ALGEBRAIC AND GEOMETRIC IDEAS IN THE THEORY OF DISCRETE OPTIMIZATION

MOS-SIAM Series on Optimization

This series is published jointly by the Mathematical Optimization Society and the Society for Industrial and Applied Mathematics. It includes research monographs, books on applications, textbooks at all levels, and tutorials. Besides being of high scientific quality, books in the series must advance the understanding and practice of optimization. They must also be written clearly and at an appropriate level for the intended audience.

Series Volumes

De Loera, Jesús A., Hemmecke, Raymond, and Köppe, Matthias, *Algebraic and Geometric Ideas in the Theory of Discrete Optimization*

Blekherman, Grigoriy, Parrilo, Pablo A., and Thomas, Rekha R., editors, *Semidefinite Optimization and Convex Algebraic Geometry*

Delfour, M. C., *Introduction to Optimization and Semidifferential Calculus*

Ulbrich, Michael, *Semismooth Newton Methods for Variational Inequalities and Constrained Optimization Problems in Function Spaces*

Biegler, Lorenz T., *Nonlinear Programming: Concepts, Algorithms, and Applications to Chemical Processes*

Shapiro, Alexander, Dentcheva, Darinka, and Ruszczyński, Andrzej, *Lectures on Stochastic Programming: Modeling and Theory*

Conn, Andrew R., Scheinberg, Katya, and Vicente, Luis N., *Introduction to Derivative-Free Optimization*

Ferris, Michael C., Mangasarian, Olvi L., and Wright, Stephen J., *Linear Programming with MATLAB*

Attouch, Hedy, Buttazzo, Giuseppe, and Michaille, Gérard, *Variational Analysis in Sobolev and* BV *Spaces: Applications to PDEs and Optimization*

Wallace, Stein W. and Ziemba, William T., editors, *Applications of Stochastic Programming*

Grötschel, Martin, editor, *The Sharpest Cut: The Impact of Manfred Padberg and His Work*

Renegar, James, *A Mathematical View of Interior-Point Methods in Convex Optimization*

Ben-Tal, Aharon and Nemirovski, Arkadi, *Lectures on Modern Convex Optimization: Analysis, Algorithms, and Engineering Applications*

Conn, Andrew R., Gould, Nicholas I. M., and Toint, Phillippe L., *Trust-Region Methods*

Algebraic and Geometric Ideas in the Theory of Discrete Optimization

Jesús A. De Loera
University of California, Davis
Davis, California

Raymond Hemmecke
Technische Universität München
München, Germany

Matthias Köppe
University of California, Davis
Davis, California

siam®
Society for Industrial and Applied Mathematics
Philadelphia

Mathematical Optimization Society
Mathematical Optimization Society
Philadelphia

10 9 8 7 6 5 4 3 2 1

Library of Congress Cataloging-in-Publication Data

De Loera, Jesús A., 1966-
Algebraic and geometric ideas in the theory of discrete optimization / Jesús A. De Loera, University of California at Davis, Davis, California, Raymond Hemmecke, Technische Universität München, München, Germany, Matthias Köppe, University of California at Davis, Davis, California.
pages cm. – (MOS-SIAM series on optimization)
Includes bibliographical references and index.
ISBN 978-1-611972-43-6
1. Combinatorial geometry. 2. Mathematical optimization–Data processing. 3. Geometry–Data processing. I. Hemmecke, Raymond, 1972- II. Köppe, Matthias, 1976- III. Title.
QA167.D43 2013
511'.1–dc23 2012033054

Contents

List of Figures ix

List of Tables xiii

List of Algorithms xv

Preface xvii

I Established Tools of Discrete Optimization 1

1 Tools from Linear and Convex Optimization 3

1.1 Convex sets and polyhedra . . . 3
1.2 Farkas' lemma and feasibility of polyhedra . . . 5
1.3 Weyl–Minkowski's representation theorem . . . 9
1.4 Decomposition of polyhedra as sums of cones and polytopes . . . 14
1.5 Faces, dimension, extreme points . . . 17
1.6 Duality of linear optimization . . . 19
1.7 Remarks on computational complexity . . . 21
1.8 The ellipsoid method and convex feasibility . . . 22
1.9 Applications of the ellipsoid method . . . 26
1.10 Notes and further references . . . 27
1.11 Exercises . . . 28

2 Tools from the Geometry of Numbers and Integer Optimization 29

2.1 Geometry of numbers in the plane . . . 29
2.2 Lattices and linear Diophantine equations . . . 34
2.3 Hermite and Smith . . . 34
2.4 Minkowski's theorems . . . 41
2.5 Gordan and Dickson . . . 43
2.6 Hilbert . . . 44
2.7 Lenstra, Lenstra, Lovász, and the shortest vector problem . . . 49
2.8 Lenstra's algorithm, integer feasibility, and optimization . . . 53
2.9 The integer hull of a polyhedron and cutting planes . . . 55
2.10 Linear versus nonlinear discrete optimization . . . 57
2.11 Notes and further references . . . 58
2.12 Exercises . . . 59

II Graver Basis Methods 61

3 Graver Bases 63
3.1 Introduction . . . 63
3.2 Graver bases and their representation property . . . 64
3.3 Optimality certificate for separable convex integer programs . . . 65
3.4 How many augmentation steps are needed? . . . 66
3.5 How do we find a Graver-best augmentation step? . . . 68
3.6 How do we find an initial feasible solution? . . . 68
3.7 Bounds for Graver basis elements . . . 69
3.8 Computation of Graver bases . . . 71
3.9 Notes and further references . . . 75
3.10 Exercises . . . 76

4 Graver Bases for Block-Structured Integer Programs 77
4.1 N-fold integer programming . . . 79
4.2 Two-stage stochastic integer programming . . . 88
4.3 N-fold 4-block decomposable integer programs . . . 93
4.4 Notes and further references . . . 101
4.5 Exercises . . . 102

III Generating Function Methods 103

5 Introduction to Generating Functions 105
5.1 The geometric series as a generating function . . . 105
5.2 Generating functions and objective functions . . . 108
5.3 Generating functions in two dimensions . . . 109
5.4 Notes and further references . . . 114
5.5 Exercises . . . 115

6 Decompositions of Indicator Functions of Polyhedra 117
6.1 Indicator functions and inclusion-exclusion . . . 117
6.2 Gram–Brianchon and Brion . . . 118
6.3 Triangulations of cones and polytopes . . . 120
6.4 Avoiding inclusion-exclusion with half-open decompositions . . . 124
6.5 Notes and further references . . . 127
6.6 Exercises . . . 127

7 Barvinok's Short Rational Generating Functions 129
7.1 Generating functions and the algorithm of Barvinok . . . 129
7.2 Specialization of the generating function . . . 135
7.3 Explicit enumeration of lattice points . . . 141
7.4 Integer linear optimization with Barvinok's algorithm . . . 142
7.5 Boolean operations on generating functions . . . 147
7.6 Integer projections . . . 149
7.7 Notes and further references . . . 152
7.8 Exercises . . . 154

8 Global Mixed-Integer Polynomial Optimization via Summation 157
8.1 Approximation algorithms and schemes . . . 157
8.2 The summation method . . . 159
8.3 FPTAS for maximizing nonnegative polynomials over integer points of polytopes . . . 159
8.4 Extension to mixed-integer optimization via discretization . . . 165
8.5 Extension to objective functions of arbitrary range . . . 172
8.6 Notes and further references . . . 176
8.7 Exercises . . . 176

9 Multicriteria Integer Linear Optimization via Integer Projection 179
9.1 Introduction . . . 179
9.2 The rational generating function encoding of all Pareto optima . . . 182
9.3 Efficiently listing all Pareto optima . . . 185
9.4 Selecting a Pareto optimum using polyhedral global criteria . . . 185
9.5 Selecting a Pareto optimum using a nonpolyhedral global criterion . . . 186
9.6 Notes and further references . . . 190
9.7 Exercises . . . 190

IV Gröbner Basis Methods 191

10 Computations with Polynomials 193
10.1 Introduction . . . 193
10.2 Univariate polynomials . . . 194
10.3 Systems of multivariate polynomial equations . . . 201
10.4 Monomial orders and the multivariate division algorithm . . . 203
10.5 Gröbner bases and Buchberger's algorithm . . . 208
10.6 Notes and further references . . . 214
10.7 Exercises . . . 214

11 Gröbner Bases in Integer Programming 217
11.1 Toric ideals and their Gröbner bases . . . 217
11.2 Toric ideals and integer programming . . . 218
11.3 Generating sets and toric ideals of lattice ideals . . . 220
11.4 Computing generating sets of lattice ideals . . . 228
11.5 Notes and further references . . . 232
11.6 Exercises . . . 233

V Nullstellensatz and Positivstellensatz Relaxations 235

12 The Nullstellensatz in Discrete Optimization 237
12.1 Motivation and examples . . . 237
12.2 Nullstellensatz and solving combinatorial systems of equations . . . 239
12.3 A simple proof of the Nullstellensatz . . . 256
12.4 Notes and further references . . . 262
12.5 Exercises . . . 263

13 Positivity of Polynomials and Global Optimization **265**
13.1 Unconstrained optimization of polynomials and sums of squares . . . 265
13.2 Positivstellensätze for semidefinite programming relaxations 273
13.3 Approximating the integer hull of combinatorial problems 277
13.4 Notes and further references 279
13.5 Exercises 280

14 Epilogue **281**
14.1 Algebraic and topological ideas in linear optimization 281
14.2 Other generating functions techniques in integer programming 283
14.3 Variations on Gomory's group relaxations 285
14.4 Connections to matrix analysis and representation theory 287

Bibliography **289**

Index **315**

List of Figures

1.1 Homogenization of a polytope. 15

2.1 Convex sets for testing primality. 30
2.2 Parallelogram in the proof of the Diophantine approximation theorem. . . 31
2.3 A (nonconvex) lattice polygon and a triangulation. 32
2.4 Convex and reflex vertices in a polygon. The vertex $\mathbf{v}_1$ is convex, while $\mathbf{v}_2$ is reflex. 32
2.5 Cases in the proof of Lemma 2.1.6. 33
2.6 A step in the proof of Lemma 2.1.6. 33
2.7 Minimal elements in a two-dimensional set of lattice points. 44
2.8 Minimal integral generating sets of two sets of lattice points. 45
2.9 For both sets S, cone(S) is not finitely generated. 45
2.10 A cone and its inclusion-minimal Hilbert basis. 46
2.11 Lattice points in a two-dimensional polyhedron covered by four shifted cones. 48
2.12 Branching on hyperplanes corresponding to approximate lattice width directions of the feasible region in a Lenstra-type algorithm [202]. 53

3.1 Augmenting an initial solution to optimality. 64

4.1 Modeling a two-stage stochastic integer multicommodity flow problem as an N-fold 4-block decomposable problem. Without loss of generality, the number of commodities and the number of scenarios are assumed to be equal. 78
4.2 Example for the map ϕ for the case of $g(A,D) = 3$ and $N = 5$. 86
4.3 The graph D_γ together with the path from $v_{0,0}$ to $v_{5,3}$ encoding the assignment from Figure 4.2. 88

5.1 A one-dimensional lattice-point set. 105
5.2 One-dimensional identity (later seen to hold in general). 107
5.3 The domains of convergence of the Laurent series. 107
5.4 Another one-dimensional identity. 108
5.5 Tiling a rational two-dimensional cone with copies of the fundamental parallelepiped. 109
5.6 The semigroup $S \subseteq \mathbb{Z}^2$ generated by $\mathbf{b}_1$ and $\mathbf{b}_2$ is a linear image of $\mathbb{Z}^2_+$. . 110
5.7 A two-dimensional cone of index 6 with its fundamental parallelepiped. Using the interior vector $\mathbf{w}$, a triangulation can be constructed. 111

5.8 A two-dimensional cone of index 6, triangulated into two unimodular cones. The integer points in the one-dimensional intersection would be counted twice, so we subtract them once (inclusion-exclusion principle). 112

5.9 A triangulation of the cone of index 5 generated by $\mathbf{b}^1$ and $\mathbf{b}^2$ into the two cones spanned by $\{\mathbf{b}^1,\mathbf{w}\}$ and $\{\mathbf{b}^2,\mathbf{w}\}$, having an index of 2 and 3, respectively. We have the inclusion-exclusion formula $g(\text{cone}\{\mathbf{b}_1,\mathbf{b}_2\};\mathbf{z}) = g(\text{cone}\{\mathbf{b}_1,\mathbf{w}\};\mathbf{z}) + g(\text{cone}\{\mathbf{b}_2,\mathbf{w}\};\mathbf{z}) - g(\text{cone}\{\mathbf{w}\};\mathbf{z})$; here the one-dimensional cone spanned by $\mathbf{w}$ needed to be subtracted. 113

5.10 A signed decomposition of the cone of index 5 generated by $\mathbf{b}^1$ and $\mathbf{b}^2$ into the two unimodular cones spanned by $\{\mathbf{b}^1,\mathbf{w}\}$ and $\{\mathbf{b}^2,\mathbf{w}\}$. We have the inclusion-exclusion formula $g(\text{cone}\{\mathbf{b}_1,\mathbf{b}_2\};\mathbf{z}) = g(\text{cone}\{\mathbf{b}_1,\mathbf{w}\};\mathbf{z}) - g(\text{cone}\{\mathbf{b}_2,\mathbf{w}\};\mathbf{z}) + g(\text{cone}\{\mathbf{w}\};\mathbf{z})$. 114

6.1 Brion's theorem. 120

6.2 Triangulations and nontriangulations. Out of these three pictures, only the left one gives a triangulation. In the middle picture, condition (i) of Definition 6.3.1 is violated. In the right picture, two full-dimensional simplices have an intersection which is another full-dimensional simplex, and thus not a face. [102] 121

6.3 A lifted configuration. 121

6.4 Lower envelope of a lifted configuration. 122

6.5 Projecting down a lifted configuration. 122

6.6 A regular and a nonregular triangulation of a point configuration. 123

6.7 Triangulation of pointed cones from triangulations of point configurations. 123

6.8 An identity, valid modulo lower-dimensional cones, corresponding to a polyhedral subdivision of a cone. [203] 125

6.9 The technique of half-open exact decomposition. The above ad hoc choice of strict inequalities (dashed lines) and weak inequalities (solid lines) appears to give an exact identity on first look. However, the apex of the cone is still counted twice. [203] 126

6.10 The technique of half-open exact decomposition 126

7.1 Integer linear optimization problem of Example 7.4.2 146

7.2 Projection of a lattice point set, codimension 1 case. The resulting lattice point set on the horizontal axis (bottom) has gaps. All nonempty fibers in this example have cardinality 1, except for the middle one, which has cardinality 2. [204] 149

7.3 Projection of a lattice point set, codimension 1 case. We remove the extra element in the fiber by taking the set difference with a shifted copy of $P \cap \mathbb{Z}^n$ in (7.29). [204] 150

7.4 A polytope and its integer projections; the same after unimodular transformation. [204] 153

8.1 Approximation properties of ℓ_k-norms. [202] 159

8.2 A sequence of optimal solutions to grid problems with two limit points, for even m and for odd m. [94] 167

8.3 The principle of grid approximation. Since we can refine the grid only in the direction of the continuous variables, we need to construct an approximating grid point $(\mathbf{x},\mathbf{z}^*)$ in the same integral slice as the target point $(\mathbf{x}^*,\mathbf{z}^*)$. [94] . . . 167
8.4 The geometry of Lemma 8.4.8. For a polynomial with a maximum total degree of 2, we construct a refinement $\frac{1}{k}\mathbb{Z}^n$ (small circles) of the standard lattice (large circles) such that $P \cap \frac{1}{k}\mathbb{Z}^n$ contains an affine image of the set $\{0,1,2\}^n$ (large dots). [94] . . . 170
8.5 Estimates in the proof of Theorem 8.1.3(a). [94] . . . 171

9.1 Strategy space. Red filled circles indicate the Pareto strategies. . . . 180
9.2 Outcome space. (a) Blue filled circles indicate the outcomes of feasible strategies; note that this set is a projection of a lattice point set and may have holes. Red filled circles indicate supported Pareto outcomes; magenta filled circles indicate nonsupported Pareto outcomes. (b) Global criteria. . . . 180
9.3 Outcome space. (a) Feasible outcomes. (b) Their discrete epigraph. . . . 182
9.4 Outcome space. (a) After erasing horizontally dominated solutions. (b) After erasing vertically dominated solutions. . . . 183
9.5 Outcome space, after erasing all dominated solutions. . . . 183
9.6 A set defining a pseudonorm with the inscribed and circumscribed cubes αB_∞ and βB_∞ (dashed lines). [98] . . . 187

11.1 The graphs (a) $\mathcal{G}(\mathcal{F}_{\mathcal{L},\mathbf{b}}, M)$ and (b) $\mathcal{G}(\mathcal{F}_{\mathcal{L},\mathbf{b}}, M')$ projected onto the (z_1,z_2)-plane. . . . 221
11.2 Reduction path between $\mathbf{z}^0$ and $\mathbf{z}^7$. . . . 223
11.3 A critical path for $(\mathbf{u},\mathbf{v})$ between $\boldsymbol{\alpha}$, $\mathbf{z}$, and $\boldsymbol{\beta}$. . . . 224
11.4 Replacing a critical path by a reduction path. . . . 225

12.1 (i) Partial 3-cycle, (ii) chordless 4-cycle. [101] . . . 246
12.2 Odd wheel. [101] . . . 247
12.3 Grötzsch graph. [101] . . . 248
12.4 Turán graph $T(5,3)$. . . . 255

List of Tables

2.1 Trace of the basis reduction algorithm (Algorithm 2.1) on the lattice of Example 2.7.6. 53

13.1 Some polynomials and their Pólya exponents 270

List of Algorithms

Algorithm 1.1	Fourier–Motzkin elimination	7
Algorithm 1.2	Ellipsoid method	25
Algorithm 2.1	Lattice basis reduction	52
Algorithm 3.1	Graver-best augmentation algorithm	66
Algorithm 3.2	Normal form algorithm	71
Algorithm 3.3	Pottier's algorithm	72
Algorithm 3.4	Project-and-lift algorithm	73
Algorithm 4.1	Graver proximity algorithm	100
Algorithm 6.1	Regular triangulation algorithm	121
Algorithm 7.1	Barvinok's algorithm, primal half-open variant	134
Algorithm 7.2	Output-sensitive enumeration with generating functions	141
Algorithm 7.3	Binary search for the optimal value	143
Algorithm 7.4	Digging algorithm	146
Algorithm 8.1	Summation method for optimizing polynomials	161
Algorithm 8.2	Approximation of the range of the objective function	173
Algorithm 10.1	Division algorithm	195
Algorithm 10.2	Extended Euclidean algorithm	196
Algorithm 10.3	Multivariate division algorithm	205
Algorithm 10.4	Buchberger's algorithm	210
Algorithm 11.1	Geometric Buchberger algorithm	226
Algorithm 11.2	Maximal decreasing path algorithm (MDPA)	227
Algorithm 11.3	Project-and-lift	230
Algorithm 12.1	NulLA algorithm	243
Algorithm 13.1	The Positivstellensatz method	275

Preface

It is undeniable that geometric ideas have been very important to the foundations of modern discrete optimization. The influence that geometric algorithms have in optimization was elegantly demonstrated in the, now classic, book *Geometric Algorithms and Combinatorial Optimization* [145] written more than 25 years ago by M. Grötschel, L. Lovász, and A. Schrijver. There, in a masterful way, we were introduced to the power that the geometry of ellipsoids, hyperplanes, convex bodies, and lattices can wield in optimization. After many years, students of integer programming today are exposed to notions such as the equivalence of separation and optimization, convex hulls, and membership, and to the many examples of successful application of these ideas such as efficient algorithms for matchings on graphs and other problems with good polyhedral characterizations [298], [299], [300] and the solution of large-scale traveling salesman problems [14]. These results were a landmark success in the theory of integer optimization.

But in just the past 15 years, there have been new developments in the understanding of the structure of polyhedra, convex sets, and their lattice points that have produced new algorithmic ideas for solving integer programs. These techniques add a new set of powerful tools for discrete optimizers and have already proved very suitable for the solution of a number of hard problems, including attempts to deal with nonlinear objective functions and constraints in discrete optimization. Unfortunately, many of these powerful tools are not yet widely known or applied. Perhaps this is because many of the developments have roots in areas of mathematics that are not normally part of the standard curriculum of students in optimization and have a much more algebraic flavor. Examples of these new tools include algebraic geometry, commutative algebra, representation theory, and number theory.

We feel that the unfamiliar technical nature of these new ideas and the lack of expository accounts have unnecessarily delayed the popularity of these techniques among those working in optimization. We decided to write a text that would not demand any background beyond what we already assume from people in mathematical programming courses. This monograph is then intended as a short, self-contained, introductory course in these new ideas and algorithms with the hope of popularizing them and inviting new applications. We were deeply inspired by the influential book [145] and we humbly try to follow in its footsteps in the hope that future generations continue to see the interdependence between beautiful mathematics and the creation of efficient optimization algorithms.

This book is meant to be used in a quick, intense course, no longer than 15 weeks. This is not a complete treatise on the subject, but rather an invitation to a set of new ideas and tools. Our aim is to popularize these new ideas among workers in optimization.

- We want to make it possible to read this book even if you are a novice of integer and linear programming (and we have taught courses with some students in that category). For this reason, we open in Part I with some of the now well-established techniques that originated before the beginning of the 1990s, a time when linear and convex programming and integer programming underwent major changes thanks to the ellipsoid method, semidefinite programming, lattice basis reduction, etc. Most of what is contained in Part I is a short summary of tools that students in optimization normally encounter in a course based on the excellent books [50, 145, 206, 259, 296] and probably should be skipped by such readers. They should go directly to the new exciting techniques in Parts II, III, IV, and V. Readers that start with Part I will add an extra three or four weeks to the course.

- Parts II, III, IV, and V form the core of this book. Roughly speaking, when the reader works in any of these sections, nonlinear, nonconvex conditions are central, making the tools of algebra necessary. When studying only these parts, the course is planned to take about 12 weeks. In fact, all the parts are quite independent from each other and each can be the focus of independent student seminars.
  - We begin in Part II with the idea of test sets and Graver bases. We show how they can be used to prove results about integer programs with linear constraints and convex objective functions.
  - Part III discusses the use of generating functions to deal with integer programs with linear constraints but with nonlinear polynomial objectives and/or with multiobjectives.
  - Part IV discusses the notion of Gröbner bases and their connection with integer programming.
  - Part V discusses the solution of global optimization problems with polynomial constraints via a sequence of linear algebra or semidefinite programming systems. These are generated based on Hilbert's Nullstellensatz and its variations.

The book contains several exercises to help students learn the material. A course based on these lectures should be suitable for advanced undergraduates with a solid mathematical background (i.e., very comfortable with proofs in linear algebra and real analysis) or for graduate students who have already taken an introductory linear programming class.

Acknowledgments. We are truly grateful to many people who helped us both on producing the research presented here and later on presenting it to a larger audience of students and colleagues.

First and foremost our collaborators and coauthors in many parts of this book were fundamental for arriving at this point; their energy and ideas show on every page. In fact several portions of the book are taken partially from our joint work. So many, many thanks for everything to Robert Hildebrand, Chris Hillar, Jon Lee, Peter N. Malkin, Susan Margulies, Mohamed Omar, Shmuel Onn, Pablo Parrilo, Uriel Rothblum, Maurice Queyranne, Christopher T. Ryan, Rüdiger Schultz, Sven Verdoolaege, Robert Weismantel, and Kevin Woods. Thanks!

Many other friends have been faithful supporters and collaborators in other closely related projects or have developed ideas of great importance to this book. We learned so much about discrete optimization from talking to Karen Aardal, Alper Atamtürk, David Avis, Egon Balas, Velleda Baldoni, Imre Bárány, Sasha Barvinok, Amitabh Basu, Nicole Berline, Dmitris Bertsimas, Lou Billera, Greg Blekherman, Sam Burer, Bill Cook, Sanjeeb Dash, Antoine Deza, Etienne de Klerk, Matthias Ehrgott, Fritz Eisenbrand, Komei Fukuda, Bernd Gärtner, Michel Goemans, João Gouveia, David Haws, Nicolai Hähnle, Martin Henk, Serkan Hoşten, Peter Huggins, Michael Joswig, Volker Kaibel, Gil Kalai,

Victor L. Klee, Steve Klee, Jean Bernard Lasserre, Monique Laurent, Jim Lawrence, Hendrik W. Lenstra, Sven Leyffer, Jeff Linderoth, Diane Maclagan, Jirka Matoušek, Nimrod Megiddo, Bernard Mourrain, Walter Morris, Jiawang Nie, Jorge Nocedal, Edwin O'Shea, Dima Pasechnik, Javier Peña, Vicky Powers, Scott Provan, Franz Rendl, Bruce Reznick, Maurice Rojas, Paco Santos, András Sebö, Bernd Sturmfels, Levent Tunçel, Tamás Terlaky, Rekha R. Thomas, Mike Todd, Frank Vallentin, Santosh Vempala, Michèle Vergne, Cynthia Vinzant, Emo Welzl, Laurence Wolsey, Yinyu Ye, Ruriko Yoshida, and Günter M. Ziegler.

We received comments, corrections, great questions, suggestions, encouragement, and help from Ilan Adler, Egon Balas, Dave Bayer, Matthias Beck, Victor Blanco, David Bremner, Winfried Bruns, Katherine Burgraf, Samantha Capozzo, Gérard Cornuéjols, Persi Diaconis, Brandon E. Dutra, Jennifer Galovich, Harvey Greenberg, Peter Gritzmann, Oktay Günlük, Christian Haase, Ilya Hicks, Dorit Hochbaum, Mark Junod, Yvonne Kemper, Eddie Kim, Bala Krishnamoorthy, Jeff Lagarias, Karla Lanzas, Adam Letchford, Quentin Louveaux, Laci Lovász, François Margot, Tyrrell McAllister, Juan Meza, Gabor Pataki, Amber Puha, Mihai Putinar, Eric Rains, Jörg Rambau, Jürgen Richter-Gebert, Carla Savage, Lex Schrijver, Markus Schweighofer, Renata Sotirov, Frank Sottile, Tamon Stephen, Seth Sullivant, Richard Tapia, Andreas Waechter, Roger Wets, Angelika Wiegele, Mark C. Wilson, Peter Winkler, Alexander Woo, David Woodruff, Doron Zeilberger, Yuriy Zinchenko, and Uri Zwick. We received help from many students that heard lectures from us on the topic. Thanks for your patience and effort! We give special thanks to Astrid Köppe for the artwork she provided for the cover.

We are truly grateful to the NSF for the financial support that made this book possible. Research and lectures about this topics were also produced with the support of the following institutions: University of California, Davis, Universität Magdeburg, Technische Universität Darmstadt, and Technische Universität München. We must stress that IMA (Institute for Mathematics and its Applications) at the University of Minnesota, the Rocky Mountains Mathematics Consortium, Banff International Research Station, MSRI (Mathematical Sciences Research Institute) at UC Berkeley, IPAM (Institute for Pure and Applied Mathematics) at UCLA, AIM (American Institute of Mathematics) at Palo Alto, MAA (Mathematical Association of America), and St. John's University deserve special acknowledgment as they allowed parts of these notes to be presented in short courses or be the focus of special workshops.

Finally, our families are very special in our lives and this project is partly theirs too, built with their love and patience in our long crazy hours and very distracted minds.

Jesús is truly grateful to his wife Ingrid who has put up with him and his difficult workaholic nature for a long, long time. Mil gracias amor mio de todo corazón, todo te lo debo a ti. Their two sons Antonio and Andrés were just little kids when the research on this book began to flourish. It is a great pleasure to see them both grow so strong in spirit and intellect. Muchas, muchas gracias hijos míos y perdón por la falta de atención, estoy orgulloso de ustedes. Mil gracias a doña Antonia y Judith, queridas madre y hermana, que siempre me quieren y me apoyan. Gracias a toda la familia y amigos en México por lo mucho que me dan.

Raymond thanks his wife Susi for her understanding and love and for the two little bright stars in their lives, Carina and Paula. You three are the best that ever happened to me!

Jesús A. De Loera, Raymond Hemmecke, Matthias Köppe

Part I

Established Tools of Discrete Optimization

Chapter 1

Tools from Linear and Convex Optimization

Convex sets and polyhedra are fundamental objects in the study of optimization. We quickly review several facts that will be used in the new results presented in this book. Discrete optimization problems are often solved, via branching, enumeration, or relaxation, by the repeated use of linear or convex methods. Here we collect some background results relevant to the book. For more details we recommend the excellent books [32, 50, 145, 297]. Readers that have already taken a course in discrete optimization most likely have been exposed to most of the material in the next two chapters and can safely proceed to Part II.

1.1 Convex sets and polyhedra

Everything that we do in this book takes place inside Euclidean n-dimensional space $\mathbb{R}^n$. We use the standard inner product which defines the standard Euclidean distance between two points.

Definition 1.1.1. A subset S of $\mathbb{R}^n$ is *convex* if for any two distinct points $\mathbf{x}_1, \mathbf{x}_2$ in S the line segment joining $\mathbf{x}_1, \mathbf{x}_2$ lies completely in S. This is equivalent to saying that $\mathbf{x} = \lambda \mathbf{x}_1 + (1-\lambda)\mathbf{x}_2$ belongs to S for all choices of λ between 0 and 1.

In general, given a finite set of points $A = \{\mathbf{x}_1, \dots, \mathbf{x}_m\}$, we say that a linear combination $\sum \gamma_i \mathbf{x}_i$ is

- an *affine combination* if $\sum_{i=1}^m \gamma_i = 1$,
- a *conic combination* if $\gamma_i \geq 0$ for all $i = 1, \dots, m$,
- a *convex combination* if it is both affine and conic.

We will assume that the empty set is also convex. Observe that the intersection of convex sets is convex too. For $A \subseteq \mathbb{R}^n$, the *convex hull* of A, denoted by conv(A), is the intersection of all the convex sets containing A. In other words, conv(A) is the smallest convex set containing A. The reader can check that the image of a convex set under a linear transformation is again a convex set. Any linear or affine subspace is a convex set too.

Recall from linear algebra that any linear functional $f : \mathbb{R}^n \to \mathbb{R}$ is given by a choice of vector $\mathbf{c} \in \mathbb{R}^n$ that gives $f(\mathbf{x}) = \mathbf{c}^\top \mathbf{x}$. For a number $\alpha \in \mathbb{R}$ we say that

$$H_\alpha = \left\{ \mathbf{x} \in \mathbb{R}^n : f(\mathbf{x}) = \alpha \right\}$$

is an *affine hyperplane*, or *hyperplane* for short. Note that a hyperplane divides $\mathbb{R}^n$ into two half spaces $H_\alpha^+ = \{\mathbf{x} \in \mathbb{R}^n : \mathbf{c}^\top \mathbf{x} \geq \alpha\}$ and $H_\alpha^- = \{\mathbf{x} \in \mathbb{R}^n : \mathbf{c}^\top \mathbf{x} \leq \alpha\}$. Half spaces are also convex sets. Another important example of a convex set is that of a polyhedron.

Definition 1.1.2. The set of solutions of a finite system of linear inequalities is called a *polyhedron*. In its general form, a polyhedron is a set of the type

$$P = \left\{\mathbf{x} \in \mathbb{R}^n : \mathbf{c}_i^\top \mathbf{x} \leq \beta_i,\ i = 1, \ldots, m\right\}$$

for some vectors $\mathbf{c}_i \in \mathbb{R}^n$ and some real numbers β_i.

Using a matrix $A \in \mathbb{R}^{m\times n}$ and a vector $\mathbf{b} \in \mathbb{R}^m$, we can write a polyhedron in the form $P = \{\mathbf{x} \in \mathbb{R}^n : A\mathbf{x} \leq \mathbf{b}\}$. Here "$\leq$" is understood componentwise.

In other words, a polyhedron in $\mathbb{R}^n$ is the intersection of finitely many half spaces. The reader can easily prove the following as a little exercise.

Lemma 1.1.3. *Every polyhedron is a convex set.*

Another important definition, closely tied to polyhedra, is that of a polytope.

Definition 1.1.4. A *polytope* is the convex hull of a finite set of points in $\mathbb{R}^n$.

Below we will prove a fundamental result, known as the Weyl–Minkowski theorem: Polytopes are bounded polyhedra and, vice versa, bounded polyhedra are polytopes.

Lemma 1.1.5. *For a set $A \subseteq \mathbb{R}^n$, conv(A) equals the set of all finite convex combinations among elements of A. In particular, for a finite set $A := \{\mathbf{a}_1, \ldots, \mathbf{a}_m\} \subseteq \mathbb{R}^n$ we have*

$$\operatorname{conv}(A) = \left\{\sum_{i=1}^{m} \gamma_i \mathbf{a}_i : \gamma_1 + \cdots + \gamma_m = 1, \gamma_1, \ldots, \gamma_m \geq 0\right\}. \tag{1.1}$$

Proof. There are two inclusions that need to be verified: Denote by B the right-hand side of equation (1.1). Clearly $A \subseteq B$, and it is easy to verify that B is convex. Therefore, we have $\operatorname{conv}(A) \subseteq B$. It remains to show that $B \subseteq \operatorname{conv}(A)$. We prove that an arbitrary element $\mathbf{x} = \sum_{i=1}^m \gamma_i \mathbf{a}_i \in B$ is also in conv(A) by induction on the number of nonzero elements r in the sum. If only $r = 1$, then clearly $\mathbf{x} \in A \subseteq \operatorname{conv}(A)$. Suppose there are $r+1$ nonzero coefficients γ_i, say

$$\mathbf{x} = \gamma_{j_1}\mathbf{a}_{j_1} + \gamma_{j_2}\mathbf{a}_{j_2} + \cdots + \gamma_{j_{r+1}}\mathbf{a}_{j_{r+1}}.$$

We construct $\beta_{j_i} = \gamma_{j_i}/(\gamma_{j_1} + \gamma_{j_2} + \cdots + \gamma_{j_r})$ and observe that these β's are all positive and sum to one. Using induction we conclude that $\mathbf{y} = \beta_{j_1}\mathbf{a}_{j_1} + \beta_{j_2}\mathbf{a}_{j_2} + \cdots + \beta_{j_r}\mathbf{a}_{j_r}$ is indeed an element of conv(A). Finally, recall that $1 - \gamma_{j_{r+1}}$ equals $\gamma_{j_1} + \gamma_{j_2} + \cdots + \gamma_{j_r}$, and therefore we have

$$\mathbf{x} = (1 - \gamma_{j_{r+1}})\mathbf{y} + \gamma_{j_{r+1}}\mathbf{a}_{j_{r+1}}.$$

Since $\mathbf{y}$ and $\mathbf{a}_{r+1}$ are in conv(A), and conv(A) is a convex set, we conclude that $\mathbf{x} \in \operatorname{conv}(A)$. □

Yet another important concept in polyhedral geometry is that of a cone.

Definition 1.1.6. A set $C \subseteq \mathbb{R}^n$ is called a *cone* if it is closed under nonnegative linear combinations, that is, $C \subseteq \mathbb{R}^n$ is a cone if for all pairs $\mathbf{x},\mathbf{y} \in C$ and all scalars $\lambda,\mu \in \mathbb{R}_+$, $\lambda\mathbf{x}+\mu\mathbf{x} \in C$.

- A cone C is called *polyhedral* if $C = \{\mathbf{x} \in \mathbb{R}^n : A\mathbf{x} \geq \mathbf{0}\}$ for some matrix A. Clearly, any polyhedral cone is a polyhedron.
- A cone C is *finitely generated* if it is the set of all conic (= nonnegative) combinations of finitely many vectors.

Below we will see that there is a strong connection between polytopes, polyhedra, and cones generated by finitely many vectors. We conclude this section with two nice structural results on convex sets.

Theorem 1.1.7 (Carathéodory's theorem). *Let $S \subseteq \mathbb{R}^n$ and $\mathbf{x} \in \mathrm{conv}(S)$. Then $\mathbf{x}$ is the convex combination of at most $n+1$ points in S.*

Proof. Take $\bar{\mathbf{x}} \in \mathrm{conv}(S)$. Consider any convex combination $\bar{\mathbf{x}} = \sum_{i=1}^m \lambda_i \mathbf{x}_i$, with $\mathbf{x}_i \in S$ and m the smallest possible. If there are more than $n+1$ summands, i.e., $m > n+1$, then there must exist an affine linear dependence among $\mathbf{x}_1,\ldots,\mathbf{x}_m$. This means that there exist $\gamma_i \in \mathbb{R}$ such that $\sum_{i=1}^m \gamma_i \mathbf{x}_i = 0$, $\sum_{i=1}^m \gamma_i = 0$, and with at least one $\gamma_i > 0$. Take $\tau = \min\{\lambda_i/\gamma_i : \gamma_i > 0\}$. Then by construction,

$$\bar{\mathbf{x}} = \sum_{i=1}^m \lambda_i \mathbf{x}_i - \tau \sum_{i=1}^m \gamma_i \mathbf{x}_i = \sum_{i=1}^m (\lambda_i - \tau\gamma_i)\mathbf{x}_i$$

is a convex combination for $\bar{\mathbf{x}}$ with less than m terms, contradicting the minimality of m. ☐

As a corollary, all triangles of a polygon in $\mathbb{R}^2$ (using the vertices of the polygon) must cover the polygon. We leave the proof of the following lemma as a nice exercise to the reader.

Lemma 1.1.8 (Radon's lemma). *If S is a set of $n+2$ points in $\mathbb{R}^n$ then S can be partitioned as $S = A \cup B$ into two disjoint subsets A and B with nonempty intersection* $\mathrm{conv}(A) \cap \mathrm{conv}(B)$.

1.2 Farkas' lemma and feasibility of polyhedra

When is a polyhedron empty? How can one solve a system of linear inequalities $A\mathbf{x} \leq \mathbf{b}$? These questions are at the heart of linear optimization and its applications. We describe a rather innocent-looking algorithm to solve any system of *inequalities* $A\mathbf{x} \leq \mathbf{b}$. Surprisingly, most of the theory of linear duality and optimality can be deduced from this very elementary process.

1.2.1 Solvability of a system of linear equations

Let us start by looking at the case of linear *equations*.

Theorem 1.2.1 (Fredholm's alternative theorem). *The system of linear equations $A\mathbf{x} = \mathbf{b}$ has a solution if and only if for each $\mathbf{y}$ with the property $\mathbf{y}^\top A = \mathbf{0}^\top$ we have $\mathbf{y}^\top \mathbf{b} = 0$.*

This theorem can be read as follows: Either $A\mathbf{x} = \mathbf{b}$ has a solution or there exists a vector $\mathbf{y}$ with the property that $\mathbf{y}^\top A = \mathbf{0}^\top$ but $\mathbf{y}^\top \mathbf{b} \neq 0$. Or in other words, $\{\mathbf{x} : A\mathbf{x} = \mathbf{b}\}$ is nonempty if and only if the set $\left\{\mathbf{y} : \mathbf{y}^\top A = \mathbf{0}^\top,\ \mathbf{y}^\top \mathbf{b} = -1\right\}$ is empty.

The proof of this fact follows directly from the Gaussian elimination procedure. Recall that the allowable row operations are

(1) interchange rows;

(2) scale rows by $\lambda \neq 0$;

(3) add a multiple of a row to another row.

Observe that any "new" row is a linear combination of the original rows, and that the "new" matrix is obtained by applying operation (1), (2), or (3) many times. Note also that all these operations can be realized by multiplying $A\mathbf{x} = \mathbf{b}$ from the left by elementary square matrices. After sufficiently many applications of the above operations we will arrive at a matrix of the form

$$(A|\mathbf{b}) \longrightarrow (D|\mathbf{d}) := \left(\begin{array}{ccccc|c|c} 1 & 0 & \cdots & 0 & & * & d_1 \\ 0 & 1 & \cdots & 0 & & * & d_2 \\ \vdots & \vdots & \ddots & \vdots & & \vdots & \vdots \\ 0 & 0 & \cdots & 1 & & * & d_r \\ \hline 0 & \cdots & 0 & \cdots & & 0 & d_{r+1} \\ \hline 0 & \cdots & 0 & \cdots & & 0 & d_{r+2} \\ \hline \vdots & \vdots & \vdots & \vdots & & \vdots & \vdots \\ \hline 0 & \cdots & 0 & \cdots & & 0 & d_m \end{array}\right).$$

Moreover, by construction, there is an invertible square matrix $Q \in \mathbb{R}^{m\times m}$ such that $QA = D$ and $Q\mathbf{b} = \mathbf{d}$.

Lemma 1.2.2. $A\mathbf{x} = \mathbf{b}$ *is solvable if and only if* $d_{r+1} = \cdots = d_m = 0$.

Proof. Gaussian elimination operations preserve solutions from one step to the next; that is, $A\mathbf{x} = \mathbf{b}$ is solvable if and only if $(QA)\mathbf{x} = Q\mathbf{b}$ is solvable for any invertible square matrix $Q \in \mathbb{R}^{m\times m}$. Hence, in particular, $A\mathbf{x} = \mathbf{b}$ is solvable if and only if $D\mathbf{x} := (QA)\mathbf{x} = Q\mathbf{b} =: \mathbf{d}$ is solvable, with D and $\mathbf{d}$ as above. Thus, $A\mathbf{x} = \mathbf{b}$ is solvable if and only if $d_{r+1} = \cdots = d_m = 0$, as claimed. □

Note that this proof includes a proof of Theorem 1.2.1. Assuming that $A\mathbf{x} = \mathbf{b}$ is *not* solvable, there is some $j > r$ with $d_j \neq 0$. Let $\mathbf{q}_j^\top$ be the jth row of Q. Then $\mathbf{q}_j^\top A = \mathbf{0}^\top$ (= the jth row of D) and $\mathbf{q}_j^\top \mathbf{b} = d_j \neq 0$. So a scalar multiple $\mathbf{y} := \lambda \mathbf{q}_j$ of $\mathbf{q}_j$ satisfies $\mathbf{y}^\top A = \mathbf{0}^\top$ and $\mathbf{y}^\top \mathbf{b} = 1$.

We can view the steps in the Gaussian elimination procedure as eliminating variables one at a time and getting a new system of equations in terms of the remaining variables $x_2, \ldots, x_n$ in such a way that any solution $(x_2, \ldots, x_n)$ to the new system can be lifted to a solution $(x_1, x_2, \ldots, x_n)$ of the original system.

1.2.2 Solvability of a system of linear inequalities

Let us now turn to the case of inequalities; that is, let us deal with the question of solving $A\mathbf{x} \leq \mathbf{b}$. This can be done by the *Fourier–Motzkin elimination* algorithm, which was first described by Fourier in the 1800s, but was rediscovered by Motzkin in the 1930s. The algorithm shows how to eliminate a single variable to obtain a new set of inequalities that describe the orthogonal projection of the polyhedron $\{\mathbf{x} : A\mathbf{x} \leq \mathbf{b}\}$ onto the hyperplane $x_1 = 0$, that is, onto the space of the remaining variables $x_2, \ldots, x_n$. Note that this already suggests that the procedure may be rather inefficient, because it is not difficult to see that some polyhedra project into lower dimensional polyhedra with many more facets than the original. For instance, imagine how certain triangular prisms could be projected into hexagons.

Given A, an $m \times n$ matrix, we will denote by $\mathbf{a}_i^\top$ the i-th row vector. The entries of A are a_{ij}.

ALGORITHM 1.1. Fourier–Motzkin elimination.

1: **input** An $m \times n$ matrix A, $A\mathbf{x} \leq \mathbf{b}, \mathbf{x} \in \mathbb{R}^n$ and a column index j, $1 \leq j \leq n$.
2: **output** A new system of inequalities $D\mathbf{x} \leq \mathbf{d}, \mathbf{x} \in \mathbb{R}^n$ where D is an $M \times n$ matrix whose j-th column is zero.
3: Partition row indices $I = \{1, \ldots, m\}$ into three sets:

$$N = \{i \in I : a_{ij} < 0\}, \quad Z = \{i \in I : a_{ij} = 0\}, \quad P = \{i \in I : a_{ij} > 0\}.$$

4: Create a new matrix D as follows:
5: **for** $i \in Z$ **do**
6: Put the row $\mathbf{a}_i^\top$ in D and b_i as the corresponding right-hand side value.
7: **for** each possible pair (s,t) with $s \in N$ and $t \in P$ **do**
8: Create a new row of D: $a_{tj}\mathbf{a}_s^\top - a_{sj}\mathbf{a}_t^\top$.
9: Create a new right-hand side in $\mathbf{d}$: $a_{tj}b_s - a_{sj}b_t$.
10: **return** $M \times n$ matrix D (where $M := |Z| + |P| \cdot |N|$) and vector $\mathbf{d}$.

Because of the growth on the number of equations during intermediate calculations, Fourier–Motzkin elimination is not a very useful algorithm in practice. Still it is rather elementary and easy to explain since it is similar to Gaussian elimination. More importantly, we can use it to prove several fundamental theorems in the theory of linear programming.

Note that in step 8 of the Fourier–Motzkin elimination procedure, we create a new row of D whose entry in the j-th column is $a_{tj}a_{sj} - a_{sj}a_{tj} = 0$. Moreover, observe that the Fourier–Motzkin elimination procedure constructs as a by-product a matrix $U_j \in \mathbb{R}_+^{M \times m}$ such that $D = U_j A$ and $\mathbf{d} = U_j \mathbf{b}$. Even more importantly, $D\mathbf{x} \leq \mathbf{d}$ describes exactly the orthogonal projection of the polyhedron $\{\mathbf{x} : A\mathbf{x} \leq \mathbf{b}\}$ onto the hyperplane $x_j = 0$, as the following theorem shows.

Theorem 1.2.3. *Algorithm* 1.1 *produces a matrix D such that there exists $\mathbf{x} \in \mathbb{R}^n$ with $A\mathbf{x} \leq \mathbf{b}$ if and only if there exists $\mathbf{x} \in \mathbb{R}^n$ with $D\mathbf{x} \leq \mathbf{d}$.*

Proof. ($\Rightarrow$) This implication is clear because the rows of D and the entries of $\mathbf{d}$ are *positive* linear combinations of the rows of A and the entries of $\mathbf{b}$. Thus when $\mathbf{x}$ satisfies $\mathbf{a}_s^\top \mathbf{x} \leq b_s$ and $\mathbf{a}_t^\top \mathbf{x} \leq b_t$, it also satisfies $a_{tj}\mathbf{a}_s^\top \mathbf{x} \leq a_{tj}b_s$ and $-a_{sj}\mathbf{a}_t^\top \mathbf{x} \leq -a_{sj}b_t$ and thus also

$$a_{tj}\mathbf{a}_s^\top \mathbf{x} - a_{sj}\mathbf{a}_t^\top \mathbf{x} \leq a_{tj}b_s - a_{sj}b_t.$$

($\Leftarrow$) We have to show how to lift a solution of $D\mathbf{x} \leq \mathbf{d}$ with $x_j = 0$ to a solution $\mathbf{y} = \mathbf{x} + \lambda \mathbf{e}_j$ of the original system $A\mathbf{y} \leq \mathbf{b}$. Now we claim that there exist numbers $L \leq U$ such that for any $\lambda \in [L,U]$, the vector $\mathbf{y} = \mathbf{x} + \lambda \mathbf{e}_j$ satisfies $A\mathbf{y} \le \mathbf{b}$. We do this in three easy steps:

1. Here are the candidates for the values of L and U: Let $\mathbf{z}$ be a solution to $D\mathbf{x} \le \mathbf{d}$ with $z_j = 0$. For any $i \in P \cup N$ let $y_i = \frac{1}{a_{ij}}(b_i - \mathbf{a}_i^\top \mathbf{z})$; that is, y_i is the amount of "slack" in the inequality $\mathbf{a}_i^\top \mathbf{z} \le b_i$. Define $U = \min\{\, y_i : i \in P \,\}$. If $P = \emptyset$, set $U = +\infty$. Moreover, define $L = \max\{\, y_i : i \in N \,\}$, and if $N = \emptyset$, simply set $L = -\infty$.

2. We want to show that $L \le U$: It is true if $P = \emptyset$ or $N = \emptyset$. Take $t \in P$ and $s \in N$ and form
$$\left.\begin{array}{c} a_{tj}\mathbf{a}_s^\top - a_{sj}\mathbf{a}_t^\top \\ a_{tj}b_s - a_{sj}b_t \end{array}\right\} \text{in } D,\mathbf{d}.$$
This corresponds to the inequality $a_{tj}\mathbf{a}_s^\top \mathbf{z} - a_{sj}\mathbf{a}_t^\top \mathbf{z} \le a_{tj}b_s - a_{sj}b_t$. Rearranging terms, we obtain
$$a_{sj}b_t - a_{sj}\mathbf{a}_t^\top \mathbf{z} \le a_{tj}b_s - a_{tj}\mathbf{a}_s^\top \mathbf{z},$$
and since $a_{sj} < 0$ and $a_{tj} > 0$, we get
$$y_t = \frac{b_t - \mathbf{a}_t^\top \mathbf{z}}{a_{tj}} \ge \frac{b_s - \mathbf{a}_s^\top \mathbf{z}}{a_{sj}} = y_s.$$
Hence, in particular, if $y_t = U$ and $y_s = L$, we get $U \ge L$ as desired.

3. We show that for all $\lambda \in [L,U]$ the vector $\mathbf{z}^\lambda = \mathbf{z} + \lambda \mathbf{e}_j$ satisfies $A\mathbf{z}^\lambda \le \mathbf{b}$. For an index $i \in Z$, the corresponding row is simply copied to $D,\mathbf{d}$, and so the inequality in question is satisfied. If $i \in P$ and thus $U < +\infty$, we have $\lambda \le U \le y_i$. Therefore, we have
$$\mathbf{a}_i^\top \mathbf{z}^\lambda = \mathbf{a}_i^\top \mathbf{z} + \lambda \mathbf{a}_i^\top \mathbf{e}_j = \mathbf{a}_i^\top \mathbf{z} + \lambda a_{ij}.$$
By definition, we have $y_i = \frac{1}{a_{ij}}(b_i - \mathbf{a}_i^\top \mathbf{z})$, and, therefore, $y_i a_{ij} + \mathbf{a}_i^\top \mathbf{z} = b_i$. Since $\lambda \le y_i$ we obtain $\mathbf{a}_i^\top \mathbf{z}^\lambda \le b_i$ as desired. An analogous argument proves the claim for any index $i \in N$. □

Example 1.2.4.

$$\left\{\begin{array}{r} x_1 - x_2 \le 0 \\ 2x_1 - 3x_2 \le -1 \\ x_1 + x_2 \le 2 \\ -2x_1 + x_2 \le 1 \end{array}\right. \to \left\{\begin{array}{r} 2x_1 \le 2 \\ -x_1 \le 1 \\ 5x_1 \le 5 \\ -4x_1 \le 2 \end{array}\right. \to \left\{\begin{array}{l} 0 \le 4 \\ 0 \le 6 \\ 0 \le 10 \\ 0 \le 30 \end{array}\right. .$$

Thus the original system must have a solution.

Note that if we repeatedly apply Algorithm 1.1 until no variables are left, that is, such that D becomes the zero matrix, and if we multiply the corresponding transformation matrices $U_1, U_2, \ldots, U_n$ together (from the left in the order that the variables are eliminated), we obtain a nonnegative matrix U such that $D = UA = O$ (the zero matrix) and $\mathbf{d} = U\mathbf{b}$. Then $A\mathbf{x} \le \mathbf{b}$ has a solution if and only if $\mathbf{d} = B\mathbf{b} \ge \mathbf{0}$. This means that the set of $\mathbf{b}$ for which $A\mathbf{x} \le \mathbf{b}$ is solvable is a polyhedral cone, namely $\{\mathbf{b} : B\mathbf{b} \ge \mathbf{0}\}$. We have just proved the following lemma.

Lemma 1.2.5. *Let A be an $m \times n$ matrix. Then there exists a nonnegative $k \times m$ matrix B with $BA = O$ and such that $A\mathbf{x} \le \mathbf{b}$ has a solution if and only if $B\mathbf{b} \ge \mathbf{0}$.*

As an immediate consequence, we obtain the well-known *Farkas' lemma*.

Corollary 1.2.6 (Farkas' lemma). *The system of inequalities* $A\mathbf{x} \le \mathbf{b}$ *has a solution if and only if the system of inequalities* $\mathbf{y}^\top A = \mathbf{0}^\top$, $\mathbf{y} \ge \mathbf{0}$, $\mathbf{y}^\top\mathbf{b} < 0$ *does not have a solution.*

Proof. It is easy to see both conditions cannot be simultaneously true because a contradiction arises: $0 = \mathbf{y}^\top A\mathbf{x} \le \mathbf{y}^\top\mathbf{b} < 0$.

Suppose $A\mathbf{x} \le \mathbf{b}$ has no solution. We wish to construct a solution $\mathbf{y}$ for the opposite system. Since the system $A\mathbf{x} \le \mathbf{b}$ has no solution, when using Fourier–Motzkin, we must obtain $O\mathbf{x} \le \mathbf{d}$ with $d_i < 0$ for some particular index i. Using Lemma 1.2.5, we obtain a nonnegative matrix B that has the property that $BA = O$, $B\mathbf{b} = \mathbf{d}$. We can define $\mathbf{y}$ as $\mathbf{y} = \mathbf{e}_i^\top B$. Thus note $\mathbf{y} \ge \mathbf{0}$, $\mathbf{y}^\top A = \mathbf{e}_i^\top BA = \mathbf{e}_i^\top O = \mathbf{0}^\top$, and $\mathbf{e}_i^\top B\mathbf{b} = \mathbf{e}_i^\top\mathbf{d} = d_i < 0$. □

Now we present a second version of Farkas' lemma that is the beginning of the theory of duality and optimality of linear programs:

Lemma 1.2.7 (Farkas' Lemma, version two). *Either* $A\mathbf{x} = \mathbf{b}$, $\mathbf{x} \ge \mathbf{0}$ *is solvable or* $\mathbf{y}^\top A \ge \mathbf{0}^\top$, $\mathbf{y}^\top\mathbf{b} < 0$ *is solvable, but not both.*

Proof. Consider the extended matrix

$$\bar{A} = \begin{pmatrix} A \\ -A \\ -I \end{pmatrix} \mathbf{x} \le \begin{pmatrix} \mathbf{b} \\ -\mathbf{b} \\ \mathbf{0} \end{pmatrix} = \bar{\mathbf{b}}.$$

By Corollary 1.2.6, either $\bar{A}\mathbf{x} \le \bar{\mathbf{b}}$ is solvable or there exists a vector $\bar{\mathbf{y}} \ge 0$ such that $\bar{\mathbf{y}}^\top\bar{A} = \mathbf{0}$, $\bar{\mathbf{y}}^\top\bar{\mathbf{b}} < 0$. The vector $\bar{\mathbf{y}}$ can be written as $(\mathbf{z}_1, \mathbf{z}_2, \mathbf{w})^\top$. This implies

$$\bar{\mathbf{y}}^\top\bar{A} = (\mathbf{z}_1, \mathbf{z}_2, \mathbf{w})^\top \begin{pmatrix} A \\ -A \\ -I \end{pmatrix} = \mathbf{z}_1^\top A - \mathbf{z}_2^\top A - \mathbf{w}^\top I = \mathbf{0}^\top.$$

Therefore $(\mathbf{z}_1 - \mathbf{z}_2)^\top A - \mathbf{w}^\top I = \mathbf{0}$, which means $(\mathbf{z}_1 - \mathbf{z}_2)^\top A \ge \mathbf{0}^\top$. Also, $\bar{\mathbf{y}}^\top\bar{\mathbf{b}} = \mathbf{z}_1^\top\mathbf{b} - \mathbf{z}_2^\top\mathbf{b} < 0$. Thus, $\mathbf{y} := \mathbf{z}_1 - \mathbf{z}_2$ is a desired solution to $\mathbf{y}^\top A \ge \mathbf{0}^\top$, $\mathbf{y}^\top\mathbf{b} < 0$. □

Note that by just a change of sign on the above calculations, we can state Farkas' lemma in another way: $A\mathbf{x} = \mathbf{b}$, $\mathbf{x} \ge \mathbf{0}$ has a solution or $\mathbf{y}^\top A \le \mathbf{0}^\top$, $\mathbf{y}^\top\mathbf{b} > 0$ has a solution.

This second version of Farkas' lemma has a clear geometric interpretation. Consider the cone $K(A)$ generated by the finitely many columns of the matrix A. Let $\mathbf{b} \in \mathbb{R}^n$. If $\mathbf{b} \notin K(A)$, then Farkas' lemma implies there exists a hyperplane $H = \{\mathbf{x} : \mathbf{a}^\top\mathbf{x} = 0\}$ that separates the cone $K(A)$ from the vector $\mathbf{b}$, that is, $K(A) \subseteq H^- = \{\mathbf{x} : \mathbf{a}^\top\mathbf{x} \le 0\}$ while $\mathbf{b} \in H^+ \setminus H = \{\mathbf{x} : \mathbf{a}^\top\mathbf{x} > 0\}$.

1.3 Weyl–Minkowski's representation theorem

In this section we study the relation between polytopes and polyhedra. Clearly standard cubes, simplices, and cross-polytopes are also polyhedra, but is this the case in general? In fact it is.

Theorem 1.3.1 (Weyl–Minkowski theorem). *Every polytope is a polyhedron. Every bounded polyhedron is a polytope.*

The Weyl–Minkowski theorem is very useful in applications to optimization. The reason is that the possible combinatorial solutions (e.g., matchings on graphs, routes in a network, etc.) are the vertices of a polytope. The Weyl–Minkowski theorem shows the existence of a different representation, one in terms of inequalities, that describes the convex hull of those vertices. From the inequality representation and linear programming we can find the optimum (e.g., the optimum matching, shortest route, etc.). We seek to find as many of those linear inequalities as possible with the goal of finding properties of the vertices.

Definition 1.3.2. A representation of a polytope by a set of linear inequalities is called an *H-representation*. On the other hand, when a polytope is given as a convex hull of a set of points, we have a *V-representation*.

The Weyl–Minkowski theorem states that all bounded polyhedra are polytopes. So they possess both a *V-representation* and an *H-representation*. These two representations give the same object, but one can be more efficient than the other; for instance, an n-cube has a V-representation with 2^n vertices but an H-representation with only $2n$ inequalities.

Before we discuss a proof of the Weyl–Minkowski theorem we need to introduce a useful operation. Given a subset A of $\mathbb{R}^n$, the *polar* of A is the set A° in $\mathbb{R}^n$ of the linear functionals whose value on A is not greater than 1; in other words,

$$A^\circ = \left\{ \mathbf{x} \in \mathbb{R}^n : \mathbf{a}^\top \mathbf{x} \leq 1 \text{ for every point } \mathbf{a} \in A \right\}.$$

Another way of thinking of the polar is as the intersection of the half spaces, one for each point $\mathbf{a} \in A$, of the form $\left\{ \mathbf{x} \in \mathbb{R}^n : \mathbf{a}^\top \mathbf{x} \leq 1 \right\}$.

Example 1.3.3. Here are some useful examples:

- For any point $\mathbf{x} \in \mathbb{R}^n$, its polar $\mathbf{x}^\circ$ is a closed half space whose bounding hyperplane is perpendicular to the vector $\mathbf{x}$ and which intersects the segment from the origin $\mathbf{0}$ to $\mathbf{x}$ at a point $\mathbf{p}$ such that the distances satisfy $d(\mathbf{0},\mathbf{p}) \cdot d(\mathbf{0},\mathbf{x}) = 1$.

- Given L, a line in $\mathbb{R}^2$ passing through the origin, what is L°? The answer is the line perpendicular to L which passes through the origin.

- If the line L does not pass through the origin, the answer is different. What is it? It is a half-line orthogonal to L that passes through the origin. To see this, simply rotate the line until it has equation $x = c$ and note that the calculation of the polar boils down to checking angles and lengths between vectors. So we must get the same answer modulo a rotation.

- Finally, what happens with a circle of radius one with center at the origin? Its polar set is the disk of radius one with center at the origin.

Lemma 1.3.4. *For any sets $A, B \subseteq \mathbb{R}^n$, the following statements are true:*

1. *The polar A° is closed, convex, and contains the origin $\mathbf{0}$.*

2. *If $A \subseteq B$, then $B^\circ \subseteq A^\circ$.*

3. *If $A = \mathrm{conv}(S)$, then $A^\circ = S^\circ$.*

Proof. For part (1) observe that A° is the intersection of closed convex sets and therefore closed and convex. Moreover, it is immediate that the origin is always in the polar. Part (2) is a simple consequence of the definition. To show part (3), note that part (2) already implies that $A^\circ \subseteq S^\circ$. Now pick $\mathbf{x} \in S^\circ$. We need to show that $\mathbf{z}^\top\mathbf{x} \leq 1$ for all $\mathbf{z} \in A$. We have that $\mathbf{z} = \sum \lambda_i \mathbf{x}_i$, with $\mathbf{x}_i \in S$ and $\sum \lambda_i = 1$, $\lambda_i \geq 0$. By linearity of the inner product, $\mathbf{z}^\top\mathbf{x} = \sum \lambda_i(\mathbf{x}_i^\top\mathbf{x}) \leq \sum \lambda_i = 1$. Thus $\mathbf{x} \in A^\circ$, implying $S^\circ \subseteq A^\circ$ and consequently $A^\circ = S^\circ$. □

The concept of polar is rather useful. We use the following lemma.

Lemma 1.3.5.

1. *If P is a polytope with $\mathbf{0} \in P$, then $(P^\circ)^\circ = P$.*
2. *Let $P \subseteq \mathbb{R}^n$ be a polytope. Then P° is a polyhedron.*

Proof.

1. Note $P \subseteq (P^\circ)^\circ$ (immediately from the definition of polar). We just need to check that $(P^\circ)^\circ \subseteq P$. Suppose $\mathbf{y} \in (P^\circ)^\circ$ but $\mathbf{y} \notin P$. Consider a hyperplane that separates P and $\mathbf{y}$; thus it is given by the equation $\mathbf{c}^\top\mathbf{x} = \alpha$ such that $\mathbf{c}^\top\mathbf{y} < \alpha$ and $\mathbf{c}^\top\mathbf{p} > \alpha$ for all $\mathbf{p} \in P$ (such a hyperplane exists, since P is compact and the distance function from $\mathbf{p}$ is a continuous function). Since $\mathbf{0} \in P$, we must have $\alpha < 0$. Therefore, letting $\mathbf{b} = \alpha^{-1}\mathbf{c}$ we get $\mathbf{b}^\top\mathbf{p} < 1$ for all $\mathbf{p} \in P$. So $\mathbf{b} \in P^\circ$. But on the other hand $\mathbf{b}^\top\mathbf{y} > 1$, which contradicts the assumption $\mathbf{y} \in (P^\circ)^\circ$.
2. We claim that if $P = \operatorname{conv}(\mathbf{a}_1, \ldots, \mathbf{a}_m)$, then $P^\circ = \left\{\mathbf{x} \in \mathbb{R}^n : \mathbf{a}_i^\top\mathbf{x} \leq 1\right\}$. Recall that Lemma 1.1.5 says that any element $\mathbf{p}$ of P is of the form $\mathbf{p} = \lambda_1\mathbf{a}_1 + \lambda_2\mathbf{a}_2 + \cdots + \lambda_m\mathbf{a}_m$ with $\lambda_i \geq 0$ and $\sum \lambda_i = 1$. Hence if $\mathbf{a}_i^\top\mathbf{x} \leq 1$ for all i, then it is true for all $\mathbf{p} \in P$, since
$$\mathbf{p}^\top\mathbf{x} = \sum \lambda_i \mathbf{a}_i^\top\mathbf{x} \leq \sum \lambda_i = 1.$$ □

An important observation is that cones have a nice shape for their polar:

Lemma 1.3.6. *Let $K \subseteq \mathbb{R}^n$ be a cone. The polar K° of cone K, is equal to*

$$\left\{\mathbf{x} \in \mathbb{R}^n : \mathbf{y}^\top\mathbf{x} \leq 0 \ \forall \mathbf{y} \in K\right\}.$$

Moreover K° is again a cone.

Proof. Clearly the polar K° is contained inside $\left\{\mathbf{x} \in \mathbb{R}^n : \mathbf{y}^\top\mathbf{x} \leq 0 \ \forall \mathbf{y} \in K\right\}$. For the other inclusion, if $\mathbf{y}^\top\mathbf{x} \leq 1$ for all $\mathbf{y} \in K$ then, in particular, $(N\mathbf{y})^\top\mathbf{x} \leq 1$, and thus $\mathbf{y}^\top\mathbf{x} \leq \frac{1}{N}$ for all positive integers N. This proves the first statement.

We claim that K° is again a cone. For this, note that $K^\circ = \bigcap_{\mathbf{y} \in K} \left\{\mathbf{x} : \mathbf{y}^\top\mathbf{x} \leq 0\right\}$. We know that $\left\{\mathbf{x} : \mathbf{y}^\top\mathbf{x} \leq 0\right\}$ is a cone, and that the intersection of cones is again a cone. Thus, K° is a cone. □

Note that a linear subspace L is a special case of a cone, and thus

$$L^\circ = L^\perp = \left\{\mathbf{x} \in \mathbb{R}^n : \mathbf{y}^\top\mathbf{x} = 0\right\}.$$

One of the most important examples of cones are those generated by finitely many vectors. In that case if $K = \mathrm{cone}(\{\mathbf{a}_1,\ldots,\mathbf{a}_m\})$, then

$$K^\circ = \left\{\mathbf{x} : \mathbf{a}_i^\top \mathbf{x} \le 0,\ i = 1,\ldots,m\right\}.$$

Note that if K is generated by the columns of a matrix A, then $K^\circ = \left\{\mathbf{x} : A^\top \mathbf{x} \le \mathbf{0}\right\}$.

Example 1.3.7. Let us look at a few examples.

1. $(\mathbb{R}^n_+)^\circ = \mathbb{R}^n_-$.
2. A linear subspace $L = \left\{\mathbf{a}^1,\ldots,\mathbf{a}^m,-(\mathbf{a}^1+\cdots+\mathbf{a}^m)\right\}$ is a finitely generated cone.
3. For a half-line $K = \mathrm{cone}(\{\mathbf{b}\})$, we have $K^\circ = \left\{\mathbf{y} : \mathbf{b}^\top \mathbf{y} \le 0\right\}$.
4. $\{\mathbf{0}\}^\circ = \mathbb{R}^n$ and $(\mathbb{R}^n)^\circ = \{\mathbf{0}\}$.
5. If $K = \{(x_1,x_2) : x_1 > 0\} \cup \{(0,0)\}$, then $K^\circ = \mathrm{cone}(\{(-1,0)\})$. But we have $(K^\circ)^\circ = \{(x_1,x_2) : x_1 \ge 0\} \supsetneq K$.

An important operation is that of a *Minkowski sum* of (convex) sets. Given sets $S_1, S_2 \subseteq \mathbb{R}^n$, their Minkowski sum $S_1 + S_2$ is the set $\{\mathbf{s}_1 + \mathbf{s}_2 : \mathbf{s}_1 \in S_1,\ \mathbf{s}_2 \in S_2\}$.

Proposition 1.3.8. *If K_1, K_2 are convex cones in $\mathbb{R}^n$, then we have the following propositions.*

1. $K_1 + K_2 = \{\mathbf{x}+\mathbf{y} : \mathbf{x} \in K_1, \mathbf{y} \in K_2\}$ *is a cone.*
2. *If $K_1 \subseteq K_2$, then $K_1^\circ \supseteq K_2^\circ$.*
3. $(K_1 + K_2)^\circ = K_1^\circ \cap K_2^\circ$.
4. $K_1^\circ + K_2^\circ \subseteq (K_1 \cap K_2)^\circ$.
5. $K \subseteq K^{\circ\circ}$.

Example 1.3.9. Example 1.3.7, part (5), shows that in Proposition 1.3.8, part (5), equality does not hold in general. Let $K_1 = \{(x_1,x_2) : x_1 > 0\} \cup \{(0,0)\} = -K_2$. Thus $K_1 \cap K_2 = \{\mathbf{0}\}$ and $(K_1 \cap K_2)^\circ = \mathbb{R}^2$. This implies $K_1^\circ + K_2^\circ = \mathbb{R} \times \{\mathbf{0}\}$, but $\mathbb{R} \times \{\mathbf{0}\} \subsetneq (K_1 \cap K_2)^\circ$.

We will show that Proposition 1.3.8, parts (3) and (4) for finitely generated cones mean that we have $K = K^{\circ\circ}$. In general, we only have $K^{\circ\circ} = \mathrm{cl}(K)$.

Theorem 1.3.10. *K is a finitely generated cone, and this implies that $K = K^{\circ\circ}$. In particular, if $K = \left\{\mathbf{x} : A^\top \mathbf{x} \le \mathbf{0}\right\}$, then K° is the cone generated by the columns of A.*

Proof. We know $K \subseteq K^{\circ\circ}$. Assume $\mathbf{y} \notin K$. We want to prove that $\mathbf{y} \notin K^{\circ\circ}$ implies $\mathbf{y} \notin K$. Note K is finitely generated and thus $K = \{\mathbf{z} : A\mathbf{x} = \mathbf{z}, \mathbf{x} \ge \mathbf{0}\}$ for some matrix A. (The columns of A generate the cone K.) $\mathbf{y} \notin K$ implies that $A\mathbf{x} = \mathbf{y}, \mathbf{x} \ge \mathbf{0}$ does not have a solution in $\mathbf{x}$. Thus Lemma 1.2.7 implies the existence of a vector $\mathbf{a} \ne \mathbf{0}$ such that $\mathbf{a}^\top \mathbf{y} > 0$ but $\mathbf{a}^\top \mathbf{x} \le 0$ for all $\mathbf{x} \in K$. (See the geometric interpretation of Lemma 1.2.7 given right after its proof.) Thus we have $\mathbf{a} \in K^\circ$, which implies $\mathbf{y} \notin K^{\circ\circ}$, as we wanted, because $\mathbf{a}^\top \mathbf{y} > 0$. This completes the proof of the first statement. For the second claim, let $\hat{K} = \text{cone}(\{\mathbf{a}^1, \ldots, \mathbf{a}^m\})$, where $\mathbf{a}^1, \ldots, \mathbf{a}^m$ are the columns of A. Clearly, by applying the definition of $\circ$ (the polar operator) and of cones, we have $\hat{K}^\circ = K$. Since $K = K^{\circ\circ}$, we conclude that $K^\circ = \hat{K} = \text{cone}(\{\mathbf{a}^1, \ldots, \mathbf{a}^m\})$. □

For our subsequent proofs we need the following immediate corollary of Lemma 1.2.5.

Corollary 1.3.11. *Given a matrix A, there exists a matrix B such that $BA \geq O$ and such that $\{\mathbf{z} : A\mathbf{x} = \mathbf{z}, \mathbf{x} \geq \mathbf{0}$ has a solution $\mathbf{x}\} = \{\mathbf{z} : B\mathbf{z} \geq \mathbf{0}\}$.*

Proof. By Lemma 1.2.5, the system $A\mathbf{x} = \mathbf{z}$, $\mathbf{x} \geq \mathbf{0}$ has a solution if and only if

$$\tilde{A}\mathbf{x} = \begin{pmatrix} A \\ -A \\ -I \end{pmatrix} \mathbf{x} \leq \begin{pmatrix} \mathbf{z} \\ -\mathbf{z} \\ 0 \end{pmatrix}$$

has a solution. This implies the existence of $\tilde{B} = (B_1|B_2|B_3) \geq O$ so that such a solution exists if and only if

$$\tilde{B} \begin{pmatrix} \mathbf{z} \\ -\mathbf{z} \\ 0 \end{pmatrix} = B_1\mathbf{z} - B_2\mathbf{z} \geq 0.$$

It follows that $(B_1 - B_2)\mathbf{z} \geq \mathbf{0}$ and $\tilde{B}\tilde{A} = O$, which implies $(B_1 - B_2)A - B_3 \geq O$ and thus $(B_1 - B_2)A \geq O$. Hence, $B = B_1 - B_2$ satisfies the claimed conditions of the corollary. □

We now state a second version of Weyl–Minkowski's theorem, this time for cones. In fact, it is equivalent to the Weyl–Minkowski theorem, Theorem 1.3.1. For this observe that $P = \text{conv}(V)$ is a polytope if and only if $\text{cone}(\{\binom{\mathbf{v}}{1} : \mathbf{v} \in V\})$ is a finitely generated cone.

Theorem 1.3.12 (Weyl). *Every finitely generated cone is polyhedral.*

Proof. Let $K = \text{cone}(\{\mathbf{a}^1, \mathbf{a}^2, \ldots, \mathbf{a}^n\}) \subseteq \mathbb{R}^m$, and let A be the matrix with columns $\mathbf{a}^1, \mathbf{a}^2, \ldots, \mathbf{a}^n$. Then we have $K = \{\mathbf{z} : A\mathbf{x} = \mathbf{z},\ \mathbf{x} \geq \mathbf{0}$ has a solution $\mathbf{x}\}$, and consequently, by Corollary 1.3.11, there exists a matrix B such that $K = \{\mathbf{z} : B\mathbf{z} \geq \mathbf{0}\}$, and thus K is a polyhedral cone. □

We remark that the matrix B does not necessarily give the minimal set of hyperplanes defining the cone. Note also that changing from one type of representation (e.g., the rays generating the cone) to the representation of the (minimal) intersection of half spaces is difficult computationally. We note the following property.

Corollary 1.3.13 (Minkowski). *Every polyhedral cone is finitely generated.*

Proof. Let K be polyhedral. Then, by Theorem 1.3.10, K° is finitely generated; thus it is polyhedral by the previous theorem. This implies that $K^{\circ\circ} = K$ is finitely generated. Note that we used the fact that K being polyhedral implies K° is finitely generated, which, by Theorem 1.3.10, gives $K^{\circ\circ} = K$. □

Observe now the following facts:

- If K is finitely generated, then K° is finitely generated (polyhedral implies finite generation).
- If K_1, K_2 is finitely generated, then $K_1 + K_2$ is finitely generated.
- $K_1 \cap K_2 = \left\{ \mathbf{z} : \binom{A^1}{A^2} \mathbf{z} \leq \mathbf{0} \right\}$ is finitely generated when K_1, K_2 are finitely generated because $K_1 \cap K_2$ is polyhedral (since $K_i = \{\mathbf{z} : A^i \mathbf{z} \leq \mathbf{0}\}$ is polyhedral), and thus $K_1 \cap K_2$ is finitely generated.

Corollary 1.3.14. *If K_1, K_2 are finitely generated, we have*

$$(K_1^\circ + K_2^\circ) = (K_1 \cap K_2)^\circ.$$

Proof. $(K_1^\circ + K_2^\circ) = ((K_1^\circ + K_2^\circ)^\circ)^\circ$, and thus we have $(K_1^{\circ\circ} \cap K_2^{\circ\circ})^\circ$ (which is true for all cones). This equals $(K_1 \cap K_2)^\circ$, as K_1, K_2 are finitely generated. □

Finally, let us remark that the Weyl–Minkowski theorem, Theorem 1.3.1, has several nice consequences.

Corollary 1.3.15. *Let P be an n-dimensional polytope in $\mathbb{R}^n$. Then the following hold:*

1. *The intersection of P with a hyperplane is a polytope. If the hyperplane passes through a point in the relative interior of P, then the intersection is an $(n-1)$-polytope.*
2. *Every projection of P is a polytope. More generally, the image of P under a linear map is another polytope.*

Proof. Part (1) follows because a polytope is a bounded polyhedron, but the intersection of a polyhedron with a hyperplane gives a polyhedron of lower dimension which is still bounded; thus the result is a polytope. For part (2), the points of P are convex combinations of vertices $\mathbf{v}_1, \ldots, \mathbf{v}_m$. Then applying a linear transformation π, we see that linearity implies that any point of the image $\pi(P)$ is a convex combination of $\pi(\mathbf{v}_1), \ldots, \pi(\mathbf{v}_m)$. □

1.4 Decomposition of polyhedra as sums of cones and polytopes

In this section we will see that cones and polytopes are, at some level, the building blocks of all polyhedra (and we will certainly exploit this structure later). We will see that every polyhedron decomposes as the Minkowski sum of cones and polytopes.

First, let us argue that all polyhedra have natural coordinates that represent them as cones. The process of *homogenization* transfers any polyhedron $K \subseteq \mathbb{R}^d$ into a cone C_K. To create this we take each point $\mathbf{a} \in K \subseteq \mathbb{R}^d$ and map it to $\tilde{\mathbf{a}} = (\mathbf{a}, 1) \in \mathbb{R}^{d+1}$. Let $C_K = \text{cone}(\{\tilde{\mathbf{a}} : \mathbf{a} \in K\})$. Note that if K is a polytope, then C_K does not contain a line. Figure 1.1 shows a picture of the homogenization of a quadrilateral, which becomes a three-dimensional cone.

We next explain a structure theorem that relates the concepts of cone, polytope, and polyhedron and gives a canonical representation of all polyhedra. This is useful from the algorithmic point of view.

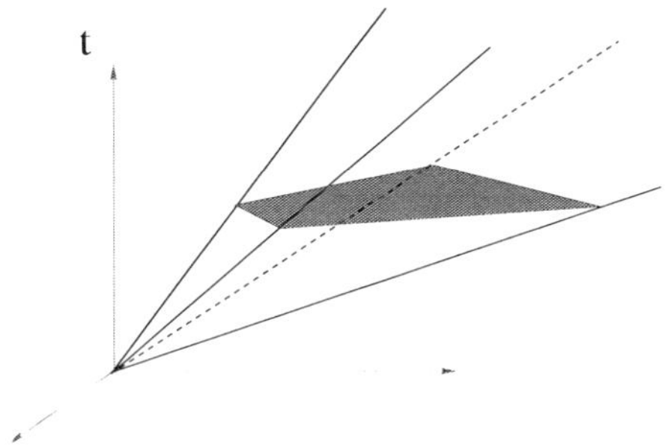

Figure 1.1. *Homogenization of a polytope.*

Definition 1.4.1. Let K be a convex cone. The set $L = K \cap (-K)$ is called the *lineality space* of K. We say that K is *pointed* if $L = K \cap (-K) = \{\mathbf{0}\}$.

Note that $L = K \cap (-K)$ is a linear subspace. For this it suffices to show that $\lambda \mathbf{x} \in L$ for $\mathbf{x} \in L$ and $\lambda < 0$, which is simple: $\lambda \mathbf{x} = (-\lambda)(-\mathbf{x}) \in L$ since K is a cone.

Theorem 1.4.2 (representation theorem for cones). *Every convex cone K is of the form $\overline{K} + L$, where L is a linear subspace and $\overline{K}$ is a pointed cone. In fact, we can take $L = K \cap (-K)$ and $\overline{K} = K \cap L^{\perp}$. Neither $\overline{K}$ nor L is uniquely specified by $K = \overline{K} + L$.*

Proof. $\mathbb{R}^n = L \oplus L^{\perp}$; i.e., $\mathbf{x} \in \mathbb{R}^n$ can be written $\mathbf{x} = \mathbf{y} + \mathbf{z}$ for $\mathbf{y} \in L$ and $\mathbf{z} \in L^{\perp}$. Take $L = K \cap (-K)$. We claim that $\mathbf{x} \in K$ if and only if $\mathbf{z} \in K$ (this is easy to prove since $\mathbf{z} = \mathbf{x} - \mathbf{y} \in K$ if $\mathbf{x} \in K$). So $\mathbf{x} \in K$ if and only if $\mathbf{z} \in K \cap L^{\perp} = \overline{K}$. We have proved $K \subseteq \overline{K} + L$. Conversely, $L \subseteq K = K \cap (-K)$, which implies

$$\overline{K} + L = (K \cap L^{\perp}) + L = (K + L) \cap (L^{\perp} + L) = K + L \subseteq K + K = K,$$

and thus $K = \overline{K} + L$, as claimed. Now we ask, why is $\overline{K}$ pointed? We verify that the lineality space is just the origin:

$$\begin{aligned} \overline{K} \cap (-\overline{K}) &= (K \cap L^{\perp}) \cap (-K \cap -L^{\perp}) \\ &= (K \cap -K) \cap (L^{\perp} \cap -L^{\perp}) = L \cap L^{\perp} = \{\mathbf{0}\}, \end{aligned}$$

since $K \cap -K = L$ and $-L^{\perp} = L^{\perp}$. □

Theorem 1.4.2 has a more precise version for polyhedra:

Theorem 1.4.3 (resolution of polyhedra). *Every nonempty polyhedron*

$$P = \left\{ \mathbf{x} \in \mathbb{R}^n : A\mathbf{x} \leq \mathbf{b} \right\}$$

is of the form $P = C + K$, where C is a nonempty polytope and $K = \{\mathbf{x} : A\mathbf{x} \leq \mathbf{0}\}$ is a convex cone.

Proof. Denoting by $\mathbf{a}_i^{\top}$ the i-th row of A, we can write

$$P = \left\{ \mathbf{x} \in \mathbb{R}^n : \mathbf{a}_i^{\top}\mathbf{x} \leq \mathbf{b}_i, i = 1, \ldots, k \right\}.$$

Let $\hat{\mathbf{a}}_i$ be the vector $\binom{\mathbf{a}_i}{-b_i} \in \mathbb{R}^{n+1}$. Then

$$\hat{K} = \left\{ \mathbf{y} \in \mathbb{R}^{n+1} : \hat{\mathbf{a}}_i^\top \mathbf{y} \leq 0, i = 1, \ldots, k, y_{n+1} \geq 0 \right\}$$

is a polyhedral cone, so it is a finitely generated cone by Weyl–Minkowski's theorem. Therefore, we get

$$\hat{K} = \operatorname{cone}\left(\left\{ \binom{\bar{\mathbf{z}}^1}{t_1}, \ldots, \binom{\bar{\mathbf{z}}^l}{t_l} \right\} \right).$$

If $t_i > 0$, replace the cone generator $\binom{\bar{\mathbf{z}}^i}{t_i}$ with $\frac{1}{t_i}\binom{\bar{\mathbf{z}}^i}{t_i}$. Then we obtain

$$\hat{K} = \operatorname{cone}\left\{ \binom{\mathbf{z}^1}{1}, \ldots, \binom{\mathbf{z}^r}{1}, \binom{\bar{\mathbf{z}}^1}{0}, \ldots, \binom{\bar{\mathbf{z}}^s}{0} \right\}.$$

If the number s of generators of the form $\binom{\bar{\mathbf{z}}^i}{0}$ is 0, include $\binom{\mathbf{0}}{0}$ as a generator of $\hat{K}$ for mere technical reasons of presentation. Denote by r the number of vectors of the form $\binom{\bar{\mathbf{z}}^i}{1}$. We have several claims to consider:

• First, we claim $\mathbf{x} \in P$ if and only if $\binom{\mathbf{x}}{1} \in \hat{K}$. This is the case because $\mathbf{x} \in P$ if and only if $\mathbf{a}_1^\top \mathbf{x} \leq \mathbf{b}_i$; the latter is equivalent to $\binom{\mathbf{a}_i}{-b_i}^\top \binom{\mathbf{x}}{1} \leq 0$, which proves the claim. As a consequence, since $P \neq \emptyset$, we can conclude that $r > 0$.

• Second, we can show analogously that $\binom{\mathbf{z}}{0} \in \hat{K}$ if and only if $A\mathbf{z} \leq \mathbf{0}$. Now define $C = \operatorname{conv}(\{\mathbf{z}^1, \ldots, \mathbf{z}^r\})$ and $K = \operatorname{cone}(\{\bar{\mathbf{z}}^1, \ldots, \bar{\mathbf{z}}^s\})$. Then $r > 0$ implies that $C \neq \emptyset$, and $s > 0$ implies that $\mathbf{0} \in K$.

• Third, we claim that $P = C + K$. For this recall that $\mathbf{x} \in P$ if and only if $\binom{\mathbf{x}}{1} \in \hat{K}$. This last membership happens if and only if

$$\binom{\mathbf{x}}{1} = \sum_{i=1}^{r} \lambda_i \binom{\mathbf{z}^i}{1} + \sum_{j=1}^{s} \mu_j \binom{\bar{\mathbf{z}}^j}{0}$$

for $\lambda_i, \mu_j \geq 0$. This is equivalent to $\mathbf{x} = \left(\sum_{i=1}^r \lambda_i \mathbf{z}^i\right) + \left(\sum_{j=1}^s \mu_j \bar{\mathbf{z}}^j\right) \in C + K$, as $\sum \lambda_i = 1$ and $\lambda_i, \mu_j \geq 0$.

• Finally, we claim that $K = \{\mathbf{x} \in \mathbb{R}^n : A\mathbf{x} \leq \mathbf{0}\}$. This holds because $\mathbf{x} \in K$ happens precisely when $\mathbf{x} = \sum \mu_j \bar{\mathbf{z}}^j$, $\mu_j \geq 0$, which is equivalent to $\binom{\mathbf{x}}{0} \in \hat{K}$. This is the same as $A\mathbf{x} \leq \mathbf{0}$. □

We remark that if $P = C + K$, where C is a polytope and K is a cone, then K is the *recession cone* of P. Note that if $\mathbf{z} \in K$, then $\lambda \mathbf{z} \in K$ for all $\lambda \geq 0$. So for $\mathbf{x} \in P$, we have $\mathbf{x} = \mathbf{y} + \bar{\mathbf{z}}$, where $\mathbf{y} \in C$ and $\bar{\mathbf{z}} \in K$ and thus $\mathbf{x} + \lambda \mathbf{z} = \mathbf{y} + (\bar{\mathbf{z}} + \lambda \mathbf{z}) \in P$, for all $\lambda \geq 0$ (this means that a half-line is strictly contained in the cone).

We conclude with a special case.

Corollary 1.4.4. *For a polyhedron P, the following are equivalent:*

1. *P is a polytope.*

2. *P is bounded.*

3. *If $P = C + K$, where C is a polytope and K is a cone, then $K = \{\mathbf{0}\}$.*

Corollary 1.4.5. *$P = \{\mathbf{x} : A\mathbf{x} \leq \mathbf{b}\}$ is bounded if and only if $K = \{\mathbf{x} : A\mathbf{x} \leq \mathbf{0}\} = \{\mathbf{0}\}$.*

Similarly, $K = \{\mathbf{x} : A\mathbf{x} \leq \mathbf{0}\}$ means that the lineality space $K \cap (-K) = \{\mathbf{x} : A\mathbf{x} = \mathbf{0}\}$.

Definition 1.4.6. A polyhedron P is *pointed* if $P = C + K$ and K is a pointed cone.

Corollary 1.4.7. *$P = \{\mathbf{x} : A\mathbf{x} \leq \mathbf{b}\}$ can be written as $P = Q + \bar{K} + L$, where Q is a polytope, $L = \{\mathbf{x} : A\mathbf{x} = \mathbf{0}\}$ is a linear subspace, and $\bar{K} = \{\mathbf{z} : A\mathbf{z} \leq \mathbf{0}\}$ is a pointed cone.*

More precisely, there are three finite sets $\{\mathbf{q}_1, \mathbf{q}_2, \ldots, \mathbf{q}_k\}$, $\{\mathbf{w}_1, \mathbf{w}_2, \ldots, \mathbf{w}_r\}$, and vector space basis $\{\mathbf{l}_1, \mathbf{l}_2, \ldots, \mathbf{l}_s\}$ such that

$$\begin{aligned} Q &= \mathrm{conv}(\{\mathbf{q}_1, \mathbf{q}_2, \ldots, \mathbf{q}_k\}), \\ K &= \mathrm{cone}(\{\mathbf{w}_1, \mathbf{w}_2, \ldots, \mathbf{w}_r\}), \\ L &= \mathrm{span}(\{\mathbf{l}_1, \mathbf{l}_2, \ldots, \mathbf{l}_s\}). \end{aligned}$$

Proof. First note that $L = \{\mathbf{x} : A\mathbf{x} = \mathbf{0}\}$, and thus $L^\perp = \mathrm{span}(\{\mathbf{a}_1, \ldots, \mathbf{a}_m\})$, where $\mathbf{a}_i^\top$, $i = 1, \ldots, m$, denote the rows of A. This implies that $L^\perp$, being a linear subspace of $\mathbb{R}^n$, is a finitely generated polyhedral cone.

By Theorem 1.4.3, we know that we can write $P = C + K$ for some polytope C and for $K = \{\mathbf{x} : A\mathbf{x} \leq \mathbf{0}\}$. Let $\bar{K} = K \cap L^\perp$. Then $P = C + \bar{K} + L$, and it remains to show that $\bar{K}$ is a pointed cone. But this follows immediately:

$$\bar{K} \cap -\bar{K} = (K \cap L^\perp) \cap (-K \cap -L^\perp) = (K \cap -K) \cap L^\perp = L \cap L^\perp = \{\mathbf{0}\}.$$

The rest follows from the Weyl–Minkowski theorem for polytopes and cones that transfers a polyhedron into a representation by vertices and rays, respectively. □

So the representation theorems can be described as saying that any polyhedron can be decomposed as the Minkowski sum of cones, polytopes, and linear subspaces.

We close this section with some useful notes for computation: first note that the polyhedron $P = \{\mathbf{x} \in \mathbb{R}^n : A\mathbf{x} \leq \mathbf{b}\}$ is bounded if and only if its *recession cone* $\{\mathbf{x} : A\mathbf{x} \leq \mathbf{0}\} = \{\mathbf{0}\}$. Similarly, P will be pointed if and only if its *lineality space* $\{\mathbf{x} : A\mathbf{x} = \mathbf{0}\} = \{\mathbf{0}\}$. Note that polyhedra of the form $P = \{\mathbf{x} : A\mathbf{x} = \mathbf{b},\ \mathbf{x} \geq \mathbf{0}\}$ are always pointed. They are bounded if and only if $\{\mathbf{x} : A\mathbf{x} = \mathbf{0},\ \mathbf{x} \geq \mathbf{0}\} = \{\mathbf{0}\}$.

1.5 Faces, dimension, extreme points

We are going to describe some of the salient characteristics of polyhedra. For this purpose, we would like to observe that all polyhedra can be put into a standard form which provides much information. Given any system of inequalities $A\mathbf{x} \leq \mathbf{b}$, it can be transformed into a new system of the form $B\bar{\mathbf{x}} = \mathbf{d}, \bar{\mathbf{x}} \geq \mathbf{0}$, with the property that one system has a solution if and only if the other system has a solution. In fact the polyhedron $\{\mathbf{x} : A\mathbf{x} \leq \mathbf{b}\}$ can be rewritten in the form $\{\bar{\mathbf{x}} : B\bar{\mathbf{x}} = \mathbf{d}, \bar{\mathbf{x}} \geq \mathbf{0}\}$.

Here is why: the inequality $\sum_{j=1}^n a_{ij} x_j \leq b_i$ can be turned into an equation by adding a slack variable $s_i \geq 0$, that is, it is equivalent to $\sum_{j=1}^n a_{ij} x_j + s_i = b_i$ and $s_i \geq 0$. Finally, note that an unrestricted variable x_j can be replaced by two nonnegative variables: $x_j = x_j^+ - x_j^-$.

Definition 1.5.1. Let S be a convex set in $\mathbb{R}^n$.

- A linear inequality $\mathbf{c}^\top\mathbf{x} \leq \alpha$ is said to be *valid* for S if every point in $\mathbf{s} \in S$ satisfies it: $\mathbf{c}^\top\mathbf{s} \leq \alpha$.

- A set $F \subseteq S$ is a *face* of S if and only if there exists a valid inequality $\mathbf{c}^\top\mathbf{x} \leq \alpha$ for S such that $F = \left\{\mathbf{x} \in S : \mathbf{c}^\top\mathbf{x} = \alpha\right\}$. In this case the hyperplane defined by $\mathbf{c}^\top\mathbf{x} = \alpha$ is a *supporting hyperplane* of S on F. Both the entire convex set S and the empty set are considered to be faces.

- For a face F of the convex set S, consider the *affine hull* aff(F) of F in $\mathbb{R}^n$, i.e., the smallest affine subspace containing F. Its dimension is called the *dimension* of F.

Definition 1.5.2. A point $\mathbf{x}$ in a convex set S is an *extreme point* of S if it is not an interior point of any line segment in S. This is equivalent to saying that when $\mathbf{x} = \lambda\mathbf{x}_1 + (1-\lambda)\mathbf{x}_2$, then either $\lambda = 1$ or $\lambda = 0$.

Proposition 1.5.3. *Every vertex (0-dimensional face) of a polyhedron is an extreme point.*

We want to be able to compute the essential features of a polyhedron or a polytope. Among them we wish to find out what are their faces or extreme points, what is their dimension, and whether the polyhedron in question is empty or not. We want to answer such questions with concrete practical algorithms.

Theorem 1.5.4. *Consider the polyhedron $P = \{\mathbf{x} : A\mathbf{x} = \mathbf{b}, \mathbf{x} \geq \mathbf{0}\}$. Suppose the m columns $\mathbf{a}_{i_1}, \mathbf{a}_{i_2}, \ldots, \mathbf{a}_{i_m}$ of the $m \times n$ matrix A are linearly independent and there exist nonnegative numbers x_{i_j} such that*

$$\mathbf{a}_{i_1}x_{i_1} + \mathbf{a}_{i_2}x_{i_2} + \cdots + \mathbf{a}_{i_m}x_{i_m} = \mathbf{b}.$$

Then the point $\bar{\mathbf{x}}$ with entry x_{i_j} in position i_j and zero elsewhere is an extreme point of the polyhedron $P = \{\mathbf{x} : A\mathbf{x} = \mathbf{b}, \mathbf{x} \geq \mathbf{0}\}$.

Proof. Suppose $\bar{\mathbf{x}}$ is not an extreme point. Then $\bar{\mathbf{x}}$ lies in the interior of a line segment in $P = \{\mathbf{x} : A\mathbf{x} = \mathbf{b}, \mathbf{x} \geq \mathbf{0}\}$. Thus $\mathbf{x} = \lambda\mathbf{u} + (1-\lambda)\mathbf{v}$ with λ between 0 and 1. But this implies, by looking at the entries of $\bar{\mathbf{x}}$ that are zero, that $\mathbf{u}, \mathbf{v}$ also have that property. Now, consider $\mathbf{y} = \bar{\mathbf{x}} - \mathbf{u}$. We have $A(\bar{\mathbf{x}} - \mathbf{u}) = \mathbf{0}$, but since the columns $\mathbf{a}_{i_1}, \mathbf{a}_{i_2}, \ldots, \mathbf{a}_{i_m}$ are linearly independent this implies that $\bar{\mathbf{x}} - \mathbf{u} = \mathbf{0}$, a contradiction. □

Theorem 1.5.5. *Suppose $\bar{\mathbf{x}} = (x_1, \ldots, x_n)^\top$ is an extreme point of a polyhedron*

$$P = \{\mathbf{x} : A\mathbf{x} = \mathbf{b}, \mathbf{x} \geq \mathbf{0}\}$$

with A an $m \times n$ matrix. Then the following statements hold true:

1. *The columns of A which correspond to positive entries of $\bar{\mathbf{x}}$ form a linearly independent set of vectors in $\mathbb{R}^m$.*

2. *At most m of the entries of $\bar{\mathbf{x}}$ can be positive, and the rest are zero.*

Proof. Suppose the columns are linearly dependent. Thus there are coefficients, not all zero, such that $\mathbf{a}_{i_1}c_{i_1} + \mathbf{a}_{i_2}c_{i_2} + \cdots + \mathbf{a}_{i_m}c_{i_m} = \mathbf{0}$. Thus we can form points $\mathbf{u}, \mathbf{v}$ whose entries are zero unless the index i_j coincides with one of those from the linear dependence, in which case we put for $\mathbf{u}$ a value $x_{i_j} - dc_{i_j}$ and for $\mathbf{v}$ a value $x_{i_j} + dc_{i_j}$. We easily verify that

$$(x_{i_1} - dc_{i_1})\mathbf{a}_{i_1} + (x_{i_2} - dc_{i_2})\mathbf{a}_{i_2} + \cdots + (x_{i_m} - dc_{i_m})\mathbf{a}_{i_m} = \mathbf{b},$$
$$(x_{i_1} + dc_{i_1})\mathbf{a}_{i_1} + (x_{i_2} + dc_{i_2})\mathbf{a}_{i_2} + \cdots + (x_{i_m} + dc_{i_m})\mathbf{a}_{i_m} = \mathbf{b}.$$

Since d is any scalar, we may choose d less than the minimum of $x_j/|c_j|$ for those indices j with $c_j \neq 0$. We have reached a contradiction because $\mathbf{x} = \mathbf{u}/2 + \mathbf{v}/2$ and both $\mathbf{u}, \mathbf{v}$ are inside the polyhedron. For the second claim of the theorem simply observe that there cannot be more than m linearly independent vectors inside $\mathbb{R}^m$. □

Definition 1.5.6. A *basis* of the system $A\mathbf{x} = \mathbf{b}$ is a list of m columns $\mathbf{a}_{i_1}, \mathbf{a}_{i_2}, \ldots, \mathbf{a}_{i_m}$ of the $m \times n$ matrix A that are linearly independent (and thus form a basis of the vector space $\mathbb{R}^m$). The corresponding variables $x_{i_1}, \ldots, x_{i_m}$ are called *basic* variables; all other variables are called *nonbasic* variables. The *basic solution* corresponding to the basis is the (unique) solution $\bar{\mathbf{x}}$ to the system $A\mathbf{x} = \mathbf{b}$ where all nonbasic variables are set to zero.

Corollary 1.5.7. *The extreme points of the polyhedron $P = \{\mathbf{x} : A\mathbf{x} = \mathbf{b}, \mathbf{x} \geq \mathbf{0}\}$ are precisely the basic feasible solutions. Similarly, every basic feasible solution of a polyhedron*

$$\{\mathbf{x} : A\mathbf{x} = \mathbf{b}, \mathbf{x} \geq \mathbf{0}\}$$

is a vertex. Thus, the sets of basic feasible solutions, vertices, and extreme points are identical.

1.6 Duality of linear optimization

We recall now some of the rich theory of linear programming (or LP for short). In its most general form an LP problem has the form

$$\max/\min\left\{\mathbf{c}^\top\mathbf{x} : A\mathbf{x} = \mathbf{b},\ B\mathbf{x} \geq \mathbf{d}\right\},$$

but it is easy to see that, either by adding slack variables or by separating equations into pairs of inequalities, one can always arrive at special forms. For example, we denote by (P) the standard form $\min\left\{\mathbf{c}^\top\mathbf{x} : A\mathbf{x} = \mathbf{b},\ \mathbf{x} \geq \mathbf{0}\right\}$.

A key point of the theory is that LPs come in pairs; to each primal linear program there is an associated *dual linear programming* problem. For example, the dual of (P) above is $\max\left\{\mathbf{b}^\top\mathbf{y} : A^\top\mathbf{y} \leq \mathbf{c}\right\}$, which, with slacks is

$$\max\left\{\mathbf{b}^\top\mathbf{y} : A^\top\mathbf{y} + \mathbf{s} = \mathbf{c},\ \mathbf{s} \geq \mathbf{0}\right\}.$$

But there are many other versions, depending on the initial form of the primal optimization problem. For example, the primal problem $\max\left\{\mathbf{c}^\top\mathbf{x} : A\mathbf{x} \leq \mathbf{b}\right\}$ has as dual

$$\min\left\{\mathbf{y}^\top\mathbf{b} : \mathbf{y}^\top A = \mathbf{c},\ \mathbf{y} \geq \mathbf{0}\right\}.$$

There are three fundamental theorems that relate the primal and dual programs. In what follows we use a third presentation of the primal and dual that we call (P) and (D):

$$\text{(P)}\quad \begin{array}{l} \max \mathbf{c}^\top\mathbf{x} \\ A\mathbf{x} \leq \mathbf{b}, \\ \mathbf{x} \geq \mathbf{0} \end{array} \qquad\qquad \text{(D)}\quad \begin{array}{l} \min \mathbf{b}^\top\mathbf{y} \\ A^\top\mathbf{y} \geq \mathbf{c}. \\ \mathbf{y} \geq \mathbf{0} \end{array}$$

First there is an easy result that is a direct consequence of both the way and the reason why duals are constructed. The dual of an LP is constructed to obtain upper bounds to the maximization problem of the primal LP. The dual variables y_i are multipliers that take linear combinations of the constraints on the primal.

Theorem 1.6.1 (weak duality). *For each pair of feasible solutions* $\mathbf{y}$ *of* (D) *and* $\mathbf{x}$ *of* (P), *they must satisfy* $\mathbf{c}^\top\mathbf{x} \leq \mathbf{b}^\top\mathbf{y}$. *In particular, when* (P) *is unbounded, it implies* (D) *is infeasible, and when* (D) *is unbounded* (P) *is infeasible.*

Proof. $\mathbf{c}^\top\mathbf{x} \leq (A^\top\mathbf{y})^\top\mathbf{x} = (\mathbf{y}^\top A)\mathbf{x} = \mathbf{y}^\top(A\mathbf{x}) \leq \mathbf{y}^\top\mathbf{b}$. □

Theorem 1.6.2 (von Neumann strong duality). *For the pairs of linear programs* (P), (D), *one and only one of the following possibilities occurs:*

1. *Neither* (P) *nor* (D) *has a feasible solution.*
2. (P) *is unbounded and* (D) *is infeasible.*
3. (D) *is unbounded and* (P) *is infeasible.*
4. *Both* (P) *and* (D) *have a feasible solution. Then both have optimal solutions,* $\mathbf{x}^*$ *and* $\mathbf{y}^*$ *and* $\mathbf{c}^\top\mathbf{x}^* = \mathbf{b}^\top\mathbf{y}^*$.

The proof of the strong duality theorem of von Neumann follows from Farkas' lemma.

Proof. The only difficult part is to prove case 4. Suppose the primal (P) has an optimal solution. We show that (D) has an optimal solution too and that the optimal objective values coincide. Suppose $\gamma = \mathbf{c}^\top\mathbf{x}^*$; then we know that the system

$$A\mathbf{x} \leq \mathbf{b}, \quad \mathbf{c}^\top\mathbf{x} \geq \gamma$$

has a nonnegative solution for sure, but $A\mathbf{x} \leq \mathbf{b}$, $\mathbf{c}^\top\mathbf{x} \geq \gamma + \epsilon$, $\epsilon > 0$ has no nonnegative solution at all. We next construct a new inequality system. Set

$$\bar{A} = \begin{pmatrix} A \\ -\mathbf{c}^\top \end{pmatrix}, \quad \bar{\mathbf{b}}_\epsilon = \begin{pmatrix} \mathbf{b} \\ -\gamma - \epsilon \end{pmatrix}.$$

In this new notation the systems become $\bar{A}\mathbf{x} \leq \bar{\mathbf{b}}_0$ and $\bar{A}\mathbf{x} \leq \bar{\mathbf{b}}_\epsilon$.

Clearly $\bar{A}\mathbf{x} \leq \bar{\mathbf{b}}_\epsilon, \mathbf{x} \geq \mathbf{0}$ is infeasible for $\epsilon > 0$. Apply Farkas' lemma in the following form, which we leave as an exercise: The system $\bar{A}\mathbf{x} \leq \mathbf{b}_\epsilon$ has no nonnegative solution if and only if there exists a nonnegative vector $\bar{\mathbf{y}} = (\mathbf{u}, z)$ such that $\bar{\mathbf{y}}^\top\bar{A} \geq \mathbf{0}$ but $\bar{\mathbf{y}}^\top\bar{\mathbf{b}}_\epsilon < \mathbf{0}$. Plugging in the expressions for $\bar{A}$ and $\bar{\mathbf{b}}_\epsilon$, we obtain $A^\top\mathbf{u} \geq z\mathbf{c}$, $\mathbf{b}^\top\mathbf{u} < z(\gamma + \epsilon)$.

But now we can apply Farkas' lemma again, this time to the system $\bar{A}\mathbf{x} \leq \bar{\mathbf{b}}_0$: $\bar{A}\mathbf{x} \leq \bar{\mathbf{b}}_0$ has a nonnegative solution if and only if every $\mathbf{y} \geq \mathbf{0}$ with $\mathbf{y}^\top\bar{A} \geq \mathbf{0}$ also satisfies $\mathbf{y}^\top\bar{\mathbf{b}}_0 \geq \mathbf{0}$. If we apply this to the vector $\bar{\mathbf{y}}$ we get $\bar{\mathbf{y}}^\top\mathbf{b}_0 \geq 0$, that is, $\mathbf{b}^\top\mathbf{u} \geq z\gamma$.

As $z\gamma \leq \mathbf{b}^\top\mathbf{u} < z(\gamma + \epsilon)$, we conclude that $z > 0$. Set $\mathbf{v} = \frac{\mathbf{u}}{z} \geq \mathbf{0}$ which yields

$$A^\top\mathbf{v} \geq \mathbf{c}, \quad \mathbf{b}^\top\mathbf{v} < \gamma + \epsilon.$$

This means that $\mathbf{v}$ is a feasible solution of the dual linear program (D) with the objective function value *strictly smaller* than $\gamma + \epsilon$. (In particular, this means that (D) is feasible.)

By weak duality, $\mathbf{c}^\top\mathbf{x} \leq \mathbf{b}^\top\mathbf{y}$, every feasible solution of (D) must have an objective function value of at least γ. As (D) is feasible and bounded, it has an optimal solution $\mathbf{y}^*$. We have $\gamma \leq \mathbf{b}^\top\mathbf{y}^* \leq \mathbf{b}^\top\mathbf{v} \leq \gamma + \epsilon$ for *all* $\epsilon > 0$. Hence, we must have $\mathbf{b}^\top\mathbf{y}^* = \gamma$. □

Corollary 1.6.3 (complementary slackness). *Given a pair of dual problems with*

$$\max\left\{\mathbf{c}^\top \mathbf{x} : A\mathbf{x} \le \mathbf{b}\right\} = \min\left\{\mathbf{y}^\top \mathbf{b} : A^\top \mathbf{y} = \mathbf{c}, \mathbf{y} \ge \mathbf{0}\right\}$$

with feasible solutions $\mathbf{x}_0, \mathbf{y}_0$, *the following are equivalent:*

1. $\mathbf{x}_0, \mathbf{y}_0$ *are optimal solutions.*
2. $\mathbf{c}^\top \mathbf{x}_0 = \mathbf{y}_0^\top \mathbf{b}$.
3. *If a component of* $\mathbf{y}_0$ *is positive, the corresponding inequalities in* $A\mathbf{x} \le \mathbf{b}$ *are satisfied by* $\mathbf{x}_0$ *with equality.*

Proof. It remains to show the equivalence (2) $\Leftrightarrow$ (3):

$$\mathbf{c}^\top \mathbf{x}_0 = \mathbf{y}_0^\top \mathbf{b} \;\Leftrightarrow\; \mathbf{y}_0^\top A\mathbf{x}_0 = \mathbf{y}_0^\top \mathbf{b} \;\Leftrightarrow\; \mathbf{y}_0^\top(\mathbf{b} - A\mathbf{x}_0) = 0 \;\Leftrightarrow\; (\mathbf{b} - A\mathbf{x}_0)_i = 0$$

if $(\mathbf{y}_0)_i > 0$. Note that $\mathbf{b} - A\mathbf{x}_0 \ge \mathbf{0}$, $\mathbf{y}_0 \ge \mathbf{0}$. □

Using the principle of complementary slackness, we can determine the optimality of a solution. If two vectors satisfy the complementary slackness, then they must be a pair of primal-dual optimal solutions. We do not give a proof here of the rest of the theorem (which also follows from Farkas' lemma).

Corollary 1.6.4. *The set of optimal solutions of the dual pair of linear programs is precisely the set of solutions of the polynomial system of equations/inequalities in the primal, dual, and slack variables:* $A\mathbf{x} = \mathbf{b}$, $A^\top \mathbf{y} - \mathbf{s} = \mathbf{c}$, $\mathbf{x}^\top \mathbf{s} = 0$, $\mathbf{x} \ge \mathbf{0}$, *and* $\mathbf{s} \ge \mathbf{0}$.

All algorithms for solving the linear programming problem are obtained by fixing some of the above five conditions of Corollary 1.6.4 and gradually adjusting the rest until they are all satisfied. Then we have reached optimality.

1.7 Remarks on computational complexity

In order to speak about the efficiency of algorithms or the hardness of algorithmic problems, we can use the standard terminology of computational complexity; see, e.g., [131]. Unless otherwise noted, we are using the standard (Turing) model of computation. Numbers are assumed to be encoded in the binary encoding scheme, so the size of integers is accounted for by their bit length. Rational numbers $p/q \in \mathbb{Q}$ are assumed to be encoded as a pair (p,q) of integers, etc.

As a direct consequence of linear programming duality (Section 1.6), we find that optimization and feasibility are equivalent operations as far as polynomial-time computability goes. This is an important observation, which we will emphasize again in the integer case in Chapter 2.

The linear programming or *linear optimization problem* (LP) is described as follows.

Given $A \in \mathbb{Q}^{m\times n}$, $\mathbf{b} \in \mathbb{Q}^m$, and $\mathbf{c} \in \mathbb{Q}^n$, find $\mathbf{x} \in \mathbb{Q}^n$ that solves

$$\min\left\{\mathbf{c}^\top \mathbf{x} : A\mathbf{x} = \mathbf{b}, \mathbf{x} \ge \mathbf{0}\right\} \tag{LP}$$

or report INFEASIBLE or UNBOUNDED.

We define the *linear feasibility problem* (LF) to be the following problem:

Given $A \in \mathbb{Q}^{m\times n}$, $\mathbf{b} \in \mathbb{Q}^m$, and $\mathbf{c} \in \mathbb{Q}^n$, find $\mathbf{x} \in \mathbb{Q}^n$ such that

$$A\mathbf{x} \le \mathbf{b} \tag{LF}$$

or report INFEASIBLE.

Lemma 1.7.1. *A polynomial-time linear optimization algorithm exists if and only if a polynomial-time linear feasibility algorithm exists.*

Proof.

($\Rightarrow$) Assuming that we can optimize in polynomial time, we wish to solve $A\mathbf{x} \le \mathbf{b}$. For this rewrite $A\mathbf{x} + I\mathbf{y} = \mathbf{b}$ for $\mathbf{x}, \mathbf{y} \ge \mathbf{0}$. One can now answer (LF) in polynomial time by solving $\min\left\{\mathbf{0}^\top\mathbf{x} : A\mathbf{x} = \mathbf{b}, \mathbf{x} \ge \mathbf{0}\right\}$.

($\Leftarrow$) Assuming we can solve systems of linear inequalities in polynomial time, we wish to solve $\min\left\{\mathbf{c}^\top\mathbf{x} : A\mathbf{x} = \mathbf{b}, \mathbf{x} \ge \mathbf{0}\right\}$. Recall the strong duality

$$\min\left\{\mathbf{c}^\top\mathbf{x} : A\mathbf{x} = \mathbf{b},\ \mathbf{x} \ge \mathbf{0}\right\} = \max\left\{\mathbf{b}^\top\mathbf{y} : \mathbf{y}^\top A \le \mathbf{c}^\top\right\}.$$

Now solving (LP) is equivalent to finding a pair of primal and dual feasible solutions with the same objective function value, that is, we have to solve

$$A\mathbf{x} = \mathbf{b},\ \mathbf{x} \ge \mathbf{0},\ \mathbf{y}^\top A \le \mathbf{c}^\top,\ \mathbf{c}^\top\mathbf{x} = \mathbf{b}^\top\mathbf{y},$$

which can be rewritten as

$$A\mathbf{x} \le \mathbf{b},\ A\mathbf{x} \ge \mathbf{b},\ \mathbf{x} \ge \mathbf{0},\ \mathbf{y}^\top A \le \mathbf{c}^\top,\ \mathbf{c}^\top\mathbf{x} - \mathbf{b}^\top\mathbf{y} \le 0,\ \mathbf{c}^\top\mathbf{x} - \mathbf{b}^\top\mathbf{y} \ge 0.$$

But this system of linear inequalities can be solved in polynomial time. □

1.8 The ellipsoid method and convex feasibility

Yudin, Nemirovskii, Shor, and Khachiyan (late 1970s) were the founders of a great geometric theory that has found many applications in the theory of discrete optimization (see, e.g., [145]). In particular, it gave the first polynomial-time algorithm for solving linear and convex optimization. In its simplest formulation it answers the geometric question: Given a polyhedron $P = \{\mathbf{x} : A\mathbf{x} \le \mathbf{b}\}$ or more generally, a convex set, *can we decide, in time polynomial in the input data, whether $P \ne \emptyset$?* The geometry is remarkably beautiful and we sketch a few key points of the algorithm.

Let K be a convex compact set in $\mathbb{R}^n$. We will assume that K is specified by a *separation oracle*: for a vector $\mathbf{y} \in \mathbb{R}^n$, either $\mathbf{y} \in K$ or there exists a hyperplane H that separates $\mathbf{y} \in H_+$ and $K \subseteq H_-$. We also assume that, if K is not empty, there exists $\mathbf{a} \in \mathbb{R}^n$ and radius r such that $B(\mathbf{a}, r) \subseteq K$, and thus there is a lower bound δ for its volume. Here $B(\mathbf{a}, r)$ is the ball with center at $\mathbf{a}$ and radius r.

Let M be a (symmetric) positive definite $n \times n$ matrix, $\mathbf{z} \in \mathbb{R}^n$. Then M describes an *ellipsoid* centered at $\mathbf{z}$ as follows:

$$E(\mathbf{z}, M) = \left\{\mathbf{x} : (\mathbf{x} - \mathbf{z})^\top M(\mathbf{x} - \mathbf{z}) \le 1\right\}.$$

Note that

$$\mathbf{x} \in E(\mathbf{z}, M) \iff \|M^{1/2}(\mathbf{x}-\mathbf{z})\| \leq 1.$$

Here $M^{1/2}$ denotes the Cholesky factor of the positive definite matrix M. Hence, putting $\mathbf{u} = M^{1/2}(\mathbf{x}-\mathbf{z})$, we have $\mathbf{x} = \mathbf{z} + M^{-1/2}\mathbf{u}$ and $\|\mathbf{u}\| \leq 1$. Therefore

$$E(\mathbf{z}, M) = \left\{\mathbf{x} = \mathbf{z} + M^{-1/2}\mathbf{u} : \mathbf{u}^\top \mathbf{u} \leq 1\right\}.$$

In other words, $E(\mathbf{z}, M) = \mathbf{z} + M^{-1/2}B(\mathbf{0}, 1)$, where $B(\mathbf{0}, 1)$ denotes the unit ball, centered at the origin. The volume of $E(\mathbf{z}, M)$ is given by

$$\mathrm{vol}(E(\mathbf{z}, M)) = \frac{\mathrm{vol}(B(\mathbf{0}, 1))}{\sqrt{\det(M)}},$$

and hence

$$\ln(\mathrm{vol}(E(\mathbf{z}, M))) = \ln(\mathrm{vol}(B(\mathbf{0}, 1))) - \frac{1}{2}\ln(\det(M)).$$

The ellipsoid method needs to generate a sequence of smaller and smaller ellipsoids that contains the convex set K. Eventually, either we find that the center of one of the ellipsoids is in K or we have reached such a small ellipsoid that K must be empty. How do we generate that sequence of ellipsoids? We are about to show the explicit formulas below, but rather than forcing the reader to suffer with a very formal proof that the formulas work we wish to derive them in a special case.

Suppose we have the unit ball centered at the origin and the half space $\{\mathbf{x} : x_1 \geq 0\}$. What is a small ellipsoid that contains their intersection? We are going to find $M = \mathrm{diag}(a_1, \ldots, a_n)$ and $\mathbf{z} = (z, 0, \ldots, 0)^\top$ such that the ellipsoid $E(\mathbf{z}, M)$ contains $B(\mathbf{0}, 1) \cap \{\mathbf{x} : x_1 \geq 0\}$ and has minimal volume. For this, note that

$$(\mathbf{x}-\mathbf{z})^\top M(\mathbf{x}-\mathbf{z}) = a_1(x_1 - z)^2 + \sum_{i=2}^{n} a_i x_i^2.$$

The unit vectors $\mathbf{e}_1, \pm\mathbf{e}_2, \ldots, \pm\mathbf{e}_n$ are on the boundary of $B(\mathbf{0}, 1) \cap \{\mathbf{x} : x_1 \geq 0\}$ and are required to lie on the boundary of the ellipsoid too. This gives

$$a_1(1-z)^2 = 1,$$
$$a_1 z^2 + a_i = 1, \; 2 \leq i \leq n.$$

From this we obtain

$$a_1 = \frac{1}{(1-z)^2}, \quad a_i = 1 - a_1 z^2 = \frac{1-2z}{(1-z)^2}, \; i = 2, \ldots, n.$$

Recall that $\mathrm{vol}(E(\mathbf{m}, M))$ is minimal if

$$\det M = \frac{(1-2z)^{n-1}}{(1-z)^{2n}}$$

is maximal, which occurs at $z = \frac{1}{n+1}$. So we take

$$a_1 = \frac{(n+1)^2}{n^2}, \quad a_i = \frac{n^2-1}{n^2}, \; i = 2, \ldots, n,$$

and the new ellipsoid will be defined by

$$M' = \frac{n^2-1}{n^2}\left(I + \frac{2}{n-1}\mathbf{e}_1\mathbf{e}_1^\top\right), \quad \mathbf{z}' = \frac{1}{n+1}\mathbf{e}_1.$$

Now we claim the following.

Lemma 1.8.1. $B(\mathbf{0},1) \cap \{\mathbf{x} : x_1 \geq 0\} \subseteq E(\mathbf{z}', M')$.

Proof. Observe that $\mathbf{y} \in E(\mathbf{z}', M')$ if and only if

$$\left(\mathbf{y} - \frac{\mathbf{e}_1}{n+1}\right)^\top \left(\frac{n^2-1}{n^2}\left(I + \frac{2\mathbf{e}_1\mathbf{e}_1^\top}{n-1}\right)\right)\left(\mathbf{y} - \frac{\mathbf{e}_1}{n+1}\right) \leq 1$$

or, equivalently,

$$\frac{n^2-1}{n^2}\|\mathbf{y}\|_2^2 + \frac{2n+2}{n^2}y_1(y_1-1) + \frac{1}{n^2} \leq 1.$$

Now we let $\mathbf{y} \in B(\mathbf{0},1) \cap \{\mathbf{x} : x_1 \geq 0\}$, then $\|\mathbf{y}\|_2^2 \leq 1$, $y_1(y_1-1) \leq 0$, and thus the above inequality is satisfied; i.e., $\mathbf{y} \in E(\mathbf{z}', M')$. □

Next, we compute the ratio of the volumes of $B(\mathbf{0},1)$ and $E(\mathbf{z}', M')$ and we claim the following.

Lemma 1.8.2. $\mathrm{vol}(E(\mathbf{z}', M')) < \mathrm{vol}(B(\mathbf{0},1)) \cdot \exp\left(\frac{-1}{2(n+1)}\right)$.

Proof. Using the fact that $1 + x \leq \mathrm{e}^x$, we have

$$\begin{aligned}
\frac{1}{\det(M')} &= 1\Big/\left(\left(\frac{n^2-1}{n^2}\right)^n\left(1+\frac{2}{n-1}\right)\right) \\
&= \left(1+\frac{1}{n^2-1}\right)^{n-1}\left(1-\frac{1}{n+1}\right)^2 \\
&\leq \exp\left(\frac{n-1}{n^2-1}\right)\exp\left(\frac{-2}{n+1}\right) = \exp\left(\frac{-1}{n+1}\right).
\end{aligned}$$

So the assertion follows by noticing that

$$\frac{\mathrm{vol}(E(\mathbf{z}', M'))}{\mathrm{vol}(B(\mathbf{0},1))} = \frac{\sqrt{\det(I)}}{\sqrt{\det(M')}} = \frac{1}{\sqrt{\det(M')}} \leq \exp\left(\frac{-1}{2(n+1)}\right). \qquad \square$$

The above lemmas show that we can indeed find a smaller ellipsoid that encloses a half-unit ball centered at the origin. Moreover the volume of the new ellipsoid is smaller than the volume of the ball.

In fact, these simple formulas also work when finding a new smaller ellipsoid that contains the intersection of an arbitrary ellipsoid and a half space. We skip the details of why the formulas work, but they are simply a repetition of the calculations in the special case above.

Consider the ellipsoid

$$E(\mathbf{z}, M) = \left\{\mathbf{x} : (\mathbf{x}-\mathbf{z})^\top M(\mathbf{x}-\mathbf{z}) \leq 1\right\} = \mathbf{z} + M^{-1/2}B(\mathbf{0},1).$$

Given $\mathbf{a}$, which defines the normal vector of a family of hyperplanes, define a new ellipsoid from its defining symmetric positive definite matrix M' and its center $\mathbf{z}'$:

$$M' = \frac{n^2-1}{n^2}\left(M + \frac{2}{n-1}\frac{\mathbf{a}\mathbf{a}^\top}{\mathbf{a}^\top M^{-1}\mathbf{a}}\right), \quad \mathbf{z}' = \mathbf{z} + \frac{1}{n+1}\frac{M^{-1}\mathbf{a}}{\sqrt{\mathbf{a}^\top M^{-1}\mathbf{a}}}.$$

One can verify that the inverse matrix of M' is given by the following expression:

$$M'^{-1} = \frac{n^2}{n^2-1}\left(M^{-1} - \frac{2}{n+1}\frac{M^{-1}\mathbf{a}\mathbf{a}^\top M^{-1}}{\mathbf{a}^\top M^{-1}\mathbf{a}}\right).$$

The key properties are the following.

Lemma 1.8.3. *The following properties are satisfied by the two ellipsoids:*

- $E(\mathbf{z}, M) \cap \left\{\mathbf{x} : \mathbf{a}^\top\mathbf{x} \le \mathbf{a}^\top\mathbf{z}\right\} \subseteq E(\mathbf{z}', M)$. *Note that the defining hyperplane of the half-space passes through the center of the ellipsoid* $E(\mathbf{z}, M)$.
- *The volume decreases:* $\mathrm{vol}(E(\mathbf{z}', M')) < \mathrm{vol}(E(\mathbf{z}, M) \cdot \exp\left(\frac{-1}{2(n+1)}\right)$.

Using this construction of enclosing ellipsoids, we can now state the famous ellipsoid method (see [145, 193]):

ALGORITHM 1.2. Ellipsoid method.

```
input A convex set K ⊆ ℝ^n given by a separation oracle; an ellipsoid E(z, M) such
  that K ⊆ E(z, M); a lower bound δ on vol(K) if K ≠ ∅.
output A vector s ∈ K or NONE EXISTS.
Set k ← 0, M^k ← M, z^k ← z.
while vol(E(z^k, M^k)) > δ do
   if z^k ∈ K then
      return s = z^k.
   else
      Find nonzero vector a such that a^T x ≤ a^T z for all x ∈ K.
      Construct the smallest volume ellipsoid E(z^{k+1}, M^{k+1}) that contains

            E(z, M) ∩ { x ∈ ℝ^n : a^T(x − z^k) ≤ 0 }.

      k ← k + 1.
return NONE EXISTS.
```

If $K \subseteq \mathbb{R}^n$ is a convex, compact, and well-specified set as in the algorithm, one can find $\mathbf{s} \in K$ using a polynomial number of calls to the separation oracle. Either the sequence of ellipsoids we construct must find a point in K, or the volumes become too small for K to be nonempty.

Theorem 1.8.4. *If we want to find a point in the set K, with $K \subseteq E(\mathbf{z}^0, M^0)$ given and* $\mathrm{vol}(S) > 0$, *then the ellipsoid method will find a point in K after at most*

$$\left\lfloor 2(n+1)\ln\left(\frac{\mathrm{vol}(E(\mathbf{z}^0, M^0))}{\mathrm{vol}(K)}\right)\right\rfloor$$

iterations.

Proof. By the recursive relation of volumes, and $K \subseteq E(\mathbf{z}^k, M^k)$ in the k-th iteration, we have

$$\mathrm{vol}(S) \leq \mathrm{vol}(E(\mathbf{z}^k, M^k)) \leq \mathrm{vol}(E(\mathbf{z}^0, M^0)) \cdot \exp\left(\frac{-k}{2(n+1)}\right).$$

The result follows by taking logarithms and rearranging. □

1.9 Applications of the ellipsoid method

Although the ellipsoid method is not good for practical computations, it has several remarkable applications in the theory of optimization.

The first famous application, due to Khachiyan [193], is that the ellipsoid method solves linear programming problems in time bounded by a polynomial in the size of the input. In feasibility terms, we can efficiently decide whether a polyhedron $P = \{\mathbf{x} : A\mathbf{x} \leq \mathbf{b}\}$ is empty or not. Here if $A \in \mathbb{Z}^{m\times n}$, the size of input is $\phi = mn\max(\log_2(A_{ij}), \log_2(b_i))$ (see [145, 193]).

Lemma 1.9.1. *If P is a polyhedron that is compact, $P \subseteq B(\mathbf{0}, 2^{4\phi^3})$, then P contains a tiny ball of radius $2^{-7\phi^3}$. Thus, if the polyhedron is nonempty, using the ellipsoid method, we will find a point in the polyhedron in polynomial time in the input size.*

A second application is in the polynomial-time solvability of some combinatorial optimization problems, such as matching problems in graphs, that may have a large polyhedral encoding: If we are given a collection of combinatorial sets $\mathcal{F}$ of a finite S, such as the matchings of a graph, or the bases of a matroid, we know there exists an inequality description of the polyhedron (or polytope) $\mathrm{conv}(\mathcal{F}) = \mathrm{conv}(\{\chi(U) : U \in \mathcal{F}\})$, the convex hull of the characteristic vectors for $\mathcal{F}$. We know such an inequality representation is always possible, thanks to Weyl–Minkowski's theorem (see Section 1.3). With that description we can solve a linear program to find the solution of the original combinatorial problem, but the problem is that the inequality representation may involve exponentially many constraints. Sometimes, we are lucky to know the full description (e.g., for matching polyhedra), but we need another strong tool to optimize efficiently.

Fortunately, Grötschel, Lovász, and Schrijver [145] showed that such problems can be solved in polynomial time if and only if the following *separation problem* is solvable in polynomial time: Given a rational vector $\mathbf{x}$ determine whether it belongs to $\mathrm{conv}(\mathcal{F})$ and, if not, find a separating hyperplane that separates $\mathbf{x}$ from $\mathrm{conv}(\mathcal{F})$. The proof of this fundamental result depends on the ellipsoid method (for details see [145]) and it has a very broad applicability.

A third application has to do with semidefinite optimization. To start we remind the reader of the following equivalent definitions of *positive semidefinite* (PSD) matrices that can be found in most linear algebra books (see, e.g., [174]):

- $\mathbf{y}^\top A\mathbf{y} \geq \mathbf{0}$ for all $\mathbf{y} \in \mathbb{R}^n$.
- All eigenvalues of A are nonnegative real numbers.
- $A = BB^\top$ for some matrix B (the Cholesky factorization of A).

The condition that a matrix A is PSD is denoted by $A \succeq O$. We must observe that the set of all PSD matrices is a convex (nonpolyhedral) cone. A *semidefinite optimization problem (or SDP for short)* is defined as the problem of finding a PSD matrix Z such that its entries satisfy a set of linear equations and inequalities.

One reason why semidefinite programs are useful is that for any integer linear programming (ILP) of the form

$$\max\left\{\mathbf{c}^\top\mathbf{x} : A\mathbf{x} \leq \mathbf{b},\ Q\mathbf{x} = \mathbf{d},\ \mathbf{x} \in \{0,1\}^n\right\},$$

there exists a corresponding SDP with linear constraints for the entries $Z_{i,j}$ and the extra condition $Z \succeq O$ that approximates the ILP solution.

The most famous example comes from combinatorial optimization. For a graph G, a *stable set* is a set of nodes without edges between them. Let x_i be a (binary) variable for each $i \in V(G)$, that is, $0 \leq x_i \leq 1$, $x_i \in \mathbb{Z}$ (or $x_i^2 - x_i = 0$). We can consider the integer optimization problem

$$\max\left\{\sum x_i : x_i + x_j \leq 1\,\forall (i,j) \in E(G),\ x_i \in \{0,1\}\ \forall i \in V(G)\right\}.$$

To finish the modeling it is useful to think nonlinearly: $x_k(x_i + x_j) \leq 1$, or $x_i x_j \geq 0$, $(1 - x_i)(1 - x_j) \geq 0$. Replace the condition $x_i + x_j \leq 1$ by $Z_{i,j} = x_i x_j$, $Z_{i,j} \geq 0$, $Z_{i,i} + Z_{j,j} - Z_{i,j} \leq 1$; also, $Z_{i,i} \geq Z_{i,j}$. Let $Z = (Z_{i,j})_{i,j} = \mathbf{x}\mathbf{x}^\top$. In conclusion, we have the following semidefinite program that gives an approximate value to the original ILP:

$$\begin{array}{ll} \max & \sum Z_{i,i} \\ \text{subject to} & Z_{i,j} \leq Z_{i,i}, \\ & 0 \leq Z_{i,i} \leq 1, \\ & Z_{i,i} + Z_{j,j} - Z_{i,j} \leq 1, \\ & Z \succeq O. \end{array}$$

It is worth mentioning that a similar nonlinear manipulation will yield a model for a general ILP: For $F := \left\{\mathbf{x} : A\mathbf{x} \leq \mathbf{b},\ \mathbf{x} \in \{0,1\}^n\right\}$ we note that multiplying by x_i and $(1 - x_i)$, and then substituting $Z = \mathbf{x}_i\mathbf{x}_j^\top$ will help define a semidefinite problem on n^2 variables whose optimal solution bounds the optimal solution of the original ILP. Later on we will see another application of semidefinite programming in nonlinear global optimization.

Theorem 1.9.2. *Any semidefinite programming problem can be solved in polynomial time on its input data size.*

Proof. It is enough to check $Z \succeq O$ in polynomial time by a separation oracle, which is implemented as follows. Compute the eigenvalues $\lambda_1 \leq \lambda_2 \leq \cdots \leq \lambda_n$ of the matrix Z and the corresponding eigenvectors. If $\lambda_1 \geq 0$, then $Z \succeq O$. Otherwise, $\lambda_1 < 0$; taking its corresponding eigenvector $\mathbf{u}_1$, we have $\mathbf{u}^\top Z\mathbf{u} = \lambda_1 \|\mathbf{u}\|^2$. □

1.10 Notes and further references

Most of this well-established material is a review in preparation for the core of the book. All missing details and proofs are given in the excellent books of Grötschel, Lovász, and Schrijver [145] and Schrijver [296]. For the background in polyhedral convexity and polyhedra we recommend [32, 298–300, 338].

1.11 Exercises

Exercise 1.11.1. Give a proof of Radon's lemma (Lemma 1.1.8).

Exercise 1.11.2. Give a proof of yet another version of Farkas' lemma:

$$\{\mathbf{x} : A\mathbf{x} \leq \mathbf{b}, \mathbf{x} \geq \mathbf{0}\} \neq \emptyset \iff \text{If } \mathbf{y}^\top A \geq \mathbf{0}^\top \text{ and } \mathbf{y} \geq \mathbf{0}, \text{ then } \mathbf{y}^\top \mathbf{b} \geq 0.$$

Exercise 1.11.3. Give a direct proof (without Weyl–Minkowski) that the d-dimensional standard cross-polytope is a polyhedron too. Write a proof of Proposition 1.3.8.

Exercise 1.11.4. Give proofs of Proposition 1.5.3 and Corollary 1.5.7. What are the extreme points of the polytope of 3×3 doubly stochastic matrices?

Exercise 1.11.5. Prove that Corollary 1.4.7 is stronger when P is a pointed polyhedron (i.e., $L = \{\mathbf{0}\}$). In that case $L = \{\mathbf{0}\}$ and the finite sets generating Q (vertices) and $\bar{K}$ (rays) are unique.

Exercise 1.11.6. Here is a demonstration of the power of semidefinite programming relaxations of integer linear programs. Given a graph G, define $\Theta(G)$ to be the optimal value of the semidefinite program

$$\begin{aligned} \max \quad & \operatorname{tr}(JZ) \\ \text{subject to} \quad & \sum Z_{i,i} = 1 \\ & Z_{i,j} = 0 \quad \text{for } (i,j) \in E(G), \\ & Z \succeq O, \end{aligned}$$

where J is a matrix with all elements 1. The number $\Theta(G)$ receives the name *Lovász theta number*. Prove that if $\alpha(G)$ is the size of the largest stable set of G, then

$$\alpha(G) \leq \Theta(G).$$

Exercise 1.11.7. Consider the max-cut problem (MAXCUT): We are given a graph G with nonnegative weights $w_{i,j}$ for each edge that connect vertices i and j, and $w_{i,j} = 0$ otherwise. Find a bipartition of V_1, V_2 of the nodes of G that maximizes the sum of the weights of edges between V_1 and V_2. Prove that MAXCUT can be approximated by an SDP.

Exercise 1.11.8. Let us return to the MAXCUT problem as discussed in the previous exercise. Letting $X_{i,j} = y_i y_j$, we have the RELAX problem

$$\begin{aligned} \text{RELAX:} \qquad \max \quad & \frac{1}{4} \sum_{(i,j)\in E} w_{i,j}(1 - X_{i,j}) \\ \text{subject to} \quad & X_{i,i} = 1, \\ & X \succeq O. \end{aligned}$$

1. Prove that the optimal values satisfy RELAX $\geq$ MAXCUT.
2. The following construction was invented by M. Goemans and D. Williamson: Solve RELAX. Let $X = V^\top V$ where $V = (\mathbf{v}_1 \ \cdots \ \mathbf{v}_n)$ is an $s \times n$ matrix. Pick a unit vector $\mathbf{r}$ uniformly at random. Partition the graph by setting $y_i = 1$ for $\mathbf{r}^\top \mathbf{v}_i \geq 0$, and $y_i = -1$ for $\mathbf{r}^\top \mathbf{v}_i < 0$.

   Prove that the expected value of the cut produced is greater than $0.87856 \cdot$ MAXCUT. In other words, MAXCUT $\geq 0.87856 \cdot$ RELAX.

Chapter 2

Tools from the Geometry of Numbers and Integer Optimization

This book is about the algebraic and geometric ideas that have influenced developments in discrete optimization. There is perhaps no better example of a topic where algebra and geometry mixed so perfectly well than the area that Hermann Minkowski named the *geometry of numbers*. Roughly speaking, this area is concerned with investigating how lattices interact with convex bodies. We list a few two-dimensional examples of this connection and preview some of the key ideas to be used later on. We regret that we can only touch upon a few topics used in the rest of the book, but fortunately one can use the excellent books [32, 34, 70, 146, 263] to learn more on this topic.

2.1 Geometry of numbers in the plane

There are many fascinating classical connections between lattice points and convex sets. In the 1800s Gauss presented the following deceptively easy problem: *Given the radius r disk* $\{(x,y) : x^2 + y^2 \leq r\}$ *with* $r \geq 1$, *compute, or at least estimate, the number of lattice points within*

$$G(\sqrt{r}) = \#\Big\{(n,m) : n,m \in \mathbb{Z},\, n^2+m^2 \leq r\Big\}.$$

For instance, $G(1) = 5, G(2) = 13, G(3) = 29$, etc.

One approximation to $G(s)$ can be obtained from an area estimation. If each lattice point $\mathbf{z}_i = (\xi, \eta)$ belongs to the square of area one,

$$Q_{\mathbf{z}} = \left\{(x,y) : |\xi - x| < \frac{1}{2},\, |\eta - y| < \frac{1}{2}\right\},$$

then each square is disjoint from the next and, therefore, the area of the slightly larger circle $x^2+y^2 = (\sqrt{r}+\frac{1}{2}\sqrt{2})^2$ provides an upper bound for $G(r)$. This implies, for $r \geq 1$, that $G(\sqrt{r}) \leq \pi(\sqrt{r}+\frac{1}{2}\sqrt{2})^2 < \pi r + 2\pi\sqrt{r}$. At the same time, because no more lattice points can lie in the "ring" of width $\frac{1}{2}\sqrt{2}$ around the circle $x^2+y^2 = r$ besides those that are on the circle, $G(\sqrt{r}) \geq \pi(\sqrt{r}-\frac{1}{2}\sqrt{2})^2 > \pi r - 2\pi\sqrt{r}$. This implies that $\big|G(\sqrt{r}) - \pi r\big| < 2\pi\sqrt{r}$. Thus, from the bounds we see $\lim_{r\to\infty} \frac{G(\sqrt{r})}{r} = \pi$. This means the function $G(\sqrt{r})$ can be used to approximate the value of π. But, what is the exact error? Surprisingly, today we still do not know the exact value of the error. Several sophisticated techniques have been developed around the problem of counting lattice points on curves (see [179]).

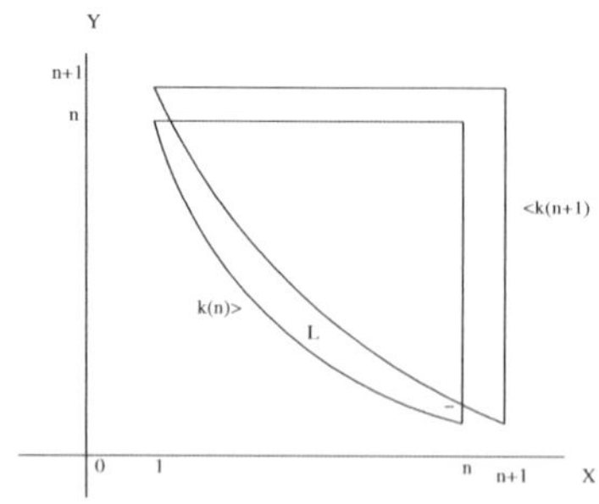

Figure 2.1. *Convex sets for testing primality.*

Our second example shows the potential applications of counting lattice points inside convex bodies. Consider the convex set bounded by a hyperbola and a square $k(n) = \left\{(x,y) : y \geq \frac{n}{x},\ 1 \leq x, y \leq n\right\}$. How many lattice points belong to $k(n)$? We can justify this as an interesting question because one can detect primality of numbers through counting.

Lemma 2.1.1. $|k(n+1) \cap \mathbb{Z}^2| - |k(n) \cap \mathbb{Z}^2| = 2n+1$ *if and only if n is a prime.*

Proof. In Figure 2.1, observe that there are only two points in the curve $xy = n$ if and only if n is a prime. Those two points are $(1,n)$ and $(n,1)$. Now let us count how many lattice points are there elsewhere in the regions.

Let L be the curvy strip band marked in the figure. However, we intend that L does not include the boundary of $k(n)$ nor that of $k(n+1)$. We claim that L has no lattice points. Suppose that L had a lattice point (c,d); then, because $c < n$, we can see that

$$\frac{n+1}{n} > d > \frac{n}{c}.$$

However, this implies that $n < cd < n+1$, which is a contradiction, since no extra integer can exist between consecutive integers. Clearly there are no lattice points in between the lines $x = n+1$, $x = n$ or in between $y = n$, $y = n+1$. Thus the only other contribution of points, besides the curve $xy = n$, must be in the line segment from $(n+1,1)$ to $(n+1,n+1)$ and the line segment from $(1,n+1)$ to $(n+1,n+1)$. Those are exactly $2n-1$ lattice points. Suppose that n is not prime; then we have that there exists a point (a,b) such that $ab = n$, i.e., a point on the lower boundary of $k(n)$, with $(a,b) \neq (n,1),(1,n)$. Thus we have more than $2n+1$ total. Now suppose that n is prime; then we have that $(n,1)$ and $(1,n)$ are the only lattice points with $xy = n$. So we get a total of $2n+1$ in the counting difference. □

Another source of counting problems for lattice points within regions of Euclidean space comes from number theory. In the 1800s researchers were interested in the problem of approximating real algebraic numbers by rational numbers and in the problem of representing numbers as the sum of squares. For example, what particular integers n can be written in the form $n = ax^2 + 2bxy + cy^2$ when x, y range over all possible integer values? Work by Hermite, Lagrange, Legendre, and of course Minkowski showed the inherent geometric form of these algebraic problems formulated as problems on lattices. One of the primary examples is given by Minkowski's first theorem, which we state in a special case without proof for now:

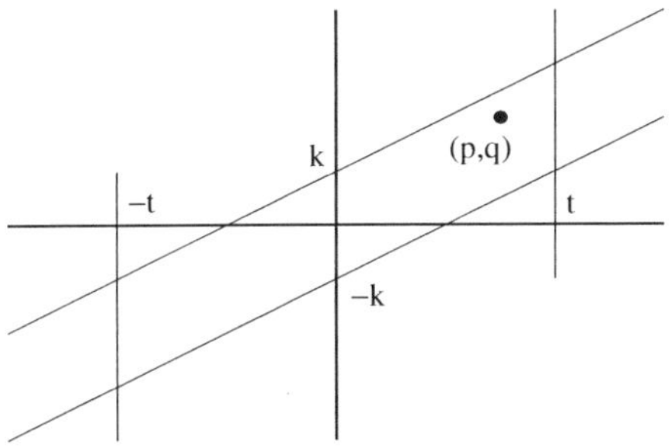

Figure 2.2. *Parallelogram in the proof of the Diophantine approximation theorem.*

Theorem 2.1.2 (Minkowski's first theorem in two dimensions). *Any convex set in* $\mathbb{R}^2$ *that has central symmetry and volume greater than* 4 *must contain an integer lattice point other than the origin* $\mathbf{0}$.

There are several applications of Minkowski's first theorem to number theory; here is an example of algebraic application of this theorem for the approximation of irrational numbers.

Corollary 2.1.3 (Diophantine approximation). *Given any real number* α *and an integer* $s > 0$, *there exist integers* p, q *such that*

$$\left| \frac{q}{p} - \alpha \right| \leq \frac{1}{|p|s}.$$

Proof. Take M to be the parallelogram bounded by the four lines

$$\begin{aligned} y - \alpha x &= k, \\ y - \alpha x &= -k, \\ x &= t, \\ x &= -t; \end{aligned}$$

see Figure 2.2. Take $t = |p|s$. This parallelogram has base $2t$, altitude $2k$, and hence Area $= (2k)(2t) = 4kt$.

If we take $k = \frac{1}{t}$, the area is just 4. By Minkowski's first theorem there must be at least one lattice point (p,q) other than $(0,0)$. This implies $|p| \leq t$. Then $\alpha p - k \leq q \leq \alpha p + k$ can be written $\alpha p - \frac{1}{t} \leq q \leq \alpha p + \frac{1}{t}$. This implies $|q - \alpha p| \leq \frac{1}{t} = \frac{1}{|p|s}$. □

So counting lattice points in planar convex regions is not an entirely trivial task. Can we at least do this more easily for polygons? Yes, we can.

Definition 2.1.4. Let $\mathbf{v}_0, \mathbf{v}_1, \ldots, \mathbf{v}_{n-1}$ be n points in the plane, such that $\mathbf{v}_i, \mathbf{v}_{i+1}, \mathbf{v}_{i+2}$ are not collinear. We denote the segments between consecutive points by $\mathbf{e}_0 = [\mathbf{v}_0, \mathbf{v}_1]$, $\mathbf{e}_1 = [\mathbf{v}_1, \mathbf{v}_2]$, $\mathbf{e}_i = [\mathbf{v}_i, \mathbf{v}_{i+1}]$, $\mathbf{e}_{n-1} = [\mathbf{v}_{n-1}, \mathbf{v}_0]$. We say that the segments $\mathbf{e}_0, \mathbf{e}_1, \ldots, \mathbf{e}_{n-1}$ form a *polygon* if the following conditions hold:

1. The intersection of each pair of segments adjacent in the cyclic ordering is the single point shared, $\mathbf{e}_i \cap \mathbf{e}_{i+1} = \mathbf{v}_{i+1}$ for all $i = 0, 1, \ldots, n-1$.

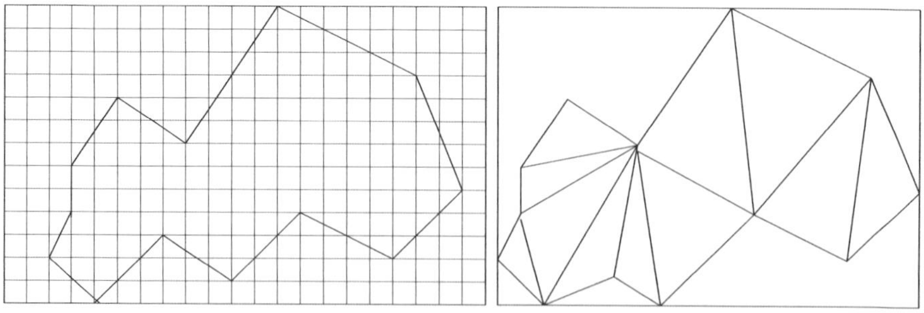

Figure 2.3. *A (nonconvex) lattice polygon and a triangulation.*

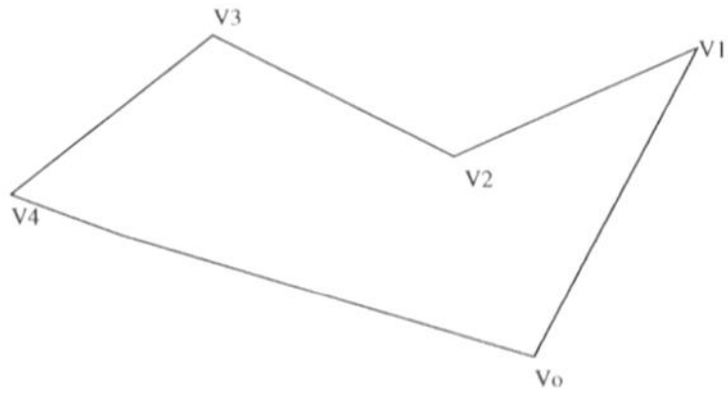

Figure 2.4. *Convex and reflex vertices in a polygon. The vertex* $\mathbf{v}_1$ *is convex, while* $\mathbf{v}_2$ *is reflex.*

2. Nonadjacent segments do not intersect, i.e., $\mathbf{e}_i \cap \mathbf{e}_j = \emptyset$ if $i \neq j \pm 1$.

The polygon is said to be *simple* if it is topologically identical to a circle, i.e., it divides the plane into two parts.

The case of counting lattice points inside a polygon (which is not necessarily a convex set) can be reduced to the case of counting lattice points inside triangles. Indeed, we can add enough extra line segments to the polygon until it decomposes into triangles (Figure 2.3). We call an extra segment, that is, one that does not cross the edges of the polygon but connects nonconsecutive vertices of the polygon, a *diagonal*. We say that $\mathbf{x}$ can *see* point $\mathbf{y}$ in polygon P if the line segment $[\mathbf{x},\mathbf{y}]$ is fully contained in P. Diagonals join vertices of the polygon P that can see each other.

Lemma 2.1.5. *Every polygon P of n vertices may be partitioned into triangles by the addition of diagonals. In fact, every triangulation of a polygon P of n vertices uses $n-3$ diagonals and consists of $n-2$ triangles.*

This lemma can be easily proved by induction on the number of edges if one can prove that any simple polygon is either a triangle or it has a diagonal. To see that this key fact is true we need to note that a vertex on a polygon P defines two angles. The angle that intersects the interior of P is called the *internal angle*. A vertex $\mathbf{v}$ of a polygon is a *reflex vertex* if its internal angle is strictly greater than π. Otherwise, it is called a *convex vertex*; see Figure 2.4.

It is not difficult to see that every simple polygon must have at least one strictly convex vertex as follows: Orient edges counterclockwise, so that the polygon is always to the left of a hypothetical walker. Order the vertices from top to bottom by their corresponding y-coordinate. Now, consider the lowest vertices from this list, next pick the rightmost

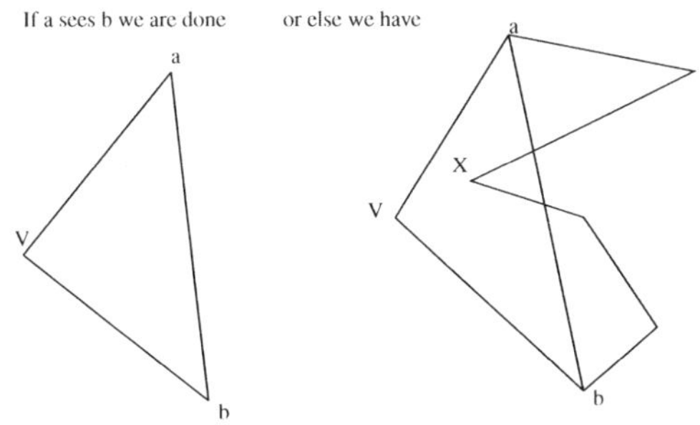

Figure 2.5. *Cases in the proof of Lemma* 2.1.6.

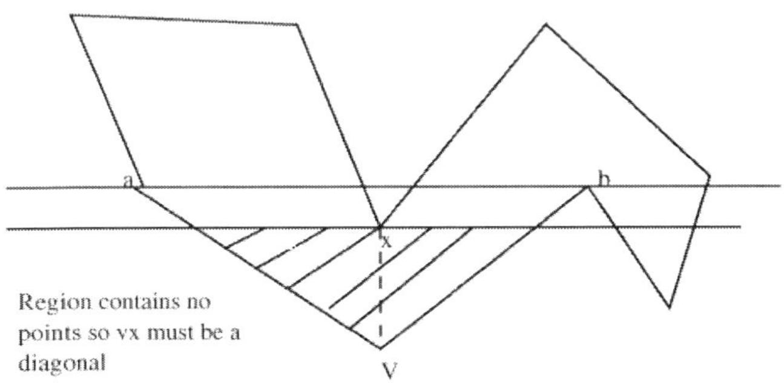

Figure 2.6. *A step in the proof of Lemma* 2.1.6.

vertex and then suppose that such vertex is reflex. However, if the vertex was reflex, then it would have contradicted the fact that we have picked the lowest rightmost vertex. So what we have is indeed a strictly convex vertex. Finally we have the next lemma:

Lemma 2.1.6. *Every simple polygon is a triangle, or it is possible to find a diagonal.*

Proof. Let $\mathbf{v}$ be a strictly convex vertex whose existence we just proved. Let $\mathbf{a}, \mathbf{b}$ be the vertices adjacent to $\mathbf{v}$. If $\mathbf{a}$ can see $\mathbf{b}$, we are done, as then $[\mathbf{a}, \mathbf{b}]$ is a diagonal (Figure 2.5). Otherwise, the triangle with vertices $\mathbf{v}, \mathbf{a}, \mathbf{b}$ must contain at least one more vertex $\mathbf{x}$ inside (possibly on its boundary). Take $\mathbf{x}$ the closest to $\mathbf{v}$ where the distance is measured perpendicular to the line $\mathbf{ab}$. Then $\mathbf{x}$ is the last vertex hit by a line L parallel to the segment $[\mathbf{a}, \mathbf{b}]$; see Figure 2.6. We claim that $[\mathbf{v}, \mathbf{x}]$ is a diagonal. The region below the line L has no points of P, so $[\mathbf{v}, \mathbf{x}]$ is fully contained in the polygon. □

Thus, assuming we can count lattice points inside segments and triangles, we can count lattice points in any polygon: First, triangulate the polygon, then count the lattice points in each triangle (with $(n-2)$ of them). Add these numbers, but subtract the number of lattice points on each diagonal. Return the result.

To conclude, we note that for lattice polygons, i.e., polygons in which every vertex is a lattice point, we can write the number of lattice points and the area of the polygon in an exact formula.

Theorem 2.1.7 (Pick's theorem; see [35]). *Given a simple closed polygon P whose vertices have integer coordinates, we have*

$$\operatorname{area}(P) = \#\textit{interior lattice points} + \frac{1}{2}\#\textit{boundary lattice points} - 1.$$

Later on we will see how to count lattice points in convex polytopes of higher dimension and how to apply that knowledge to problems in optimization.

2.2 Lattices and linear Diophantine equations

We are interested in studying lattices and Diophantine linear systems because of their relations to integer programming. There are three fundamental problems on lattice points that we touch upon now:

1. Solvability of systems of linear Diophantine equations.
2. Conditions for the existence of convex lattice-free bodies.
3. Finding shortest vectors in a lattice.

Definition 2.2.1. A *lattice* $\mathcal{L}$ is a subset of $\mathbb{R}^m$ which is a an additive Abelian subgroup of $\mathbb{R}^m$ under standard vector addition, and there are linearly independent vectors $\mathbf{b}_1, \mathbf{b}_2, \ldots, \mathbf{b}_n$ where $\mathcal{L} = \left\{ \sum_{i=1}^{n} \lambda_i \mathbf{b}_i : \lambda_1, \ldots, \lambda_n \in \mathbb{Z} \right\}$. The vectors $\mathbf{b}_1, \ldots, \mathbf{b}_n$ form a *basis* for the lattice.

We can use a more compact notation. Given an $m \times n$ matrix A of linearly independent columns, it has a lattice associated to it which we will denote by $\mathcal{L}(A) = \{ A\mathbf{x} : \mathbf{x} \in \mathbb{Z}^n \}$. The most important lattice we will use is of course the canonical integer lattice $\mathbb{Z}^n$. What makes lattices interesting is their combination of algebraic and metric properties. A lattice has infinitely many different bases, for example, $\mathbb{Z}^n$ is of course generated not only by the standard unit vectors but also by any vectors $\mathbf{a}_1, \ldots, \mathbf{a}_n$ whose associated $n \times n$ matrix has determinant plus or minus one. Such matrices will be very important for our study.

Now we shift our attention to a practical application of lattices. We consider the problem of solving linear systems of p equations in n unknowns with integral coefficients; this time the solutions must be integral vectors too.

2.3 Hermite and Smith

In linear algebra (over fields like the real or the complex numbers) you have been introduced to useful special reductions of matrices, like matrices in row-echelon form to find the solution. As we work over the integer numbers, which do not form a field but only a ring (not every nonzero integer number has a multiplicative inverse), we need other types of special canonical forms that explicitly deal with the integral nature of the problem. Two such types of matrices are matrices in Hermite normal form and those in Smith normal form.

Definition 2.3.1. A square nonsingular integral matrix C is *unimodular* if $\left|\det(C)\right| = 1$.

From basic linear algebra we can conclude that a unimodular matrix C has the nice property that its inverse C^{-1} is also an integral matrix.

Lemma 2.3.2. *If A is an integral $m \times n$ matrix and C is a unimodular $n \times n$ matrix, then $\mathcal{L}(AC) = \mathcal{L}(A)$.*

Proof. By substituting $\mathbf{x} = C\mathbf{w}$ we have

$$\begin{aligned}\mathcal{L}(A) &= \left\{ A\mathbf{x} : \mathbf{x} \in \mathbb{Z}^n \right\} \\ &= \left\{ AC\mathbf{w} : C\mathbf{w} \in \mathbb{Z}^n \right\}.\end{aligned}$$

The result follows by showing that $\{C\mathbf{w} : \mathbf{w} \in \mathbb{Z}^n\} = \mathbb{Z}^n$. Since C is an integral matrix, $\mathbf{w} \in \mathbb{Z}^n$ implies $C\mathbf{w} \in \mathbb{Z}^n$. As C is unimodular, C^{-1} is an integral matrix. Thus, for any $\mathbf{x} \in \mathbb{Z}^n$ we find $\mathbf{w}$ such that $\mathbf{x} = C\mathbf{w}$ by multiplying both sides by C^{-1}: $\mathbf{w} = C^{-1}\mathbf{x} \in \mathbb{Z}^n$. □

Definition 2.3.3 (Hermite normal form). An $m \times n$ matrix is in *Hermite normal form* if it is of the form $(H \mid O)$ with H being a lower-triangular matrix with strictly positive entries on the diagonal and all entries d_{ij} of H with $j < i$ are nonnegative and strictly smaller than the element d_{ii} of the diagonal of H in the same row.

Next we show that any $m \times n$ matrix A of full row rank can be brought into Hermite normal form by performing only elementary integer column operations (change of sign, addition of integer multiples of a column to another column, interchanging two columns). One can show (but we omit this here) that there is in fact a *unique* such matrix, which we will denote by HNF(A), the Hermite normal form of A. Given A, the matrix HNF(A) can be computed in polynomial time [190, 248, 296].

Theorem 2.3.4. *If A is an $m \times n$ integer matrix with* $\operatorname{rank}(A) = m$, *then there exists an $n \times n$ unimodular matrix C such that*

1. $AC = (H \mid O) = \mathrm{HNF}(A)$,
2. *$H^{-1}A$ is an integer matrix.*

Proof. We will use elementary column operations:

1. Interchange columns j and k.
2. Multiply column j by -1.
3. Add $\lambda \in \mathbb{Z}$ times column k to column j.

We describe an algorithm to find the Hermite normal form of A. Suppose at the k-th step of the transformation, A is rewritten as

$$A_k = \begin{pmatrix} H_k & O \\ F_k & G_k \end{pmatrix}$$

where we have H_k in Hermite normal form already. We create a new matrix as follows: By using operations 1 and 2 we can make sure the first row on G_K has only nonnegative entries $g_1, g_2, \ldots, g_{m-k}$. Note that at least one nonzero entry must be present; otherwise the matrix A was not full-row rank. If $g_i > g_j$ for some $i < j$, multiply the j-th column by $-\lfloor g_i/g_j \rfloor$

and add the result to the i-th column. We are applying an operation of type 3. This has the effect of reducing the values of the entries keeping them nonnegative. We can repeat these operations, using the Euclidean algorithm, to replace g_i with the greatest common divisor of g_i, g_j. Eventually we just end up with a single nonzero entry on that first row of G_k, which is the greatest common g divisor of $g_1, g_2, \ldots, g_{m-k}$. By permuting columns, move that entry to be either the first entry in G_k or the entry in position $(k+1, k+1)$ of the former A_k.

All we have to do now is to make sure the entries of F_k are positive and smaller than g. For this we can make sure the entries of the first row of F_k, $f_1, f_2, \ldots, f_k$ are positive by adding multiplies of the first column of the new G_k (the $(k+1)$st column of the entire matrix) to columns in F_k. This does not alter the current values in H_k because the corresponding entries that are added are zero. Finally by adding $-\left\lfloor \frac{f_i}{g} \right\rfloor$ we can be sure the entries are smaller than g.

Finally, note that the elementary operations are easily achieved by right-multiplication with elementary unimodular matrices. Note that we have written $\mathrm{HNF}(A) = AU_1 \cdots U_s$, where the U_i are $n \times n$ matrices that encode the column operations performed on A. Note that we get the matrix $C := I_n U_1 \cdots U_s \in \mathbb{Z}^{n \times n}$ from the identity matrix I_n by performing the same column operations on I_n as on A. The determinants of each U_i and hence also of C are ± 1. Therefore, C is a unimodular matrix. □

Solving a linear Diophantine system of equations

We show how to use the Hermite normal form of a matrix to find a general integer solution to a system of linear Diophantine equations $A\mathbf{z} = \mathbf{b}$, $\mathbf{z} \in \mathbb{Z}^n$. We will explain the general construction first in an example:

Example 2.3.5. $A = \left(\begin{smallmatrix} 1 & 1 & 1 & 1 \\ 1 & 5 & 10 & 25 \end{smallmatrix}\right)$ and $\mathbf{b} = \left(\begin{smallmatrix} 10 \\ 110 \end{smallmatrix}\right)$. The general solution to the system $A\mathbf{z} = \mathbf{b}$, $\mathbf{z} \in \mathbb{Z}^n$ will look similar to the general solution in the nonintegral situation:

$$\{\mathbf{z} : A\mathbf{z} = \mathbf{b},\ \mathbf{z} \in \mathbb{Z}^n\} = \left\{\mathbf{z} : \mathbf{z} = \mathbf{z}_0 + \sum_{i=1}^{k} \lambda_i \mathbf{v}_i,\ \lambda_1, \ldots, \lambda_k \in \mathbb{Z}\right\}$$

for some particular solution $\mathbf{z}_0$ to $A\mathbf{z} = \mathbf{b}$, $\mathbf{z} \in \mathbb{Z}^n$, and for some lattice basis $\{\mathbf{v}_1, \ldots, \mathbf{v}_k\}$ for the lattice $\ker_{\mathbb{Z}^n}(A)$.

We first compute the Hermite normal form $\mathrm{HNF}(A) = (H\,|\,O)$ of A:

$$\left(\begin{smallmatrix} 1 & 1 & 1 & 1 \\ 1 & 5 & 10 & 25 \end{smallmatrix}\right) \rightarrow \left(\begin{smallmatrix} 1 & 0 & 0 & 0 \\ 1 & 4 & 9 & 24 \end{smallmatrix}\right) \rightarrow \left(\begin{smallmatrix} 1 & 0 & 0 & 0 \\ 1 & 4 & 1 & 0 \end{smallmatrix}\right) \rightarrow \left(\begin{smallmatrix} 1 & 0 & 0 & 0 \\ 1 & 1 & 4 & 0 \end{smallmatrix}\right) \rightarrow \left(\begin{smallmatrix} 1 & 0 & 0 & 0 \\ 0 & 1 & 0 & 0 \end{smallmatrix}\right).$$

Thus

$$\mathrm{HNF}(A) = \begin{pmatrix} 1 & 0 & 0 & 0 \\ 0 & 1 & 0 & 0 \end{pmatrix}.$$

For our example, we obtain

$$C = \begin{pmatrix} 0 & 1 & -5 & 5 \\ 2 & -2 & 9 & -6 \\ -1 & 1 & -4 & 0 \\ 0 & 0 & 0 & 1 \end{pmatrix}.$$

Let $\mathbf{y} = C^{-1}\mathbf{z}$ and hence $\mathbf{z} = C\mathbf{y}$. As C and C^{-1} are integral matrices, we have $\mathbf{y} \in \mathbb{Z}^n$ if and only if $\mathbf{z} \in \mathbb{Z}^n$. Consequently, there is an integer solution $\mathbf{z}_0$ to $A\mathbf{z} = \mathbf{b}$ if and only if there is an integer solution $\mathbf{y}_0$ to $\mathrm{HNF}(A)\mathbf{y}=\mathbf{b}$, since $A\mathbf{z}=\mathrm{HNF}(A)C^{-1}\mathbf{z}=\mathrm{HNF}(A)\mathbf{y}$. As $\mathrm{HNF}(A)=(H\,|\,O)$ for some lower triangular matrix H, an integer solution $\mathbf{y}_0$ to $\mathrm{HNF}(A)\mathbf{y}=\mathbf{b}$ is quickly found. In particular, an integer solution $\mathbf{y}_0$ exists if and only if $H^{-1}\mathbf{b}\in\mathbb{Z}^d$. To see this, split $\mathbf{y}$ into $\binom{\mathbf{y}'}{\mathbf{y}''}$ and observe that $\mathbf{b}=\mathrm{HNF}(A)\mathbf{y}=H\mathbf{y}'$. Therefore, any solution $\mathbf{y}_0$ to $\mathrm{HNF}(A)\mathbf{y}=\mathbf{b}$ has $\mathbf{y}_0'=H^{-1}\mathbf{b}$ and $\mathbf{y}_0''\in\mathbb{Z}^{n-d}$ arbitrary. Once we have found an integer solution $\mathbf{y}_0$, $\mathbf{z}_0=C\mathbf{y}_0$ gives an integer solution to $A\mathbf{z}=\mathbf{b}$. In our example, we first solve $H\mathbf{y}=\binom{10}{110}$, which gives $\mathbf{y}_0=(10,110,0,0)^\top$ as an integer solution. Next, we compute $\mathbf{z}_0=C\mathbf{y}_0=(110,-200,100,0)^\top$. A quick check verifies $A\mathbf{z}_0=\binom{10}{110}$.

We can now extract a lattice basis for $\ker_{\mathbb{Z}^4}(A)$ from C. The last two columns of $\mathrm{HNF}(A)$ are zero. Thus the last two columns of C,

$$\left\{(-5,9,-4,0)^\top,(5,-6,0,1)^\top\right\},$$

form a lattice basis of $\ker_{\mathbb{Z}^4}(A)$. Consequently, the general solution to our example $A\mathbf{z}=\mathbf{b}$, $\mathbf{z}\in\mathbb{Z}^4$ is given by

$$\mathbf{z}=(110,-200,100,0)^\top+\lambda_1(-5,9,-4,0)^\top+\lambda_2(5,-6,0,1)^\top.$$

The following theorem gives the general procedure for finding solutions of Diophantine systems of linear equations.

Theorem 2.3.6. *Let $S=\{\mathbf{x}\in\mathbb{Z}^n : A\mathbf{x}=\mathbf{b}\}$ and let H and $C=(C_1|C_2)$ be as in Theorem 2.3.4, with C_1 an $n\times m$ matrix and C_2 an $n\times(n-m)$ matrix. Then the following hold:*

1. *$S\neq\emptyset$ if and only if $\mathbf{b}\in\mathcal{L}(A)$.*
2. *$S\neq\emptyset$ if and only if $H^{-1}\mathbf{b}\in\mathbb{Z}^m$.*
3. *If $S\neq\emptyset$, every solution is of the form $\mathbf{x}=C_1H^{-1}\mathbf{b}+C_2\mathbf{z}$, $\mathbf{z}\in\mathbb{Z}^{n-m}$.*

Proof. We have

$$\begin{aligned}S&=\left\{\mathbf{x}\in\mathbb{Z}^n : A\mathbf{x}=\mathbf{b}\right\}\\&=\left\{\mathbf{x}:\mathbf{x}=C\mathbf{w},AC\mathbf{w}=\mathbf{b},\mathbf{w}\in\mathbb{Z}^n\right\}\\&=\left\{\mathbf{x}:\mathbf{x}=C_1\mathbf{w}_1+C_2\mathbf{w}_2,H\mathbf{w}_1=\mathbf{b},\mathbf{w}_1\in\mathbb{Z}^m,\mathbf{w}_2\in\mathbb{Z}^{n-m}\right\}\\&=\left\{\mathbf{x}:\mathbf{x}=C_1H^{-1}\mathbf{b}+C_2\mathbf{w}_2,H^{-1}\mathbf{b}\in\mathbb{Z}^m,\mathbf{w}_2\in\mathbb{Z}^{n-m}\right\}.\end{aligned}$$

□

Example 2.3.7. Find the solutions (if any) of

$$\begin{aligned}2x_1+6x_2+x_3&=7,\\4x_1+7x_2+7x_3&=4.\end{aligned}$$

Here we have

$$A=\begin{pmatrix}2&6&1\\4&7&7\end{pmatrix},\quad C=\begin{pmatrix}4&3&7\\-1&-1&-2\\-1&0&-2\end{pmatrix},$$

$$H=\begin{pmatrix}1&0\\2&5\end{pmatrix},\quad H^{-1}=\frac{1}{5}\begin{pmatrix}5&0\\-2&1\end{pmatrix}.$$

The given system of equations has a solution because

$$H^{-1}\begin{pmatrix}7\\4\end{pmatrix}=\begin{pmatrix}1&0\\-2/5&1/5\end{pmatrix}\begin{pmatrix}7\\4\end{pmatrix}=\begin{pmatrix}7\\-2\end{pmatrix}.$$

All solutions look like

$$\begin{pmatrix}x_1\\x_2\\x_3\end{pmatrix}=\begin{pmatrix}4&3\\-1&-1\\-1&0\end{pmatrix}\begin{pmatrix}7\\-2\end{pmatrix}+\begin{pmatrix}-7\\2\\2\end{pmatrix}w=\begin{pmatrix}22\\-5\\-7\end{pmatrix}+\begin{pmatrix}-7\\2\\2\end{pmatrix}w$$

for $w \in \mathbb{Z}$.

Example 2.3.8. Suppose instead that $\mathbf{b} = \binom{8}{4}$. Then

$$H^{-1}\mathbf{b}=\frac{1}{5}\begin{pmatrix}5&0\\-2&1\end{pmatrix}\begin{pmatrix}8\\4\end{pmatrix}=\begin{pmatrix}8\\\frac{-12}{5}\end{pmatrix}.$$

This implies that there is no integral solution.

We can now easily prove the following characterization of the solutions, which was suggested by the previous example:

Lemma 2.3.9. *If $\tilde{\mathbf{x}} \in \mathbb{Z}^n$ is a solution of $A\mathbf{x} = \mathbf{b}$, then all other solutions are of the form $\mathbf{x} = \tilde{\mathbf{x}} + \boldsymbol{\epsilon}$, where $\boldsymbol{\epsilon} \in \mathbb{Z}^n$ is a solution of the homogenous system $A\boldsymbol{\epsilon} = 0$. More precisely, if we let $\mathrm{HNF}(A) = AC$ and let $\mathbf{c}_1,\dots,\mathbf{c}_n$ denote the columns of C, then the last $n-d$ columns of C, $\mathbf{c}_{d+1},\dots,\mathbf{c}_n$, generate $\mathcal{L} = \ker_{\mathbb{Z}^n}(A)$ over $\mathbb{Z}$.*

Proof. Computing $AC = \mathrm{HNF}(A) = (H\,|\,O)$, we see that $A\mathbf{c}_i = \mathbf{0}$ for $i = d+1,\dots,n$, and thus $\mathbf{c}_{d+1},\dots,\mathbf{c}_n \in \ker_{\mathbb{Z}^n}(A)$. Let $\mathbf{z} \in \ker_{\mathbb{Z}^n}(A)$ and consider $\mathbf{y} = C^{-1}\mathbf{z}$. Then $\mathrm{HNF}(A)\mathbf{y} = \mathbf{0}$. By the arguments above, we conclude that $\mathbf{y}' = H^{-1}\mathbf{0} = \mathbf{0}$ and $\mathbf{y}'' \in \mathbb{Z}^{n-d}$ arbitrary, where $\mathbf{y} = \binom{\mathbf{y}'}{\mathbf{y}''}$. Therefore, $\mathbf{y} = \sum_{i=d+1}^{n} y_i \mathbf{e}_i$ is an integer linear combination of the unit vectors $\mathbf{e}_{d+1},\dots,\mathbf{e}_n$. We conclude that $\mathbf{z} = C\mathbf{y} = \sum_{i=d+1}^{n} y_i C\mathbf{e}_i = \sum_{i=d+1}^{n} y_i \mathbf{c}_i$ is an integer linear combination of the columns $\mathbf{c}_{d+1},\dots,\mathbf{c}_n$ of C, and the claim is proved. □

There is another more specialized format for matrices:

Definition 2.3.10 (Smith normal form). A square matrix $S \in \mathbb{Z}^{n\times n}$ is in *Smith normal form* if it is a diagonal matrix with strictly positive entries on the diagonal and with $s_{ii} \mid s_{jj}$ whenever $i < j$.

Just like with the Hermite normal form, every nonsingular square matrix $A \in \mathbb{Z}^{n\times n}$ can be brought into Smith normal form by performing only elementary integer column *and* row operations. We will not prove it here, but the Smith normal form is uniquely determined. We write $\mathrm{SNF}(A)$ for the Smith normal form of A. As $\mathrm{SNF}(A)$ is obtained from A by elementary row and column operations, there are unimodular matrices $U, V \in \mathbb{Z}^{n\times n}$ such that $\mathrm{SNF}(A) = VAU$. Note that also $\mathrm{SNF}(A)$ can be computed in polynomial time by the Kannan–Bachem algorithm [190].

Enumerating lattice points in fundamental parallelepipeds

Let $A \in \mathbb{Z}^{n\times n}$ be a nonsingular matrix. Its columns generate a sublattice $\mathcal{L}(A)$ of $\mathbb{Z}^n$.

Definition 2.3.11. Let $A = \{\mathbf{a}_1, \mathbf{a}_2, \ldots, \mathbf{a}_n\}$ be a basis of $\mathcal{L}(A)$. The half-open parallelepiped

$$\Pi_A = \left\{ \sum_{i=1}^{n} \lambda_i \mathbf{a}_i : 0 \le \lambda_i < 1 \right\}$$

is called the *fundamental parallelepiped* of $\mathcal{L}(A)$ (spanned by A).

Theorem 2.3.12. *Let $\mathcal{L} \subseteq \mathbb{R}^n$ be a lattice and A a basis of $\mathcal{L}$. Let Π_A be the corresponding fundamental parallelepiped. Then for any $\mathbf{x} \in \mathbb{R}^n$, there exists a unique pair $\mathbf{v} \in \mathcal{L}$, $\mathbf{y} \in \Pi$ such that $\mathbf{x} = \mathbf{v} + \mathbf{y}$.*

In other words, every point $\mathbf{x}$ can be shifted by a lattice vector into the fundamental parallelepiped. We introduce a notation for this canonical representative: $\mathbf{x} \bmod \mathcal{L}(A)$.

Corollary 2.3.13. *Let Π be a fundamental parallelepiped of a lattice $\mathcal{L} \subseteq \mathbb{R}^n$. Then the translates $\mathbf{v} + \Pi$ of Π by the vectors $\mathbf{v} \in \mathcal{L}$ cover $\mathbb{R}^n$ without overlapping.*

***Proof of Theorem* 2.3.12.** Let $\mathbf{a}_1, \mathbf{a}_2, \ldots, \mathbf{a}_n$ be a basis of $\mathcal{L}$. Then $\mathbf{a}_1, \mathbf{a}_2, \ldots, \mathbf{a}_n$ also span $\mathbb{R}^n$ as a vector space. Let $\mathbf{x} \in \mathbb{R}^n$. Then $\mathbf{x} = \lambda_1 \mathbf{a}_1 + \lambda_2 \mathbf{a}_2 + \cdots + \lambda_n \mathbf{a}_n$, where the coefficients λ_i are real numbers. Take $\lfloor \lambda \rfloor$ to be the integer part of λ, that is, the largest integer that does not exceed λ. Take $\{\lambda\} = \lambda - \lfloor \lambda \rfloor$ to be the fractional part of λ. Setting $\mathbf{v} = \sum \lfloor \lambda_i \rfloor \mathbf{a}_i$ and $\mathbf{y} = \sum \{\lambda_i\} \mathbf{a}_i$, we have $\mathbf{v} \in \mathcal{L}$ and $\mathbf{y} \in \Pi$.

To prove the uniqueness, suppose there are two representations $\mathbf{x} = \mathbf{v}_1 + \mathbf{y}_1 = \mathbf{v}_2 + \mathbf{y}_2$. Then $\mathbf{y}_1 - \mathbf{y}_2 = \mathbf{v}_1 - \mathbf{v}_2$, and thus $\mathbf{y}_1 - \mathbf{y}_2$ is a nonzero element of the lattice $\mathcal{L}$. So $\mathbf{y}_1 - \mathbf{y}_2 = r_1 \mathbf{a}_1 + r_2 \mathbf{a}_2 + \cdots + r_n \mathbf{a}_n$ with integer coefficients r_i, but on the other hand $|r_i| < 1$, and thus $r_i = 0$. □

We will see below that the number of lattice points in Π_A, $|\Pi_A \cap \mathbb{Z}^n|$, is always equal to $\det(A)$. Our goal is to enumerate this set of lattice points.

Lemma 2.3.14. *Let $A \in \mathbb{Z}^{n\times n}$ be a regular matrix and let Π_A be its fundamental parallelepiped. Then $|\Pi_A \cap \mathbb{Z}^n| = |\det(A)|$.*

Proof. Let $V, U \in \mathbb{Z}^{n\times n}$ be unimodular matrices such that $D = \mathrm{SNF}(A) = VAU$ and let $\mathcal{L}(D)$ denote the sublattice of $\mathbb{Z}^n$ spanned by the columns of D. Now consider the map $\phi : \mathbb{R}^n \to \mathbb{R}^n$ given by $\phi(\mathbf{x}) = V\mathbf{x}$ and take any element $\mathbf{a} \in \mathcal{L}(A)$. Then there is some $\mathbf{x} \in \mathbb{Z}^n$ such that $\mathbf{a} = A\mathbf{x}$. But then $\mathbf{a} = (AU)(U^{-1}\mathbf{x})$, where $\mathbf{y} := U^{-1}\mathbf{x} \in \mathbb{Z}^n$ as U is unimodular. Thus we get $\phi(\mathbf{a}) = V\mathbf{a} = VAU\mathbf{y} = D\mathbf{y} \in \mathcal{L}(D)$. Conversely, let $\mathbf{b} \in \mathcal{L}(D)$. Then $\mathbf{b} = VAU\mathbf{y}$ for some $\mathbf{y} \in \mathbb{Z}^n$. As $U\mathbf{y} \in \mathbb{Z}^n$, we obtain $\phi^{-1}(\mathbf{b}) = A(U\mathbf{y}) \in \mathcal{L}(A)$. Hence, ϕ gives a one-to-one correspondence between the lattices $\mathcal{L}(A)$ and $\mathcal{L}(D)$ and their fundamental parallelepipeds, where $\mathbf{a} + \mathcal{L}(A)$ is mapped to $\phi(\mathbf{a}) + \mathcal{L}(D)$, and we conclude that $|\Pi_A \cap \mathbb{Z}^n| = |F_D \cap \mathbb{Z}^n|$. The latter, however, is $|F_D \cap \mathbb{Z}^n| = |\det(D)|$, since D is a diagonal matrix, and thus F_D is just a (half-open) rectangular box. This, however, implies

$$|\Pi_A \cap \mathbb{Z}^n| = |F_D \cap \mathbb{Z}^n| = |\det(D)| = |\det(V)\det(A)\det(U)| = |\det(A)|,$$

as claimed. □

This proof shows even more. The few missing details are left to the reader as a little exercise.

Lemma 2.3.15. *Let $A \in \mathbb{Z}^{n\times n}$ be a regular matrix and let U be a unimodular matrix. Then $\mathcal{L}(AU) = \mathcal{L}(A)$ and $|\det(AU)| = |\det(A)|$. Moreover, every basis of $\mathcal{L}(A)$ can be obtained from A via multiplication by a unimodular matrix U.*

This lemma states that any basis of a lattice $\mathcal{L}$ has the same determinant in absolute value. Hence, we can put $\det(\mathcal{L}) := |\det(A)|$ for any basis A of $\mathcal{L}$. The number $\det(\mathcal{L})$ is called the *determinant of the lattice* $\mathcal{L}$. It is an invariant of the lattice: The fundamental parallelepipeds are different for different bases of the lattice, but the volumes of these parallelepipeds are always the same.

Lemma 2.3.16. *Let $A \in \mathbb{Z}^{n\times n}$ be a nonsingular matrix, let Π_A be its fundamental parallelepiped, and let $D = \mathrm{SNF}(A) = VAU$. Moreover, let $\mathbf{d} \in \mathbb{Z}^n$ such that $D = \mathrm{diag}(\mathbf{d})$. Then we have*

$$\Pi_A = \left\{ V^{-1}\mathbf{y} \bmod \mathcal{L}(A) : \mathbf{0} \le \mathbf{y} \le \mathbf{d} - \mathbf{1} \right\}.$$

Proof. We have to show that $V^{-1}\mathbf{y}_1 + \mathcal{L}(A)$ and $V^{-1}\mathbf{y}_2 + \mathcal{L}(A)$ are disjoint for any two different $\mathbf{y}_1, \mathbf{y}_2$ with $\mathbf{0} \le \mathbf{y}_1, \mathbf{y}_2 \le \mathbf{d} - \mathbf{1}$. If these sets are not disjoint, we must have $V^{-1}\mathbf{y}_1 - V^{-1}\mathbf{y}_2 \in \mathcal{L}(A)$, that is, $V^{-1}\mathbf{y}_1 - V^{-1}\mathbf{y}_2 = A\mathbf{x}$ for some $\mathbf{x} \in \mathbb{Z}^n$. Multiplying by V, we obtain

$$\mathbf{y}_1 - \mathbf{y}_2 = VA\mathbf{x} = VAU(U^{-1}\mathbf{x}) = \mathrm{SNF}(A)(U^{-1}\mathbf{x}).$$

As U is unimodular, we have $\mathbf{z} := U^{-1}\mathbf{x} \in \mathbb{Z}^n$ and, consequently, $\mathbf{y}_1 - \mathbf{y}_2 = VAU\mathbf{z} = \mathrm{SNF}(A)\mathbf{z} \in \mathcal{L}(D)$. Since $\mathbf{0} \le \mathbf{y} \le \mathbf{d} - \mathbf{1}$, this is possible only if $\mathbf{y}_1 - \mathbf{y}_2 = \mathbf{0}$, contradicting $\mathbf{y}_1 \ne \mathbf{y}_2$. ☐

There is a more abstract way to look at the above results. One elegant algebraic way to think about lattices of $\mathbb{Z}^n$ is as Abelian subgroups of $\mathbb{Z}^n$ and their associated quotients. Given an integer lattice we can think of the lattice points in the fundamental parallelepiped as coset representatives of the quotient group:

Definition 2.3.17. Let $\mathcal{L}$ be a lattice and let $\mathcal{L}_0 \subseteq \mathcal{L}$. If the subset $\mathcal{L}_0$ is also a lattice, we say that $\mathcal{L}_0$ is a *sublattice* of $\mathcal{L}$.

Since $\mathcal{L}_0$ is a subgroup, we can define the quotient group, whose elements are the equivalence classes of $\mathcal{L}$ under the following relation: We let $\mathbf{x} \sim \mathbf{y}$ if and only if $\mathbf{x} - \mathbf{y} \in \mathcal{L}_0$. Then the equivalence classes are sets of the form $\mathbf{y} + \mathcal{L}_0$ for some $\mathbf{y} \in \mathcal{L}$. How many elements are there in this group? $|\mathcal{L} : \mathcal{L}_0|$ is the index. The quotient group is a finitely generated Abelian group. In this setup we can rewrite the existence of the Smith normal form as follows:

Lemma 2.3.18. *Let $\mathcal{L}_0$ be a sublattice of $\mathcal{L} \subseteq \mathbb{R}^n$. Then there exists a basis $\mathbf{u}_1, \dots, \mathbf{u}_n$ of $\mathcal{L}$ and a basis $\mathbf{v}_1, \dots, \mathbf{v}_n$ of $\mathcal{L}_0$ such that $\mathbf{v}_i = m_i \mathbf{u}_i$ for $m_i \in \mathbb{Z}$ and $m_i \mid m_j$ for all $j \ge i$ (the transition matrix is diagonal).*

What we have proved above can be rephrased as the following theorem.

Theorem 2.3.19. *Let $\mathcal{L}_0$ be a sublattice of $\mathcal{L} \subseteq \mathbb{R}^n$. The index $|\mathcal{L} : \mathcal{L}_0|$ is equal to the following numbers:*

1. *The number* $|\mathcal{L} \cap \Pi|$ *of points of* $\mathcal{L}$ *inside the fundamental parallelepiped* Π *of some basis of* $\mathcal{L}_0$.
2. *The absolute value of the determinant* $\det(M)$ *of the matrix* $M = (m_{ij})$

$$\begin{aligned}
\mathbf{v}_1 &= m_{11}\mathbf{u}_1 + m_{12}\mathbf{u}_2 + \cdots + m_{1n}\mathbf{u}_n,\\
\mathbf{v}_2 &= m_{21}\mathbf{u}_1 + m_{22}\mathbf{u}_2 + \cdots + m_{2n}\mathbf{u}_n,\\
&\vdots\\
\mathbf{v}_n &= m_{n1}\mathbf{u}_1 + m_{n2}\mathbf{u}_2 + \cdots + m_{nn}\mathbf{u}_n,
\end{aligned}$$

where $\mathbf{u}_1, \ldots, \mathbf{u}_n$ *is a basis of* $\mathcal{L}$ *and* $\mathbf{v}_1, \ldots, \mathbf{v}_n$ *is a basis of* $\mathcal{L}_0$.
3. *The ratio* $\det(\mathcal{L}_0) : \det(\mathcal{L})$.

2.4 Minkowski's theorems

When can we be sure that a convex set contains a nonzero lattice point? In general this problem is NP-hard, but sufficient conditions for some special, but very useful, geometric situations are available.

We begin with Blichfeldt's theorem. The theorem says that if a region of $\mathbb{R}^n$ has more volume than the volume of the fundamental parallelepiped of the lattice $\det(\mathcal{L})$, then we can translate X in such a way that it will contain one or more nonzero lattice points inside.

Theorem 2.4.1 (Blichfeldt's theorem). *Let* $\mathcal{L}$ *be a lattice,* $X \subseteq \mathbb{R}^n$ *a measurable set, and* $m \in \mathbb{Z}_{>0}$ *a number. Suppose* $\mathrm{vol}(X) > m\det(\mathcal{L})$. *Then there exist* $m+1$ *points* $\mathbf{x}_1, \mathbf{x}_2, \ldots, \mathbf{x}_{m+1}$ *inside* X *such that the nonzero vectors* $\mathbf{x}_i - \mathbf{x}_j \in \mathcal{L}$ *for all* $i \neq j$.

Proof. Let Π be a fundamental parallelepiped of $\mathcal{L}$. We know from Theorem 2.3.12 and Corollary 2.3.13 that the set X is covered completely by copies of Π, $\Pi + \mathbf{v}$, $\mathbf{v} \in \mathcal{L}$. Therefore, $\bigcup_{\mathbf{v}\in\mathcal{L}} (\Pi + \mathbf{v}) \cap X = X$. This implies

$$\sum_{\mathbf{v}\in\mathcal{L}} \mathrm{vol}[(\Pi + \mathbf{v}) \cap X] = \mathrm{vol}(X) > m\det(\mathcal{L}),$$

and since

$$\sum_{\mathbf{v}\in\mathcal{L}} \mathrm{vol}(\Pi \cap (X - \mathbf{v})) = \sum_{\mathbf{v}\in\mathcal{L}} \mathrm{vol}[(\Pi + \mathbf{v}) \cap X],$$

we must have

$$\sum_{\mathbf{v}\in\mathcal{L}} \mathrm{vol}(\Pi \cap (X - \mathbf{v})) > m\det(\mathcal{L}).$$

Now we define the indicator function

$$I_{\mathbf{v}}(\mathbf{x}) = [\Pi \cap (X - \mathbf{v})](\mathbf{x}) = \begin{cases} 1 & \text{if } \mathbf{x} \in \Pi \cap (X - \mathbf{v}),\\ 0 & \text{otherwise.} \end{cases}$$

Let $S(\mathbf{x}) = \sum_{\mathbf{v}\in\mathcal{L}} I_{\mathbf{v}}(\mathbf{x})$.

$$\int_\Pi S(\mathbf{x})\,\mathrm{d}\mathbf{x} = \sum_{\mathbf{v}\in\mathcal{L}} \int_\Pi I_{\mathbf{v}}(\mathbf{x})\,\mathrm{d}\mathbf{x} = \sum_{\mathbf{v}\in\mathcal{L}} \mathrm{vol}[(X - \mathbf{v}) \cap \Pi] > m\det(L) = m\,\mathrm{vol}(\Pi).$$

Therefore, at least one point $\mathbf{z}$ belongs to the $m+1$ sets $\Pi \cap (X - \mathbf{v}_1), \Pi \cap (X - \mathbf{v}_2), \ldots$. This means $\mathbf{z} \in (X - \mathbf{v}_1) \cap (X - \mathbf{v}_2) \cap \cdots \cap (X - \mathbf{v}_{m+1})$. This implies that $\mathbf{z} = \mathbf{x}_1 - \mathbf{v}_1 = \mathbf{x}_2 - \mathbf{v}_2$, and so $\mathbf{x}_1, \mathbf{x}_2, \ldots, \mathbf{x}_{m+1}$ are the points we wanted. □

Theorem 2.4.1 can be used to derive Minkowski's first theorem:

Theorem 2.4.2 (Minkowski's first theorem). *Let $\mathcal{L}$ be any lattice and let $K \subseteq \mathbb{R}^n$ be a convex set that is centrally symmetric ($-\mathbf{x} \in K$ if $\mathbf{x} \in K$).*

(a) *If* $\operatorname{vol}(K) > 2^n \det(\mathcal{L})$*, then K contains a nonzero lattice point.*

(b) *If* $\operatorname{vol}(K) \geq 2^n \det(\mathcal{L})$ *and K is compact, then K contains a nonzero lattice point.*

Proof. *Part* (a). Let $X = \frac{1}{2}K = \left\{\frac{\mathbf{x}}{2} : \mathbf{x} \in K\right\}$. Then $\operatorname{vol}(X) = 2^{-n}\operatorname{vol}(K) > \det(\mathcal{L})$. Now Blichfeldt's theorem, Theorem 2.4.1, implies that there exists a pair $\mathbf{x}, \mathbf{y} \in X$ such that $\mathbf{x} - \mathbf{y} \in \mathcal{L} \setminus \{0\}$. Finally, $2\mathbf{x}, 2\mathbf{y} \in K$, and since K is centrally symmetric, $-2\mathbf{y} \in K$. This implies that $\mathbf{u} = \frac{1}{2}(2\mathbf{x}) + \frac{1}{2}(-2\mathbf{y}) \in K$ since K is convex.

Part (b). In the compact case, we consider the dilated convex sets $\rho_i K$ for a sequence of $\rho_i > 1$ with $\rho_i \to 1$. Part (a) of the theorem then implies that there exists a sequence of nonzero lattice points $\mathbf{x}_i \in \rho_i K \cap \mathcal{L}$. A compactness argument implies that the sequence has a limit point $\mathbf{x}$, but on the other hand the discreteness of the lattice implies that $\mathbf{0} \neq \mathbf{x} \in K \cap \mathcal{L}$. □

Following are some of many important applications of Minkowski's theorem. We leave the last two as exercises:

Length of the shortest nonzero vector in a lattice. Computing shortest nonzero lattice vectors is a fundamental problem with many applications in integer programming (more about this in Section 2.7). Minkowski's theorem allows us to find estimates on the length of a shortest lattice vector.

Theorem 2.4.3. *Let $\mathcal{L} \subseteq \mathbb{R}^n$ be a lattice; then there exists a vector $\mathbf{u} \in \mathcal{L} \setminus \{\mathbf{0}\}$ such that*

$$\|\mathbf{u}\| \leq \frac{2}{\sqrt{\pi}} \left(\Gamma\left(1 + \frac{n}{2}\right)\right)^{1/n} (\det(\mathcal{L}))^{1/n} \approx \sqrt{\frac{2n}{e\pi}} (\det(\mathcal{L}))^{1/n},$$

where

$$\Gamma(s) = \int_0^\infty x^{s-1} e^{-x}\, dx.$$

Proof. Consider the ball $B_r(\mathbf{0}) = \{\mathbf{x} : \|\mathbf{x}\| \leq r\}$. Then $\operatorname{vol}(B_r(\mathbf{0})) = \frac{r^n \pi^{n/2}}{\Gamma(1+\frac{n}{2})}$. The ball $B_r(\mathbf{0})$ is compact and centrally symmetric, which implies that there exists a nonzero vector if $\operatorname{vol}(B_r(\mathbf{0})) > 2^n \det(\mathcal{L})$, which holds if

$$r \geq 2(\det(\mathcal{L}))^{1/n} \left(\Gamma\left(1 + \frac{n}{2}\right)\right)^{1/n} \pi^{-1/2}. \qquad \square$$

The same can be done for the shortest vector with respect to other norms.

Theorem 2.4.4. *Let $\mathcal{L} \subseteq \mathbb{R}^n$ be a lattice; then there exists a vector $\mathbf{u} \in \mathcal{L} \setminus \{\mathbf{0}\}$ such that*

$$\|\mathbf{u}\|_\infty \leq (\det(\mathcal{L}))^{1/n}.$$

Proof. Take the cube $Q = \{\mathbf{x} : \|\mathbf{x}\|_\infty \le (\det(\mathcal{L}))^{1/n}\}$. It has volume $2^n \det(\mathcal{L})$. By Minkowski's first theorem, there exists $\mathbf{0} \neq \mathbf{u} \in Q \cap \mathcal{L}$. □

Applications in number theory. Number theorists have been interested in the representation of numbers as sums of squares. Lagrange proved the following result, which can also be proved using Minkowski's theorem.

Theorem 2.4.5. *Every positive integer n can be expressed as a sum of four squares,*

$$n = x_1^2 + x_2^2 + x_3^2 + x_4^2,$$

where x_i are nonnegative integers.

Another result generalizes Corollary 2.1.3.

Theorem 2.4.6. *Let $\alpha_1, \alpha_2, \ldots, \alpha_n$ be any n irrational numbers. Then there exist infinitely many sets of integers $p_1, p_2, \ldots, p_n, p$ with $p \ge 1$ such that*

$$\left|\alpha_1 - \frac{p_1}{p}\right| < \frac{1}{p^{\frac{n+1}{n}}}, \ldots, \left|\alpha_n - \frac{p_n}{p}\right| < \frac{1}{p^{\frac{n+1}{n}}}.$$

Let us conclude with a counterpart to Minkowski's first theorem. There is much more to learn about this topic; see [32, 34, 70, 146, 263].

Theorem 2.4.7 (Minkowski–Hlawka theorem). *Suppose $n > 1$ and $M \subseteq \mathbb{R}^n$ is a bounded measurable set. Let $d > \mathrm{vol}(M)$. Then there exists a lattice $\mathcal{L}$ such that $d = \det(\mathcal{L})$ and M does not contain any lattice points other than the origin.*

2.5 Gordan and Dickson

In this section, we introduce two equivalent versions of the Gordan–Dickson lemma. Whereas the *set version* is suitable to show finiteness of certain sets of lattice points, the *sequence version* is suitable to show termination of many algorithms.

Let us start with the sequence version. It is a generalization of the simple fact that every strictly decreasing sequence of nonnegative integers is finite.

Lemma 2.5.1 (Gordan–Dickson lemma, sequence version). *Let $\{\mathbf{p}_1, \mathbf{p}_2, \ldots\}$ be a sequence of points in $\mathbb{Z}_+^n$ such that $\mathbf{p}_i \not\le \mathbf{p}_j$ whenever $i < j$. Then this sequence is finite.*

Lemma 2.5.2 (Gordan–Dickson lemma, set version). *Every infinite set $S \subseteq \mathbb{Z}_+^n$ contains only finitely many $\le$-minimal points.*

In Exercises 2.12.4, 2.12.5, and 2.12.6, you will be asked to prove Lemma 2.5.1 and to show that the two presented versions of the Gordan–Dickson lemma are in fact equivalent. Let us give a small example for Lemma 2.5.2.

Example 2.5.3. For $n = 1$, Lemma 2.5.2 states that every infinite set of nonnegative integers contains only finitely many smallest elements. In fact, when $n = 1$, there is exactly one such minimal integer number.

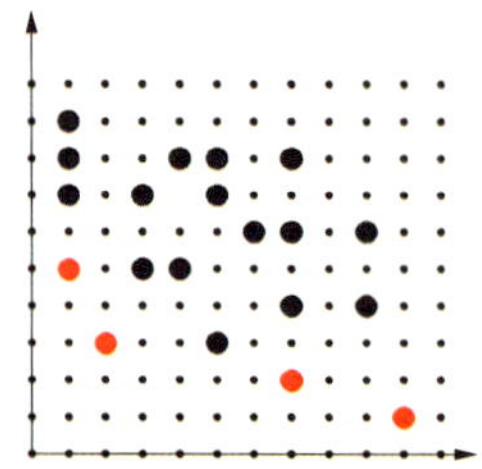

Figure 2.7. *Minimal elements in a two-dimensional set of lattice points.*

For $n = 2$, the situation becomes more complicated. However, there are still only finitely many $\leq$-minimal elements, as claimed in Lemma 2.5.2; see Figure 2.7.

For our purposes, we need to extend the Gordan–Dickson lemma from $\mathbb{Z}_+^n$ to $\mathbb{Z}^n$. For this, let us generalize the partial ordering $\leq$ on $\mathbb{R}_+^n$ to a partial ordering $\sqsubseteq$ on $\mathbb{R}^n$.

Definition 2.5.4. We call $\mathbf{u}, \mathbf{v} \in \mathbb{R}^n$ *sign-compatible* if $u^{(j)} v^{(j)} \geq 0$ for all components $j = 1, \ldots, n$.

For $\mathbf{u}, \mathbf{v} \in \mathbb{R}^n$, we say that $\mathbf{u} \sqsubseteq \mathbf{v}$ if $\mathbf{u}$ and $\mathbf{v}$ are sign-compatible and if $|u^{(j)}| \leq |v^{(j)}|$ for all components $j = 1, \ldots, n$; that is, if $\mathbf{u}$ belongs to the same orthant as $\mathbf{v}$ and if its components are not greater in absolute value than the corresponding components of $\mathbf{v}$.

Example 2.5.5. Checking the definition, we see that $(1, -1, 0) \sqsubseteq (2, -1, 4)$. On the other hand, we have $(1, -1, 0) \not\sqsubseteq (2, 1, 4)$, since the signs of the second components disagree.

For $\mathbf{v} \in \mathbb{Z}^n$, we define $\mathbf{v}^+$ and $\mathbf{v}^-$ as the vectors with components $\max(0, \mathbf{v}^{(j)})$ and $\max(0, -\mathbf{v}^{(j)})$, respectively. With this, it can be checked easily that $\mathbf{u} \sqsubseteq \mathbf{v}$ if and only if $(\mathbf{u}^+, \mathbf{u}^-) \leq (\mathbf{v}^+, \mathbf{v}^-)$; see Exercise 2.12.7. This simple correspondence readily implies an extension of both versions of the Gordan–Dickson lemma to the $\sqsubseteq$-situation.

Lemma 2.5.6 (Gordan–Dickson lemma, $\sqsubseteq$-version).

- *Every sequence $\{\mathbf{p}_1, \mathbf{p}_2, \ldots\}$ of points in $\mathbb{Z}^n$ such that $\mathbf{p}_i \not\sqsubseteq \mathbf{p}_j$ whenever $i < j$ is finite.*
- *Every infinite set $S \subseteq \mathbb{Z}^n$ contains only finitely many $\sqsubseteq$-minimal points.*

2.6 Hilbert

Let us show a nice finiteness result on the representation of lattice points in rational polyhedra and cones. For this, we define the notion of an integral generating set.

Definition 2.6.1. Let $S \subseteq \mathbb{Z}^n$. Then we call $T \subseteq S$ an *integral generating set* of S if, for every $\mathbf{s} \in S$, there exists a finite integer linear combination $\mathbf{s} = \sum \alpha_i \mathbf{t}_i$, with $\mathbf{t}_i \in T$ and $\alpha_i \in \mathbb{Z}_+$. We call T an *integral basis* of S if T is an inclusion-minimal integral generating set.

Note that an integral generating set of S is allowed to contain elements from S only! See Figure 2.8 for two examples.

Figure 2.8. *Minimal integral generating sets of two sets of lattice points.*

Figure 2.9. *For both sets S,* cone(S) *is not finitely generated.*

The following theorem characterizes all sets $S \subseteq \mathbb{Z}^n$ that possess *finite* integral generating sets, i.e., such that every point of S can be expressed as a nonnegative linear integer combination of a *finite* subset of S.

Theorem 2.6.2. *Let S be a set of lattice points in $\mathbb{Z}^n$.*

(a) *S has a finite integral generating set if and only if $C = \operatorname{cone}(S)$ is a rational polyhedral cone.*

(b) *If the cone $C = \operatorname{cone}(S)$ is rational and pointed, then there exists a unique integral basis of S.*

A proof appears in [158].

Example 2.6.3. For $S_1 = \{(x,y) \in \mathbb{Z}^2_+ : x \geq y^2\}$ and $S_2 = \{(x,y) \in \mathbb{Z}^2_+ : y \geq 1\}$, we easily see that $\operatorname{cone}(S_i) = \mathbb{R}^2_+ \setminus \{(x,0) : x > 0\}$ is not a rational polyhedral cone; see Figure 2.9. Consequently, there do not exist finite integral generating sets for these sets.

In the special situation that S consists of the lattice points in a rational polyhedral cone, Theorem 2.6.2 implies the following well-known fact about rational polyhedral cones; see Figure 2.10.

Corollary 2.6.4. *For every rational polyhedral cone C, the set $C \cap \mathbb{Z}^n$ possesses a finite integral generating set, a so-called* Hilbert basis *of C. If C is pointed, there is a unique inclusion-minimal Hilbert basis of C.*

Proof. As C is a rational polyhedral cone, there exist generators $\mathbf{v}_1, \ldots, \mathbf{v}_k \in \mathbb{Z}^n$ with $C =$

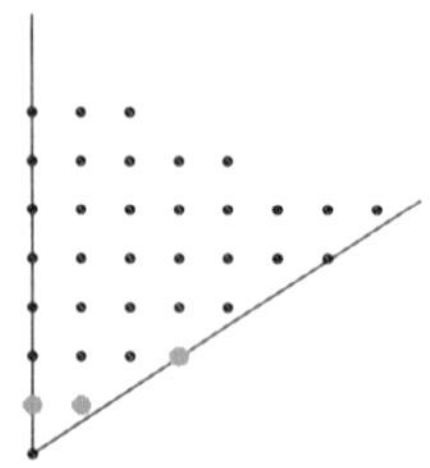

Figure 2.10. *A cone and its inclusion-minimal Hilbert basis.*

cone($\mathbf{v}_1,\ldots,\mathbf{v}_k$) by Corollary 1.3.13. Now consider the set

$$F := \left\{ \mathbf{x} : \mathbf{x} = \sum_{i=1}^{k} \lambda_i \mathbf{v}_i : 0 \le \lambda_i < 1,\, i = 1,\ldots,k \right\},$$

which is clearly bounded, as the ℓ_1-norms of the points in F are bounded by

$$\|\mathbf{x}\|_1 \le \sum_{i=1}^{k} \lambda_i \|\mathbf{v}_i\|_1 < \sum_{i=1}^{k} \|\mathbf{v}_i\|_1.$$

Therefore, $F \cap \mathbb{Z}^n$ is finite. We claim that $H := (F \cap \mathbb{Z}^n) \cup \{\mathbf{v}_1,\ldots,\mathbf{v}_k\}$ is an integral generating set for $C \cap \mathbb{Z}^n$. For this, take an arbitrary $\mathbf{z} \in C \cap \mathbb{Z}^n$. Clearly, as $\mathbf{z} \in C$, we can write $\mathbf{z} = \sum_{i=1}^{k} \lambda_i \mathbf{v}_i$ for some $\lambda_1,\ldots,\lambda_k \ge 0$. From this, we get

$$\mathbf{z} = \sum_{i=1}^{k} \lfloor \lambda_i \rfloor \mathbf{v}_i + \underbrace{\sum_{i=1}^{k} (\lfloor \lambda_i \rfloor - \lambda_i) \mathbf{v}_i}_{\in F},$$

giving a desired representation of $\mathbf{z}$ as a nonnegative integer linear combination of elements in H, and the first part of the corollary is proved.

To show the second part, assume that C is a *pointed* rational polyhedral cone. Thus, there exists some $\mathbf{c} \in \mathbb{R}^n$ such that $C \subseteq \left\{ \mathbf{x} \in \mathbb{R}^n : \mathbf{c}^\top \mathbf{x} \ge 0 \right\}$ and $C \cap \left\{ \mathbf{x} \in \mathbb{R}^n : \mathbf{c}^\top \mathbf{x} = 0 \right\} = \{\mathbf{0}\}$. Consider now the set

$$R := \left\{ \mathbf{x} \in C \cap \mathbb{Z}^n \setminus \{\mathbf{0}\} : \nexists\, \mathbf{y},\mathbf{z} \in C \cap \mathbb{Z}^n \setminus \{\mathbf{0}\} \text{ with } \mathbf{x} = \mathbf{y} + \mathbf{z} \right\}$$

of all lattice points in C that *cannot* be written as the sum of two nonzero lattice points in C. As no element in R can be represented as a (nontrivial) sum of two lattice points in C, all elements from R must belong to any Hilbert basis of C. In particular, $R \subseteq H$ for our Hilbert basis constructed above. Consequently, R is always finite. It remains to prove that R itself is already a Hilbert basis of C.

Assume on the contrary that not all lattice points in C can be written as a nonnegative integer linear combination of elements in R. Let $\mathbf{x} \in C \cap \mathbb{Z}^n$ be such a point with minimal value $\mathbf{c}^\top \mathbf{x} > 0$. As $\mathbf{x} \in R$ cannot hold, we can write $\mathbf{x}$ as $\mathbf{x} = \mathbf{y} + \mathbf{z}$ for some $\mathbf{y},\mathbf{z} \in C \cap \mathbb{Z}^n \setminus \{\mathbf{0}\}$. We conclude that $\mathbf{c}^\top \mathbf{y}, \mathbf{c}^\top \mathbf{z} < \mathbf{c}^\top \mathbf{x}$, and therefore $\mathbf{y}$ and $\mathbf{z}$ can be represented as nonnegative integer linear combinations of elements in R. Together, this gives a valid representation of $\mathbf{x} = \mathbf{y} + \mathbf{z}$, a contradiction. Hence, R is already a Hilbert basis of C, and therefore it is the unique inclusion-minimal Hilbert basis of C. □

Note that this proof in fact shows that the unique inclusion-minimal Hilbert basis consists exactly of all *indecomposable* lattice points in C. Moreover, the first part of the proof gives an algorithm to compute a Hilbert basis of any rational cone C:

- *Triangulate* C into *simplicial* cones, that is, into cones generated by linearly independent vectors (see Section 6.3 for how to do this in arbitrary dimension).
- Use Lemma 2.3.16 to enumerate all lattice points in the fundamental parallelepipeds F of each simplicial cone, and collect them into a set H.
- Remove all those lattice points $\mathbf{x}$ from H such that there exists some $\mathbf{y} \in H$ with $\mathbf{x} - \mathbf{y} \in C$; that is, $\mathbf{x}$ can be written as a nontrivial sum of two other lattice points, $\mathbf{y}$ and $\mathbf{x} - \mathbf{y}$, of C.

This approach is implemented in the software package `Normaliz` [64].

Solving a linear Diophantine system of inequalities

In this section, we use Hilbert bases to show an integer analogue to Theorem 1.4.3, which states that every rational polyhedron P can be written as the Minkowski sum $P = Q + C$ of a convex rational polytope Q and a rational polyhedral cone C. The integer analogue is as follows.

Lemma 2.6.5. *Let $P = \{\mathbf{z} : A\mathbf{z} \leq \mathbf{b},\ \mathbf{z} \in \mathbb{R}^n\}$ be a rational polyhedron and let*

$$C = \left\{\mathbf{z} : A\mathbf{z} \leq \mathbf{0},\ \mathbf{z} \in \mathbb{R}^n\right\}$$

be its recession cone. Then there exists a polytope $Q \subseteq \mathbb{R}^n$ such that

$$(P \cap \mathbb{Z}^n) = (Q \cap \mathbb{Z}^n) + (C \cap \mathbb{Z}^n).$$

It should be noted that, in general, the polytope Q from Theorem 1.4.3 and the polytope Q from its integer analogue, Lemma 2.6.5, are different.

Example 2.6.6. Let $P = \{z : z \geq 1/2,\ z \in \mathbb{R}\}$ so $C = \mathbb{R}_+$. Then we have $Q = \{1/2\}$ in the continuous and $Q = \{1\}$ in the integer setting.

Before we prove Lemma 2.6.5, let us start with an example where this integer representation is useful.

Example 2.6.7. Let us have a look at the polyhedron P defined by the linear system

$$\begin{aligned} x - y &\leq 2, \\ -3x + y &\leq 1, \\ x + y &\geq 1, \\ y &\geq 0, \end{aligned}$$

and let us solve this system over $\mathbb{Z}$. All the integer solutions (x, y) to this linear system are of the form

$$\begin{pmatrix} x \\ y \end{pmatrix} = \begin{pmatrix} 1 \\ 0 \end{pmatrix} + \alpha_1 \begin{pmatrix} 1 \\ 1 \end{pmatrix} + \alpha_2 \begin{pmatrix} 1 \\ 2 \end{pmatrix} + \alpha_3 \begin{pmatrix} 1 \\ 3 \end{pmatrix},$$

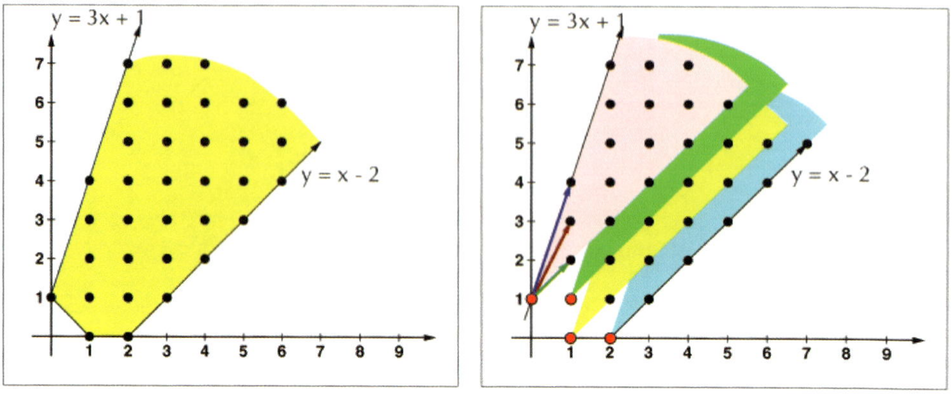

Figure 2.11. *Lattice points in a two-dimensional polyhedron covered by four shifted cones.*

or

$$\binom{x}{y} = \binom{2}{0} + \alpha_1\binom{1}{1} + \alpha_2\binom{1}{2} + \alpha_3\binom{1}{3},$$

or

$$\binom{x}{y} = \binom{0}{1} + \alpha_1\binom{1}{1} + \alpha_2\binom{1}{2} + \alpha_3\binom{1}{3},$$

or

$$\binom{x}{y} = \binom{1}{1} + \alpha_1\binom{1}{1} + \alpha_2\binom{1}{2} + \alpha_3\binom{1}{3}$$

for some nonnegative integers α_1, α_2, and α_3. From Figure 2.11, we see that every lattice point of P is covered at least once. Moreover, every covered lattice point belongs to P. Finally, note that $\binom{1}{1}, \binom{1}{2}, \binom{1}{3}$ form the minimal Hilbert basis of the recession cone C of P, which is described by

$$\begin{aligned} x - y &\leq 0, \\ -3x + y &\leq 0, \\ x + y &\geq 0, \\ y &\geq 0. \end{aligned}$$

Thus

$$P \cap \mathbb{Z}^n = \left\{\binom{1}{0}, \binom{2}{0}, \binom{0}{1}, \binom{1}{1}\right\} + \mathbb{Z}_+\binom{1}{1} + \mathbb{Z}_+\binom{1}{2} + \mathbb{Z}_+\binom{1}{3}$$

gives a nice finite representation of all lattice points in P.

Such a finite representation does in fact exist for every rational polyhedron.

Lemma 2.6.8. *Let $P = \{\mathbf{z} \in \mathbb{R}^n : A\mathbf{z} \leq \mathbf{b}\}$ be a rational polyhedron and let*

$$C = \left\{\mathbf{z} \in \mathbb{R}^n : A\mathbf{z} \leq \mathbf{0}\right\}$$

be its recession cone. If $P \cap \mathbb{Z}^n \neq \emptyset$, then there exist finitely many points $\mathbf{z}_1, \ldots, \mathbf{z}_k \in P \cap \mathbb{Z}^n$ and $\mathbf{h}_1, \ldots, \mathbf{h}_s \in C \cap \mathbb{Z}^n$ such that every solution $\mathbf{z} \in P \cap \mathbb{Z}^n$ can be written as

$$\mathbf{z} = \mathbf{z}_i + \sum_{j=1}^{s} \alpha_j \mathbf{h}_j$$

for some $i \in \{1,\dots,k\}$ *and with* $\alpha_j \in \mathbb{Z}_+$ *for all* $j = 1,\dots,s$. *Moreover, the vectors* $\mathbf{h}_1,\dots,\mathbf{h}_s$ *form a Hilbert basis of* C.

Proof. Consider the rational polyhedral cone

$$\bar{P} := \left\{ (\mathbf{z},u) \in \mathbb{R}^{n+1} : A\mathbf{z} - \mathbf{b}u \le \mathbf{0},\ u \ge 0,\ \mathbf{z} \in P \right\}.$$

Note that $\mathbf{z} \in P$ if and only if $(\mathbf{z},1) \in \bar{P}$ and $\mathbf{z} \in C$ if and only if $(\mathbf{z},0) \in \bar{P}$. Consider a (finite) Hilbert basis H of $\bar{P}$ and let $\{\mathbf{z}_1,\dots,\mathbf{z}_k\} := \{\mathbf{z} : (\mathbf{z},1) \in H\}$ and $\{\mathbf{h}_1,\dots,\mathbf{h}_s\} := \{\mathbf{h} : (\mathbf{h},0) \in H\}$. We show that the vectors $\mathbf{z}_i$ and $\mathbf{h}_j$ have the required properties.

Let $\mathbf{z} \in P \cap \mathbb{Z}^n$. Then $(\mathbf{z},1) \in \bar{P} \cap \mathbb{Z}^{n+1}$ and consequently

$$(\mathbf{z},1) = \sum \alpha_j \mathbf{g}_j, \tag{2.1}$$

with $\alpha_j \in \mathbb{Z}_+$ and $\mathbf{g}_j \in H$ for all j. As the last component of each element in $\bar{P}$ is nonnegative, the last component of each $\mathbf{g}_j$ with $\alpha_j > 0$ in equation (2.1) must be 0 or 1. Moreover, exactly *one* such $\mathbf{g}_i$ with $\alpha_i > 0$ has a last component equal to 1 and necessarily $\alpha_i = 1$, and all the other vectors $\mathbf{g}_j$, $j \neq i$, must have a zero as their last component. Thus,

$$(\mathbf{z},1) = (\mathbf{z}_i,1) + \sum_{j=1}^{s} \alpha_j(\mathbf{h}_j,0)$$

for some $i \in \{1,\dots,k\}$. By deleting the last component, we get the desired representation of $\mathbf{z}$.

The vectors $\mathbf{h}_1,\dots,\mathbf{h}_s$ form a Hilbert basis of C, since with $\mathbf{z} \in C \cap \mathbb{Z}^n$ we have $(\mathbf{z},0) \in \bar{P}$, and consequently all $\mathbf{g}_j$ with $\alpha_j > 0$ in the representation $(\mathbf{z},0) = \sum \alpha_j \mathbf{g}_j$ must have a zero as their last component and therefore are among the vectors $\mathbf{h}_1,\dots,\mathbf{h}_s$. □

The representation from Lemma 2.6.8 immediately proves Lemma 2.6.5. By construction, the polytope $Q = \operatorname{conv}(\mathbf{z}_1,\dots,\mathbf{z}_k)$ with $\mathbf{z}_1,\dots,\mathbf{z}_k$ as defined in the proof of Lemma 2.6.8 has the desired property claimed in Lemma 2.6.5.

2.7 Lenstra, Lenstra, Lovász, and the shortest vector problem

The problem of the shortest nonzero vector of a lattice $\mathcal{L}$ with respect to a given norm is called the *shortest vector problem* (SVP)indexSVP (shortest vector problem). It is, in its variant for the ℓ_2-norm, one of the most well-studied algorithmic problems related to lattices.

The problem is NP-hard when the dimension is not fixed—even to approximate within a constant factor [247]—and it is still difficult for the case of "average" lattices; see [9] and references therein.

But it appears to be much easier to solve than a general integer program. Therefore, it makes sense to use algorithms for solving the SVP, exactly or approximately, as subroutines for solving integer programs. Most famously, H.W. Lenstra's polynomial-time integer linear programming algorithm for problems in fixed dimension [231] relies on it to find thin directions for branching on hyperplanes. Frank and Tardos [129] showed how the computation of approximate shortest vectors, providing a simultaneous Diophantine approximation of linear objective functions, turn polynomial-time algorithms into *strongly* polynomial-time algorithms for some combinatorial problems with polyhedral characterizations (see also [145, 296]).

Until recently, the best deterministic algorithm for solving SVP was due to Kannan [188], with time-complexity $2^{O(n\log n)}$. Probabilistic algorithms, such as the sieving method by Ajtai, Kumar, and Sivakumar [10], achieve single exponential time, $2^{O(n)}$; the exponent was improved by Micciancio and Voulgaris [249] to give a run time of $2^{3.199n}$. In 2010, Micciancio and Voulgaris [250] gave the first deterministic single exponential algorithms for SVP, and other lattice problems, using Voronoi cell computations. Dadush, Peikert, and Vempala [80] extend this result, using so-called M-ellipsoid coverings, to arbitrary norms, including the ℓ_∞-norm, which is relevant to us in Section 7.1.

We do not present any of the abovementioned algorithms in this book, but we do present one famous algorithm to achieve fairly short vectors with a good size upper bound guarantee. This is an algorithm invented by Lenstra, Lenstra, and Lovász [230], originally for solving the problem of factoring a polynomial into irreducibles. It is known today as the *LLL algorithm*. All known deterministic algorithms for SVP rely on the LLL algorithm as a subroutine.

The beginning of the LLL algorithm goes back to the elementary Gram–Schmidt orthogonalization algorithm: From an initial basis $\mathbf{b}_1,\dots,\mathbf{b}_n$ we recursively construct a new basis as follows: First, set $\mathbf{b}_1^* = \mathbf{b}_1$, then recursively compute

$$\mathbf{b}_k^* = \mathbf{b}_k - \sum_{j=1}^{k-1} \mu_{kj}\mathbf{b}_j^*, \quad \text{where} \quad \mu_{kj} = \frac{\mathbf{b}_j^{*\top}\mathbf{b}_k}{\|\mathbf{b}_j^*\|^2}.$$

Theorem 2.7.1. *The Gram–Schmidt vectors have the following properties:*

1. *For each k, the first k vectors generate the same subspace as the first k original vectors; i.e., if $U_k = \operatorname{span}\{\mathbf{b}_1,\mathbf{b}_2,\dots,\mathbf{b}_k\}$, then $U_k = \operatorname{span}\{\mathbf{b}_1^*,\mathbf{b}_2^*,\dots,\mathbf{b}_k^*\}$.*

2. *The vectors are shorter than or of the same length as the original ones, $\|\mathbf{b}_i^*\| \le \|\mathbf{b}_i\|$.*

3. *The lattices generated have the same determinant, i.e.,*

$$\det(\mathbf{b}_1,\mathbf{b}_2,\dots,\mathbf{b}_n) = \det(\mathbf{b}_1^*,\mathbf{b}_2^*,\dots,\mathbf{b}_n^*).$$

4. *The vectors we generate are pairwise orthogonal, i.e., $\mathbf{b}_i^{*\top}\mathbf{b}_j^* = 0$ for $i \ne j$.*

Example 2.7.2. Given vectors

$$\begin{pmatrix}2\\4\\0\end{pmatrix}, \quad \begin{pmatrix}6\\7\\0\end{pmatrix}, \quad \begin{pmatrix}1\\7\\1\end{pmatrix},$$

we can compute

$$\mathbf{b}_1^* = \begin{pmatrix} 2 \\ 4 \\ 0 \end{pmatrix},$$

$$\mathbf{b}_2^* = \begin{pmatrix} 6 \\ 7 \\ 0 \end{pmatrix} - 2\begin{pmatrix} 2 \\ 4 \\ 0 \end{pmatrix} = \begin{pmatrix} 2 \\ -1 \\ 0 \end{pmatrix},$$

$$\mathbf{b}_3^* = \begin{pmatrix} 1 \\ 7 \\ 1 \end{pmatrix} - \frac{3}{2}\begin{pmatrix} 2 \\ 4 \\ 0 \end{pmatrix} - \frac{-5}{5}\begin{pmatrix} 2 \\ -1 \\ 0 \end{pmatrix} = \begin{pmatrix} 0 \\ 0 \\ 1 \end{pmatrix}.$$

Lemma 2.7.3. *Let $\mathcal{L}$ be a lattice with basis $\mathbf{b}_1, \ldots, \mathbf{b}_n$. Then for all $\mathbf{x} \in \mathcal{L} \setminus \{\mathbf{0}\}$,*

$$\|\mathbf{x}\| \geq \min\left\{\|\mathbf{b}_1^*\|, \|\mathbf{b}_2^*\|, \ldots, \|\mathbf{b}_n^*\|\right\}.$$

Proof. We have

$$\mathbf{x} = \sum \lambda_i \mathbf{b}_i = \sum \lambda_i \left(\mathbf{b}_i^* + \sum \mu_{ij} \mathbf{b}_j^*\right),$$

and thus

$$\|\mathbf{x}\|^2 = \mathbf{x}^\top \mathbf{x} = \lambda_i^2 \mathbf{b}_i^{*\top} \mathbf{b}_i^* + \text{nonnegative terms} \geq \lambda_s^2 \mathbf{b}_s^{*\top} \mathbf{b}_s^*,$$

where s is the index of the vector of the smallest length. □

To summarize, the Gram–Schmidt vectors $\mathbf{b}_1^*, \ldots, \mathbf{b}_n^*$ have favorable properties. Unfortunately, in general they do not form a basis of the lattice $\mathcal{L}$. In fact, most lattices do not have an orthogonal basis. We next define a notion of *reduced* bases, which are "almost" orthogonal. The Gram–Schmidt orthogonalization plays a key role in this definition.

Definition 2.7.4. A basis $\mathbf{g}_1, \ldots, \mathbf{g}_n$ of a lattice $\mathcal{L}$ is said to be *reduced* if its Gram–Schmidt vectors $\mathbf{g}_1^*, \ldots, \mathbf{g}_n^*$ and multipliers μ_{kj} satisfy

$$\left|\mu_{kj}\right| \leq \frac{1}{2} \qquad \text{for } 1 \leq j < k \leq n, \tag{2.2}$$

$$\|\mathbf{g}_j^*\|^2 \leq 2\|\mathbf{g}_{j+1}^*\|^2 \quad \text{for } j = 1, \ldots, n-1. \tag{2.3}$$

Here are some reasons why we like reduced bases.

Theorem 2.7.5. *Let $\det(\mathcal{L})$ be the determinant of the lattice $\mathcal{L}$ and let $\mathbf{g}^1, \ldots, \mathbf{g}^n$ be a reduced basis. Then we have*

$$\|\mathbf{g}_1\| \leq 2^{\frac{n-1}{2}} \min\{\,\|\mathbf{x}\| : \mathbf{x} \in \mathcal{L},\ \mathbf{x} \neq \mathbf{0}\}, \tag{2.4}$$

$$\|\mathbf{g}_1\| \leq 2^{\frac{n-1}{4}} \det(\mathcal{L})^{\frac{1}{n}}. \tag{2.5}$$

Thus the first basis vector, $\mathbf{g}^1$, is an approximate shortest nonzero lattice vector. The approximation factor is $2^{\frac{n-1}{2}}$. The estimate (2.5) should be compared with the bound from Minkowski's first theorem (Theorem 2.4.3).

Proof. We have

$$(\det(\mathcal{L}))^2 = \prod_{j=1}^{n} \|\mathbf{g}_j^*\|^2 \geq \prod_{j=1}^{n} 2^{-(j-1)} \|\mathbf{g}_1\|^{2n} = 2^{\frac{-n(n-1)}{2}} \|\mathbf{g}_1\|^{2n}.$$

Therefore, $\|\mathbf{g}_j\|^2 \geq 2^{-(j-1)}\|\mathbf{g}_1\|^2 \geq 2^{-(n-1)}\|\mathbf{g}_1\|^2$. Hence, by Lemma 2.7.3, we are done. □

We now present the algorithm that computes a reduced basis of a lattice $\mathcal{L}$ from an arbitrary basis. For $\mu \in \mathbb{R}$, we write $\lceil \mu \rfloor = \lfloor \mu + \frac{1}{2} \rfloor$ for the integer nearest to μ.

ALGORITHM 2.1. Lattice basis reduction.

input basis $\mathbf{b}_1, \ldots, \mathbf{b}_n \in \mathbb{Q}^n$ of a lattice $\mathcal{L}$.
output a reduced basis $\mathbf{g}_1, \ldots, \mathbf{g}_n$ of the lattice $\mathcal{L}$.
for $i = 1, \ldots, n$ **do**
  $\mathbf{g}_i \leftarrow \mathbf{b}_i$.
Compute the Gram–Schmidt orthogonal basis (GSO) $\mathbf{g}_1^*, \ldots, \mathbf{g}_n^* \in \mathbb{Q}^n$ and multipliers $\mu_{kj} \in \mathbb{Q}$.
$i \leftarrow 2$.
while $i \leq n$ **do**
  for $j = i-1, i-2, \ldots, 1$ **do**
    $\mathbf{g}_i \leftarrow \mathbf{g}_i - \lceil \mu_{ij} \rfloor \mathbf{g}_j$, update the GSO (replacement step).
    if $i > 1$ and $\|\mathbf{g}_{i-1}^*\|^2 > 2\|\mathbf{g}_i^*\|^2$ **then**
      Exchange $\mathbf{g}_{i-1}$ and $\mathbf{g}_i$ and update the GSO, $i \leftarrow i-1$.
    else
      $i \leftarrow i+1$.
return $\mathbf{g}_1, \ldots, \mathbf{g}_n$.

The algorithm runs in polynomial time. We omit the proof of termination and polynomial-time complexity. It can be found in [296].

Example 2.7.6. We will go through the following example, which is also illustrated in Table 2.1. Given vectors $\mathbf{g}_1 = \binom{12}{2}$ and $\mathbf{g}_2 = \binom{13}{4}$, we can compute

$$\mathbf{g}_1^* = \begin{pmatrix} 12 \\ 2 \end{pmatrix}, \qquad \mathbf{g}_2^* = \begin{pmatrix} 13 \\ 4 \end{pmatrix} - \frac{\binom{13}{4}^\top \binom{12}{2}}{\|\binom{12}{2}\|^2} \begin{pmatrix} 12 \\ 2 \end{pmatrix} = \begin{pmatrix} -11/37 \\ 66/37 \end{pmatrix}.$$

We can compute: $\binom{13}{4} - \binom{12}{2} = \binom{1}{2}$. Thus we have $\binom{12}{2}, \binom{1}{2}$ as a new basis. Then we need to exchange because $12^2 + 2^2 > 2\left(\left(\frac{-11}{37}\right)^2 + \left(\frac{66}{37}\right)^2\right) \approx 6.540$. We update the Gram–Schmidt form:

$$\mathbf{g}_1^* = \begin{pmatrix} 1 \\ 2 \end{pmatrix}, \qquad \mathbf{g}_2^* = \begin{pmatrix} 12 \\ 2 \end{pmatrix} - \frac{\binom{12}{2}^\top \binom{1}{2}}{\|\binom{1}{2}\|^2} \begin{pmatrix} 1 \\ 2 \end{pmatrix} = \begin{pmatrix} 44/5 \\ -22/5 \end{pmatrix}.$$

Then we find our new $\mathbf{g}_2 = \binom{12}{2} - 3\binom{1}{2} = \binom{9}{-4}$. Can we stop now? We must check: $1^2 + 2^2 \leq 2\left(\left(\frac{44}{5}\right)^2 + \left(\frac{-22}{5}\right)^2\right) = 193.6$? Yes! Our reduced basis is $\binom{1}{2}, \binom{9}{-4}$. It turns out that the first basis vector, $\binom{1}{2}$, is actually the shortest nonzero vector of the lattice.

Table 2.1. *Trace of the basis reduction algorithm (Algorithm 2.1) on the lattice of Example 2.7.6.*

$(\mathbf{g}_1 \ \ \mathbf{g}_2)$	$\begin{pmatrix}\mu_{11} & \cdot \\ \mu_{21} & \mu_{22}\end{pmatrix}$	$(\mathbf{g}_1^* \ \ \mathbf{g}_2^*)$	Action
$\begin{pmatrix}12 & 13\\ 2 & 4\end{pmatrix}$	$\begin{pmatrix}1 & \cdot\\ \frac{41}{37} & 1\end{pmatrix}$	$\begin{pmatrix}12 & -\frac{11}{37}\\ 2 & \frac{66}{37}\end{pmatrix}$	row 2 ← row 2 + row 1
$\begin{pmatrix}12 & 1\\ 2 & 2\end{pmatrix}$	$\begin{pmatrix}1 & \cdot\\ \frac{4}{37} & 1\end{pmatrix}$	$\begin{pmatrix}12 & -\frac{11}{37}\\ 2 & \frac{66}{37}\end{pmatrix}$	exchange rows 1 and 2
$\begin{pmatrix}1 & 12\\ 2 & 2\end{pmatrix}$	$\begin{pmatrix}1 & \cdot\\ \frac{16}{5} & 1\end{pmatrix}$	$\begin{pmatrix}1 & \frac{44}{5}\\ 2 & -\frac{22}{5}\end{pmatrix}$	row 2 ← row 2 − row 1
$\begin{pmatrix}1 & 9\\ 2 & -4\end{pmatrix}$	$\begin{pmatrix}1 & \cdot\\ \frac{1}{5} & 1\end{pmatrix}$	$\begin{pmatrix}1 & \frac{44}{5}\\ 2 & -\frac{22}{5}\end{pmatrix}$	

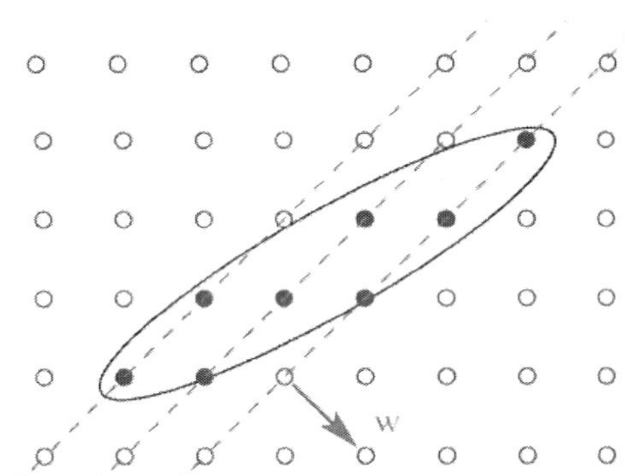

Figure 2.12. *Branching on hyperplanes corresponding to approximate lattice width directions of the feasible region in a Lenstra-type algorithm* [202].

2.8 Lenstra's algorithm, integer feasibility, and optimization

When the dimension is fixed, the integer linear optimization problem can be solved efficiently using variants of Lenstra's famous algorithm [231] for integer programming. Lenstra-type algorithms are algorithms for solving feasibility problems. However, similar to the linear optimization case in Section 1.6, optimization and feasibility are equivalent operations as far as polynomial-time computability goes. The *integer linear optimization problem* (ILP) is described as follows:

Given $A \in \mathbb{Z}^{m\times n}$, $\mathbf{b} \in \mathbb{Z}^m$, and $\mathbf{c} \in \mathbb{Z}^n$, find $\mathbf{x} \in \mathbb{Z}^n$ that solves

$$\min\left\{\mathbf{c}^\top\mathbf{x} : A\mathbf{x} \le \mathbf{b},\ \mathbf{x} \ge \mathbf{0},\ \mathbf{x} \in \mathbb{Z}^n\right\} \qquad \text{(ILP)}$$

or report INFEASIBLE or UNBOUNDED.

We define the *integer linear feasibility problem* (ILF) to be the following problem:

Given $A \in \mathbb{Z}^{m\times n}$, $\mathbf{b} \in \mathbb{Z}^m$, and $\mathbf{c} \in \mathbb{Z}^n$, find $\mathbf{x} \in \mathbb{Z}^n$ such that

$$A\mathbf{x} \leq \mathbf{b}, \quad \mathbf{x} \geq \mathbf{0} \tag{ILF}$$

or report INFEASIBLE.

We let $P = \{\mathbf{x} : A\mathbf{x} \leq \mathbf{b},\ \mathbf{x} \geq \mathbf{0}\}$. To prove the equivalence, we consider the following family of integer feasibility problems associated with the integer minimization problem:

$$\exists \mathbf{x} \in P_\alpha \cap \mathbb{Z}^n \quad \text{where } P_\alpha = \{\mathbf{x} \in P : \mathbf{c}^\top \mathbf{x} \leq \alpha\} \quad \text{for } \alpha \in \mathbb{Z}. \tag{2.6}$$

The unbounded case can be easily eliminated; we omit the details. Then from linear optimization, a bound R on all components of feasible solutions $\mathbf{x}$, $|x_i| \leq R$, can be effectively computed [145]. The binary encoding size (bit length) of R is bounded polynomially. From this we find bounds for the range of the objective function $\mathbf{c}^\top\mathbf{x}$ over $\mathbf{x} \in P$, whose bit lengths are also bounded polynomially. Then a binary search for the right value of α solves the optimization problem in polynomial time.

A Lenstra-type algorithm uses *branching on hyperplanes* (Figure 2.12) to obtain polynomial-time complexity in fixed dimension. Note that only the binary encoding size of the bound R, but not R itself, is bounded polynomially. Thus, multiway branching on the values of a single variable x_i will create an exponentially large number of subproblems. Instead, a Lenstra-type algorithm computes a primitive lattice vector $\mathbf{d} \in \mathbb{Z}^n$ such that there are only a few lattice hyperplanes $\mathbf{d}^\top\mathbf{x} = \gamma$ (with $\gamma \in \mathbb{Z}$) that can have a nonempty intersection with P_α.

Definition 2.8.1. The *width* of P_α in the direction $\mathbf{d}$, defined as

$$\operatorname{width}_{\mathbf{d}}(P_\alpha) = \max\{\mathbf{d}^\top\mathbf{x} : \mathbf{x} \in P_\alpha\} - \min\{\mathbf{d}^\top\mathbf{x} : \mathbf{x} \in P_\alpha\}, \tag{2.7}$$

essentially gives the number of these lattice hyperplanes. A *lattice width direction* is a minimizer of the width among all directions $\mathbf{d} \in \mathbb{Z}^n$. The *lattice width*, denoted by $\operatorname{width}(P_{\mathbf{d}})$, gives the corresponding width.

Any polynomial upper bound on the lattice width of a polytope without integer points will yield a polynomial-time algorithm in fixed dimension. In fact, much better bounds are available. There exist bounds that depend only on the dimension, but not on the polytope (or, more generally, convex body).

Theorem 2.8.2 (Khinchin's flatness theorem [195]). *There exists a function $f^*(n)$, depending only on the dimension, such that if $K \subset \mathbb{R}^n$ is convex and $w(K) > f^*(n)$, then K contains an integer point.*

The best known bound for $f^*(n)$ is $O(n^{4/3}\log^{O(1)}(n))$, and it is conjectured that the bound for ellipsoids, $f^*(n) = \Theta(n)$, also holds for general convex bodies [24].

Exact and approximate lattice width directions $\mathbf{d}$ can be constructed using geometry of numbers techniques. We refer to the excellent tutorial [122] and the classic references cited therein. The standard technique to deal with the feasible region P_α is to apply ellipsoidal rounding. By applying a variant of the ellipsoid method (cf. Section 1.8), the *shallow-cut* ellipsoid method, one finds concentric, proportional inscribed and circumscribed ellipsoids that differ by some factor β that depends only on the dimension n. Then any η-approximate lattice width direction for the ellipsoids gives a $\beta\eta$-approximate lattice width direction for P_α.

Approximate lattice width directions for ellipsoids, in turn, can be computed using approximate algorithms for the SVP of a certain lattice $\mathcal{L}$ in the ℓ_2-norm (Section 2.7). Indeed, Lenstra's original algorithm just uses a basis vector of an LLL-reduced basis of $\mathcal{L}$ as this approximation.

2.9 The integer hull of a polyhedron and cutting planes

Given the rational convex polyhedron $P = \{\mathbf{x} : A\mathbf{x} \le \mathbf{b}\}$, we are interested in the convex hull of the lattice points in P, denoted by P_I. This new polytope is called the *integer hull.*

The polyhedral combinatorics of integer hulls, of course, is the basis for strong cutting plane algorithms and branch-and-cut technology. If we had a complete facet description of P_I, we could just solve the integer optimization problem as a linear program. We only outline a few key points, as the focus of this book is on other methods. For more information we refer the reader to the books by Schrijver [296] and Korte and Vygen [206].

It is clear that finding a full description is hard; in fact, even deciding whether the integer hull is nonempty is an NP-hard problem. Next we give a more detailed picture of the complexity of the integer hull. A *subdeterminant* of the given integer matrix A is $\det(B)$ for some square submatrix B of A (defined by arbitrary sets of row and column indices). We write $\Xi(A)$ for the maximum absolute value among all the subdeterminants of A.

Theorem 2.9.1. *Let A be an integral $m \times n$ matrix and $\mathbf{b} \in \mathbb{R}^m$, $\mathbf{c} \in \mathbb{R}^n$ arbitrary vectors. Let $P := \{\mathbf{x} : A\mathbf{x} \le \mathbf{b}\}$ and suppose that $P_I \ne \emptyset$.*

1. *Suppose $\mathbf{y}$ is an optimal solution of $\max\left\{\mathbf{c}^\top\mathbf{x} : \mathbf{x} \in P\right\}$. Then there exists an optimal solution $\mathbf{z}$ of $\max\left\{\mathbf{c}^\top\mathbf{x} : \mathbf{x} \in P_I\right\}$ with $\|\mathbf{z}-\mathbf{y}\|_\infty \le n\,\Xi(A)$.*
2. *Suppose $\mathbf{y}$ is a feasible but nonoptimal solution of $\max\left\{\mathbf{c}^\top\mathbf{x} : \mathbf{x} \in P_I\right\}$. Then there exists a feasible solution $\mathbf{z} \in P_I$ with $\mathbf{c}^\top\mathbf{z} > \mathbf{c}^\top\mathbf{y}$ and $\|\mathbf{z}-\mathbf{y}\|_\infty \le n\,\Xi(A)$.*
3. *Let $\mathbf{c}$ be some vector. Then $\max\left\{\mathbf{c}^\top\mathbf{x} : \mathbf{x} \in P\right\}$ is bounded if and only if $\max\left\{\mathbf{c}^\top\mathbf{x} : \mathbf{x} \in P_I\right\}$ is bounded.*
4. *There exists an integral matrix M whose entries have absolute values bounded above by $n^{2n}\,\Xi(A)^n$, such that for each vector $\mathbf{b} \in \mathbb{Q}^m$ there exists a vector $\mathbf{d}$ with*
$$\{\mathbf{x} : A\mathbf{x} \le \mathbf{b}\}_I = \{\mathbf{x} : M\mathbf{x} \le \mathbf{d}\}.$$
Thus the integer hulls of the parametric polyhedra admit a parametric description.

When the dimension is fixed, P_I has only a polynomial number of vertices. This was first shown for the knapsack polytope by Shevchenko [311] and Hayes and Larman [151], and the following improved result was given by Cook et al. [74].

Theorem 2.9.2. *Let $P = \{\mathbf{x} : A\mathbf{x} \le \mathbf{b}\} \subseteq \mathbb{R}^n$ be a rational polyhedron with $A \in \mathbb{Q}^{m\times n}$ and let ϕ be the largest binary encoding size of any of the rows of the system $A\mathbf{x} \le \mathbf{b}$. Then the number of vertices of P_I is at most $2m^n(6n^2\phi)^{n-1}$.*

Moreover, Hartmann [149] gave an algorithm for enumerating all the vertices, which runs in polynomial time in fixed dimension.

A different systematic way for computing the inequalities of P_I from those determining P is the Gomory–Chvátal integer rounding procedure, which goes back to the work of Ralph Gomory in the 1950s [138].

Definition 2.9.3. Let $P = \{\mathbf{x} : A\mathbf{x} \leq \mathbf{b}\}$ be a polyhedron. Then we define the *Gomory–Chvátal closure*

$$P' := \bigcap_{P \subseteq H} H_I,$$

where the intersection ranges over all rational affine half spaces $H = \left\{\mathbf{x} : \mathbf{c}^\top \mathbf{x} \leq \delta\right\}$ containing P, and H_I is the integer hull of H. This operation can be iterated. We set $P^{(0)} := P$ and $P^{(i+1)} := \left(P^{(i)}\right)'$. Then $P^{(i)}$ is called the *i-th Gomory–Chvátal closure* of P.

Theorem 2.9.4. *Let $P = \{\mathbf{x} : A\mathbf{x} \leq \mathbf{b}\}$ be a rational polyhedron.*

1. *The Gomory–Chvátal closure P' equals*

$$\left\{\mathbf{x} : \mathbf{u}^\top A\mathbf{x} \leq \left\lfloor \mathbf{u}^\top \mathbf{b} \right\rfloor \forall \mathbf{u} \geq \mathbf{0} \text{ with } \mathbf{u}^\top A \text{ integral}\right\}.$$

2. *P' is a polyhedron; i.e., there is a* finite *set of inequalities that describes the Gomory–Chvátal closure.*

Theorem 2.9.5 (Chvátal (1973), Schrijver (1980)). *For each rational polyhedron P, there exists a positive integer number t such that $P^{(t)} = P_I$.*

The smallest t for which we have finally reached the integer hull is a measure of the complexity. Unfortunately, as Chvátal showed, there is no fixed t that works. There are several fascinating results about the closure operation; for example, given a rational polyhedron P and some rational vector $\mathbf{x}$, deciding whether $\mathbf{x}$ is outside the Gomory–Chvátal closure P' is NP-complete [120]. However, in fixed dimension the inequality description of the Gomory–Chvátal closure can be computed in polynomial time [59].

The very best scenario for optimization is, of course, when the input rational polyhedron P is *integral*, i.e., $P = P_I$.

Theorem 2.9.6 (Hoffman (1974), Edmonds and Giles (1977)). *Let P be a rational polyhedron. Then the following statements are equivalent:*

1. *P is integral.*
2. *Each face of P contains integral vectors.*
3. *Each minimal face of P contains integral vectors.*
4. *Each supporting hyperplane contains integral vectors.*
5. *Each rational supporting hyperplane contains integral vectors.*
6. $\max\left\{\mathbf{c}^\top \mathbf{x} : \mathbf{x} \in P\right\}$ *is attained by an integral vector for each* $\mathbf{c}$ *for which the maximum is finite.*
7. $\max\left\{\mathbf{c}^\top \mathbf{x} : \mathbf{x} \in P\right\}$ *is an integer for each integral* $\mathbf{c}$ *for which the maximum is finite.*

A very special case when the polyhedron will be integral from the beginning, and in fact integral for all integral right-hand side vectors, is when the defining matrix is *totally unimodular*. We say A is totally unimodular if each subdeterminant of A is 0, +1, or −1. Such matrices play an important role in the theory of combinatorial optimization and have been studied extensively. We know from work by Hoffman and Kruskal that an integral matrix A is totally unimodular if and only if, for all integral vectors **b** and **c**, both optima in the equation of strong LP duality,

$$\max\left\{\mathbf{c}^\top\mathbf{x} : A\mathbf{x} \leq \mathbf{b}, \mathbf{x} \geq \mathbf{0}\right\} = \min\left\{\mathbf{y}^\top\mathbf{b} : \mathbf{y} \geq \mathbf{0}, \mathbf{y}^\top A \geq \mathbf{c}\right\}$$

are attained by integral vectors (if they are finite). But when is a matrix totally unimodular?

Theorem 2.9.7 (Ghouila–Houri (1962)). *A matrix $A = (a_{ij}) \in \mathbb{Z}^{m\times n}$ is totally unimodular if and only if for every $R \subseteq \{1,\dots,m\}$ there is a partition $R = R_1 \cup R_2$ such that for all $j = 1,\dots,n$, we have*

$$\sum_{i\in R_1} a_{ij} - \sum_{i\in R_2} a_{ij} \in \{-1,0,1\}.$$

Using the above result the reader can verify that the incidence matrix of an undirected graph G is totally unimodular if and only if G is bipartite. Similarly, the incidence matrix of any digraph is totally unimodular. A celebrated deep result of Seymour says that each totally unimodular matrix can be obtained from network matrices and the 5×5 special matrix R_{10}, by applying elementary operations, dualization, 1-sums, 2-sums, and 3-sums (see more details in [295]). The key consequence is a polynomial-time test for total unimodularity of matrices.

2.10 Linear versus nonlinear discrete optimization

As mentioned above, techniques from polyhedral combinatorics have been very successful when dealing with a variety of integer *linear* programming problems. There are various situations in which we obtain excellent algorithms coming from good polyhedral characterizations. For example, there are important theoretical results on the efficiency of algorithms for matchings on graphs and matroid problems [298–300]. On the practical side, even though no polynomiality result may be available, very efficient branch-and-cut algorithms have been obtained, as showcased by the solution of large-scale traveling salesman problems [14].

Despite these good results there are limitations for models where only linear constraints are used. One is more generally interested in the solution of the following mixed-integer nonlinear optimization model:

$$\begin{array}{lll} \max & f(\mathbf{x}) & \\ \text{subject to} & g_i(\mathbf{x}) \leq 0, & i = 1,2,\dots,k, \\ & h_j(\mathbf{x}) = 0, & j = 1,2,\dots,m, \\ & \mathbf{x} \in \mathbb{R}^{n_1} \times \mathbb{Z}^{n_2}. & \end{array} \qquad \text{(MOAOP)}$$

Here the objective function f and the constraint functions g_i, h_j are assumed to be *arbitrary* polynomial functions on $\mathbb{R}^n$ ($n = n_1 + n_2$) with rational coefficients. (Of course, even more general models are possible.)

But, how difficult is this problem (MOAOP)? We mentioned that in order to speak about the hardness of algorithmic problems, we want to use the terminology of computational complexity [131]. Already the traditional integer linear optimization problem is NP-hard. Unfortunately, the problem (MOAOP) is even harder to solve. In fact, it is known that there does not exist *any* algorithm to solve it. This follows from the obvious relation to the following famous problem.

Problem 2.10.1 (Hilbert's 10th problem). Find an algorithm to decide, given polynomials $h_i \in \mathbb{Z}[x_1,\dots,x_n]$, whether there exists a vector $\mathbf{x} \in \mathbb{Z}^n$ such that

$$h_1(\mathbf{x}) = 0, \ \dots, \ h_n(\mathbf{x}) = 0.$$

Hilbert's 10th problem was answered in the negative, showing that the decision problem is incomputable, by Matiyasevich [242], based on earlier work by Davis, Putnam, and Robinson; see also [244].

Thus it will be necessary to assume some restrictions on the polynomials in (MOAOP) to be able to prove good complexity results. For example, incomputability results no longer apply when finite bounds for all variables are known; in this case, a trivial enumeration approach gives a finite algorithm. However, the hardness remains. With nonlinearity, hardness results arise even in small fixed dimension, as do encoding issues for solutions. As a perhaps surprising example, the problem of maximizing a degree 4 polynomial over the lattice points of a polygon is NP-hard. For a discussion of negative complexity results, such as NP-hardness and inapproximability, we refer to the recent survey [202].

One of the purposes of this book is to show that under the right restrictions we will be able to prove positive complexity results.

2.11 Notes and further references

Most of the material we presented here is found in any of the excellent books [50, 145, 206, 259, 296, 297].

Our presentation of the LLL algorithm took some material from [134, 259]. The LLL algorithm is a practical algorithm with many high-profile applications. Related to its use in solving the SVP and its use in Lenstra's algorithm, it also makes an appearance in the lattice-based reformulation of integer programs [2–4, 209].

Lenstra's algorithm for integer programming in fixed dimension (Section 2.8) appeared in [231] and was closely tied to the LLL algorithm. Since then many improvements and extensions have been found. Kannan first observed that SVP could be used to minimize the number of branching directions in Lenstra's algorithm [187]. Our exposition follows Eisenbrand's excellent tutorial in presenting this idea in the context of flatness directions [122]. The first Lenstra-type algorithm for convex integer minimization was announced by Khachiyan [194]. The case of quasi-convex polynomial functions appeared in [152] and was later improved in [168] and, building upon Kannan's improvement and ideas, in [198].

The best complexity of a Lenstra-type algorithm so far appears in [80]. In that paper, an "any-norm" shortest vector algorithm is developed. By using this shortest vector algorithm directly with a norm related to (2.7), rather than using the approximation by an ellipsoid first, Dadush, Peikert, and Vempala obtain an algorithm whose running time dependence on n is $\mathrm{O}\big(n^{4/3}\log^{\mathrm{O}(1)} n\big)^n$. See [80] for details and a comparison with the complexity of other methods.

While the binary search method used in Section 2.8 gives an easy way to focus on the integer feasibility problem, this approach does not necessarily give the best complexity. One of the techniques in Eisenbrand's fast integer programming algorithm in fixed dimension (assuming a fixed number of constraints also) is the use of a parametric integer feasibility problem instead, which in turn leads to the approximate parametric lattice width problem [121].

2.12 Exercises

Exercise 2.12.1. Suppose you are given a line $L = \{(x,y) : y = ax + b\}$. Prove that there are only five possibilities:

1. L has rational slope and no lattice points.
2. L has rational slope and infinitely many lattice points.
3. L has irrational slope and no lattice points.
4. L has irrational slope and exactly one lattice point.
5. L is a line parallel to one of the axes (x or y) with infinitely many lattice points or none, depending on constant $x = k$ or $y = k$.

Exercise 2.12.2. Prove that one can triangulate any lattice polygon into *primitive triangles*, i.e., triangles having no lattice points except their vertices. Note that from Lemma 2.1.5 it is enough to prove that every lattice triangle can be divided into primitive lattice triangles. Prove that every primitive triangle has area $\frac{1}{2}$. This can be used to prove Pick's theorem (see Theorem 2.1.7).

Exercise 2.12.3. Use Minkowski's first theorem (see Theorem 2.4.2) to give proofs of Theorems 2.4.5 and 2.4.6. Show with examples that Minkowski's first theorem fails if any of the conditions is relaxed.

Exercise 2.12.4. Prove Lemma 2.5.1. Hint: Use induction on the number n of variables.

Exercise 2.12.5. Show that Lemma 2.5.1 implies Lemma 2.5.2.

Exercise 2.12.6. Show that Lemma 2.5.2 implies Lemma 2.5.1.

Exercise 2.12.7. Show that, for $\mathbf{u},\mathbf{v} \in \mathbb{Z}^n$, we have $\mathbf{u} \sqsubseteq \mathbf{v}$ if and only if $(\mathbf{u}^+,\mathbf{u}^-) \leq (\mathbf{v}^+,\mathbf{v}^-)$.

Exercise 2.12.8. Prove Lemma 2.5.6.

Exercise 2.12.9. Using `4ti2` (see [1]) or by hand, solve the system

$$\begin{aligned} x - 2y &\leq 2, \\ -2x + y &\leq 1, \\ x + y &\geq 1, \\ y &\geq 0 \end{aligned}$$

over $\mathbb{Z}$. By drawing a diagram, verify that your computed solution is correct.

Exercise 2.12.10. Assuming the incomputability of Hilbert's 10th problem (Problem 2.10.1), prove that the feasibility problem corresponding to (MOAOP) is incomputable.

Part II

Graver Basis Methods

Chapter 3

Graver Bases

3.1 Introduction

Let $A \in \mathbb{Z}^{m\times n}$, $\mathbf{b} \in \mathbb{Z}^m$, $\mathbf{l},\mathbf{u} \in \mathbb{Z}^n$, and an objective function $f : \mathbb{R}^n \to \mathbb{R}$ be given. In this chapter, we show how to solve integer programs (IPs) of the form

$$(\mathrm{IP})_{A,\mathbf{b},\mathbf{l},\mathbf{u},f} : \qquad \min\left\{ f(\mathbf{z}) : A\mathbf{z} = \mathbf{b}, \mathbf{l} \leq \mathbf{z} \leq \mathbf{u}, \mathbf{z} \in \mathbb{Z}^n \right\}$$

via a simple augmentation procedure: Given an initial feasible solution, we repeatedly search for augmenting directions until we reach an optimal solution; see Figure 3.1.

To design such an augmentation scheme, we employ an *optimality certificate* or *test set* that either declares our current solution to be optimal or that returns a direction to a better solution. More formally, a set $\mathcal{T} \subseteq \mathbb{Z}^n$ is called an *optimality certificate* (or *test set*) for $(\mathrm{IP})_{A,\mathbf{b},\mathbf{l},\mathbf{u},f}$ if, for every nonoptimal feasible solution $\mathbf{z}_0$ of $(\mathrm{IP})_{A,\mathbf{b},\mathbf{l},\mathbf{u},f}$, there exists a vector $\mathbf{t} \in \mathcal{T}$ and some positive integer α such that

1. $\mathbf{z}_0 + \alpha\mathbf{t}$ is feasible, and
2. $f(\mathbf{z}_0 + \alpha\mathbf{t}) < f(\mathbf{z}_0)$.

The vector $\mathbf{t} \in \mathcal{T}$ (or $\alpha\mathbf{t}$) is called an *improving* or *augmenting vector* or *direction*.

Clearly, once we have an optimality certificate $\mathcal{T}$ for $(\mathrm{IP})_{A,\mathbf{b},\mathbf{l},\mathbf{u},f}$ available, any feasible solution $\mathbf{z}_0$ to $(\mathrm{IP})_{A,\mathbf{b},\mathbf{l},\mathbf{u},f}$ can be iteratively augmented to optimality, provided that a finite optimum exists. If $\mathcal{T}$ is finite, one may test every single $\mathbf{t} \in \mathcal{T}$ as to whether it is augmenting. Hence the name *test set*. However, for the augmentation scheme to work, $\mathcal{T}$ need not be finite. It is sufficient to have an algorithm that either constructs an augmenting vector from $\mathcal{T}$ for the given feasible solution or declares that no such vector exists in $\mathcal{T}$.

Often, there are several augmenting vectors in $\mathcal{T}$ for a given nonoptimal solution. Therefore, the question arises of whether the augmenting vectors can be chosen in such a way that only a polynomial number of augmentation steps (in the encoding length of the input data) are needed in order to reach an optimal solution.

In this chapter, we will introduce the *Graver basis* (or *Graver test set*) associated to a matrix A. This set turns out to be a finite optimality certificate that gives improving directions for a whole family of problems, namely for all $(\mathrm{IP})_{A,\mathbf{b},\mathbf{l},\mathbf{u},f}$ with fixed matrix A and with arbitrary choices of the integer vectors $\mathbf{b}$, $\mathbf{l}$, $\mathbf{u}$, and of a separable convex function f. Herein, we call $f : \mathbb{R}^n \to \mathbb{R}$ a *separable convex function* if it can be written as $f(\mathbf{z}) =$

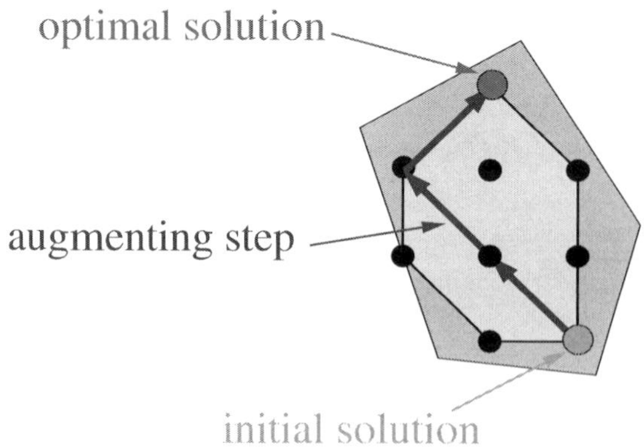

Figure 3.1. *Augmenting an initial solution to optimality.*

$\sum_{i=1}^{n} f_i(z_i)$ for some convex functions $f_i : \mathbb{R} \to \mathbb{R}$. Note that linear functions $\mathbf{c}^\top \mathbf{z}$ are separable convex functions (with $f_i(z_i) = c_i z_i$).

We wish to point out here that any such universal optimality certificate can depend only on the given matrix A, since all other data are allowed to vary. Moreover, the vectors in such an optimality certificate must belong to the lattice $\ker_{\mathbb{Z}^n}(A) = \{\mathbf{z} \in \mathbb{Z}^n : A\mathbf{z} = \mathbf{0}\}$, since adding $\alpha\mathbf{t}$ to a feasible solution $\mathbf{z}_0$ must lead to another feasible solution $\mathbf{z}_0 + \alpha\mathbf{t}$ with $A(\mathbf{z}_0 + \alpha\mathbf{t}) = \mathbf{b}$. We will see that the Graver basis of A indeed allows us to reach an optimal solution via only polynomially many augmentation steps.

3.2 Graver bases and their representation property

Definition 3.2.1. For any given matrix $A \in \mathbb{Z}^{m\times n}$, we define the *Graver basis* $\mathcal{G}(A)$ of A to be the set of all $\sqsubseteq$-minimal elements in $\ker_{\mathbb{Z}^n}(A) \setminus \{\mathbf{0}\}$.

(For the definition of $\sqsubseteq$ see Definition 2.5.4 on page 44.) By the Gordan–Dickson lemma, Lemma 2.5.6, $\mathcal{G}(A)$ is always finite. For example, the Graver basis of the matrix $A = (1\ 2\ 3)$ is

$$\mathcal{G}(A) = \{\pm(2,-1,0), \pm(3,0,-1), \pm(1,1,-1), \pm(1,-2,1), \pm(0,3,-2)\}.$$

Note that $\mathcal{G}(A)$ is always symmetric, that is, if $\mathbf{g} \in \mathcal{G}(A)$, then also $-\mathbf{g} \in \mathcal{G}(A)$. The following lemma states that our definition above is indeed equivalent to Graver's original definition [144] (Exercise 3.10.1).

Lemma 3.2.2. *For any given matrix $A \in \mathbb{Z}^{m\times n}$ and for every orthant $\mathbb{O}_j$ of $\mathbb{R}^n$, let H_j denote the unique inclusion-minimal Hilbert basis of the pointed rational polyhedral cone $\mathbb{O}_j \cap \ker_{\mathbb{R}^n}(A)$. Then $\mathcal{G}(A)$ is the union of $H_j \setminus \{\mathbf{0}\}$ taken over all 2^n orthants $\mathbb{O}_j$ of $\mathbb{R}^n$.*

This lemma immediately implies that Graver bases have the following nice representation property with respect to $\ker_{\mathbb{Z}^n}(A)$. In fact, it can be shown (Exercise 3.10.2) that $\mathcal{G}(A)$ is the unique inclusion-minimal subset of $\ker_{\mathbb{Z}^n}(A)$ with this representation property, giving yet another possible definition for $\mathcal{G}(A)$.

Lemma 3.2.3. *Every vector $\mathbf{z} \in \ker_{\mathbb{Z}^n}(A)$ can be written as a finite sign-compatible nonnegative integer linear combination $\mathbf{z} = \sum_{i=1}^{r} \alpha_i \mathbf{g}_i$ with $r \leq 2n-2$, $\alpha_i \in \mathbb{Z}_+$, $\mathbf{g}_i \in \mathcal{G}(A)$,* *and* $\mathbf{g}_i \sqsubseteq \mathbf{z}$ *for all* i.

Proof. Choose any $\mathbf{z} \in \ker_{\mathbb{Z}^n}(A)$. Then $\mathbf{z}$ belongs to some orthant $\mathbb{O}_j$ and thus, by Lemma 3.2.2, $\mathbf{z}$ can be written as a finite sign-compatible nonnegative integer linear combination $\mathbf{z} = \sum_{i=1}^{r} \alpha_i \mathbf{g}_i$ using only Hilbert basis elements $\mathbf{g}_i \in H_j \setminus \{\mathbf{0}\} \subseteq \mathcal{G}(A)$. Clearly, we then also have $\mathbf{g}_i \sqsubseteq \mathbf{z}$ for all i.

A result of Sebö [306] implies that there is always such a sign-compatible representation using at most $r \leq 2n-2$ summands $\alpha_i \mathbf{g}_i$. □

This fundamental property of Graver bases implies the following important fact that we will need later (Exercise 3.10.4).

Lemma 3.2.4. *Let* $\mathbf{z}_0$ *and* $\mathbf{z}_1$ *be feasible solutions to* $A\mathbf{z} = \mathbf{b}$, $\mathbf{l} \leq \mathbf{z} \leq \mathbf{u}$. *Moreover, let* $\mathbf{z}_1 - \mathbf{z}_0 = \sum_{i=1}^{r} \alpha_i \mathbf{g}_i$ *be a nonnegative integer linear sign-compatible decomposition into Graver basis elements* $\mathbf{g}_i \in \mathcal{G}(A)$. *Then for all choices of* $\beta_1, \dots, \beta_r \in \mathbb{Z}$ *with* $0 \leq \beta_j \leq \alpha_j$, $j = 1, \dots, r$, *the vector* $\mathbf{z}_0 + \sum_{i=1}^{r} \beta_i \mathbf{g}_i$ *is also a feasible solution to* $A\mathbf{z} = \mathbf{b}$, $\mathbf{l} \leq \mathbf{z} \leq \mathbf{u}$.

3.3 Optimality certificate for separable convex integer programs

In the following, we consider the minimization of a separable convex function over the lattice points in a polyhedron. Recall that a convex function $f : \mathbb{R}^n \to \mathbb{R}$ is called *separable* if it can be written as $f(\mathbf{z}) := \sum_{j=1}^{n} f_j(z_j)$ for convex functions $f_1, \dots, f_n : \mathbb{R} \to \mathbb{R}$. Such functions satisfy the following nice property (Exercise 3.10.4).

Lemma 3.3.1. *Let* $f(\mathbf{z}) := \sum_{j=1}^{n} f_j(z_j)$ *be separable convex, let* $\mathbf{z} \in \mathbb{R}^n$, *and let* $\mathbf{g}_1, \dots, \mathbf{g}_r \in \mathbb{R}^n$ *be vectors with the same sign pattern from* $\{\leq 0, \geq 0\}^n$; *that is, they belong to a common orthant of* $\mathbb{R}^n$. *Then we have*

$$f\left(\mathbf{z} + \sum_{i=1}^{r} \alpha_i \mathbf{g}_i\right) - f(\mathbf{z}) \geq \sum_{i=1}^{r} \alpha_i (f(\mathbf{z} + g_i) - f(\mathbf{z}))$$

for arbitrary integers $\alpha_1, \dots, \alpha_r \in \mathbb{Z}_+$.

This property together with the representation property of Graver bases, Lemma 3.2.3, implies that Graver bases are optimality certificates for separable convex integer minimization problems. With this we mean that if $\mathbf{z}_0$ is a nonoptimal feasible solution, then there exists an element $\mathbf{g} \in \mathcal{G}(A)$ and a step length $\alpha \in \mathbb{Z}_+$ such that $\mathbf{z}_0 + \alpha \mathbf{g}$ is feasible with $f(\mathbf{z}_0 + \alpha \mathbf{g}) < f(\mathbf{z}_0)$. Clearly, if no such pair $\mathbf{g} \in \mathcal{G}(A)$ and $\alpha \in \mathbb{Z}_+$ exists, $\mathbf{z}_0$ must be optimal.

Lemma 3.3.2. *The set* $\mathcal{G}(A)$ *is an optimality certificate for* $(IP)_{A,\mathbf{b},\mathbf{l},\mathbf{u},f}$ *for any vectors* $\mathbf{b} \in \mathbb{Z}^m$, $\mathbf{l}, \mathbf{u} \in \mathbb{Z}^n$ *and for any separable convex function* f.

Proof. Let $\mathbf{z}_0$ denote a feasible solution of the problem and assume that there is some other feasible solution $\mathbf{z}_1$ with $f(\mathbf{z}_1) < f(\mathbf{z}_0)$. Then $\mathbf{z}_1 - \mathbf{z}_0 \in \ker(A)$ which implies that

$$\mathbf{z}_1 - \mathbf{z}_0 = \sum_{i=1}^{r} \alpha_i \mathbf{g}_i$$

for some positive integers $\alpha_1,\ldots,\alpha_r$ and for some vectors $\mathbf{g}_1,\ldots,\mathbf{g}_r \in \mathcal{G}(A)$ such that $\mathbf{g}_i \sqsubseteq \mathbf{z}_1-\mathbf{z}_0$. By Lemma 3.2.4, we know that for all i, $\mathbf{z}_0+\mathbf{g}_i$ is a feasible solution to $(\mathrm{IP})_{A,\mathbf{b},\mathbf{l},\mathbf{u},f}$. Moreover, since all $\mathbf{g}_i$ have the same sign pattern, we can apply Lemma 3.3.1 and get

$$\begin{aligned} 0 > f(\mathbf{z}_1)-f(\mathbf{z}_0) &= f(\mathbf{z}_0+(\mathbf{z}_1-\mathbf{z}_0))-f(\mathbf{z}_0)\\ &= f\left(\mathbf{z}_0+\sum_{i=1}^{r}\alpha_i\mathbf{g}_i\right)-f(\mathbf{z}_0)\\ &\geq \sum_{i=1}^{r}\alpha_i(f(\mathbf{z}_0+\mathbf{g}_i)-f(\mathbf{z}_0)). \end{aligned}$$

Thus, we must have $f(\mathbf{z}_0+\mathbf{g}_i)-f(\mathbf{z}_0)<0$ for at least one index i. Consequently, $\mathbf{z}_0+\mathbf{g}_i$ is a feasible solution with a strictly smaller objective function value than $\mathbf{z}_0$ and our claim is proved. □

Note that it suffices that the function f is given only by a *comparison oracle*, which, when queried on $\mathbf{x},\mathbf{y}\in\mathbb{Z}^n$, decides whether $f(\mathbf{x})<f(\mathbf{y})$, $f(\mathbf{x})=f(\mathbf{y})$, or $f(\mathbf{x})>f(\mathbf{y})$.

3.4 How many augmentation steps are needed?

Once we have an optimality certificate *and* a feasible solution $\mathbf{z}_0$ to the problem $(\mathrm{IP})_{A,\mathbf{b},\mathbf{l},\mathbf{u},f}$ the iterative augmentation procedure can be applied. But how many augmentation steps do we need if we use directions from the Graver basis $\mathcal{G}(A)$?

Several techniques have been found to turn the augmentation algorithm into an efficient algorithm, bounding the number of augmentation steps polynomially. Three such speedup techniques are known in the literature: For 0/1 integer linear problems, a simple *bit-scaling technique* suffices [304]. For general integer linear problems, one can use the *directed augmentation technique* [303], in which one uses Graver basis elements $\mathbf{v}\in\mathcal{G}(E)$ that are improving directions for the nonlinear functions $\mathbf{c}^\top\mathbf{v}^+ + \mathbf{d}^\top\mathbf{v}^-$, which are adjusted during the augmentation algorithm. For separable convex integer minimization problems, one can use the *Graver-best augmentation technique* [160], which we will present below. Herein, one uses an augmentation vector that is at least as good as the best augmentation step of the form $\gamma\mathbf{g}$ with $\gamma\in\mathbb{Z}_+$ and $\mathbf{g}\in\mathcal{G}(A)$.

In order to get a good bound for the number of augmentation steps, let us spend a bit more effort on finding the next augmentation step. Instead of finding just *any* pair $\mathbf{g}\in\mathcal{G}(A)$, $\alpha\in\mathbb{Z}_{>0}$ with $\mathbf{z}_0+\alpha\mathbf{g}$ feasible and $f(\mathbf{z}_0+\alpha\mathbf{g})<f(\mathbf{z}_0)$, find a *best* such pair, that is, such that $\mathbf{z}_0+\alpha\mathbf{t}$ is feasible and such that $f(\mathbf{z}_0+\alpha\mathbf{t})$ is minimal among all such choices. We call such a best step a *Graver-best augmentation step*. Note that using a *best improvement* or *greedy* augmentation step is a common strategy. This leads to the following adapted augmentation algorithm.

ALGORITHM 3.1. Graver-best augmentation algorithm.

input A, $\mathbf{b}$, $\mathbf{l}$, $\mathbf{u}$, a function $f:\mathbb{R}^n\to\mathbb{R}$ given by a comparison oracle, a finite test set $\mathcal{T}$ for $(\mathrm{IP})_{A,\mathbf{b},\mathbf{l},\mathbf{u},f}$, a feasible solution $\mathbf{z}_0$ to $(\mathrm{IP})_{A,\mathbf{b},\mathbf{l},\mathbf{u},f}$.
output an optimal solution $\mathbf{z}_{\min}$ of $(\mathrm{IP})_{A,\mathbf{b},\mathbf{l},\mathbf{u},f}$.
while there are $\mathbf{t}\in\mathcal{T}$, $\alpha\in\mathbb{Z}_{>0}$ with $\mathbf{z}_0+\alpha\mathbf{t}$ feasible and $f(\mathbf{z}_0+\alpha\mathbf{t})<f(\mathbf{z}_0)$ **do**
  Among all such pairs $\mathbf{t}\in\mathcal{T}$, $\alpha\in\mathbb{Z}_{>0}$ choose one with $f(\mathbf{z}_0+\alpha\mathbf{t})$ minimal.
  $\mathbf{z}_0\leftarrow\mathbf{z}_0+\alpha\mathbf{t}$.
return $\mathbf{z}_0$.

In our case, with a separable convex objective function f, applying only such Graver-best augmentation steps will guarantee that the number of augmentation steps to an optimum point is polynomially bounded in the binary encoding length of the input data.

Theorem 3.4.1. *There exists an algorithm that, given a matrix A along with its Graver basis $\mathcal{G}(A)$, vectors $\mathbf{z}_0 \in \mathbb{Z}^n$, $\mathbf{l},\mathbf{u} \in \mathbb{Z}^n$ with $\mathbf{l} \leq \mathbf{z}_0 \leq \mathbf{u}$, a separable convex function f mapping $\mathbb{Z}^n$ to $\mathbb{Z}$ given by a comparison oracle, and a number M with $|f(\mathbf{z})| \leq M$ for all solutions $\mathbf{z}$ to $A\mathbf{z} = A\mathbf{z}_0$, $\mathbf{l} \leq \mathbf{z} \leq \mathbf{u}$, solves the integer program $(IP)_{A,A\mathbf{z}_0,\mathbf{l},\mathbf{u},f}$ in time polynomial in the binary encoding lengths of $\mathcal{G}(A)$, $\mathbf{z}_0$, $\mathbf{l}$, $\mathbf{u}$, M.*

Proof. To prove this result, we strengthen the proof to Lemma 3.3.2. Let $\mathbf{z}_{\min}$ be an optimal solution to $(\mathrm{IP})_{A,A\mathbf{z}_0,\mathbf{l},\mathbf{u},f}$. Consider

$$\mathbf{z}_{\min} - \mathbf{z}_0 = \sum_{i=1}^{r} \alpha_i \mathbf{g}_i,$$

for some positive integers $\alpha_1,\ldots,\alpha_r$ and for some vectors $\mathbf{g}_1,\ldots,\mathbf{g}_r \in \mathcal{G}(A)$ such that $\alpha_i \mathbf{g}_i \sqsubseteq \mathbf{z}_1 - \mathbf{z}_0$. Note that by Lemma 3.2.3, we can assume $r \leq 2n-2$. Moreover, by Lemma 3.2.4, $\mathbf{z}_0 + \alpha_i \mathbf{g}_i$ is a feasible solution to $(\mathrm{IP})_{A,A\mathbf{z}_0,\mathbf{l},\mathbf{u},f}$ for all i. Moreover, since all $\alpha_i \mathbf{g}_i$ have the same sign pattern, we can apply Lemma 3.3.1 to get

$$\begin{aligned} 0 > f(\mathbf{z}_{\min}) - f(\mathbf{z}_0) &= f(\mathbf{z}_0 + (\mathbf{z}_{\min} - \mathbf{z}_0)) - f(\mathbf{z}_0) \\ &= f\left(\mathbf{z}_0 + \sum_{i=1}^{r} \alpha_i \mathbf{g}_i\right) - f(\mathbf{z}_0) \\ &\geq \sum_{i=1}^{r} [f(\mathbf{z}_0 + \alpha_i \mathbf{g}_i) - f(\mathbf{z}_0)]. \end{aligned}$$

Multiplying through by -1, we obtain

$$\sum_{i=1}^{r} [f(\mathbf{z}_0) - f(\mathbf{z}_0 + \alpha_i \mathbf{g}_i)] \geq f(\mathbf{z}_0) - f(\mathbf{z}_{\min}) > 0.$$

The crucial observation is now that there is some i such that $f(\mathbf{z}_0) - f(\mathbf{z}_0 + \alpha_i \mathbf{g}_i)$ is not only positive but is at least as big as the arithmetic mean:

$$f(\mathbf{z}_0) - f(\mathbf{z}_0 + \alpha_i \mathbf{g}_i) \geq \frac{1}{r} \sum_{i=1}^{r} [f(\mathbf{z}_0) - f(\mathbf{z}_0 + \alpha_i \mathbf{g}_i)] \geq \frac{1}{2n-2} [f(\mathbf{z}_0) - f(\mathbf{z}_{\min})].$$

This implies that

$$f(\mathbf{z}_0) - f(\mathbf{z}_0 + \alpha \mathbf{g}) \geq f(\mathbf{z}_0) - f(\mathbf{z}_0 + \alpha_i \mathbf{g}_i) \geq \frac{1}{2n-2} [f(\mathbf{z}_0) - f(\mathbf{z}_{\min})]$$

holds for a Graver-best augmentation step $\alpha \mathbf{g}$. This means that a Graver-best augmentation step is at least as good as $1/(2n-2)$ times the maximum possible augmentation $f(\mathbf{z}_0) - f(\mathbf{z}_{\min})$ in objective value. This so-called *geometric improvement* implies that only $\mathrm{O}((2n-2)\log(M)) = \mathrm{O}(n\log(M))$ augmentation steps are needed in order to reach an optimal solution; see [8]. The claim now follows, since by Lemma 3.5.1, such a Graver-best augmentation step can be found in time polynomial in the binary encoding lengths of $\mathcal{G}(A)$ and of the input data. □

3.5 How do we find a Graver-best augmentation step?

As we have seen in the previous section, Graver-best augmentation steps allow us to augment any feasible solution to optimality in polynomially many augmentation steps. Let us now deal with the question of constructing such a Graver-best augmentation step. For this, we consider each $\mathbf{g} \in \mathcal{G}(A)$ independently and construct for each $\mathbf{g}$ a best possible step length $\alpha_{\mathbf{g}}$. Then we only need to find the vector $\alpha_{\mathbf{g}}\mathbf{g}$ that minimizes $f(\mathbf{z}_0 + \alpha_{\mathbf{g}}\mathbf{g})$.

Note that once we fix the search direction $\mathbf{g}$, we are left with a one-dimensional convex integer minimization problem in α.

Lemma 3.5.1. *Let $f\colon \mathbb{R} \to \mathbb{R}$ be a convex function given by a comparison oracle. Then for any given numbers $l,u \in \mathbb{Z}$, the one-dimensional minimization problem* $\min\{f(\alpha) : l \leq \alpha \leq u\}$ *can be solved by* $O(\log(u-l))$ *calls to the comparison oracle. That is, this problem is (oracle-) polynomial-time solvable.*

Proof. If the interval $[l,u]$ contains at most two integers, return l or u as the minimum, depending on the values of $f(l)$ and $f(u)$. If the interval $[l,u]$ contains at least three integers, consider the integers $\lfloor (l+u)/2 \rfloor - 1$, $\lfloor (l+u)/2 \rfloor$, $\lfloor (l+u)/2 \rfloor + 1 \in [l,u]$ and exploit convexity of f to bisect the interval $[l,u]$ as follows:

If $f(\lfloor (l+u)/2 \rfloor - 1) < f(\lfloor (l+u)/2 \rfloor)$ holds, then the minimum of f must be attained in the interval $[l, \lfloor (l+u)/2 \rfloor]$. If, on the other hand, $f(\lfloor (l+u)/2 \rfloor) > f(\lfloor (l+u)/2 \rfloor + 1)$, then the minimum of f must be attained in the interval $[\lfloor (l+u)/2 \rfloor + 1, u]$. If neither holds, the minimum of f is attained at the point $\alpha = \lfloor (l+u)/2 \rfloor$.

Clearly, after $O(\log(u-l))$ bisection steps, the minimization problem is solved. □

This lemma implies that by looking at the directions $\mathbf{g} \in \mathcal{G}(A)$ one by one, we can find a Graver-best augmentation step $\alpha\mathbf{g}$ in time polynomial in the encoding lengths of the input data and of $\mathcal{G}(A)$.

3.6 How do we find an initial feasible solution?

Finding an initially feasible solution can be done via a process similar to phase I of the simplex method. First, we find an integer solution $\mathbf{z}_0$ to $A\mathbf{z} = \mathbf{b}$. This can be done in polynomial time using the Hermite normal form; see Section 2.3. Therefore, the encoding length of $\mathbf{z}_0$ must be polynomial in the encoding lengths of A and $\mathbf{b}$. Now we only have to observe the following.

Lemma 3.6.1. *Let $A \in \mathbb{Z}^{m\times n}$, $\mathbf{b} \in \mathbb{Z}^m$, $\mathbf{l},\mathbf{u} \in \mathbb{Z}^n$, and let $\mathbf{z}_0$ be an integer solution to $A\mathbf{z} = \mathbf{b}$. Then an integer solution to $A\mathbf{z} = \mathbf{b}, \mathbf{l} \leq \mathbf{z} \leq \mathbf{u}$ can be found by solving the integer program*

$$\min\left\{ \|(\mathbf{z}-\mathbf{l})^-\|_1 + \|(\mathbf{u}-\mathbf{z})^-\|_1 : A\mathbf{z} = \mathbf{b}, \bar{\mathbf{l}} \leq \mathbf{z} \leq \bar{\mathbf{u}} \right\}, \tag{3.1}$$

where $\bar{\mathbf{l}} := \min\{\mathbf{l}, \mathbf{z}_0\}$ and $\bar{\mathbf{u}} := \max\{\mathbf{u}, \mathbf{z}_0\}$ are defined componentwise.

Proof. Clearly, this integer program has an optimal solution of value 0 if and only if there exists an integer solution to $A\mathbf{z} = \mathbf{b}, \mathbf{l} \leq \mathbf{z} \leq \mathbf{u}$. □

More important is the following consequence.

Corollary 3.6.2. *Let $A \in \mathbb{Z}^{m\times n}$, $\mathbf{b} \in \mathbb{Z}^m$, and $\mathbf{l},\mathbf{u} \in \mathbb{Z}^n$. Then in time polynomial in the encoding lengths of the input data and of $\mathcal{G}(A)$, we can either find an integer solution to $A\mathbf{z} = \mathbf{b}, \mathbf{l} \leq \mathbf{z} \leq \mathbf{u}$ or can assert that no such solution exists.*

Proof. First, we can find an integer solution $\mathbf{z}_0$ in polynomial time using the Hermite normal form. Then, by solving the integer program (3.1), we can find a feasible solution (if it exists). This integer program, however, has a separable convex objective function and thus can be solved via polynomially many augmentations (with respect to the binary encoding lengths of $A, \mathbf{b}, \mathbf{l}, \mathbf{u}, \mathbf{z}, f$) of $\mathbf{z}_0$ along Graver basis directions. $\square$

3.7 Bounds for Graver basis elements

In this section we prove a few bounds on the maximal 1-norm of a Graver basis element. They will be used later in Section 4.3.1.

First let us state the following folklore fact on Graver bases (see, for example, Corollary 3.2 in [154]), whose proof we leave to the reader as a little exercise.

Lemma 3.7.1 (aggregation). *Let $G = (F\ \mathbf{f}\ \mathbf{f})$ be a matrix with two identical columns. Then the Graver bases of $(F\ \mathbf{f})$ and G are related as follows:*

$$\mathcal{G}(G) = \{(\mathbf{u}, v, w) : vw \geq 0, (\mathbf{u}, v+w) \in \mathcal{G}((F\ \mathbf{f}))\} \cup \{\pm(\mathbf{0}, 1, -1)\}.$$

The Graver basis of $G' = (F\ \mathbf{f}\ -\mathbf{f})$ can be obtained by reversing the sign of the last component in any element of $\mathcal{G}(G)$.

In particular, this means that the maximum ℓ_1-norm of Graver basis elements does not change if we repeat columns.

Corollary 3.7.2. *Let G be a matrix obtained from a matrix F by repeating columns. Then*

$$\max\{\,\|\mathbf{v}\|_1 : \mathbf{v} \in \mathcal{G}(G)\} = \max\{2, \max\{\,\|\mathbf{v}\|_1 : \mathbf{v} \in \mathcal{G}(F)\}\}.$$

We continue with the following result, which can be found, for instance, in [265, Lemma 3.20].

Lemma 3.7.3 (determinant bound). *Let $A \in \mathbb{Z}^{m\times n}$ be a matrix of rank r and let $\Delta(A)$ denote the maximum absolute value of subdeterminants of A. Then $\max\{\,\|\mathbf{v}\|_1 : \mathbf{v} \in \mathcal{G}(A)\} \leq (n-r)(r+1)\Delta(A)$. Moreover, $\Delta(A) \leq (\sqrt{m}M)^m$, where M is the maximum absolute value of an entry of A.*

As a corollary of Lemma 3.7.3 and the aggregation technique (Corollary 3.7.2), we obtain the following result.

Corollary 3.7.4 (determinant bound, aggregated). *Let $A \in \mathbb{Z}^{m\times n}$ be a matrix of rank r, let d be the number of different columns in A, and let M be the maximum absolute value of an entry of A. Then*

$$\max\{\,\|\mathbf{v}\|_1 : \mathbf{v} \in \mathcal{G}(A)\} \leq (d-r)(r+1)(\sqrt{m}M)^m.$$

For matrices with only one row ($m = r = 1$), there are only $2M+1$ different columns, and so this bound simplifies to $4M^2$. However, a tighter bound is known for this special case. The following lemma is a straightforward consequence of Theorem 2 in [114].

Lemma 3.7.5 (primitive partition identity bound). *Let $A \in \mathbb{Z}^{1\times n}$ be a matrix consisting* *of only one row and let M be an upper bound on the absolute values of the entries of A. Then we have* $\max\{\|\mathbf{v}\|_1 : \mathbf{v} \in \mathcal{G}(A)\} \leq 2M - 1$.

Let us now prove some more general degree bounds on Graver bases that we will use in Section 4.3.1.

Lemma 3.7.6 (Graver basis length bound for stacked matrices). *Let $L \in \mathbb{Z}^{d\times n}$ and let $F \in \mathbb{Z}^{m\times n}$. Moreover, put $E := \binom{F}{L}$. Then we have*

$$\max\{\|\mathbf{v}\|_1 : \mathbf{v} \in \mathcal{G}(E)\} \leq \max\{\|\boldsymbol{\lambda}\|_1 : \boldsymbol{\lambda} \in \mathcal{G}(F\cdot\mathcal{G}(L))\}\cdot\max\{\|\mathbf{v}\|_1 : \mathbf{v} \in \mathcal{G}(L)\}.$$

Proof. Let $\mathbf{v} \in \mathcal{G}(E)$. Then $\mathbf{v} \in \ker(L)$ implies that $\mathbf{v}$ can be written as a nonnegative integer linear sign-compatible sum $\mathbf{v} = \sum \lambda_i \mathbf{g}_i$ using Graver basis vectors $\mathbf{g}_i \in \mathcal{G}(L)$. Adding zero components if necessary, we can write $\mathbf{v} = \mathcal{G}(L)\boldsymbol{\lambda}$. We now claim that $\mathbf{v} \in \mathcal{G}(E)$ implies $\boldsymbol{\lambda} \in \mathcal{G}(F\cdot\mathcal{G}(L))$.

First, observe that $\mathbf{v} \in \ker(F)$ implies $F\mathbf{v} = F\cdot(\mathcal{G}(L)\boldsymbol{\lambda}) = (F\cdot\mathcal{G}(L))\boldsymbol{\lambda} = \mathbf{0}$ and thus, $\boldsymbol{\lambda} \in \ker(F\cdot\mathcal{G}(L))$. If $\boldsymbol{\lambda} \notin \mathcal{G}(F\cdot\mathcal{G}(L))$, then it can be written as a sign-compatible sum $\boldsymbol{\lambda} = \boldsymbol{\mu} + \boldsymbol{\nu}$ with $\boldsymbol{\mu}, \boldsymbol{\nu} \in \ker(F\cdot\mathcal{G}(L))$. But then

$$\mathbf{v} = (\mathcal{G}(L)\boldsymbol{\mu}) + (\mathcal{G}(L)\boldsymbol{\nu})$$

gives a sign-compatible decomposition of $\mathbf{v}$ into vectors $\mathcal{G}(L)\boldsymbol{\mu}, \mathcal{G}(L)\boldsymbol{\nu} \in \ker(E)$, contradicting the minimality property of $\mathbf{v} \in \mathcal{G}(E)$. Hence, $\boldsymbol{\lambda} \in \mathcal{G}(F\cdot\mathcal{G}(L))$.

From $\mathbf{v} = \sum \lambda_i \mathbf{g}_i$ with $\mathbf{g}_i \in \mathcal{G}(L)$ and $\boldsymbol{\lambda} \in \mathcal{G}(F\cdot\mathcal{G}(L))$, the desired estimate follows. □

We will employ the following simple corollary.

Corollary 3.7.7. *Let $L \in \mathbb{Z}^{d\times n}$ and let $\mathbf{a}^\top \in \mathbb{Z}^n$ be a row vector. Moreover, put $E := \begin{pmatrix}\mathbf{a}^\top \\ L\end{pmatrix}$. Then we have*

$$\max\{\|\mathbf{v}\|_1 : \mathbf{v} \in \mathcal{G}(E)\} \leq \left(2\cdot\max\left\{|\mathbf{a}^\top\mathbf{v}| : \mathbf{v} \in \mathcal{G}(L)\right\} - 1\right)\cdot\max\{\|\mathbf{v}\|_1 : \mathbf{v} \in \mathcal{G}(L)\}.$$

In particular, if $M := \max\{|a^{(i)}| : i = 1,\ldots,n\}$, then

$$\max\{\|\mathbf{v}\|_1 : \mathbf{v} \in \mathcal{G}(E)\} \leq 2nM\left(\max\{\|\mathbf{v}\|_1 : \mathbf{v} \in \mathcal{G}(L)\}\right)^2.$$

Proof. By Lemma 3.7.6, we already get

$$\max\{\|\mathbf{v}\|_1 : \mathbf{v} \in \mathcal{G}(E)\} \leq \max\{\|\boldsymbol{\lambda}\|_1 : \boldsymbol{\lambda} \in \mathcal{G}(\mathbf{a}^\top\cdot\mathcal{G}(L))\}\cdot\max\{\|\mathbf{v}\|_1 : \mathbf{v} \in \mathcal{G}(L)\}.$$

Now, observe that $\mathbf{a}^\top\cdot\mathcal{G}(L)$ is a $1\times|\mathcal{G}(L)|$ matrix. Thus, the degree bound of primitive partition identities, Lemma 3.7.5, applies, which gives

$$\max\{\|\boldsymbol{\lambda}\|_1 : \boldsymbol{\lambda} \in \mathcal{G}(\mathbf{a}^\top\cdot\mathcal{G}(L))\} \leq 2\cdot\max\left\{|\mathbf{a}^\top\mathbf{v}| : \mathbf{v} \in \mathcal{G}(L)\right\} - 1,$$

and thus the first claim is proved. The second claim is a trivial consequence of the first. □

Let us now extend this corollary to a form that we need in Section 4.3.1.

Corollary 3.7.8. *Let $L \in \mathbb{Z}^{d\times n}$ and let $F \in \mathbb{Z}^{m\times n}$. Let the entries of F be bounded by M in absolute value. Moreover, put $E := \binom{F}{L}$. Then we have*

$$\max\{\|\mathbf{v}\|_1 : \mathbf{v} \in \mathcal{G}(E)\} \leq (2nM)^{2^m-1}\left(\max\{\|\mathbf{v}\|_1 : \mathbf{v} \in \mathcal{G}(L)\}\right)^{2^m}.$$

Proof. This claim follows by simple induction, adding one row of F at a time, and by using the second inequality of Corollary 3.7.7 to bound the sizes of the intermediate Graver bases in comparison to the Graver basis of the matrix with one fewer row of F. □

3.8 Computation of Graver bases

In this section we present two algorithms to compute Graver bases. In fact, these algorithms compute all $\sqsubseteq$-minimal elements in $\mathcal{L}\setminus\{\mathbf{0}\}$ for any given sublattice $\mathcal{L}\subseteq\mathbb{Z}^n$. Clearly, if we choose $\mathcal{L} = \ker_{\mathbb{Z}^n}(A)$ for some integer matrix A, these algorithms compute exactly the Graver basis $\mathcal{G}(A)$ of A.

The first algorithm due to Pottier [275] is a so-called *completion procedure* that starts with any lattice basis G of $\mathcal{L}$ and that iteratively adds new vectors to G until every $\mathbf{z}\in\mathcal{L}$ can be written as a nonnegative integer linear *sign-compatible* representation $\mathbf{z} = \sum \lambda_i \mathbf{g}_i$ with $\lambda_i \in \mathbb{Z}_+$, $\mathbf{g}_i \in G$, and $\mathbf{g}_i \sqsubseteq \mathbf{z}$ for all i. If G has this representation property, then we can conclude that G contains all $\sqsubseteq$-minimal elements (see Exercise 3.10.2). Finally, we simply have to remove all elements $\mathbf{v}\in G$ that are not $\sqsubseteq$-minimal, which is the case if there is some other (and different) $\mathbf{u}\in G$ such that $\mathbf{u}\sqsubseteq\mathbf{v}$. Sturmfels and Thomas [319] presented an algorithm to compute Graver bases by computing Gröbner bases of certain toric/lattice ideals. This algebraic algorithm, however, can be translated one-to-one into Pottier's geometric algorithm.

The second algorithm is the currently fastest algorithm to compute the $\sqsubseteq$-minimal elements in $\mathcal{L}$. It is based on the project-and-lift idea from [153]. The main idea of this approach is to project $\mathcal{L}$ down to a lower-dimensional space by deleting components and to compute all $\sqsubseteq$-minimal elements F' of the resulting lattice $\pi(\mathcal{L})\subseteq\mathbb{Z}^d$. As π is chosen to be injective, we can uniquely append the components that were projected out to each element in F' and obtain a generating set $F\subseteq\mathbb{Z}^n$ for $\mathcal{L}$ that has the sign-compatible representation property already on d components. This set F is then used as input to Pottier's algorithm. Then the knowledge about $\sqsubseteq$-minimality of the elements in $\pi(F)$ is exploited to simplify and to speedup Pottier's algorithm tremendously.

Pottier's algorithm

Let $F\subseteq\mathcal{L}\subseteq\mathbb{Z}^n$ generate $\mathcal{L}$ over $\mathbb{Z}$. Before we present Pottier's algorithm to compute all $\sqsubseteq$-minimal elements in $\mathcal{L}\setminus\{\mathbf{0}\}$, let us have a look at the following little algorithm.

ALGORITHM 3.2. Normal form algorithm.

1: **input** vector $\mathbf{s}\in\mathcal{L}$, set $G\subseteq\mathcal{L}$.
2: **output** vector $\mathbf{r}$ = normalForm($\mathbf{s}, G$) $\in\mathcal{L}$ such that $\mathbf{s} = \sum\alpha_i\mathbf{g}_i + \mathbf{r}$ with $\alpha_i\in\mathbb{Z}_{>0}$, $\mathbf{g}_i, \mathbf{r}\sqsubseteq\mathbf{s}$, and $\mathbf{g}_i\in G$ for all i; and $\mathbf{g}\not\sqsubseteq\mathbf{r}$ for all $\mathbf{g}\in G$.
3: $\mathbf{r}\leftarrow\mathbf{s}$.

while $\exists \mathbf{g} \in G$ with $\mathbf{g} \sqsubseteq \mathbf{r}$ **do**
  $\mathbf{r} \leftarrow \mathbf{r} - \mathbf{g}$.
return r.

Clearly, if $\mathbf{r} = \mathbf{0}$ is returned, $\mathbf{s}$ has successfully been written as a nonnegative integer linear combination of elements from G all having the same sign pattern as $\mathbf{s}$. If $\mathbf{r} \neq \mathbf{0}$, such a representation could still be possible, but the normal form algorithm did not succeed in producing such a representation. However, it has successfully produced such a representation of $\mathbf{s}$ by the elements in $G \cup \{\mathbf{r}\}$. This knowledge will be enough for our purposes.

ALGORITHM 3.3. Pottier's algorithm.

input a generating set F of $\mathcal{L} \subseteq \mathbb{Z}^n$.
output a set $G \subseteq \mathcal{L}$ containing all $\sqsubseteq$-minimal elements in $\mathcal{L} \setminus \{\mathbf{0}\}$.
$G \leftarrow F \cup (-F)$.
$C \leftarrow \bigcup_{\mathbf{f},\mathbf{g} \in G} \{\mathbf{f}+\mathbf{g}\}$.
while $C \neq \emptyset$ **do**
  $\mathbf{s} \leftarrow$ an element in C.
  $C \leftarrow C \setminus \{\mathbf{s}\}$.
  $\mathbf{r} \leftarrow$ normalForm($\mathbf{s}, G$).
  if $\mathbf{r} \neq \mathbf{0}$ **then**
    $C \leftarrow C \cup \{\mathbf{r}+\mathbf{g} : \mathbf{g} \in G\}$.
    $G \leftarrow G \cup \{\mathbf{r}\}$.
return G.

Lemma 3.8.1. *Algorithm 3.3 terminates and computes a set G^{ret} containing all $\sqsubseteq$-minimal elements of $\mathcal{L} \setminus \{\mathbf{0}\}$.*

Proof. The algorithm terminates, since the sequence of new elements $\{\mathbf{r}_1, \mathbf{r}_2, \ldots\}$ that are added to the initial set $F \cup (-F)$ satisfies $\mathbf{r}_i \not\sqsubseteq \mathbf{r}_j$ whenever $i < j$. By the $\sqsubseteq$-version of the Gordan–Dickson lemma, Lemma 2.5.6, any such sequence must be finite.

Let G^{ret} denote the set returned by Algorithm 3.3 and let $\mathbf{g}$ be a $\sqsubseteq$-minimal element of $\mathcal{L} \setminus \{\mathbf{0}\}$. Since $F \cup (-F) \subseteq G^{\text{ret}}$, there exists a positive integer linear combination $\mathbf{g} = \sum \alpha_i \mathbf{g}_i$ with $\alpha_i \in \mathbb{Z}_{>0}$ and $\mathbf{g}_i \in G^{\text{ret}}$ for all i. Among all such representations of $\mathbf{g}$ choose one such that $\sum \alpha_i \|\mathbf{g}_i\|_1$ is minimal. Clearly, by the triangle inequality, we have $\|\mathbf{g}\|_1 \leq \sum \alpha_i \|\mathbf{g}_i\|_1$ with equality if and only if all $\mathbf{g}_i$ have the same sign pattern in $\{\leq 0, \geq 0\}^n$ as $\mathbf{g}$.

If $\|\mathbf{g}\|_1 = \sum \alpha_i \|\mathbf{g}_i\|_1$ holds for our minimal representation, nothing is left to show. Thus, assume that $\|\mathbf{g}\|_1 < \sum \alpha_i \|\mathbf{g}_i\|_1$. Then there must exist two summands that have a different sign pattern. Without loss of generality, assume that these are $\mathbf{g}_1$ and $\mathbf{g}_2$. During the run of the algorithm, $\mathbf{g}_1 + \mathbf{g}_2$ was considered as $\mathbf{s} \in C$ and was written (in the normal form algorithm) as $\mathbf{g}_1 + \mathbf{g}_2 = \sum \beta_j \mathbf{g}'_j + \mathbf{r}$ with $\beta_j \in \mathbb{Z}_{>0}$, $\mathbf{g}'_j, \mathbf{r} \in G^{\text{ret}}$ for all j, and where $\mathbf{r}$ and all $\mathbf{g}'_j$ have the same sign pattern as $\mathbf{g}_1 + \mathbf{g}_2$. Thus, $\sum \beta_j \|\mathbf{g}'_j\|_1 + \|\mathbf{r}\|_1 = \|\mathbf{g}_1 + \mathbf{g}_2\|_1 < \|\mathbf{g}_1\|_1 + \|\mathbf{g}_2\|_1$, since $\mathbf{g}_1$ and $\mathbf{g}_2$ do not have the same sign pattern. Thus, the representation

$$\mathbf{g} = \sum \beta_j \mathbf{g}'_j + \mathbf{r} + (\alpha_1 - 1)\mathbf{g}_1 + (\alpha_2 - 1)\mathbf{g}_2 + \sum_{i>2} \alpha_i \mathbf{g}_i$$

satisfies

$$\sum \beta_j \|\mathbf{g}'_j\|_1 + \|\mathbf{r}\|_1 + (\alpha_1 - 1)\|\mathbf{g}_1\|_1 + (\alpha_2 - 1)\|\mathbf{g}_2\|_1 + \sum_{i>2} \alpha_i \|\mathbf{g}_i\|_1 < \sum \alpha_i \|\mathbf{g}_i\|_1,$$

contradicting minimality of $\sum \alpha_i \|\mathbf{g}_i\|_1$.

Consequently, we must have $\mathbf{g} = \sum \alpha_i \mathbf{g}_i$ and $\mathbf{g}_i \sqsubseteq \mathbf{g}$ for all i. As $\mathbf{g}$ is $\sqsubseteq$-minimal, this is only possible if the representation is trivial, that is, if $\mathbf{g} = \mathbf{g}_1 \in G^{\text{ret}}$. This proves our claim. □

A simple consequence is the following.

Corollary 3.8.2. *Algorithm* 3.3 *applied to* $\mathcal{L} = \ker_{\mathbb{Z}^n}(A)$ *computes the Graver basis* $\mathcal{G}(A)$.

Note that Algorithm 3.3 has two major disadvantages:

- The set G^{ret} might contain many elements of $\mathcal{L}$ that are not $\sqsubseteq$-minimal.
- The computation of the normal form of $\mathbf{s}$ with respect to G is very costly.

Let us now show how to get rid of these problems.

Project-and-lift approach

Now we present the currently fastest algorithm to compute the $\sqsubseteq$-minimal elements in $\mathcal{L}$. It is based on the project-and-lift idea from [153]. Let d be the dimension of $\mathcal{L}$ and assume without loss of generality that the projection map $\pi\colon \mathbb{Z}^n \to \mathbb{Z}^d$ that maps every vector $\mathbf{v} \in \mathbb{Z}^n$ onto its first d components is injective; that is, if $\pi(\mathbf{u}) = \pi(\mathbf{v})$ then $\mathbf{u} = \mathbf{v}$. Clearly, this can be achieved easily by suitably permuting components of the elements in $\mathcal{L}$.

Observe that $\pi(\mathcal{L})$ is a sublattice of $\mathbb{Z}^d$. Assume that we have already computed the $\sqsubseteq$-minimal elements in $\pi(\mathcal{L}) \setminus \{\mathbf{0}\}$ and that they are collected in a set $F' \subseteq \mathbb{Z}^d$. As π is injective and since $F' \subseteq \pi(\mathcal{L})$, there is a unique set $F \subseteq \mathcal{L}$ with $F' = \pi(F)$. Note that by construction F is symmetric, that is, $-F = F$. We will now use F to start Algorithm 3.3 and use the knowledge about $\pi(F)$ to simplify and speedup the algorithm. We obtain the following project-and-lift algorithm.

ALGORITHM 3.4. Project-and-lift algorithm.

input set F of $\mathcal{L} \subseteq \mathbb{Z}^n$, such that $\pi(F)$ are the $\sqsubseteq$-minimal elements in $\pi(\mathcal{L}) \setminus \{\mathbf{0}\}$.
output set $G \subseteq \mathcal{L}$ containing all $\sqsubseteq$-minimal elements in $\mathcal{L} \setminus \{\mathbf{0}\}$.
$G \leftarrow F$.
$C \leftarrow \bigcup_{\mathbf{f},\mathbf{g} \in G} \{\mathbf{f} + \mathbf{g}\}$.
while $C \neq \emptyset$ **do**
    $\mathbf{s} \leftarrow$ an element in C with $\|\pi(\mathbf{s})\|_1 = \min\{\|\pi(\mathbf{t})\|_1 : \mathbf{t} \in C\}$.
    $C \leftarrow C \setminus \{\mathbf{s}\}$.
    if $\nexists \mathbf{v} \in G$ with $\mathbf{v} \sqsubseteq \mathbf{s}$ **then**
        $C \leftarrow C \cup \{\mathbf{s} + \mathbf{g} : \mathbf{g} \in G$ where $\pi(\mathbf{s})$ and $\pi(\mathbf{g})$ are sign compatible$\}$.
        $G \leftarrow G \cup \{\mathbf{s}\}$.
return G.

Lemma 3.8.3. *Algorithm* 3.4 *terminates and computes a set* G^{ret} *that contains* exactly *all* $\sqsubseteq$*-minimal elements of* $\mathcal{L}\setminus\{\mathbf{0}\}$.

Proof. The algorithm terminates, since the sequence of new elements $\{\mathbf{s}_1,\mathbf{s}_2,\dots\}$ that are added to the initial set F satisfies $\mathbf{s}_i \not\sqsubseteq \mathbf{s}_j$ whenever $i<j$. By the $\sqsubseteq$-version of the Gordan–Dickson lemma, Lemma 2.5.6, any such sequence must be finite.

Let G^{ret} denote the set returned by Algorithm 3.4. First observe that the new elements $\{\mathbf{s}_1,\mathbf{s}_2,\dots\}$ added to G^{ret} are added by increasing norm $\|\pi(\mathbf{s}_1)\|_1 \leq \|\pi(\mathbf{s}_2)\|_1 \leq \dots$, since $\|\pi(\mathbf{s}+\mathbf{g})\|_1 > \|\pi(\mathbf{s})\|_1$ for all new elements added to C.

Now assume that G^{ret} does not contain all $\sqsubseteq$-minimal elements of $\mathcal{L}\setminus\{\mathbf{0}\}$. Among those $\sqsubseteq$-minimal elements missing in G^{ret}, let $\mathbf{g}$ be one with $\|\pi(\mathbf{g})\|_1$ minimal. That is, if $\mathbf{v}$ is $\sqsubseteq$-minimal in $\mathcal{L}\setminus\{\mathbf{0}\}$ with $\|\pi(\mathbf{v})\|_1 < \|\pi(\mathbf{g})\|_1$, then $\mathbf{v}$ must belong to G^{ret}. By construction of F and since $F\subseteq G^{\text{ret}}$, $\pi(\mathbf{g})$ can be written as a positive integer linear combination $\mathbf{g}=\sum\alpha_i\mathbf{g}_i$ with $\alpha_i\in\mathbb{Z}_{>0}$, $\mathbf{g}_i\in G^{\text{ret}}$, and $\pi(\mathbf{g}_i)\sqsubseteq\pi(\mathbf{g})$ for all i. Among all such representations of $\mathbf{g}$ consider one such that $\sum\alpha_i\|\mathbf{g}_i\|_1$ is minimal.

If $\|\mathbf{g}\|_1=\sum\alpha_i\|\mathbf{g}_i\|_1$ holds for our minimal representation, we must have $\mathbf{g}=\sum\alpha_i\mathbf{g}_i$ and $\mathbf{g}_i\sqsubseteq\mathbf{g}$ for all i. As $\mathbf{g}$ is $\sqsubseteq$-minimal, this is possible only if the representation is trivial, that is, if $\mathbf{g}=\mathbf{g}_1\in G^{\text{ret}}$, in contradiction to our assumption. Thus, we have $\|\mathbf{g}\|_1 < \sum\alpha_i\|\mathbf{g}_i\|_1$ and therefore there must exist two summands that have a different sign pattern. Without loss of generality, assume that these are $\mathbf{g}_1$ and $\mathbf{g}_2$. During the run of the algorithm, $\mathbf{g}_1+\mathbf{g}_2$ was considered as $\mathbf{s}\in C$ and therefore $\mathbf{g}=\mathbf{g}_1+\mathbf{g}_2$ cannot hold, since $\mathbf{g}\notin G^{\text{ret}}$. Thus, we must have $\|\pi(\mathbf{g}_1+\mathbf{g}_2)\|_1 < \|\pi(\mathbf{g})\|_1$ since $\pi(\mathbf{g}_1),\pi(\mathbf{g}_2)\sqsubseteq\pi(\mathbf{g})$ by construction. Since by our assumption on $\mathbf{g}$ all $\sqsubseteq$-minimal elements $\mathbf{v}$ of $\mathcal{L}\setminus\{\mathbf{0}\}$ with $\|\pi(\mathbf{v})\|_1 < \|\pi(\mathbf{g})\|_1$ belong to G^{ret}, we can write $\mathbf{g}_1+\mathbf{g}_2=\sum\beta_j\mathbf{g}'_j$ with $\beta_j\in\mathbb{Z}_{>0}$, $\mathbf{g}'_j\in G^{\text{ret}}$ for all j, and where all $\mathbf{g}'_j$ have the same sign pattern as $\mathbf{g}_1+\mathbf{g}_2$. Thus, $\sum\beta_j\|\mathbf{g}'_j\|_1=\|\mathbf{g}_1+\mathbf{g}_2\|_1 < \|\mathbf{g}_1\|_1+\|\mathbf{g}_2\|_1$, since $\mathbf{g}_1$ and $\mathbf{g}_2$ do not have the same sign pattern. Thus, the representation

$$\mathbf{g}=\sum\beta_j\mathbf{g}'_j+(\alpha_1-1)\mathbf{g}_1+(\alpha_2-1)\mathbf{g}_2+\sum_{i>2}\alpha_i\mathbf{g}_i$$

satisfies

$$\sum\beta_j\|\mathbf{g}'_j\|_1+(\alpha_1-1)\|\mathbf{g}_1\|_1+(\alpha_2-1)\|\mathbf{g}_2\|_1+\sum_{i>2}\alpha_i\|\mathbf{g}_i\|_1 < \sum\alpha_i\|\mathbf{g}_i\|_1,$$

contradicting minimality of $\sum\alpha_i\|\mathbf{g}_i\|_1$. Consequently, G^{ret} contains all $\sqsubseteq$-minimal elements in $\mathcal{L}\setminus\{\mathbf{0}\}$. It remains to show that G^{ret} contains no additional vector.

By construction, all elements in $\pi(F)$ are $\sqsubseteq$-minimal in $\pi(\mathcal{L})\setminus\{\mathbf{0}\}$ and thus also all elements in F are $\sqsubseteq$-minimal in $\mathcal{L}\setminus\{\mathbf{0}\}$. Therefore, no unnecessary vector has entered G^{ret} from the input F. Assume now that some vector $\mathbf{s}$ was added to G^{ret} that is not $\sqsubseteq$-minimal in $\pi(\mathcal{L})\setminus\{\mathbf{0}\}$. Thus, there is some $\sqsubseteq$-minimal element $\mathbf{g}\in\mathcal{L}\setminus\{\mathbf{0}\}$ with $\mathbf{g}\sqsubseteq\mathbf{s}$. As the new vectors are added to G^{ret} by increasing 1-norm of their projection under π, $\mathbf{g}$ has to be present in $G\subseteq G^{\text{ret}}$ at the time when $\mathbf{s}$ got added to G. This is a contradiction to the fact that at the time at which $\mathbf{s}$ was added to G, no $\mathbf{v}\in G$ with $\mathbf{v}\sqsubseteq\mathbf{s}$ existed. $\square$

The huge speedup of Algorithm 3.4 compared to Algorithm 3.3 stems from the fact that instead of performing full normal form computations, we only do reduction tests $\mathbf{v}\sqsubseteq\mathbf{s}$. Moreover, as elements $\mathbf{s}$ are chosen in a suitable order, no unnecessary vector has been computed which would lead to more vectors $\mathbf{s}+\mathbf{g}$ being checked unnecessarily. This latter statement implies the following important observation. There exists an enumeration algorithm for the elements of a Graver basis that runs in polynomial total time (for a discussion of output-sensitive complexity analysis, see Section 7.3).

Corollary 3.8.4. *Algorithm* 3.4 *runs in polynomial total time.*

3.9 Notes and further references

Graver bases were first introduced by Jack Graver in 1975 [144]. He already showed that they provide an optimality certificate for integer linear programming. However, he did not provide an algorithm to compute them.

Later, Herbert Scarf introduced another type of optimality certificate, the so-called *neighbors of the origin* [291, 293]. The definition of these neighbors, however, requires genericity assumptions on the matrix A and the cost vector $\mathbf{c}$ which typically imply that the matrix A has to be irrational. This genericity assumption was later relaxed to allow also a definition for rational matrices A [294, 324]. Neighbors of the origin provide only augmenting directions for fixed matrix A *and* fixed cost vector $\mathbf{c}$.

In 1991, Pasqualina Conti and Carlo Traverso introduced yet another optimality certificate that provides augmenting directions for fixed matrix A *and* fixed cost vector $\mathbf{c}$. They translated the geometric problem into the language of commutative algebra where a solution via *Gröbner bases* was already known. As there is extensive literature on Gröbner bases and a variety of fast implementations in computer algebra systems to compute them, this seminal paper sparked a renewed interest in using optimality certificates (or, as they were called back then, test sets) for the solution of integer programs. This approach will be presented in Chapters 10 (short introduction to Gröbner bases) and 11 (Gröbner bases in integer programming).

In the 1990s, following the paper by Conti and Traverso, quite a few papers on test sets and augmentation algorithms appeared (see, for example, [51, 52, 76, 162, 163, 175, 178, 294, 303, 304, 317, 319, 320, 322–326, 334]). In 2004, Murota, Saito, and Weismantel proved that Graver bases even provide optimality certificates for integer separable convex minimization problems [258]. In fact, also in this situation polynomially many augmentation steps via Graver basis directions suffice to reach an optimal solution [160]. A similar result was shown for certain integer convex maximization problems [99].

In the last decade, more emphasis was put on integer programs with structure, for example, stochastic integer programs [16, 157] or N-fold integer programs [96, 159, 177, 289], and some major progress was made by exploiting these structures. Both cases were combined into a common generalization: N-fold 4-block decomposable IPs, over which separable convex integer minimization is possible in polynomial time [159, 161]. These results will be presented in Chapter 4. Other structured problems like independence systems were considered in [227, 228].

There are only a few papers on the actual computation of Graver bases [275, 319]. In Section 3.8, we described the currently fastest method to compute Graver bases. This algorithm is based on the project-and-lift algorithm for the computation of Hilbert bases as presented in [153]. The project-and-lift algorithms to compute Hilbert bases of cones and Graver bases of lattices are implemented in the software package `4ti2` [1].

Graver bases are not only useful for separable convex integer minimization, but can also be used to maximize a convex function $f(W\mathbf{z})$ over the lattice points in a polytope $P = \{\mathbf{z} : A\mathbf{z} = \mathbf{b}, \mathbf{z} \geq \mathbf{0}\}$, where f is convex and where W is a $d \times n$ matrix [99, 266]. The parameter d is considered to be fixed. For the solution, one exploits the property that the edge directions of conv$(P \cap \mathbb{Z}^n)$ are all contained in $\mathcal{G}(A)$, which allows the enumeration of all vertices of $\operatorname{conv}(\{ W\mathbf{z} : \mathbf{z} \in P \cap \mathbb{Z}^n \})$. This leads to a polynomial-time algorithm over N-fold matrices. (See Section 4.1 for an introduction to N-fold IPs.)

Graver bases also allow the minimization of quadratic and higher degree polynomial functions which lie in suitable cones. These cones always include all separable convex polynomials, but this is a strict inclusion in general; see [229].

For another nice introductory book on Graver bases see [265].

3.10 Exercises

Exercise 3.10.1. Prove Lemma 3.2.2.

Exercise 3.10.2. Let $\mathcal{L} \subseteq \mathbb{Z}^n$ be a sublattice of $\mathbb{Z}^n$. Moreover, let $G \subseteq \mathcal{L}$ be a finite set of vectors such that every $\mathbf{z} \in \mathcal{L}$ can be written as a sign-compatible nonnegative integer linear combination $\mathbf{z} = \sum \alpha_i \mathbf{g}_i$ with $\alpha_i \in \mathbb{Z}_+$, $\mathbf{g}_i \in G$, and $\mathbf{g}_i \sqsubseteq \mathbf{z}$ for all i. Then G contains all $\sqsubseteq$-minimal elements of $\mathcal{L} \setminus \{\mathbf{0}\}$.

In particular, if $\mathcal{L} = \ker_{\mathbb{Z}^n}(A)$ for some integer matrix A, then G contains the Graver basis $\mathcal{G}(A)$.

Exercise 3.10.3. Prove Lemma 3.2.4.

Exercise 3.10.4. Prove Lemma 3.3.1. (Hint: Prove the result first for the case $n = 1$ and then exploit separability to lift it to general n.)

Exercise 3.10.5. Consider the optimization problem

$$\min \left\{ \begin{pmatrix} 0 & 1 & 0 & 2 \end{pmatrix} \mathbf{z} : \begin{pmatrix} 1 & 1 & 1 & 1 \\ 1 & 2 & 3 & 4 \end{pmatrix} \mathbf{z} = \begin{pmatrix} 10 \\ 21 \end{pmatrix},\ \mathbf{z} \in \mathbb{Z}^4_+ \right\}.$$

(a) With

$$A = \begin{pmatrix} 1 & 1 & 1 & 1 \\ 1 & 2 & 3 & 4 \end{pmatrix},$$

compute

- the Hermite normal form $\mathrm{HNF}(A) = (H \,|\, O)$ of A, and
- the transformation matrix U with $\mathrm{HNF}(A) = AU$.

(b) Find

- a lattice basis F for $\ker_{\mathbb{Z}^4}(A)$, and
- a solution $\mathbf{z}_0$ to $A\mathbf{z} = \binom{10}{21}$, $\mathbf{z} \in \mathbb{Z}^4$.

(c) Use `4ti2` to compute the Graver basis $\mathcal{G}(A)$ of A.

(d) Using the Graver basis $\mathcal{G}(A)$, turn the integer point $\mathbf{z}_0$ into a feasible solution $\mathbf{z}_1$ of $A\mathbf{z} = \binom{10}{21}$, $\mathbf{z} \in \mathbb{Z}^4_+$.

(e) Using the Graver basis $\mathcal{G}(A)$, augment the feasible solution $\mathbf{z}_1$ until an optimal solution $\mathbf{z}_{\min}$ is reached.

Exercise 3.10.6. Prove Lemma 3.7.3.

Chapter 4

Graver Bases for Block-Structured Integer Programs

In this chapter, we study a certain family of block-structured separable convex integer minimization problems. The constraint matrix of these problems is *N-fold 4-block decomposable* as follows:

$$\begin{pmatrix} C & D \\ B & A \end{pmatrix}^{(N)} := \begin{pmatrix} C & D & D & \cdots & D \\ B & A & O & & O \\ B & O & A & & O \\ \vdots & & & \ddots & \\ B & O & O & & A \end{pmatrix}$$

for some given $N \in \mathbb{Z}_+$ and N copies of A, B, and D.

N-fold 4-block decomposable matrices arise in many contexts and have been studied in various special cases, three of which are particularly relevant. We denote by O a zero matrix of compatible dimensions and by $\cdot$ a matrix with no columns or no rows. Some interesting special cases are as follows:

(i) For $B = \cdot$ and $C = \cdot$ we recover the problem matrix $\left(\begin{smallmatrix} \cdot & D \\ \cdot & A \end{smallmatrix}\right)^{(N)}$ of a so-called N-fold integer program (IP). Here, if we let A be the node-edge incidence matrix of a given network and set D to be the identity matrix, then the resulting N-fold IP is a multicommodity network flow problem. Or, if we let A be the node-edge incidence matrix of the complete bipartite graph $K_{L,M}$ and set D to be the identity matrix, then the resulting N-fold IP is the three-way transportation problem of dimensions $L \times M \times N$. We will see that separable convex N-fold IPs can be solved in polynomial time, provided that the matrices A and D are fixed [96, 160]; see Section 4.1.

(ii) For $C = \cdot$ and $D = \cdot$ we recover the problem matrix $\left(\begin{smallmatrix} \cdot & \cdot \\ B & A \end{smallmatrix}\right)^{(N)}$ of a two-stage stochastic integer optimization problem. Then, B is the matrix associated with the first-stage decision variables and A is associated with the decision in stage two. The number of occurrences of blocks of the matrix A reflects all the possible scenarios that pop up once a first-stage decision has been made; see Section 4.2.

(iii) For totally unimodular matrices C, A, their so-called 1-sum $\left(\begin{smallmatrix} C & O \\ O & A \end{smallmatrix}\right)$ is totally unimodular. Similarly, total unimodularity is preserved under the so-called 2-sum and 3-sum composition [296, 308]. For example, for matrices C and A, column vector $\mathbf{a}$, and row vector $\mathbf{b}^\top$ of appropriate dimensions, the 2-sum of $(C \ \mathbf{a})$ and $\begin{pmatrix} \mathbf{b}^\top \\ A \end{pmatrix}$ gives $\begin{pmatrix} C & \mathbf{a}\mathbf{b}^\top \\ O & A \end{pmatrix}$. The

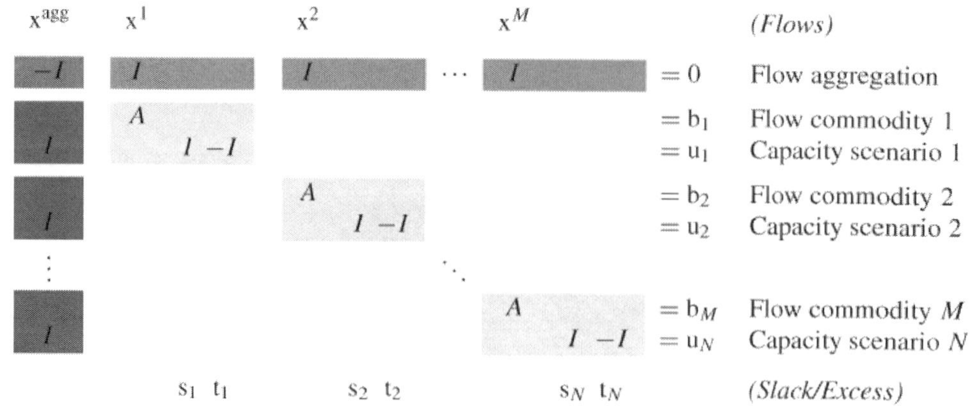

Figure 4.1. *Modeling a two-stage stochastic integer multicommodity flow problem as an N-fold 4-block decomposable problem. Without loss of generality, the number of commodities and the number of scenarios are assumed to be equal.*

2-sum of $\begin{pmatrix} C & \mathbf{a}\mathbf{b}^\top & \mathbf{a} \\ O & A & O \end{pmatrix}$ and $\begin{pmatrix} \mathbf{b}^\top \\ B \end{pmatrix}$ creates the matrix

$$\begin{pmatrix} C & \mathbf{a}\mathbf{b}^\top & \mathbf{a}\mathbf{b}^\top \\ O & A & O \\ O & O & A \end{pmatrix},$$

which is the 2-fold 4-block decomposable matrix $\begin{pmatrix} C & \mathbf{a}\mathbf{b}^\top \\ O & A \end{pmatrix}^{(2)}$. Repeated application of certain 1-sum, 2-sum, and 3-sum compositions leads to a particular family of N-fold 4-block decomposable matrices with special structure regarding the matrices B and D.

(iv) The general case appears in stochastic integer programs with second-order dominance relations [137] and stochastic integer multicommodity flows. See [159] for further details of the model as an N-fold 4-block decomposable problem. To give one example, consider a stochastic integer multicommodity flow problem, introduced in [252, 276]. Let M integer (in contrast to continuous) commodities be transported over a given network. While we assume that supply and demand are deterministic, we assume that the upper bounds for the capacities per edge are uncertain and given initially only via some probability distribution. In a first stage we have to decide how to transport the M commodities over the given network without knowing the true capacities per edge. Then, after observing the true capacities per edge, penalties have to be paid if the capacity is exceeded. Assuming that we have knowledge about the probability distributions of the uncertain upper bounds, we wish to minimize the costs for the integer multicommodity flow plus the expected penalties to be paid for exceeding capacities. To solve this problem, we discretize as usual the probability distribution for the uncertain upper bounds into N scenarios. Doing so, we obtain a (typically large-scale) (two-stage stochastic) integer programming problem as shown in Figure 4.1. Herein, A is the node-edge incidence matrix of the given network, I is an identity matrix of appropriate size, and the columns containing $-I$ correspond to the penalty variables. We deal with this general case of N-fold 4-block decomposable matrices in Section 4.3.

4.1 N-fold integer programming

One fundamental combinatorial optimization problem is the integer two-way transportation problem

$$\min\left\{\mathbf{c}^\top\mathbf{z} : \sum_i z_{i,j} = r_j,\ \sum_j z_{i,j} = c_i,\ \mathbf{0} \le \mathbf{z} \le \mathbf{u},\ \mathbf{z} \in \mathbb{Z}^{M\times N}\right\},$$

which models the transport of a commodity from M factories to N stores at minimum costs. The underlying problem matrix is totally unimodular, and hence any vertex of the two-way transportation polytope is integral for any integer row and column sums r_j and c_i. Hence, we can drop integrality and solve the integer two-way transportation problem as a linear program in polynomial time [296]. Let us now consider the next more general case of three-way transportation problems. Interestingly, as the following theorem shows, every integer linear program is polynomial-time equivalent to an $L \times M \times N$ integer three-way transportation problem. Thus, these integer three-way transportation problems already capture all the hardness of integer linear programming, even if one of the three dimensions is fixed to $L = 3$.

Theorem 4.1.1 (universality theorem [86]). *Every (bounded) integer programming problem* $\min\left\{\mathbf{c}^\top\mathbf{y} : \mathbf{y} \in \mathbb{Z}^k_+,\ V\mathbf{y} = \mathbf{v}\right\}$ *is polynomial-time equivalent to an IP of the form*

$$\min\left\{\mathbf{c}^\top\mathbf{z} : \sum_i z_{i,j,k} = r_{j,k},\ \sum_j z_{i,j,k} = s_{i,k},\ \sum_k z_{i,j,k} = t_{i,j},\ \mathbf{0} \le \mathbf{z} \le \mathbf{u},\ \mathbf{z} \in \mathbb{Z}^{3\times M\times N}\right\}.$$

If we keep M and N fixed, we deal with integer linear programs in fixed dimension $3MN$, and thus these IPs can be solved in polynomial time in the encoding length of the remaining input using Lenstra's algorithm. If we let M and N vary, this family of problems is NP-hard to solve since it can model any integer linear program. But what about the intermediate cases, in which we keep M fixed and let N vary? As we will see below, these integer three-way transportation problems can be solved in polynomial time. To show this, we consider slightly more general IPs, so-called N-fold integer programs. As we saw it is a special case of N-fold 4-block decomposition.

N-fold integer programs have problem matrices of a very special form:

$$\begin{pmatrix} \cdot & D \\ \cdot & A \end{pmatrix}^{(N)} = \begin{pmatrix} D & D & \cdots & D \\ A & O & \cdots & O \\ O & A & \cdots & O \\ \vdots & \vdots & \ddots & \vdots \\ O & O & \cdots & A \end{pmatrix},$$

where N copies of given fixed integer matrices A and D with the same number of columns, of dimensions $d_A \times n$ and $d_D \times n$, respectively, are used. If $D = I_n$ is simply the identity matrix, we use the shorthand $A^{(N)} := \begin{pmatrix} \cdot & I_n \\ \cdot & A \end{pmatrix}^{(N)}$. The matrix $\begin{pmatrix} \cdot & D \\ \cdot & A \end{pmatrix}^{(N)}$ is called an *N-fold matrix*. We are interested in the complexity of solving the IP

$$\min\left\{ f(\mathbf{z}) : \begin{pmatrix} \cdot & D \\ \cdot & A \end{pmatrix}^{(N)}\mathbf{z} = \mathbf{b},\ \mathbf{l} \le \mathbf{z} \le \mathbf{u},\ \mathbf{z} \in \mathbb{Z}^{Nn}\right\} \tag{4.1}$$

for linear and, more generally, for separable convex functions f. Clearly, if we make no further assumptions, this problem is NP-hard to solve already for $N = 1$ and linear f.

Thus, we will assume that A and D are kept fixed, but N is allowed to vary. As we will see below, for fixed matrices A and D, the sizes of the Graver bases of $\left(\begin{smallmatrix}\cdot & D\\ \cdot & A\end{smallmatrix}\right)^{(N)}$ increase only polynomially with N. Consequently, with the techniques from the previous chapter, problem (4.1) is solvable in polynomial time! In fact, for separable convex piecewise linear objectives, we construct a *Graver-based dynamic programming* approach for the solution of the optimization problem, whose running time will increase only cubically with N. Let us start by exhibiting a fundamental structural result on the Graver bases of $\left(\begin{smallmatrix}\cdot & D\\ \cdot & A\end{smallmatrix}\right)^{(N)}$.

4.1.1 Graver complexity

Let $A \in \mathbb{Z}^{d_A \times n}$ and $D \in \mathbb{Z}^{d_D \times n}$ be fixed, and let N vary. As N increases, the Graver bases $\mathcal{G}\left(\left(\begin{smallmatrix}\cdot & D\\ \cdot & A\end{smallmatrix}\right)^{(N)}\right)$ become bigger and bigger. However, the structure of the matrix $\left(\begin{smallmatrix}\cdot & D\\ \cdot & A\end{smallmatrix}\right)^{(N)}$ implies a structure on the elements of $\mathcal{G}\left(\left(\begin{smallmatrix}\cdot & D\\ \cdot & A\end{smallmatrix}\right)^{(N)}\right)$. So to say, the Graver basis elements have a bounded *complexity*. Let us motivate the notion of *Graver complexity* in the special case $D = I_n$.

For this, consider the set of all 3×3 tables/arrays whose entries are filled with integer numbers in such a way that the sums along each row and along each column are 0. These arrays correspond exactly to the lattice points inside $\ker(A_{3\times 3})$, where $A_{3\times 3}$ denotes the problem matrix of the 3×3 transportation problem. One particular example is the table

$$\begin{pmatrix} 1 & -1 & 0 \\ -1 & 3 & -2 \\ 0 & -2 & 2 \end{pmatrix}.$$

If we encode the nine entries of the table as $z_1, \dots, z_9$, then the set of 3×3 tables coincides with the integer vectors in the kernel of the matrix

$$A_{3\times 3} = \begin{pmatrix} 1 & 0 & 0 & 1 & 0 & 0 & 1 & 0 & 0 \\ 0 & 1 & 0 & 0 & 1 & 0 & 0 & 1 & 0 \\ 0 & 0 & 1 & 0 & 0 & 1 & 0 & 0 & 1 \\ 1 & 1 & 1 & 0 & 0 & 0 & 0 & 0 & 0 \\ 0 & 0 & 0 & 1 & 1 & 1 & 0 & 0 & 0 \\ 0 & 0 & 0 & 0 & 0 & 0 & 1 & 1 & 1 \end{pmatrix} = \begin{pmatrix} I_3 & I_3 & I_3 \\ A & O & O \\ O & A & O \\ O & O & A \end{pmatrix} = (1\ 1\ 1)^{(3)},$$

that is, with all $\mathbf{z} \in \mathbb{Z}^9$ satisfying $A_{3\times 3}\mathbf{z} = \mathbf{0}$. The Graver basis of $A_{3\times 3}$ consists of all $\sqsubseteq$-minimal *nonzero* tables among them. The particular 3×3 table above does not belong to the Graver basis of $A_{3\times 3}$, since

$$\begin{pmatrix} 1 & -1 & 0 \\ -1 & 1 & 0 \\ 0 & 0 & 0 \end{pmatrix} \sqsubseteq \begin{pmatrix} 1 & -1 & 0 \\ -1 & 3 & -2 \\ 0 & -2 & 2 \end{pmatrix}.$$

Using the computer program `4ti2` [1], we find that the following 15 vectors (and their negatives) constitute the Graver basis of $A_{3\times 3}$:

$$\begin{pmatrix}1&-1&0\\-1&1&0\\0&0&0\end{pmatrix},\quad \begin{pmatrix}1&0&-1\\-1&0&1\\0&0&0\end{pmatrix},\quad \begin{pmatrix}0&-1&1\\0&1&-1\\0&0&0\end{pmatrix},$$

$$\begin{pmatrix}1&-1&0\\0&0&0\\-1&1&0\end{pmatrix},\quad \begin{pmatrix}1&0&-1\\0&0&0\\-1&0&1\end{pmatrix},\quad \begin{pmatrix}0&1&-1\\0&0&0\\0&-1&1\end{pmatrix},$$

$$\begin{pmatrix}0&0&0\\1&-1&0\\-1&1&0\end{pmatrix},\quad \begin{pmatrix}0&0&0\\1&0&-1\\-1&0&1\end{pmatrix},\quad \begin{pmatrix}0&0&0\\0&1&-1\\0&-1&1\end{pmatrix},$$

$$\begin{pmatrix}1&-1&0\\-1&0&1\\0&1&-1\end{pmatrix},\quad \begin{pmatrix}1&-1&0\\0&1&-1\\-1&0&1\end{pmatrix},\quad \begin{pmatrix}0&1&-1\\1&-1&0\\-1&0&1\end{pmatrix},$$

$$\begin{pmatrix}0&1&-1\\-1&0&1\\1&-1&0\end{pmatrix},\quad \begin{pmatrix}1&0&-1\\-1&1&0\\0&-1&1\end{pmatrix},\quad \begin{pmatrix}1&0&-1\\0&-1&1\\-1&1&0\end{pmatrix}.$$

However, there is an obvious symmetry group $S_3 \times S_3$ acting on the set of 3×3 tables whose elements transform a given table $\mathbf{v} \in \ker(A_{3\times 3})$ into another table $\mathbf{w} \in \ker(A_{3\times 3})$ by permuting rows or columns. If we take these symmetries into account, we see that among these 15 elements there are in fact only two essentially different elements:

$$\begin{pmatrix}1&-1&0\\-1&1&0\\0&0&0\end{pmatrix} \quad\text{and}\quad \begin{pmatrix}1&-1&0\\-1&0&1\\0&1&-1\end{pmatrix}.$$

It should be clear that for larger or higher-dimensional tables, this difference in sizes becomes far more striking, since the symmetry groups are much bigger.

Interestingly enough, the Graver basis elements of $N \times 3$ arrays, $N \geq 4$, also fall into only two symmetry classes:

$$\begin{pmatrix}1&-1&0\\-1&1&0\\0&0&0\\0&0&0\\\vdots&\vdots&\vdots\\0&0&0\end{pmatrix} \quad\text{and}\quad \begin{pmatrix}1&-1&0\\-1&0&1\\0&1&-1\\0&0&0\\\vdots&\vdots&\vdots\\0&0&0\end{pmatrix}.$$

This example encourages us to decompose vectors $\mathbf{z}$ into parts corresponding to copies of A, that is, we consider vectors $\mathbf{z} = \left(\mathbf{z}^{(1)},\ldots,\mathbf{z}^{(N)}\right)$ with $\mathbf{z}^{(i)} \in \mathbb{Z}^n$ for $i = 1,\ldots,N$. The vectors $\mathbf{z}^{(i)}$ are called *building blocks*, *bricks*, or *layers* of $\mathbf{z}$. The number $\operatorname{type}(\mathbf{z}) := |\{i : \mathbf{z}^{(i)} \neq \mathbf{0}\}|$ of nonzero bricks $\mathbf{z}^{(i)} \in \mathbb{Z}^n$ of $\mathbf{z}$ is called the *type* of $\mathbf{z}$.

As we saw above, each Graver basis element of $A_{3\times 3}$ was of type 2 or 3. The same holds true for $A_{N\times 3}$. In fact, the following amazing finiteness result holds for an arbitrary matrix A [289].

Theorem 4.1.2. *For all matrices $A \in \mathbb{Z}^{d_A \times n}$, there exists a constant $g(A)$ such that for all N, the Graver basis of $A^{(N)}$ consists of vectors of type at most $g(A)$.*

The smallest such constant $g(A)$ is called the *Graver complexity* of A. Thus, the Graver complexity of $A = \begin{pmatrix} 1 & 1 & 1 \end{pmatrix}$ is 3. Theorem 4.1.2 holds in fact for any matrix $D \in \mathbb{Z}^{d_D \times n}$ [177], not just for $D = I_n$.

Theorem 4.1.3. *For all matrices $A \in \mathbb{Z}^{d_A \times n}$ and $D \in \mathbb{Z}^{d_D \times n}$, there exists a constant $g(A, D)$ such that for all N, the Graver basis of $\left(\begin{smallmatrix} \cdot & D \\ \cdot & A \end{smallmatrix}\right)^{(N)}$ consists of vectors of type at most $g(A, D)$.*

Proof. Let $\mathbf{u} = \left(\mathbf{u}^{(1)}, \ldots, \mathbf{u}^{(N)}\right) \in \mathcal{G}\left(\left(\begin{smallmatrix} \cdot & D \\ \cdot & A \end{smallmatrix}\right)^{(N)}\right)$. Then $\mathbf{u}^{(i)} \in \ker(A)$ for all i. Thus, for all $i = 1, \ldots, N$, we can write down nonnegative integer linear sign-compatible representations $\mathbf{u}^{(i)} = \sum_{j=1}^{k_i} \lambda_{i,j} \mathbf{g}_{i,j}$ with $\mathbf{g}_{i,j} \in \mathcal{G}(A)$. We claim that for

$$N' = \sum_{i=1}^{N} \sum_{j=1}^{k_i} \lambda_{i,j} \geq N,$$

the vector

$$\mathbf{u}' = (\underbrace{\mathbf{g}_{1,1}, \ldots, \mathbf{g}_{1,1}}_{\lambda_{1,1} \text{ times}}, \underbrace{\mathbf{g}_{1,2}, \ldots, \mathbf{g}_{1,2}}_{\lambda_{1,2} \text{ times}}, \ldots, \underbrace{\mathbf{g}_{N,k_N}, \ldots, \mathbf{g}_{N,k_N}}_{\lambda_{N,k_N} \text{ times}})$$

belongs to $\mathcal{G}\left(\left(\begin{smallmatrix} \cdot & D \\ \cdot & A \end{smallmatrix}\right)^{(N')}\right)$. Clearly, if there was some nontrivial vector

$$\mathbf{v}' = (\mathbf{v}'_{1,1,1}, \ldots, \mathbf{v}'_{1,1,\lambda_{1,1}}, \mathbf{v}'_{1,2,1}, \ldots, \mathbf{v}'_{1,2,\lambda_{1,2}}, \ldots, \mathbf{v}'_{N,k_N,1}, \ldots, \mathbf{v}'_{N,k_N,\lambda_{N,k_N}})$$

with $\mathbf{v}' \in \ker\left(\left(\begin{smallmatrix} \cdot & D \\ \cdot & A \end{smallmatrix}\right)^{(N')}\right)$ and $\mathbf{v}' \sqsubseteq \mathbf{u}'$, we can construct $\mathbf{v} = \left(\mathbf{v}^{(1)}, \ldots, \mathbf{v}^{(N)}\right) \in \ker\left(\left(\begin{smallmatrix} \cdot & D \\ \cdot & A \end{smallmatrix}\right)^{(N)}\right)$ with

$$\mathbf{v}^{(i)} = \sum_{j=1}^{k_i} \sum_{m=1}^{\lambda_{i,j}} \mathbf{v}'_{i,j,m}.$$

By construction, $\mathbf{v} \sqsubseteq \mathbf{u}$ with $\mathbf{v} \notin \{\mathbf{0}, \mathbf{u}\}$, contradicting $\mathbf{u} \in \mathcal{G}\left(\left(\begin{smallmatrix} \cdot & D \\ \cdot & A \end{smallmatrix}\right)^{(N)}\right)$. Thus, as type($\mathbf{u}$) $\leq$ type($\mathbf{u}'$), it suffices to bound (for varying N) the type of only those vectors $\mathbf{u} \in \mathcal{G}\left(\left(\begin{smallmatrix} \cdot & D \\ \cdot & A \end{smallmatrix}\right)^{(N)}\right)$ whose bricks $\mathbf{u}^{(i)}$ all belong to $\mathcal{G}(A)$.

Consider now $\mathbf{u} \in \mathcal{G}\left(\left(\begin{smallmatrix} \cdot & D \\ \cdot & A \end{smallmatrix}\right)^{(N)}\right)$ with $\mathbf{u}^{(i)} \in \mathcal{G}(A)$ for all $i = 1, \ldots, N$. Moreover, let $\mathcal{G}(A) = \left\{\mathbf{g}_1, \ldots, \mathbf{g}_k\right\}$. By abuse of notation, we write

$$D\mathcal{G}(A) = \begin{pmatrix} D\mathbf{g}_1 & D\mathbf{g}_2 & \cdots & D\mathbf{g}_k \end{pmatrix}$$

for the matrix whose columns are $D\mathbf{g}_1, \ldots, D\mathbf{g}_k$. Let $\mathbf{u}$ consist of λ_1 bricks $\mathbf{g}_1$, λ_2 bricks $\mathbf{g}_2$, ..., and λ_k bricks $\mathbf{g}_k$. Without loss of generality, we may assume that $N = \sum_{i=1}^{k} \lambda_i$, that is, we drop any (uninteresting) $\mathbf{0}$-brick from $\mathbf{u}$. As any permutation of bricks of the $\sqsubseteq$-minimal vector $\mathbf{u}$ still leads to a $\sqsubseteq$-minimal vector $\mathbf{u}' \in \ker\left(\left(\begin{smallmatrix} \cdot & D \\ \cdot & A \end{smallmatrix}\right)^{(N)}\right)$ (why?), we may assume that $\mathbf{u}$ has the form

$$\mathbf{u} = (\underbrace{\mathbf{g}_1, \ldots, \mathbf{g}_1}_{\lambda_1 \text{ times}}, \underbrace{\mathbf{g}_2, \ldots, \mathbf{g}_2}_{\lambda_2 \text{ times}}, \ldots, \underbrace{\mathbf{g}_k, \ldots, \mathbf{g}_k}_{\lambda_k \text{ times}}).$$

As $\mathbf{u} \in \ker\left(\left(\begin{smallmatrix} \cdot & D \\ \cdot & A \end{smallmatrix}\right)^{(N)}\right)$, we have $\left(\begin{smallmatrix} \cdot & D \\ \cdot & A \end{smallmatrix}\right)^{(N)}\mathbf{u} = \mathbf{0}$. The latter is equivalent to

$$\sum_{i=1}^{k} D(\lambda_i \mathbf{g}_i) = \sum_{i=1}^{k} (D\mathbf{g}_i)\lambda_i = D\mathcal{G}(A)\begin{pmatrix} \lambda_1 \\ \vdots \\ \lambda_k \end{pmatrix} =: D\mathcal{G}(A)\boldsymbol{\lambda},$$

that is, $\boldsymbol{\lambda} \in \mathbb{Z}_+^k$ belongs to $\ker\big(D\mathcal{G}(A)\big)$. Let us now assume that there exists some other solution $\boldsymbol{\mu} \in \mathbb{Z}_+^k$ with $D\mathcal{G}(A)\boldsymbol{\mu} = \mathbf{0}$ and $\boldsymbol{\mu} \leq \boldsymbol{\lambda}$, $\boldsymbol{\mu} \notin \{\mathbf{0}, \boldsymbol{\lambda}\}$. Consider now the vector

$$\mathbf{v} = (\underbrace{\mathbf{g}_1, \dots, \mathbf{g}_1}_{\mu_1 \text{ times}}, \underbrace{\mathbf{0}, \dots, \mathbf{0}}_{\lambda_1 - \mu_1 \text{ times}}, \dots, \underbrace{\mathbf{g}_k, \dots, \mathbf{g}_k}_{\mu_k \text{ times}}, \underbrace{\mathbf{0}, \dots, \mathbf{0}}_{\lambda_k - \mu_k \text{ times}}).$$

By construction, $\mathbf{v} \in \ker\left(\left(\begin{smallmatrix} \cdot & D \\ \cdot & A \end{smallmatrix}\right)^{(N)}\right)$ and $\mathbf{v} \sqsubseteq \mathbf{u}$, contradicting $\sqsubseteq$-minimality of $\mathbf{u}$.

We conclude that $\mathbf{u} \in \mathcal{G}\left(\left(\begin{smallmatrix} \cdot & D \\ \cdot & A \end{smallmatrix}\right)^{(N)}\right)$ implies that $\boldsymbol{\lambda} \in \mathbb{Z}^n$ is a $\leq$-minimal lattice point in the pointed rational cone $C = \{\mathbf{x} : D\mathcal{G}(A)\mathbf{x} = \mathbf{0}, \mathbf{x} \geq \mathbf{0}\}$; that is, $\boldsymbol{\lambda}$ belongs to the (unique) inclusion-minimal Hilbert basis of C. Let $g(A,D)$ be the maximum of the 1-norms of the elements in the inclusion-minimal Hilbert basis of C. By construction, $g(A,D)$ is independent of N and $\text{type}(\mathbf{u}) = \|\boldsymbol{\lambda}\|_1 \leq g(A,D)$. This proves our claim. □

This proof is in fact constructive. Note that $D\mathcal{G}(A)$ contains columns for $\mathbf{g} \in \mathcal{G}(A)$ and for $-\mathbf{g} \in \mathcal{G}(A)$. Let $D\bar{\mathcal{G}}(A)$ be a submatrix of $D\mathcal{G}(A)$ where we choose only one of the two columns for each pair $\mathbf{g}, -\mathbf{g} \in \mathcal{G}(A)$. Then we have the following fact:

Lemma 4.1.4. *The maximum ℓ_1-norm of the elements in the inclusion-minimal Hilbert basis of the cone $\{\mathbf{x} : D\mathcal{G}(A)\mathbf{x} = \mathbf{0}, \mathbf{x} \geq \mathbf{0}\}$ is equal to the maximum ℓ_1-norm of the elements in $\mathcal{G}(D\bar{\mathcal{G}}(A))$.*

This result follows immediately from the construction in the proof of Theorem 4.1.3 and by applying Lemma 3.7.1.

Example 4.1.5. Let us compute the Graver complexity $g(A) = g(A, I_3)$ for the matrix $A = \begin{pmatrix} 1 & 1 & 1 \end{pmatrix}$. First, we compute the Graver basis of A and get

$$\mathcal{G}(A) = \{\pm(1,-1,0), \pm(1,0,-1), \pm(0,1,-1)\}.$$

As $D = I_3$, we set up a new matrix $A' := D\bar{\mathcal{G}}(A)$, where the columns of $\bar{\mathcal{G}}(A)$ consist of the elements in $\mathcal{G}(A)$, but with only one column for each pair of vectors $\mathbf{v}, -\mathbf{v} \in \mathcal{G}(A)$. Thus, let

$$A' := D\bar{\mathcal{G}}(A) = \begin{pmatrix} 1 & 1 & 0 \\ -1 & 0 & 1 \\ 0 & -1 & -1 \end{pmatrix}.$$

Now, we compute the Graver basis $\mathcal{G}(A')$ and obtain $\mathcal{G}(A') = \{\pm(1,-1,1)\}$. Thus, we get

$$g(A) = \max\left\{\|\mathbf{z}\|_1 : \mathbf{z} \in \mathcal{G}(A')\right\} = 3.$$

Hence, the Graver bases of $3 \times N$ transportation matrices do not get more complicated than the Graver basis of the 3×3 transportation matrix. We just have $N-3$ additional $\mathbf{0}$-bricks.

How can we compute $g(A) = 3$ using the software `4ti2`? If we stored the matrix $A = (1\ 1\ 1)$ into a file, say A3:

```
1 3
1 1 1
```

then we could compute the Graver complexity of A as follows:

```
graver A3
output --transpose A3.gra
graver A3.gra.tra
output --degree A3.gra.tra.gra
```

Herein, the call "`graver A3`" produces (one half of) the Graver basis of A, then "`output -transpose A3.gra`" constructs A', then "`graver A3.gra.tra`" produces (one half of) the Graver basis of A', and finally, the highest listed norm we get by calling "`output -degree A3.gra.tra.gra`" is the desired Graver degree.

4.1.2 Polynomial-size Graver bases

The existence of a finite number $g(A,D)$ that bounds the type of any element in $\mathcal{G}\left(\left(\begin{smallmatrix} \cdot & D \\ \cdot & A \end{smallmatrix}\right)^{(N)}\right)$ already implies that $\mathcal{G}\left(\left(\begin{smallmatrix} \cdot & D \\ \cdot & A \end{smallmatrix}\right)^{(N)}\right)$ is of polynomial size in N.

Theorem 4.1.6. *Fix any matrices $A \in \mathbb{Z}^{d_A \times n}$ and $D \in \mathbb{Z}^{d_D \times n}$. Then there is a polynomial-time algorithm that, given N, computes the Graver basis $\mathcal{G}\left(\left(\begin{smallmatrix} \cdot & D \\ \cdot & A \end{smallmatrix}\right)^{(N)}\right)$ of the N-fold matrix $\left(\begin{smallmatrix} \cdot & D \\ \cdot & A \end{smallmatrix}\right)^{(N)}$. In particular, the cardinality and the binary encoding length of $\mathcal{G}\left(\left(\begin{smallmatrix} \cdot & D \\ \cdot & A \end{smallmatrix}\right)^{(N)}\right)$ are bounded by a polynomial function of N.*

Proof. Let $g := g(A,D)$ be the Graver complexity of A and D, and consider any $N \geq g$. We show that the Graver basis of $\left(\begin{smallmatrix} \cdot & D \\ \cdot & A \end{smallmatrix}\right)^{(N)}$ is the union of $\binom{N}{g}$ suitably embedded copies of the Graver basis of $\left(\begin{smallmatrix} \cdot & D \\ \cdot & A \end{smallmatrix}\right)^{(g)}$. Consider any g indices $1 \leq k_1 < \cdots < k_g \leq N$ and define a map $\phi_{k_1,\ldots,k_g}$ from $\mathbb{Z}^{gn}$ to $\mathbb{Z}^{Nn}$ by sending $\mathbf{x} = \left(\mathbf{x}^{(1)},\ldots,\mathbf{x}^{(g)}\right)$ to $\mathbf{y} = \left(\mathbf{y}^{(1)},\ldots,\mathbf{y}^{(N)}\right)$ defined by $\mathbf{y}^{(k_t)} := \mathbf{x}^{(t)}$ for $t = 1,\ldots,g$, and $\mathbf{y}^{(i)} := \mathbf{0}$ for all other bricks of $\mathbf{y}$.

We claim that the Graver basis of $\left(\begin{smallmatrix} \cdot & D \\ \cdot & A \end{smallmatrix}\right)^{(N)}$ is the union of the images of the Graver basis of $\left(\begin{smallmatrix} \cdot & D \\ \cdot & A \end{smallmatrix}\right)^{(g)}$ under the $\binom{N}{g}$ maps $\phi_{k_1,\ldots,k_g}$ for all $1 \leq k_1 < \cdots < k_g \leq N$, that is,

$$\mathcal{G}\left(\left(\begin{smallmatrix} \cdot & D \\ \cdot & A \end{smallmatrix}\right)^{(N)}\right) = \bigcup_{1 \leq k_1 < \cdots < k_g \leq N} \phi_{k_1,\ldots,k_g}\left(\mathcal{G}\left(\left(\begin{smallmatrix} \cdot & D \\ \cdot & A \end{smallmatrix}\right)^{(g)}\right)\right). \tag{4.2}$$

To see this, recall first that the Graver basis of a matrix M is the set of all $\sqsubseteq$-minimal elements in $\ker_{\mathbb{Z}}(M) \setminus \{\mathbf{0}\}$. Thus, if $\mathbf{x} = \left(\mathbf{x}^{(1)},\ldots,\mathbf{x}^{(g)}\right) \in \mathcal{G}\left(\left(\begin{smallmatrix} \cdot & D \\ \cdot & A \end{smallmatrix}\right)^{(g)}\right)$, then $\mathbf{x}$ is $\sqsubseteq$-minimal in $\ker_{\mathbb{Z}}\left(\left(\begin{smallmatrix} \cdot & D \\ \cdot & A \end{smallmatrix}\right)^{(N)}\right) \setminus \{\mathbf{0}\}$, implying that $\phi_{k_1,\ldots,k_g}(\mathbf{x})$ is also $\sqsubseteq$-minimal in $\ker_{\mathbb{Z}}\left(\left(\begin{smallmatrix} \cdot & D \\ \cdot & A \end{smallmatrix}\right)^{(N)}\right) \setminus \{\mathbf{0}\}$, and hence $\phi_{k_1,\ldots,k_g}(\mathbf{x}) \in \mathcal{G}\left(\left(\begin{smallmatrix} \cdot & D \\ \cdot & A \end{smallmatrix}\right)^{(N)}\right)$. This establishes that the right-hand side of equation (4.2) is contained in the left-hand side. Conversely, consider any $\mathbf{y} \in \mathcal{G}\left(\left(\begin{smallmatrix} \cdot & D \\ \cdot & A \end{smallmatrix}\right)^{(N)}\right)$. Then, by Theorem 4.1.3, the type of $\mathbf{y}$ is at most g, so there are indices $1 \leq k_1 < \cdots < k_g \leq N$ such that all nonzero components of $\mathbf{y}$ are among those of the reduced vector $\mathbf{x} := \left(\mathbf{y}^{(k_1)},\ldots,\mathbf{y}^{(k_g)}\right)$, and therefore $\mathbf{y} = \phi_{k_1,\ldots,k_g}(\mathbf{x})$. Now, $\mathbf{y} \in \mathcal{G}\left(\left(\begin{smallmatrix} \cdot & D \\ \cdot & A \end{smallmatrix}\right)^{(N)}\right)$ implies that $\mathbf{y}$ is $\sqsubseteq$-minimal in $\ker_{\mathbb{Z}}\left(\left(\begin{smallmatrix} \cdot & D \\ \cdot & A \end{smallmatrix}\right)^{(N)}\right) \setminus \{\mathbf{0}\}$, and therefore $\mathbf{x}$ is $\sqsubseteq$-minimal in $\ker_{\mathbb{Z}}\left(\left(\begin{smallmatrix} \cdot & D \\ \cdot & A \end{smallmatrix}\right)^{(g)}\right) \setminus \{\mathbf{0}\}$ and hence $\mathbf{x} \in \mathcal{G}\left(\left(\begin{smallmatrix} \cdot & D \\ \cdot & A \end{smallmatrix}\right)^{(g)}\right)$, showing that $\mathbf{y} \in \phi_{k_1,\ldots,k_g}\left(\mathcal{G}\left(\left(\begin{smallmatrix} \cdot & D \\ \cdot & A \end{smallmatrix}\right)^{(g)}\right)\right)$. This establishes that the left-hand side of Equation (4.2) is contained in the right-hand side. Thus, the Graver basis of $\left(\begin{smallmatrix} \cdot & D \\ \cdot & A \end{smallmatrix}\right)^{(N)}$ is indeed given by Equation (4.2).

Since A and D are fixed and hence $g = g(A,D)$ is constant, the g-fold matrix $\left(\begin{smallmatrix}\cdot & D\\ \cdot & A\end{smallmatrix}\right)^{(g)}$ is also fixed and so the cardinality and bit size of its Graver basis $\mathcal{G}\left(\left(\begin{smallmatrix}\cdot & D\\ \cdot & A\end{smallmatrix}\right)^{(g)}\right)$ are constant as well. It follows from Equation (4.2) that

$$|\mathcal{G}(\left(\begin{smallmatrix}\cdot & D\\ \cdot & A\end{smallmatrix}\right)^{(N)})| \leq \binom{N}{g} |\mathcal{G}(\left(\begin{smallmatrix}\cdot & D\\ \cdot & A\end{smallmatrix}\right)^{(g)})| \in O(N^g). \tag{4.3}$$

Further, each element of $\mathcal{G}\left(\left(\begin{smallmatrix}\cdot & D\\ \cdot & A\end{smallmatrix}\right)^{(N)}\right)$ is a vector $\phi_{k_1,\dots,k_g}(\mathbf{x}) \in \mathbb{Z}^{Nn}$ obtained from some $\mathbf{x} \in \mathcal{G}\left(\left(\begin{smallmatrix}\cdot & D\\ \cdot & A\end{smallmatrix}\right)^{(g)}\right)$ (of constant bit size) by appending zero components, and therefore is of linear bit size $O(N)$, showing that the bit size of the entire Graver basis $\mathcal{G}\left(\left(\begin{smallmatrix}\cdot & D\\ \cdot & A\end{smallmatrix}\right)^{(N)}\right)$ is $O(N^{g+1})$. Finally, it is clear that the $\binom{N}{g} = O(N^g)$ images $\phi_{k_1,\dots,k_g}\left(\mathcal{G}\left(\left(\begin{smallmatrix}\cdot & D\\ \cdot & A\end{smallmatrix}\right)^{(g)}\right)\right)$ and their union $\mathcal{G}\left(\left(\begin{smallmatrix}\cdot & D\\ \cdot & A\end{smallmatrix}\right)^{(N)}\right)$ can be computed in polynomial time in N, completing the proof. □

Example 4.1.7. Consider the matrices $A = (1\ 1)$ and $D = I_2$. The Graver complexity of the pair A and D is $g(A,D) = 2$. The 2-fold matrix and its Graver basis, consisting only of two antipodal vectors, are

$$\left(\begin{smallmatrix}\cdot & D\\ \cdot & A\end{smallmatrix}\right)^{(2)} = \begin{pmatrix} 1 & 0 & 1 & 0\\ 0 & 1 & 0 & 1\\ 1 & 1 & 0 & 0\\ 0 & 0 & 1 & 1 \end{pmatrix}, \qquad \mathcal{G}(\left(\begin{smallmatrix}\cdot & D\\ \cdot & A\end{smallmatrix}\right)^{(2)}) = \{\pm(1,-1,-1,1)\}.$$

By Theorem 4.1.6, the Graver basis of the 4-fold matrix $\left(\begin{smallmatrix}\cdot & D\\ \cdot & A\end{smallmatrix}\right)^{(4)}$ can be computed by taking the union of the images of the $6 = \binom{4}{2}$ maps $\phi_{k_1,k_2}\colon \mathbb{Z}_+^{2\cdot 2} \longrightarrow \mathbb{Z}_+^{4\cdot 2}$ for $1 \leq k_1 < k_2 \leq 4$, and we obtain

$$\left(\begin{smallmatrix}\cdot & D\\ \cdot & A\end{smallmatrix}\right)^{(4)} = \begin{pmatrix} 1 & 0 & 1 & 0 & 1 & 0 & 1 & 0\\ 0 & 1 & 0 & 1 & 0 & 1 & 0 & 1\\ 1 & 1 & 0 & 0 & 0 & 0 & 0 & 0\\ 0 & 0 & 1 & 1 & 0 & 0 & 0 & 0\\ 0 & 0 & 0 & 0 & 1 & 1 & 0 & 0\\ 0 & 0 & 0 & 0 & 0 & 0 & 1 & 1 \end{pmatrix}$$

and

$$\mathcal{G}\left(\left(\begin{smallmatrix}\cdot & D\\ \cdot & A\end{smallmatrix}\right)^{(4)}\right) = \left\{\begin{array}{c} \pm(1,-1,-1,1,0,0,0,0)\\ \pm(1,-1,0,0,-1,1,0,0)\\ \pm(1,-1,0,0,0,0,-1,1)\\ \pm(0,0,1,-1,-1,1,0,0)\\ \pm(0,0,1,-1,0,0,-1,1)\\ \pm(0,0,0,0,1,-1,-1,1) \end{array}\right\}.$$

Note that Theorems 3.4.1 and 4.1.6 now imply that for fixed matrices A and D, N-fold IPs can be solved in polynomial time.

Theorem 4.1.8. *Fix any integer matrices A and D of appropriate sizes. Then there is a polynomial-time algorithm that, given any N, integer vectors* $\mathbf{b}$, $\mathbf{l}$, $\mathbf{u}$, $\mathbf{c}$, *a separable convex function f mapping $\mathbb{Z}^n$ to $\mathbb{Z}$ given by a comparison oracle, and a number M with $|f(\mathbf{z})| \leq M$ for all feasible solutions* $\mathbf{z}$, *solves the corresponding N-fold integer programming problem*

$$\min\left\{ f(\mathbf{x}) : \left(\begin{smallmatrix}\cdot & D\\ \cdot & A\end{smallmatrix}\right)^{(N)}\mathbf{x} = \mathbf{b},\ \mathbf{l} \leq \mathbf{x} \leq \mathbf{u},\ \mathbf{x} \in \mathbb{Z}^{Nn} \right\}. \tag{4.4}$$

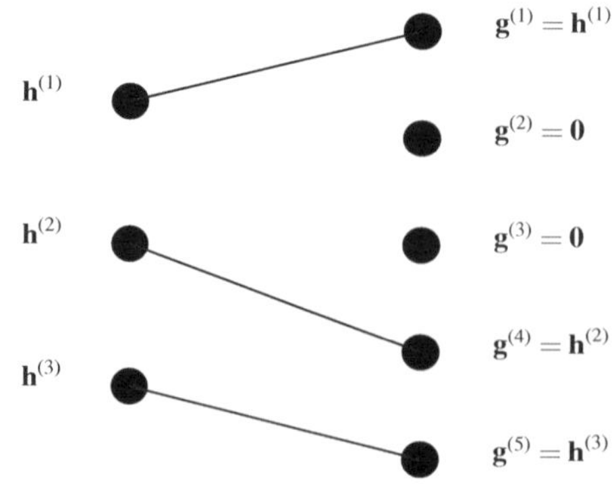

Figure 4.2. *Example for the map ϕ for the case of $g(A,D) = 3$ and $N = 5$.*

The running time of this algorithm is governed by the size of the Graver basis $\mathcal{G}\left(\left(\begin{smallmatrix} D \\ A \end{smallmatrix}\right)^{(N)}\right)$, that is, by $O\left(N^{g(A,D)}\right)$. In the following, we show how to reduce this running time to $O(N^3)$ when the objective function is separable convex and piecewise linear.

4.1.3 Graver-based dynamic programming for N-fold IPs

If the objective function is separable convex and piecewise affine linear, the N-fold IP can be solved much faster than in $O\left(N^{g(A,D)}\right)$ many steps. For this, let $f(\mathbf{z}) = \sum_{i=1}^{n} f^i(\mathbf{z}^{(i)}) = \sum_{i=1}^{n} \sum_{j=1}^{t} f_j^i(z_j^i)$ with each f_j^i univariate convex. Moreover, we also assume that for some fixed p, each f_j^i is p-piecewise affine linear, that is, the interval between the lower bound $\mathbf{l}_j^i$ and upper bound $\mathbf{u}_j^i$ is partitioned into at most p intervals with integer endpoints, and the restriction of f_j^i to each interval k is an affine-linear function $(\mathbf{w}_{j,k}^{(i)})^\top z_j^{(i)} + a_{j,k}^{(i)}$ with all $\mathbf{w}_{j,k}^{(i)}, a_{j,k}^{(i)}$ being integer. Note that the binary encoding length of f is the sum of the binary encoding lengths of all interval endpoints and $\mathbf{w}_{j,k}^{(i)}, a_{j,k}^{(i)}$ needed to describe it. The binary encoding length L of the input for the nonlinear N-fold IP is the binary encoding length of $f, \mathbf{b}, \mathbf{l}, \mathbf{u}$.

Theorem 4.1.9. *For any fixed p and for any pair of fixed integer matrices A and D (with the same number of columns), there is an algorithm that, given N, integer vectors $\mathbf{b}, \mathbf{l}, \mathbf{u}$, and separable convex p-piecewise affine linear f, solves in time $O(N^3 L)$ the N-fold integer programming problem*

$$\min\left\{ f(\mathbf{z}) : \left(\begin{smallmatrix} D \\ A \end{smallmatrix}\right)^{(N)} \mathbf{z} = \mathbf{b},\ \mathbf{l} \leq \mathbf{z} \leq \mathbf{u},\ \mathbf{z} \in \mathbb{Z}^{Nn} \right\}.$$

To prove this theorem, we have to understand the structure of $\mathcal{G}\left(\left(\begin{smallmatrix} D \\ A \end{smallmatrix}\right)^{(N)}\right)$. By the mapping ϕ from the proof of Theorem 4.1.6, every Graver basis vector $\mathbf{g} = \left(\mathbf{g}^{(1)}, \ldots, \mathbf{g}^{(N)}\right) \in \mathcal{G}\left(\left(\begin{smallmatrix} D \\ A \end{smallmatrix}\right)^{(N)}\right)$ arises from some Graver basis vector $\mathbf{h} = \left(\mathbf{h}^{(1)}, \ldots, \mathbf{h}^{(g(A,D))}\right) \in \mathcal{G}\left(\left(\begin{smallmatrix} D \\ A \end{smallmatrix}\right)^{(g(A,D))}\right)$ by suitably inserting $N - g(A,D)$ zero bricks into $\mathbf{h}$. (See Figure 4.2 for an example.)

In fact, from each fixed $\mathbf{h} \in \mathcal{G}\left(\left(\begin{smallmatrix} D \\ A \end{smallmatrix}\right)^{(\mathrm{type}(\mathbf{h}))}\right)$, we can obtain many such lifted vectors and we denote the set of all $\mathbf{g} \in \mathcal{G}\left(\left(\begin{smallmatrix} D \\ A \end{smallmatrix}\right)^{(N)}\right)$ constructible from $\mathbf{h}$ by $\mathcal{G}_{\mathbf{h}}\left(\left(\begin{smallmatrix} D \\ A \end{smallmatrix}\right)^{(N)}\right)$.

Recall that $\mathbf{g}$ gives an augmenting vector for a feasible solution $\mathbf{z}_0$, if $\mathbf{z}_0+\mathbf{g}$ is feasible with $f(\mathbf{z}_0+\mathbf{g}) < f(\mathbf{z}_0)$. However, if the objective function f is separable convex and piecewise linear, we do not have to check every possible $\mathbf{g}\in\mathcal{G}_{\mathbf{h}}\left(\left(\begin{smallmatrix}\cdot & D\\ \cdot & A\end{smallmatrix}\right)^{(N)}\right)$ one by one as to whether it is augmenting, but we can find a feasible step $\mathbf{g}\in\mathcal{G}_{\mathbf{h}}\left(\left(\begin{smallmatrix}\cdot & D\\ \cdot & A\end{smallmatrix}\right)^{(N)}\right)$ with smallest value of $f(\mathbf{z}_0+\mathbf{g})$ by solving a suitable shortest path problem.

Lemma 4.1.10. *Let $p\in\mathbb{Z}_+$. For any fixed $\mathbf{h}\in\mathcal{G}\left(\left(\begin{smallmatrix}\cdot & D\\ \cdot & A\end{smallmatrix}\right)^{(\text{type}(\mathbf{h}))}\right)$, there is an algorithm that, given N, integer vectors $\mathbf{z}_0,\mathbf{l},\mathbf{u}$ with $\mathbf{l}\leq\mathbf{z}_0\leq\mathbf{u}$, and separable convex p-piecewise affine-linear f, computes in time $O(N^2)$ among all combinations of $\gamma\in\mathbb{Z}_+$ and $\mathbf{g}\in\mathcal{G}_{\mathbf{h}}\left(\left(\begin{smallmatrix}\cdot & D\\ \cdot & A\end{smallmatrix}\right)^{(N)}\right)$ one that minimizes $f(\mathbf{z}_0+\gamma\mathbf{g})$.*

Proof. First let us define the following set $\Gamma\subseteq\mathbb{Z}_+$:

- $0\in\Gamma$.
- For $i=1,\dots,\text{type}(\mathbf{h})$ and $j=1,\dots,N$, add to Γ the largest value of $\gamma\in\mathbb{Z}_+$ such that $\mathbf{l}^{(j)}\leq\mathbf{z}_0^{(j)}+\gamma\mathbf{h}^{(i)}\leq\mathbf{u}^{(j)}$ holds.
- For $i=1,\dots,\text{type}(\mathbf{h})$ $j=1,\dots,N$, and $k=1,\dots,n$, add all nonnegative integers $\gamma,\gamma+1$ to Γ for which ${\mathbf{z}_0}_k^{(j)}+\gamma\mathbf{h}_k^{(i)}$ and ${\mathbf{z}_0}_k^{(j)}+(\gamma+1)\mathbf{h}_k^{(i)}$ belong to different affine-linear pieces of $f_k^{(j)}$.

Clearly, $|\Gamma|\in O(pN)=O(N)$, as p is assumed to be a constant. Write $\Gamma=\{0,\gamma_1,\dots,\gamma_r\}$ with $0=:\gamma_0<\gamma_1<\dots<\gamma_r$. By construction, for all $\mathbf{g}\in\mathcal{G}_{\mathbf{h}}\left(\left(\begin{smallmatrix}\cdot & D\\ \cdot & A\end{smallmatrix}\right)^{(N)}\right)$ the following holds true:

- For all $s=1,\dots,r$, and for all (real values of) $\gamma\in[\gamma_{s-1},\gamma_s]$, the vector $\mathbf{z}_0+\gamma\mathbf{g}$ is either always feasible (that is, $\mathbf{l}\leq\mathbf{z}_0+\gamma\mathbf{g}\leq\mathbf{u}$) or always infeasible.
- For all $s=1,\dots,r$, either we have $\gamma_{s-1}+1=\gamma_s$ or the function f is affine linear over the line segment $[\mathbf{z}_0+\gamma_{s-1}\mathbf{g},\mathbf{z}_0+\gamma_s\mathbf{g}]$ (or even both conditions hold).

Let

$$s(\gamma)=\min\left\{f(\mathbf{z}_0)+\gamma\mathbf{g}:\mathbf{l}\leq\mathbf{z}_0+\gamma\mathbf{g}\leq\mathbf{u},\ \mathbf{g}\in\mathcal{G}_{\mathbf{h}}\left(\left(\begin{smallmatrix}\cdot & D\\ \cdot & A\end{smallmatrix}\right)^{(N)}\right)\right\}.$$

We are interested in $\min\{s(\gamma):\gamma\in\mathbb{Z}_+\}$, which is in fact the same as $\min\{s(\gamma):\gamma\in[0,\gamma_r]\cap\mathbb{Z}\}$. Note that in each interval $[\gamma_{s-1},\gamma_s]$, $s=1,\dots,r$, the function $s(\gamma)$ is obtained by taking the minimum over affine-linear functions (one for each $\mathbf{g}\in\mathcal{G}_{\mathbf{h}}\left(\left(\begin{smallmatrix}\cdot & D\\ \cdot & A\end{smallmatrix}\right)^{(N)}\right)$) and thus, $s(\gamma)$ is a piecewise affine-linear concave function that takes on its minimum in one of its endpoints γ_{s-1} or γ_s. Consequently, we have that

$$\min\{s(\gamma):\gamma\in\mathbb{Z}_+\}=\min\{s(\gamma):\gamma\in\Gamma\}.$$

It remains to show that we can find $s(\gamma)$ for any fixed γ (in Γ), together with a direction $\mathbf{g}\in\mathcal{G}_{\mathbf{h}}\left(\left(\begin{smallmatrix}\cdot & D\\ \cdot & A\end{smallmatrix}\right)^{(N)}\right)$ in which this value $s(\gamma)$ is attained.

Let us now fix $\gamma\in\mathbb{Z}_+$ and construct the following directed graph $D_\gamma=(V,E)$: The node set V of D_γ consists of a special source node $v_{0,0}$ and N layers of $\text{type}(\mathbf{h})+1$ nodes each. The nodes in the k-th layer, $k=1,\dots,N$, are labeled by $v_{k,0},\dots,v_{k,\text{type}(\mathbf{h})}$. For all $k=0,\dots,N-1$, and for all $i=0,\dots,\text{type}(\mathbf{h})$, there are arcs $v_{k,i}\to v_{k+1,i}$ and $v_{k,i}\to v_{k+1,i+1}$ provided that both nodes of an arc exists. (So, there is no arc $v_{0,1}\to v_{1,1}$, as there is no node $v_{0,1}\in V$.) By construction, there is now a one-to-one correspondence between the directed paths from $v_{0,0}$ to $v_{N,\text{type}(\mathbf{h})}$ and all $\mathbf{g}\in\mathcal{G}_{\mathbf{h}}\left(\left(\begin{smallmatrix}\cdot & D\\ \cdot & A\end{smallmatrix}\right)^{(N)}\right)$. (See Figure 4.3

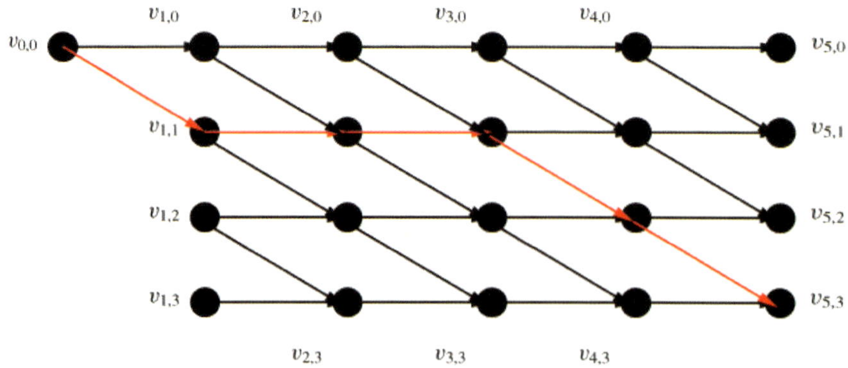

Figure 4.3. *The graph D_γ together with the path from $v_{0,0}$ to $v_{5,3}$ encoding the assignment from Figure 4.2.*

for an example.) If the path contains the arc $v_{k,i} \to v_{k+1,i}$, then $\mathbf{g}^{(k+1)} = \mathbf{0}$, and if the path walks along an arc $v_{k,i} \to v_{k+1,i+1}$, then $\mathbf{g}^{(k+1)} = \mathbf{h}^{(i+1)}$. Clearly, given $\mathbf{g} \in \mathcal{G}\left(\left(\begin{smallmatrix} \cdot & D \\ \cdot & A \end{smallmatrix}\right)^{(N)}\right)$, we can reconstruct such a path uniquely.

For given feasible solution $\mathbf{z}_0$, bounds $\mathbf{l}, \mathbf{u}$, and step length $\gamma \in \mathbb{Z}_+$, we now change this graph as follows: By construction and the interpretation of the arcs, it makes sense

- to remove any arc $v_{k,i} \to v_{k+1,i+1}$ if $\mathbf{l}^{(k+1)} \le \mathbf{z}^{(k+1)} + \gamma \mathbf{h}^{(i+1)} \le \mathbf{u}^{(k+1)}$ is not satisfied,
- to assign weights 0 to arcs $v_{k,i} \to v_{k+1,i}$ (as they encode $\mathbf{g}^{(k+1)} = \mathbf{0}$) and weights $f^{(k+1)}(\mathbf{z}_0^{(k+1)} + \gamma \mathbf{h}^{(i+1)}) - f^{(k+1)}(\mathbf{z}_0^{(k+1)})$ to arcs $v_{k,i} \to v_{k+1,i+1}$ (as they encode $\mathbf{g}^{(k+1)} = \mathbf{h}^{(i+1)}$).

It is now evident that a shortest path from $v_{0,0}$ to $v_{N,\text{type}(\mathbf{h})}$ in this graph encodes a vector $\gamma \mathbf{g}$ with $\mathbf{g} \in \mathcal{G}_{\mathbf{h}}\left(\left(\begin{smallmatrix} \cdot & D \\ \cdot & A \end{smallmatrix}\right)^{(N)}\right)$ such that $\mathbf{z}_0 + \gamma \mathbf{g}$ is feasible and $f(\mathbf{z}_0 + \alpha \mathbf{g}) - f(\mathbf{z}_0)$ is smallest among all such $\mathbf{g}$ (and for fixed γ). Applying Moore–Bellman–Ford's algorithm to find such a shortest path from $v_{0,0}$ to $v_{N,\text{type}(\mathbf{h})}$, we see that it can be computed in $O(N)$ steps [297]. Thus, we can find

$$\min\{s(\gamma) : \gamma \in \mathbb{Z}_+\} = \min\{s(\gamma) : \gamma \in \Gamma\}$$

(and a solution $\gamma \mathbf{g}$ attaining it) in $O(NL \cdot N) = O(N^2 L)$ many steps. This completes the proof. □

4.2 Two-stage stochastic integer programming

The two-stage integer linear stochastic program is the optimization problem

$$\min\left\{\mathbf{c}^\top \mathbf{x} + Q(\mathbf{x}) : M\mathbf{x} = \mathbf{a},\ \mathbf{x} \in \mathbb{Z}_+^m\right\}, \tag{4.5a}$$

where

$$Q(\mathbf{x}) := \int_{\mathbb{R}^s} \Phi(\boldsymbol{\xi} - T\mathbf{x})\,\mu(\mathrm{d}\boldsymbol{\xi}) \tag{4.5b}$$

and

$$\Phi(\mathbf{h}) := \min\left\{\mathbf{d}^\top \mathbf{y} : W\mathbf{y} = \mathbf{h},\ \mathbf{y} \in \mathbb{Z}_+^n\right\}. \tag{4.5c}$$

Herein, the measure μ is a Borel probability measure on $\mathbb{R}^s$. The model (4.5) arises in decision making under uncertainty. Given an optimization problem with random data where parts of the decisions (the first-stage variables $\mathbf{x}$) have to be taken before and parts (the second-stage variables $\mathbf{y}$) are taken after the realizations of the random data are known, the purpose of (4.5) is to minimize the sum of the direct costs $\mathbf{c}^\top\mathbf{x}$ and the expected costs of optimal decisions in the second stage. The model has a multistage extension where a multistage process of alternating decision and observation replaces the two-stage process assumed above. For further details on the modeling background we refer the reader to [53, 186, 279].

Under mild assumptions, all ingredients in the above model are well defined [301]: We assume that $W(\mathbb{R}^n) = \mathbb{R}^s$ and $\left\{\mathbf{u} \in \mathbb{R}^s : W^\top\mathbf{u} \leq \mathbf{d}\right\} \neq \emptyset$. Standard existence results for mixed-integer linear programs then imply that the problem $\min\left\{\mathbf{d}^\top\mathbf{y} : W\mathbf{y} = \mathbf{h},\ \mathbf{y} \in \mathbb{Z}^n_+\right\}$ is solvable for any $\mathbf{h} \in \mathbb{R}^s$ [259]. If, moreover, μ has a finite first moment, i.e., $\int_{\mathbb{R}^s} \|\boldsymbol{\xi}\|\, \mu(d\boldsymbol{\xi}) < \infty$, then the integral in (4.5b) is finite and $Q(\mathbf{x}) \in \mathbb{R}$ for all $\mathbf{x} \in \mathbb{R}^m$ [301].

Algorithmically, (4.5) provides some challenges since the integral defining $Q(\mathbf{x})$ is multidimensional and its integrand is given only implicitly. Therefore, computations have almost exclusively worked with discrete measures μ, tacitly assuming that, if necessary, continuous measures can be approximated by discrete ones. If we assume that μ is a discrete probability measure with finitely many realizations (or scenarios) $\boldsymbol{\xi}^{(1)}, \ldots, \boldsymbol{\xi}^{(N)}$ and probabilities $\pi^{(1)}, \ldots, \pi^{(N)}$, then (4.5) is equivalent to the integer linear program

$$\min\left\{\mathbf{c}^\top\mathbf{x} + \sum_{i=1}^{N} \pi^{(i)}\mathbf{d}^\top\mathbf{y}^{(i)} : M\mathbf{x} = \mathbf{a},\ \mathbf{x} \in \mathbb{Z}^m_+,\ T\mathbf{x} + W\mathbf{y}^{(i)} = \boldsymbol{\xi}^{(i)},\ \mathbf{y}^{(i)} \in \mathbb{Z}^n_+\ \forall i\right\}.$$

As the number N of scenarios is large in general, this problem is large scale and not amenable to general purpose integer linear programming solvers. This has motivated research into decomposition algorithms. The latter have a long tradition in stochastic linear programming without integer requirements [53, 167, 186, 279, 286].

In the following, we present a decomposition approach that does not decompose the problem itself, but rather exploits the structure of the problem matrix to decompose its Graver basis and to use those parts to efficiently reconstruct an augmenting vector if it exists.

4.2.1 Graver bases for two-stage stochastic IPs

Before we continue, note that we can think of the constraint $M\mathbf{x} = \mathbf{a}$ as being written as $M\mathbf{x} + O\mathbf{y}^{(i)} = \mathbf{a}$, $i = 1, \ldots, N$. Thus, instead of using T and W, we can also use $B := \binom{M}{T}$ and $A := \binom{O}{W}$ as blocks of the problem matrix. Although this inclusion of copies of the constraint $M\mathbf{x} = \mathbf{a}$ enlarges the problem matrix, the problem itself remains the same and the presentation is simplified. Writing down the problem matrix for two-stage stochastic programs with N scenarios, we obtain

$$\begin{pmatrix} B & A & & \\ B & & A & \\ \vdots & & & \ddots \\ B & & & & A \end{pmatrix},$$

which is simply the special case $\left(\begin{smallmatrix} \cdot & \cdot \\ B & A \end{smallmatrix}\right)^{(N)}$ of an N-fold 4-block decomposable matrix.

As in the previous section, we will assume that the matrices A and B are fixed. Again, as for N-fold IPs, we can split any vector $\mathbf{z} = \left(\mathbf{x}, \mathbf{y}^{(1)}, \ldots, \mathbf{y}^{(N)}\right)^\top$ into a brick $\mathbf{x}$ corresponding to the variables of the first-stage decision, and into N bricks $\mathbf{y}^{(1)}, \ldots, \mathbf{y}^{(N)}$. However, in contrast to the N-fold IP-situation, the sizes of the Graver bases of $\left(\begin{smallmatrix} \cdot & \cdot \\ B & A \end{smallmatrix}\right)^{(N)}$ are not bounded polynomially in N, as the following simple example shows.

Example 4.2.1. For $B = (1)$ and $A = (1\ 1)$, the 2^N vectors $\left(\mathbf{x}, \mathbf{y}^{(1)}, \ldots, \mathbf{y}^{(N)}\right)^\top$ with $\mathbf{x} = (1)$ and with $\mathbf{y}^{(i)} \in \left\{(-1,0)^\top, (0,-1)^\top\right\}$, $i = 1, \ldots, N$, all belong to $\mathcal{G}\left(\left(\begin{smallmatrix} \cdot & \cdot \\ B & A \end{smallmatrix}\right)^{(N)}\right)$.

Although the sizes of the Graver bases increase exponentially with N, the structure of the matrix $\left(\begin{smallmatrix} \cdot & \cdot \\ B & A \end{smallmatrix}\right)^{(N)}$ leads to a structure on the elements in $\mathcal{G}\left(\left(\begin{smallmatrix} \cdot & \cdot \\ B & A \end{smallmatrix}\right)^{(N)}\right)$, which we can exploit to construct in $O(N^2)$ many steps an augmentation vector for any given nonoptimal feasible solution $\mathbf{z}_0$ that is at least as good as the best Graver basis augmentation step $\alpha\mathbf{g}$.

Let us first observe that $\left(\mathbf{x}, \mathbf{y}^{(1)}, \ldots, \mathbf{y}^{(N)}\right)^\top \in \ker\left(\left(\begin{smallmatrix} \cdot & \cdot \\ B & A \end{smallmatrix}\right)^{(N)}\right)$ if and only if $\left(\mathbf{x}, \mathbf{y}^{(i)}\right)^\top \in \ker\left(\left(\begin{smallmatrix} \cdot & \cdot \\ B & A \end{smallmatrix}\right)^{(1)}\right)$ for $i = 1, \ldots, N$. This implies that by permuting the second-stage bricks we do not leave $\ker\left(\left(\begin{smallmatrix} \cdot & \cdot \\ B & A \end{smallmatrix}\right)^{(N)}\right)$. Also as $\sqsubseteq$-minimality is preserved under such a permutation, we can construct from one Graver basis element many others by simply permuting bricks. In other words, we can efficiently store exponentially many Graver basis elements by simply remembering the bricks they are composed of. This motivates the following decomposition approach.

Let $U \subseteq \mathbb{Z}^{n_B}$ be the set of all $\mathbf{u} \in \mathbb{Z}^{n_B}$ such that there exist some $N \in \mathbb{Z}_+$ and some Graver basis element $\left(\mathbf{u}, \mathbf{v}^{(1)}, \ldots, \mathbf{v}^{(N)}\right)^\top \in \mathcal{G}\left(\left(\begin{smallmatrix} \cdot & \cdot \\ B & A \end{smallmatrix}\right)^{(N)}\right)$. Moreover, for each $\mathbf{u} \in U$, let $V_\mathbf{u}$ be the set of all $\mathbf{v} \in \mathbb{Z}^{n_A}$ that appear as a second-stage brick of some Graver basis element $\left(\mathbf{u}, \mathbf{v}^{(1)}, \ldots, \mathbf{v}^{(N)}\right)^\top \in \mathcal{G}\left(\left(\begin{smallmatrix} \cdot & \cdot \\ B & A \end{smallmatrix}\right)^{(N)}\right)$ for some $N \in \mathbb{Z}_+$. In the following, we will show that for any fixed integer matrices $A \in \mathbb{Z}^{d \times n_A}$ and $B \in \mathbb{Z}^{d \times n_B}$, the sets U and $V_\mathbf{u}$, $\mathbf{u} \in U$, are all finite and we will see how we can use these sets to efficiently reconstruct augmentation vectors. We start by exhibiting some structure on $V_\mathbf{u}$:

Lemma 4.2.2. *Given fixed integer matrices $A \in \mathbb{Z}^{d \times n_A}$ and $B \in \mathbb{Z}^{d \times n_B}$, we have the following:*

(a) $V_\mathbf{0} = \mathcal{G}(A)$.

(b) *For $\mathbf{u} \in U \setminus \{\mathbf{0}\}$, $V_\mathbf{u}$ coincides with the set of $\sqsubseteq$-minimal solutions to $A\mathbf{v} = -B\mathbf{u}$.*

$V_\mathbf{0} = \mathcal{G}(A)$ is readily seen, since any Graver basis vector $\left(\mathbf{u}, \mathbf{v}^{(1)}, \ldots, \mathbf{v}^{(N)}\right)^\top \in \mathcal{G}\left(\left(\begin{smallmatrix} \cdot & \cdot \\ B & A \end{smallmatrix}\right)^{(N)}\right)$ with $\mathbf{u} = \mathbf{0}$ decomposes into the parts for each second-stage brick. Moreover, for $\mathbf{u} \in U \setminus \{\mathbf{0}\}$, each brick in $V_\mathbf{u}$ must be a $\sqsubseteq$-minimal solution to $A\mathbf{v} = -B\mathbf{u}$, which can be seen as follows. Suppose on the contrary that there exists $\left(\mathbf{u}, \mathbf{v}^{(1)}, \ldots, \mathbf{v}^{(N)}\right)^\top \in \mathcal{G}\left(\left(\begin{smallmatrix} \cdot & \cdot \\ B & A \end{smallmatrix}\right)^{(N)}\right)$ for some $N \in \mathbb{Z}_+$, whose second-stage bricks $\mathbf{v}^{(1)}, \ldots, \mathbf{v}^{(N)}$ are not all $\sqsubseteq$-minimal solutions to $A\mathbf{v} = -B\mathbf{u}$. Then we can find vectors $\bar{\mathbf{v}}^{(i)} \sqsubseteq \mathbf{v}^{(i)}$, being $\sqsubseteq$-minimal solutions to $A\mathbf{v} = -B\mathbf{u}$, for $i = 1, \ldots, N$. Then, clearly, $\mathbf{0} \neq \left(\mathbf{u}, \bar{\mathbf{v}}^{(1)}, \ldots, \bar{\mathbf{v}}^{(N)}\right)^\top \sqsubseteq \left(\mathbf{u}, \mathbf{v}^{(1)}, \ldots, \mathbf{v}^{(N)}\right)^\top$. As the last relation is strict, this contradicts $\left(\mathbf{u}, \mathbf{v}^{(1)}, \ldots, \mathbf{v}^{(N)}\right)^\top \in \mathcal{G}\left(\left(\begin{smallmatrix} \cdot & \cdot \\ B & A \end{smallmatrix}\right)^{(N)}\right)$. The simple proof that $V_\mathbf{u}$ actually *coincides* with the set of $\sqsubseteq$-minimal solutions to $A\mathbf{v} = -B\mathbf{u}$ is left to the reader as a little exercise. Note that the $\sqsubseteq$-version of the Gordan–Dickson lemma, Lemma 2.5.6, now implies that each $V_\mathbf{u}$ is finite.

In order to show that also U is finite, let us associate with each $\mathbf{u} \in U$ the following monomial ideal:

$$J_{\mathbf{u}} := \left\langle \mathbf{x}^{(\mathbf{u}^+,\mathbf{u}^-,\mathbf{v}^+,\mathbf{v}^-)} : \mathbf{v} \in V_{\mathbf{u}} \right\rangle \subseteq \mathbb{K}[x_1,\ldots,x_{2n_A+2n_B}].$$

The crucial observation now is the following.

Lemma 4.2.3. *For all pairs* $\mathbf{u},\bar{\mathbf{u}} \in U \setminus \{\mathbf{0}\}$, $J_{\bar{\mathbf{u}}} \subseteq J_{\mathbf{u}}$ *implies* $\bar{\mathbf{u}} = \mathbf{u}$. *That is, any two such ideals are incomparable with respect to inclusion.*

Proof. Assume to the contrary that there exist bricks $\mathbf{u},\bar{\mathbf{u}} \in U \setminus \{\mathbf{0}\}$ with $J_{\bar{\mathbf{u}}} \subseteq J_{\mathbf{u}}$ and $\bar{\mathbf{u}} \neq \mathbf{u}$. Now choose an arbitrary $N \in \mathbb{Z}_+$ and look at an arbitrary Graver basis element in $\mathcal{G}(\left(\begin{smallmatrix} \cdot & \cdot \\ B & A\end{smallmatrix}\right)^{(N)})$ of the form $\left(\bar{\mathbf{u}},\bar{\mathbf{v}}^{(1)},\ldots,\bar{\mathbf{v}}^{(N)}\right)^\top$. We construct a vector $\left(\mathbf{u},\mathbf{v}^{(1)},\ldots,\mathbf{v}^{(N)}\right)^\top \sqsubseteq \left(\bar{\mathbf{u}},\bar{\mathbf{v}}^{(1)},\ldots,\bar{\mathbf{v}}^{(N)}\right)^\top$ that contradicts the $\sqsubseteq$-minimality of $\left(\bar{\mathbf{u}},\bar{\mathbf{v}}^{(1)},\ldots,\bar{\mathbf{v}}^{(N)}\right)^\top$.

For $i = 1,\ldots,N$, the monomial $\mathbf{x}^{(\bar{\mathbf{u}}^+,\bar{\mathbf{u}}^-,(\bar{\mathbf{v}}^{(i)})^+,(\bar{\mathbf{v}}^{(i)})^-)}$ belongs to $J_{\bar{\mathbf{u}}}$ and thus also to $J_{\mathbf{u}}$. Consequently, there is some $\mathbf{x}^{(\mathbf{u}^+,\mathbf{u}^-,(\mathbf{v}^{(i)})^+,(\mathbf{v}^{(i)})^-)} \in J_{\mathbf{u}}$ with $\mathbf{v}^{(i)} \in V_{\mathbf{u}}$ and with

$$\mathbf{x}^{(\mathbf{u}^+,\mathbf{u}^-,(\mathbf{v}^{(i)})^+,(\mathbf{v}^{(i)})^-)} \,\Big|\, \mathbf{x}^{(\bar{\mathbf{u}}^+,\bar{\mathbf{u}}^-,(\bar{\mathbf{v}}^{(i)})^+,(\bar{\mathbf{v}}^{(i)})^-)},$$

that is, with

$$\left(\mathbf{u}^+,\mathbf{u}^-,\left(\mathbf{v}^{(i)}\right)^+,\left(\mathbf{v}^{(i)}\right)^-\right) \leq \left(\bar{\mathbf{u}}^+,\bar{\mathbf{u}}^-,\left(\bar{\mathbf{v}}^{(i)}\right)^+,\left(\bar{\mathbf{v}}^{(i)}\right)^-\right),$$

or, in other words, with $\left(\mathbf{u},\mathbf{v}^{(i)}\right) \sqsubseteq \left(\bar{\mathbf{u}},\bar{\mathbf{v}}^{(i)}\right)$. In this way we have constructed all the bricks of a desired vector $\left(\mathbf{u},\mathbf{v}^{(1)},\ldots,\mathbf{v}^{(N)}\right)^\top \sqsubseteq \left(\bar{\mathbf{u}},\bar{\mathbf{v}}^{(1)},\ldots,\bar{\mathbf{v}}^{(N)}\right)^\top$ that contradicts $\sqsubseteq$-minimality of $\left(\bar{\mathbf{u}},\bar{\mathbf{v}}^{(1)},\ldots,\bar{\mathbf{v}}^{(N)}\right)^\top$. □

Finiteness of U now follows immediately from the following result.

Theorem 4.2.4 (Maclagan [238]). *Every collection $\mathcal{I}$ of monomial ideals in a polynomial ring $\mathbb{K}[x_1,\ldots,x_s]$ contains at most finitely many inclusion-maximal elements.*

Applying the result now to $\mathcal{I} = \{\, J_{\mathbf{u}} : \mathbf{u} \in U \,\}$, we obtain our main finiteness result, originally proved in [157].

Theorem 4.2.5 (Graver basis for stochastic IPs). *Let $A \in \mathbb{Z}^{d\times n_A}$ and $B \in \mathbb{Z}^{d\times n_B}$, and let $\mathcal{G} = \mathcal{G}\left(\left(\begin{smallmatrix} \cdot & \cdot \\ B & A\end{smallmatrix}\right)^{(N)}\right)$. There exist numbers $g,\xi,\eta \in \mathbb{Z}_+$ depending only on A and B but not on N such that the following holds:*

(a) *For every $N \in \mathbb{Z}_+$ and for every $\mathbf{v} \in \mathcal{G}$, we have $\|\mathbf{v}\|_\infty \leq g$; i.e., the components of $\mathbf{v}$ are bounded by g in absolute value.*

(b) *As a corollary, $\|\mathbf{v}\|_1 \leq (n_B + Nn_A)g$ for all $\mathbf{v} \in \mathcal{G}$.*

(c) *More precisely, there exists a finite set $U \subseteq \mathbb{Z}^{n_B}$ of cardinality $|U| \leq \xi$, and for each $\mathbf{u} \in U$ there exists a finite set $V_{\mathbf{u}} \subseteq \mathbb{Z}^{n_A}$ of cardinality $|V_{\mathbf{u}}| \leq \eta$ such that the elements $\mathbf{g} \in \mathcal{G}$ take the form $\mathbf{g} = (\mathbf{u},\mathbf{v}^{(1)},\ldots,\mathbf{v}^{(N)})$, with $\mathbf{u} \in U$ and $\mathbf{v}^{(1)},\ldots,\mathbf{v}^{(N)} \in V_{\mathbf{u}}$.*

Remark 4.2.6. The finiteness of the numbers g, ξ, η comes from a saturation result in commutative algebra; see Theorem 4.2.4. Concrete bounds on these numbers are unfortunately not available. However, for given matrices A and B, the finite sets U and $V_{\mathbf{u}}$ for $\mathbf{u} \in U$ can be computed using the completion algorithm in [157, section 3.3]. Thus, the numbers g, ξ, η are effectively computable.

4.2.2 Reconstruction of an augmenting vector

Let us abbreviate $\mathcal{H}(B,A) := \{(\mathbf{u}, V_{\mathbf{u}}) : \mathbf{u} \in U\}$, and let us show how we can use this finite set of bricks in order to efficiently solve the two-stage stochastic integer linear programming problem as the number N of scenarios grows. In fact, we will prove the following result.

Theorem 4.2.7. *Let $A \in \mathbb{Z}^{d \times n_A}$ and $B \in \mathbb{Z}^{d \times n_B}$ be fixed integer matrices. Then the two-stage integer linear programming problem*

$$\min\left\{ \mathbf{c}^\top \mathbf{z} : \left(\begin{smallmatrix} \cdot & \cdot \\ B & A \end{smallmatrix}\right)^{(N)} \mathbf{z} = \mathbf{b}, \mathbf{l} \leq \mathbf{z} \leq \mathbf{u}, \mathbf{z} \in \mathbb{Z}^{n_B + N n_A} \right\}$$

can be solved in time $O(N^3 L)$, where L denotes the binary encoding length of the input data, that is, of the objective function, the right-hand side vector, and the lower and bound vectors for the variables.

Proof. Let us assume that we are given a problem instance together with a possibly nonoptimal feasible solution $\mathbf{z}_0$. From Chapter 3 we know that only $O((n_B + N n_A)L) = O(NL)$ many Graver-best augmentation steps are needed to solve the two-stage stochastic IP, and thus we only need to show how to find an augmentation step for a given feasible solution $\mathbf{z}_0$ whose improvement is at least as good as that of a Graver-best augmentation step.

Let us start by constructing an augmentation step $\gamma \mathbf{g}$ for fixed step length $\gamma \in \mathbb{Z}_+$. It is surprisingly simple! The crucial point is that there are only constantly many first-stage bricks $\mathbf{u}$ in $\mathcal{H}(B,A)$ that have to be considered for the desired vector $\mathbf{g} := \left(\bar{\mathbf{u}}, \mathbf{v}^{(1)}, \ldots, \mathbf{v}^{(N)}\right)^\top$. But once we fix such a first-stage brick $\mathbf{u}$ for $\mathbf{g}$, the search for the N second-stage bricks $\left\{\mathbf{v}^{(1)}, \ldots, \mathbf{v}^{(N)}\right\}$ of $\mathbf{g}$ decomposes into N *independent* problems, each of which can be solved in constant time: We simply find the best brick $\mathbf{v}^{(i)}$ from the constant-size set $V_{\mathbf{u}}$ for each of the N positions so that $\mathbf{z}_0 + \gamma \mathbf{g}$ is feasible (that is, satisfies the bounds) and such that it has a smallest objective value among all feasible choices. If for any $i = 1, \ldots, N$ the construction of $\mathbf{v}^{(i)}$ is not possible (as any of the bricks from $V_{\mathbf{u}}$ would lead to a violation of the bounds), there is no augmenting vector $\gamma \mathbf{g}$ for $\mathbf{z}_0$ having first-stage brick $\mathbf{u}$. If such a vector $\mathbf{g}$ can be constructed, but satisfies $\mathbf{c}^\top \mathbf{g} \geq 0$, there is again no augmenting vector $\gamma \mathbf{g}$ for $\mathbf{z}_0$ having first-stage brick $\mathbf{u}$, as even the best choices for the $\mathbf{v}^{(i)}$ lead to a vector that does not improve the objective function. Finally, if such a vector $\mathbf{g}$ can be constructed and satisfies $\mathbf{c}^\top \mathbf{g} < 0$, then $\gamma \mathbf{g}$ is an augmenting step for $\mathbf{z}_0$. As there are only constantly many possible $\mathbf{u}$ in $\mathcal{H}(B,A)$, we can construct an augmentation step $\gamma \mathbf{g}$ (for given step length $\gamma \in \mathbb{Z}_+$) in time $O(N)$, or we obtain a certificate that no such vector exists.

But how many step lengths $\gamma \in \mathbb{Z}_+$ do we have to consider for fixed $\mathbf{u}$? For this, observe that if γ and $\gamma + 1$ lead to the same vector $\mathbf{g}$, then we need not consider γ. But how could we know that *before* constructing $\mathbf{g}$ for both cases? Again this is simple: If for every $i = 1, \ldots, N$, the set of applicable bricks $\mathbf{v} \in V_{\mathbf{u}}$ is the same for γ and $\gamma + 1$, that is, if $\mathbf{l}^{(i)} \leq \mathbf{z}_0{}^{(i)} + \gamma \mathbf{v} \leq \mathbf{u}^{(i)}$ implies $\mathbf{l}^{(i)} \leq \mathbf{z}_0{}^{(i)} + (\gamma + 1)\mathbf{v} \leq \mathbf{u}^{(i)}$, then the constructed brick $\mathbf{v}^{(i)}$ of $\mathbf{g}$ is the same for γ and for $\gamma + 1$. Hence, we only need to consider $\gamma \in \mathbb{Z}_+$, such that there is some $i \in \{1,\dots,N\}$, and $\mathbf{v} \in V_{\mathbf{u}}$, such that $\mathbf{v}$ is applicable at second-stage position i for step length γ but not for $\gamma+1$. As $V_{\mathbf{u}}$ has constant size, this situation can appear only $O(N)$ times.

We conclude that an optimal solution can be found in $O(NL \cdot N \cdot N) = O(N^3 L)$ many steps. □

4.3 N-fold 4-block decomposable integer programs

In this section, we generalize the polynomial-time algorithms for N-fold IPs and for two-stage stochastic IPs to N-fold 4-block decomposable IPs for separable convex functions f. We assume that the following approximate continuous convex optimization oracle is available:

Problem 4.3.1 (approximate continuous convex optimization). Given the problem data A, B, C, D, N, $\mathbf{l}, \mathbf{u}$, $\mathbf{b}$ and a number $\epsilon \in \mathbb{Q}_{>0}$, find a feasible solution $\mathbf{r}_\epsilon \in \mathbb{Q}^{n_B+Nn_A}$ for the continuous relaxation

$$(\mathrm{CP})_{N,\mathbf{b},\mathbf{l},\mathbf{u},f}: \qquad \min\left\{ f(\mathbf{r}) : \begin{pmatrix} C & D \\ B & A \end{pmatrix}^{(N)} \mathbf{r} = \mathbf{b},\ \mathbf{l} \le \mathbf{r} \le \mathbf{u},\ \mathbf{r} \in \mathbb{R}^{n_B+Nn_A} \right\},$$

such that there exists an optimal solution $\hat{\mathbf{r}}$ to $(\mathrm{CP})_{N,\mathbf{b},\mathbf{l},\mathbf{u},f}$ with

$$\|\hat{\mathbf{r}} - \mathbf{r}_\epsilon\|_\infty \le \epsilon,$$

or report INFEASIBLE or UNBOUNDED.

Theorem 4.3.2. *Let $A \in \mathbb{Z}^{d_A \times n_A}$, $B \in \mathbb{Z}^{d_A \times n_B}$, $C \in \mathbb{Z}^{d_C \times n_B}$, $D \in \mathbb{Z}^{d_C \times n_A}$ be fixed. For given $N \in \mathbb{Z}_+$, let $\mathbf{l} \in (\mathbb{Z} \cup \{-\infty\})^{n_B+Nn_A}$, $\mathbf{u} \in (\mathbb{Z} \cup \{+\infty\})^{n_B+Nn_A}$, and $\mathbf{b} \in \mathbb{Z}^{d_C+Nd_A}$, let $f: \mathbb{R}^{n_B+Nn_A} \to \mathbb{R}$ be a separable convex function that takes integer values on $\mathbb{Z}^{n_B+Nn_A}$, and denote by $\hat{f}$ an upper bound on the maximum of $|f|$ over the feasible region of the N-fold 4-block decomposable convex integer minimization problem*

$$(\mathrm{IP})_{N,\mathbf{b},\mathbf{l},\mathbf{u},f}: \qquad \min\left\{ f(\mathbf{z}) : \begin{pmatrix} C & D \\ B & A \end{pmatrix}^{(N)} \mathbf{z} = \mathbf{b},\ \mathbf{l} \le \mathbf{z} \le \mathbf{u},\ \mathbf{z} \in \mathbb{Z}^{n_B+Nn_A} \right\}.$$

We assume that the objective function f is given by an evaluation oracle and an approximate continuous convex optimization oracle for $(\mathrm{CP})_{N,\mathbf{b},\mathbf{l},\mathbf{u},f}$.

Then there exists an algorithm that finds an optimal solution to $(\mathrm{IP})_{N,\mathbf{b},\mathbf{l},\mathbf{u},f}$ or decides that $(\mathrm{IP})_{N,\mathbf{b},\mathbf{l},\mathbf{u},f}$ is infeasible or unbounded and that runs in time polynomial in N and in the binary encoding length of $\mathbf{l}$, $\mathbf{u}$, $\mathbf{b}$, and $\hat{f}$.

To prove this result, we combine Graver basis techniques with a proximity result developed by Hochbaum and Shanthikumar [170] in the context of their so-called proximity-scaling technique. This allows us to first use the approximate continuous convex optimization oracle to find a point in whose proximity the optimal integer solution has to lie. The integer problem restricted to this neighborhood is then efficiently solvable with primal (augmentation) algorithms using Graver bases, which will find the optimal integer solution in a polynomial number of steps.

Let us start by bounding the ℓ_1-norm of Graver basis elements of $\begin{pmatrix} C & D \\ B & A \end{pmatrix}^{(N)}$.

4.3.1 Bounds for Graver basis elements of N-fold 4-block decomposable matrices

In this section we prove the following structural result.

Theorem 4.3.3. *If* $A \in \mathbb{Z}^{d_A \times n_A}$, $B \in \mathbb{Z}^{d_A \times n_B}$, $C \in \mathbb{Z}^{d_C \times n_B}$, $D \in \mathbb{Z}^{d_C \times n_A}$ *are fixed matrices, then* $\max\left\{ \|\mathbf{v}\|_1 : \mathbf{v} \in \mathcal{G}\left(\left(\begin{smallmatrix} C & D \\ B & A \end{smallmatrix}\right)^{(N)}\right)\right\}$ *is bounded by a polynomial in* N.

We note that in the special case of N-fold IPs, the ℓ_1-norm is bounded by a constant (depending only on the fixed problem matrices and not on N), and in the special case of two-stage stochastic IPs, the ℓ_1-norm is bounded linearly in N. This fact demonstrates that N-fold 4-block IPs are much richer and more difficult to solve than the two special cases.

To get a bound for the ℓ_1-norms of the Graver basis elements of $\left(\begin{smallmatrix} C & D \\ B & A \end{smallmatrix}\right)^{(N)}$, we combine the bounds from Theorem 4.2.5 with Corollary 3.7.8.

Proposition 4.3.4 (Graver basis length bound for 4-block IPs). *Let* $A \in \mathbb{Z}^{d_A \times n_A}$, $B \in \mathbb{Z}^{d_A \times n_B}$, $C \in \mathbb{Z}^{d_C \times n_B}$, $D \in \mathbb{Z}^{d_C \times n_A}$ *be given matrices. Moreover, let* M *be a bound on the absolute values of the entries in* C *and* D, *and let* $g \in \mathbb{Z}_+$ *be the number from Theorem* 4.2.5. *Then for any* $N \in \mathbb{Z}_+$ *we have*

$$\begin{aligned} &\max\left\{\|\mathbf{v}\|_1 : \mathbf{v} \in \mathcal{G}\left(\left(\begin{smallmatrix} C & D \\ B & A \end{smallmatrix}\right)^{(N)}\right)\right\} \\ &\leq (2(n_B + Nn_A)M)^{2^{d_C}-1}\left(\max\left\{\|\mathbf{v}\|_1 : \mathbf{v} \in \mathcal{G}\left(\left(\begin{smallmatrix} \cdot & \cdot \\ B & A \end{smallmatrix}\right)^{(N)}\right)\right\}\right)^{2^{d_C}} \\ &\leq (2(n_B + Nn_A)M)^{2^{d_C}-1}((n_B + Nn_A)g)^{2^{d_C}}. \end{aligned}$$

If A, B, C, D *are fixed matrices, then* $\max\left\{\|\mathbf{v}\|_1 : \mathbf{v} \in \mathcal{G}\left(\left(\begin{smallmatrix} C & D \\ B & A \end{smallmatrix}\right)^{(N)}\right)\right\}$ *is bounded by* $O(N^{2^{d_C}+1})$, *a polynomial in* N.

Proof. The first claim is a direct consequence of Theorem 4.2.5 and Corollary 3.7.8 with $L = \left(\begin{smallmatrix} \cdot & \cdot \\ B & A \end{smallmatrix}\right)^{(N)}$, $F = \left(\begin{smallmatrix} C & D \\ \cdot & \cdot \end{smallmatrix}\right)^{(N)}$, and $E = \left(\begin{smallmatrix} C & D \\ B & A \end{smallmatrix}\right)^{(N)}$. The polynomial bound for fixed matrices A, B, C, D, and varying N follows by observing that n_A, n_B, d_C, M, g are constants, as they depend only on the fixed matrices A, B, C, D. □

We now complement this bound with another useful bound, which is given by the following result [161]:

Proposition 4.3.5 (alternative length bound for 4-block IPs). *Let* $A \in \mathbb{Z}^{d_A \times n_A}$, $B \in \mathbb{Z}^{d_A \times n_B}$, $C \in \mathbb{Z}^{d_C \times n_B}$, $D \in \mathbb{Z}^{d_C \times n_A}$ *be given matrices. Moreover, let* M *be a bound on the absolute values of the entries in* C *and* D, *and let* $g, \xi, \eta \in \mathbb{Z}_+$ *be the numbers, depending on* A *and* B, *from Theorem* 4.2.5. *Then for any* $N \in \mathbb{Z}_+$ *we have*

$$\begin{aligned} &\max\left\{\|\mathbf{v}\|_1 : \mathbf{v} \in \mathcal{G}\left(\left(\begin{smallmatrix} C & D \\ B & A \end{smallmatrix}\right)^{(N)}\right)\right\} \\ &\leq \xi \cdot (N+\eta)^{\eta} \cdot d_C \cdot \left(\sqrt{d_C}(n_B + Nn_A)M\right)^{d_C} \cdot (n_B + Nn_A)g. \end{aligned}$$

If A, B, C, D *are fixed matrices, then* $\max\left\{\|\mathbf{v}\|_1 : \mathbf{v} \in \mathcal{G}\left(\left(\begin{smallmatrix} C & D \\ B & A \end{smallmatrix}\right)^{(N)}\right)\right\}$ *is bounded by* $O(N^{d_C+\eta})$, *a polynomial in* N.

Proof. Let $L = \begin{pmatrix} \cdot & \cdot \\ B & A \end{pmatrix}^{(N)}$ and

$$F = \begin{pmatrix} C & D \\ \cdot & \cdot \end{pmatrix}^{(N)} = (C\ D\ \dots\ D).$$

First of all, Theorem 4.2.5 (b) gives the bound

$$\|\mathbf{v}\|_1 \leq (n_B + Nn_A)g \quad \text{for} \quad \mathbf{v} \in \mathcal{G}(L), \tag{4.6}$$

where g is a constant that depends only on A and B.

We now consider the matrix $F \cdot \mathcal{G}(L)$. Each column of it is given by

$$F\mathbf{v} = C\mathbf{x} + D\sum_{i=1}^{N} \mathbf{y}^{(i)} \quad \text{with} \quad \mathbf{v} = (\mathbf{x}, \mathbf{y}^{(1)}, \dots, \mathbf{y}^{(N)}) \in \mathcal{G}(L).$$

By Theorem 4.2.5 (c), there are at most $\xi = \mathrm{O}(1)$ different vectors $\mathbf{x}$, and for each $\mathbf{x}$ at most $\eta = \mathrm{O}(1)$ different vectors $\mathbf{y}^{(i)}$. We now determine the number σ of different sums $\mathbf{s} = \sum_{i=1}^{N} \mathbf{y}^{(i)}$ that can arise from these choices. This number is bounded by the number of weak compositions of N into η nonnegative integer parts: $\sigma \leq \binom{N+\eta-1}{\eta-1} \leq (N+\eta)^\eta = \mathrm{O}(N^\eta)$. Thus $F\mathcal{G}(L)$ has at most $d := \xi \cdot \sigma \leq \xi \cdot (N+\eta)^\eta = \mathrm{O}(N^\eta)$ different columns.

Using the bound on the entries of C and D, we find that the maximum absolute value of the entries of $F\mathcal{G}(L)$ is bounded by $(n_B + Nn_A)M$.

We now determine a length bound for the elements $\boldsymbol{\lambda}$ of $\mathcal{G}(F \cdot \mathcal{G}(L))$. By Corollary 3.7.4, we find that

$$\begin{aligned} \|\boldsymbol{\lambda}\|_1 &\leq d \cdot d_C \cdot \left(\sqrt{d_C}(n_B + Nn_A)M\right)^{d_C} \\ &\leq \xi \cdot (N+\eta)^\eta \cdot d_C \cdot \left(\sqrt{d_C}(n_B + Nn_A)M\right)^{d_C}. \end{aligned} \tag{4.7}$$

Combining the two bounds (4.6) and (4.7) using Corollary 3.7.6 then gives the result. □

Either of the two bounds implies Theorem 4.3.3.

Remark 4.3.6. Comparing the two results is difficult because bounds for the finite number $\eta(A, B)$ are unknown. However, one should expect that the bound of Proposition 4.3.5 is better for matrices with large upper blocks $\begin{pmatrix} C & D \\ \cdot & \cdot \end{pmatrix}$, whereas the bound of Proposition 4.3.4 is better for matrices with large lower blocks $\begin{pmatrix} \cdot & \cdot \\ B & A \end{pmatrix}$.

4.3.2 Constructing a feasible solution

Now that we have polynomially bounded the size of $\mathcal{G}\left(\begin{pmatrix} C & D \\ B & A \end{pmatrix}^{(N)}\right)$, let us show how to construct a feasible solution to the problem $(\mathrm{IP})_{N,\mathbf{b},\mathbf{l},\mathbf{u},f}$ in polynomial time.

Let $N \in \mathbb{Z}_+$, $\mathbf{l}, \mathbf{u} \in \mathbb{Z}^{n_B + Nn_A}$, $\mathbf{b} \in \mathbb{Z}^{d_C + Nd_A}$. First, construct an integer solution to the system $\begin{pmatrix} C & D \\ B & A \end{pmatrix}^{(N)} \mathbf{z} = \mathbf{b}$. This can be done in polynomial time using the Hermite normal form of $\begin{pmatrix} C & D \\ B & A \end{pmatrix}^{(N)}$. (See Section 2.3 for the definition and computation of the Hermite normal form of a matrix.) Then we turn it into a feasible solution (satisfying $\mathbf{l} \leq \mathbf{z} \leq \mathbf{u}$) by a sequence of at most $\mathrm{O}(Nd_A)$ many integer linear programs (with the same problem matrix $\begin{pmatrix} C & D \\ B & A \end{pmatrix}^{(N)}$, but with bounds $\tilde{\mathbf{l}}, \tilde{\mathbf{u}}$ adjusted so that the current solution is feasible) with auxiliary objective functions that move the components of $\mathbf{z}$ into the direction of the given original bounds $\mathbf{l},\mathbf{u}$; see Lemma 3.6.1. This step is similar to phase I of the simplex method in linear programming.

In order to solve these auxiliary integer linear programs with polynomially many augmentation steps, we use the speedup provided by the directed augmentation procedure [303]. This procedure requires us to repeatedly find, for certain vectors $\mathbf{c}$ and $\mathbf{d}$ that it constructs, an augmentation vector $\mathbf{v}$ with respect to the (separable convex) piecewise linear function $h(\mathbf{v}) = \mathbf{c}^\top \mathbf{v}^+ + \mathbf{d}^\top \mathbf{v}^-$.

Consequently, we only need to show how to find, for a given solution $\mathbf{z}_0$ that is feasible for $(\text{IP})_{N,\mathbf{b},\tilde{\mathbf{l}},\tilde{\mathbf{u}},h}$, an augmenting Graver basis element $\mathbf{v} \in \mathcal{G}\left(\left(\begin{smallmatrix} C & D \\ B & A \end{smallmatrix}\right)^{(N)}\right)$ for a separable convex piecewise linear function $h(\mathbf{v})$ in polynomial time in N and in the binary encoding length of $\mathbf{z}_0$, $\mathbf{c}$, and $\mathbf{d}$.

Let us now assume that we are given a solution $\mathbf{z}_0 = (\mathbf{x}_0, \mathbf{y}_0^{(1)}, \dots, \mathbf{y}_0^{(N)})$ that is feasible for $(\text{IP})_{N,\mathbf{b},\tilde{\mathbf{l}},\tilde{\mathbf{u}},h}$ and that we wish to decide whether there exists another feasible solution $\mathbf{z}_1$ with $h(\mathbf{z}_1 - \mathbf{z}_0) < 0$. As Graver bases provide an optimality certificate for separable convex integer minimization problems, Lemma 3.3.2, it suffices to decide whether there exists some vector $\mathbf{v} = (\bar{\mathbf{x}}, \bar{\mathbf{y}}^{(1)}, \dots, \bar{\mathbf{y}}^{(N)})$ in the Graver basis of $\left(\begin{smallmatrix} C & D \\ B & A \end{smallmatrix}\right)^{(N)}$ such that $\mathbf{z}_0 + \mathbf{v}$ is feasible and $h(\mathbf{v}) < 0$. By Proposition 4.3.4 or 4.3.5, the ℓ_1-norm of $\mathbf{v}$ is bounded polynomially in N. Thus, since n_B is constant, there is only a polynomial number of candidates for the $\bar{\mathbf{x}}$-part of $\mathbf{v}$. Since the bounds given by Propositions 4.3.4 and 4.3.5 are effectively computable (cf. Remark 4.2.6), we can actually list all possible vectors $\bar{\mathbf{x}}$ that satisfy these bounds.

For each such candidate $\bar{\mathbf{x}}$, we can find a best possible choice for $\bar{\mathbf{y}}^{(1)}, \dots, \bar{\mathbf{y}}^{(N)}$ by solving the following N-fold IP:

$$
\min \left\{ h(\mathbf{v}) : \begin{array}{c} \left(\begin{smallmatrix} C & D \\ B & A \end{smallmatrix}\right)^{(N)} (\mathbf{z}_0 + \mathbf{v}) = \mathbf{b}, \\ \tilde{\mathbf{l}} \le (\mathbf{z}_0 + \mathbf{v}) \le \tilde{\mathbf{u}}, \\ \mathbf{v} = (\bar{\mathbf{x}}, \bar{\mathbf{y}}^{(1)}, \dots, \bar{\mathbf{y}}^{(N)}) \in \mathbb{Z}^{n_B + N n_A} \end{array} \right\}
$$

$$
= \min \left\{ h \begin{pmatrix} \bar{\mathbf{x}} \\ \bar{\mathbf{y}}^{(1)} \\ \vdots \\ \bar{\mathbf{y}}^{(N)} \end{pmatrix} : \begin{array}{c} \left(\begin{smallmatrix} \cdot & D \\ \cdot & A \end{smallmatrix}\right)^{(N)} \begin{pmatrix} \bar{\mathbf{y}}^{1} \\ \vdots \\ \bar{\mathbf{y}}^{N} \end{pmatrix} = \mathbf{b} - \left(\begin{smallmatrix} C & D \\ B & A \end{smallmatrix}\right)^{(N)} \mathbf{z}_0 - \left(\begin{smallmatrix} C & \cdot \\ B & \cdot \end{smallmatrix}\right)^{(N)} \bar{\mathbf{x}}, \\ \tilde{\mathbf{l}} - \mathbf{z}_0 \le \begin{pmatrix} \bar{\mathbf{x}} \\ \bar{\mathbf{y}}^{(1)} \\ \vdots \\ \bar{\mathbf{y}}^{(N)} \end{pmatrix} \le \tilde{\mathbf{u}} - \mathbf{z}_0, \\ \bar{\mathbf{y}}^{(1)}, \dots, \bar{\mathbf{y}}^{(N)} \in \mathbb{Z}^{n_A} \end{array} \right\}
$$

for given $\mathbf{z}_0 = (\mathbf{x}_0, \mathbf{y}_0^{(1)}, \dots, \mathbf{y}_0^{(N)})$ and $\bar{\mathbf{x}}$. As shown in the second line, this problem does indeed simplify to a separable convex N-fold IP with problem matrix $\left(\begin{smallmatrix} \cdot & D \\ \cdot & A \end{smallmatrix}\right)^{(N)}$ because $\mathbf{z}_0 = (\mathbf{x}_0, \mathbf{y}_0^{(1)}, \dots, \mathbf{y}_0^{(N)})$ and $\bar{\mathbf{x}}$ are fixed. Since the matrices A and D are fixed, each such N-fold IP is solvable in polynomial time by Theorem 4.1.8. In fact, as shown in Theorem 4.1.9, because the function h is "2-piecewise affine," this problem can be solved in time $O(N^3 L)$ by Graver-based dynamic programming, where L denotes the binary encoding length of $\mathbf{c}$, $\mathbf{d}$, $\tilde{\mathbf{l}}$, $\tilde{\mathbf{u}}$, $\mathbf{z}_0$, and $\bar{\mathbf{x}}$.

If the N-fold IP is infeasible, there does not exist an augmenting vector using the particular choice of $\bar{\mathbf{x}}$. If it is feasible, let $\mathbf{v} = (\bar{\mathbf{x}}, \bar{\mathbf{y}}^{(1)}, \dots, \bar{\mathbf{y}}^{(N)})$ be an optimal solution. Now if we have $h(\mathbf{v}) \ge 0$, then no augmenting vector can be constructed using this particular choice of $\bar{\mathbf{x}}$. If, on the other hand, we have $h(\mathbf{v}) < 0$, then $\mathbf{v}$ is a desired augmenting vector for $\mathbf{z}_0$ and we can stop.

As we solve polynomially many polynomially solvable N-fold IPs, one for each choice of $\bar{\mathbf{x}}$, an optimality certificate or a desired augmentation step can be computed in polynomial time and the claim follows.

To construct Graver-best augmenting steps for augmenting the feasible solution that we have just found, we need to introduce some more machinery that restricts the bounds in which we need to search for an optimal solution to $(\mathrm{IP})_{N,\mathbf{b},\mathbf{l},\mathbf{u},f}$.

4.3.3 Using Hochbaum–Shanthikumar's proximity results

Hochbaum and Shanthikumar [170] present an algorithm for nonlinear separable convex integer minimization problems for matrices with small subdeterminants. The algorithm is based on the so-called proximity-scaling technique. It is pseudopolynomial in the sense that the running time depends polynomially on the absolute value of the largest subdeterminant of the problem matrix. The results of the paper [170] cannot be directly applied to our situation, since the subdeterminants of N-fold 4-block decomposable matrices typically grow exponentially in N. In the following we adapt a lemma from [170] that establishes proximity of optimal solutions of the integer problem and its continuous relaxation; we do not use the scaling technique, however.

We consider the separable convex integer minimization problem

$$\min\left\{ f(\mathbf{z}) : E\mathbf{z} = \mathbf{b},\ \mathbf{l} \leq \mathbf{z} \leq \mathbf{u},\ \mathbf{z} \in \mathbb{Z}^n \right\}. \tag{4.8}$$

Then we have the following result.

Theorem 4.3.7 (proximity). *Let $\hat{\mathbf{r}}$ be an optimal solution of the continuous relaxation of* (4.8),

$$\min\{ f(\mathbf{r}) : E\mathbf{r} = \mathbf{b},\ \mathbf{l} \leq \mathbf{r} \leq \mathbf{u},\ \mathbf{r} \in \mathbb{R}^n \}. \tag{4.9}$$

Then there exists an optimal solution $\mathbf{z}^*$ of the integer optimization problem (4.8) *with*

$$\|\hat{\mathbf{r}} - \mathbf{z}^*\|_\infty \leq n \cdot \max\left\{ \|\mathbf{v}\|_\infty : \mathbf{v} \in \mathcal{G}(E) \right\}.$$

We remark that we actually just need a bound on the circuits of E. A vector in $\ker(E)$ is called a *circuit* of E if its support is inclusion minimal among all elements in $\ker(E)$ and its components are integer and relatively prime. The set of circuits forms a subset of the Graver basis of E.

Hochbaum and Shanthikumar [170] prove a version of this result where the maximum of the absolute values of the subdeterminants of E appears on the right-hand side. The following proof from [161] is almost identical.

Proof. Let $\hat{\mathbf{z}}$ be an optimal solution of the integer optimization problem (4.8). Since $\hat{\mathbf{z}}$ is a feasible solution to the continuous relaxation, there exists a conformal (orthant-compatible) decomposition of $\hat{\mathbf{r}} - \hat{\mathbf{z}}$ into rational multiples of the circuits of E,

$$\hat{\mathbf{r}} - \hat{\mathbf{z}} = \sum_{i=1}^{n} \alpha_i \mathbf{u}^i, \quad \alpha_i \geq 0,\ \mathbf{u}^i \in \mathcal{C}(E),$$

where, due to Carathéodory's theorem, at most n circuits are needed. Then

$$\hat{\mathbf{r}} - \hat{\mathbf{z}} = \sum_{i=1}^{n} \lfloor \alpha_i \rfloor \mathbf{u}^i + \sum_{i=1}^{n} \beta_i \mathbf{u}^i,$$

setting $\beta_i = \alpha_i - \lfloor \alpha_i \rfloor$. Now we define

$$\mathbf{r}^* = \hat{\mathbf{z}} + \sum_{i=1}^{n} \beta_i \mathbf{u}^i \quad \text{and} \quad \mathbf{z}^* = \hat{\mathbf{z}} + \sum_{i=1}^{n} \lfloor \alpha_i \rfloor \mathbf{u}^i.$$

Since the vectors $\mathbf{u}^i$ lie in the kernel of matrix E, both $\mathbf{z} = \mathbf{z}^*$ and $\mathbf{z} = \mathbf{r}^*$ satisfy the equation $E\mathbf{z} = \mathbf{b}$. Moreover, since both $\hat{\mathbf{r}}$ and $\hat{\mathbf{z}}$ lie within the lower and upper bounds and the vectors $\mathbf{u}^i$ lie in the same orthant as $\hat{\mathbf{r}} - \hat{\mathbf{z}}$, $\mathbf{z}^*$ and $\mathbf{r}^*$ also lie within the lower and upper bounds. Thus, $\mathbf{r}^*$ is a feasible solution to the continuous relaxation of (4.8). Since $\mathbf{z}^*$ is also an integer vector, it is a feasible solution to the integer optimization problem (4.8).

We can write

$$\hat{\mathbf{r}} - \hat{\mathbf{z}} = [\mathbf{r}^* - \hat{\mathbf{z}}] + [\mathbf{z}^* - \hat{\mathbf{z}}].$$

Then we use an important superadditivity property of separable convex functions, Lemma 3.3.1, which gives

$$f(\hat{\mathbf{r}}) - f(\hat{\mathbf{z}}) \geq [f(\mathbf{r}^*) - f(\hat{\mathbf{z}})] + [f(\mathbf{z}^*) - f(\hat{\mathbf{z}})] \tag{4.10}$$

or, equivalently,

$$f(\hat{\mathbf{r}}) - f(\mathbf{r}^*) \geq f(\mathbf{z}^*) - f(\hat{\mathbf{z}}). \tag{4.11}$$

Since $\hat{\mathbf{r}}$ is an optimal solution to the continuous relaxation and $\mathbf{r}^*$ is a feasible solution to it, the left-hand side is nonpositive, and so $f(\mathbf{z}^*) \leq f(\hat{\mathbf{z}})$. Thus, since $\mathbf{z}^*$ is a feasible solution to (4.8), it is, in fact, another optimal solution of the integer optimization problem and $f(\mathbf{z}^*) = f(\hat{\mathbf{z}})$.

We now verify the proximity of $\mathbf{z}^*$ to $\hat{\mathbf{r}}$. From the definition of $\mathbf{z}^*$, we immediately get

$$\begin{aligned}
\left\|\hat{\mathbf{r}} - \mathbf{z}^*\right\|_\infty &= \left\|[\hat{\mathbf{r}} - \hat{\mathbf{z}}] + [\hat{\mathbf{z}} - \mathbf{z}^*]\right\|_\infty \\
&= \left\|\textstyle\sum_{i=1}^n \alpha_i \mathbf{u}^i - \sum_{i=1}^n \lfloor \alpha_i \rfloor \mathbf{u}^i\right\|_\infty \\
&= \left\|\textstyle\sum_{i=1}^n \beta_i \mathbf{u}^i\right\|_\infty \\
&\leq n \cdot \max\{\|\mathbf{u}^j\|_\infty : j = 1, \dots, n\} \\
&\leq n \cdot \max\left\{\|\mathbf{v}\|_\infty : \mathbf{v} \in \mathcal{G}(E)\right\}.
\end{aligned}$$

This concludes the proof. □

As an immediate corollary, we obtain the following result.

Corollary 4.3.8. *Let* $\epsilon \geq 0$ *and let* $\hat{\mathbf{r}}$ *be an optimal solution to the continuous relaxation* (4.9). *Setting*

$$\begin{aligned}
\mathbf{l}' &= \max\{\mathbf{l}, \lfloor \hat{\mathbf{r}} - (n \cdot \ell)\mathbf{1} \rfloor\}, \\
\mathbf{u}' &= \min\{\mathbf{u}, \lceil \hat{\mathbf{r}} + (n \cdot \ell)\mathbf{1} \rceil\},
\end{aligned}$$

where $\ell = \max\left\{\|\mathbf{v}\|_\infty : \mathbf{v} \in \mathcal{G}(E)\right\}$, *we have*

$$\begin{aligned}
&\min\{f(\mathbf{z}) : E\mathbf{z} = \mathbf{b},\ \mathbf{l} \leq \mathbf{z} \leq \mathbf{u},\ \mathbf{z} \in \mathbb{Z}^n\} \\
&\qquad = \min\{f(\mathbf{z}) : E\mathbf{z} = \mathbf{b},\ \mathbf{l}' \leq \mathbf{z} \leq \mathbf{u}',\ \mathbf{z} \in \mathbb{Z}^n\}.
\end{aligned} \tag{4.12}$$

Later we will use a simple modification of Corollary 4.3.8, using an ϵ-approximate optimal solution to the continuous relaxation (4.9).

For $E = \left(\begin{smallmatrix} C & D \\ B & A \end{smallmatrix}\right)^{(N)}$, we can control the size of ℓ using Proposition 4.3.4 or 4.3.5 and thus obtain an equivalent IP with small (polynomial-sized) bounds.

We note that though the bounds are small, the dimension is still variable, and so the problem cannot be solved efficiently with elementary techniques such as dynamic programming. In the following we show how to solve this IP with Graver basis techniques.

4.3.4 Graver-best augmentation for the restricted problem

Theorem 4.3.7 allows us to restrict the bounds for an optimal solution to $(\text{IP})_{N,\mathbf{b},\mathbf{l},\mathbf{u},f}$. In the restricted problem, however, no long augmentation steps are possible, and therefore it is possible to efficiently construct a Graver-best augmentation vector. Using this observation, we prove the following theorem.

Theorem 4.3.9. *Let $A \in \mathbb{Z}^{d_A \times n_A}$, $B \in \mathbb{Z}^{d_A \times n_B}$, $C \in \mathbb{Z}^{d_C \times n_B}$, $D \in \mathbb{Z}^{d_C \times n_A}$ be fixed matrices. Then there exists an algorithm that, given $N \in \mathbb{Z}_+$, $\mathbf{c} \in \mathbb{Z}^{kn_B + kNn_A}$, $\mathbf{b} \in \mathbb{Z}^{d_C + Nd_A}$, $\mathbf{l}, \mathbf{u} \in \mathbb{Z}^{n_B + Nn_A}$, a feasible solution $\mathbf{z}_0$, and a comparison oracle for the function $f : \mathbb{R}^{n_B + Nn_A} \to \mathbb{R}$, finds an optimal solution to*

$$\min\left\{ f(\mathbf{z}) : \left(\begin{smallmatrix} C & D \\ B & A \end{smallmatrix}\right)^{(N)} \mathbf{z} = \mathbf{b},\ \mathbf{l}' \le \mathbf{z} \le \mathbf{u}',\ \mathbf{z} \in \mathbb{Z}^{n_B + Nn_A} \right\}$$

and that runs in time that is polynomially bounded in N, in $k := \|\mathbf{u}' - \mathbf{l}'\|_\infty$, and in the binary encoding length of $\mathbf{b}$, $\mathbf{c}$, and $\hat{f}$.

Proof. By the Graver-best speed-up technique (see Chapter 3), it suffices to show that for a given feasible solution $\mathbf{z}_0$, we can construct a vector $\gamma \mathbf{g}$, where $\gamma \in \mathbb{Z}_+$ and $\mathbf{g} \in \mathcal{G}\left(\left(\begin{smallmatrix} C & D \\ B & A \end{smallmatrix}\right)^{(N)}\right)$, such that $\mathbf{z}_0 + \gamma \mathbf{g}$ is feasible, and γ and $\mathbf{g}$ minimize $f(\mathbf{z}_0 + \gamma \mathbf{g})$ among all possible choices. It actually suffices to construct any vector $\mathbf{v}$ such that $\mathbf{z}_0 + \mathbf{v}$ is feasible and $f(\mathbf{z}_0 + \mathbf{v}) \le f(\mathbf{z}_0 + \gamma \mathbf{g})$.

Write $\mathbf{z}_0 = (\mathbf{x}_0, \mathbf{y}_0^{(1)}, \dots, \mathbf{y}_0^{(N)})$ and let $\mathbf{v} = (\bar{\mathbf{x}}, \dots)$ be any vector in the Graver basis of $\left(\begin{smallmatrix} C & D \\ B & A \end{smallmatrix}\right)^{(N)}$. By Proposition 4.3.4 or 4.3.5, the ℓ_1-norm of $\mathbf{v}$ is bounded polynomially in N. Thus, since n_B is constant, there is only a polynomial number of candidates for the $\bar{\mathbf{x}}$-part of $\mathbf{v}$. Since the bounds given by Proposition 4.3.4 and 4.3.5 are effectively computable (cf. Remark 4.2.6), we can actually list all possible vectors $\bar{\mathbf{x}}$ that satisfy these bounds.

For each such vector $\bar{\mathbf{x}}$, we now consider all vectors of the form $(\gamma \bar{\mathbf{x}}, \bar{\mathbf{y}}^{(1)}, \dots, \bar{\mathbf{y}}^{(N)})$ as candidate augmentation vectors, not just multiples $\gamma \mathbf{v}$ of Graver basis elements.

In the special case $\bar{\mathbf{x}} = \mathbf{0}$, this is equivalent to the construction of a Graver-best augmentation vector for the N-fold IP with the problem matrix $\left(\begin{smallmatrix} \cdot & D \\ \cdot & A \end{smallmatrix}\right)^{(N)}$, which can be done in polynomial time, by Theorem 4.1.8.

Otherwise, if $\bar{\mathbf{x}} \ne \mathbf{0}$, we determine the largest step length $\hat{\gamma} \in \mathbb{Z}_+$ such that $\mathbf{x}_0 + \hat{\gamma} \bar{\mathbf{x}}$ lies within the bounds $\mathbf{l}', \mathbf{u}'$. Certainly $\hat{\gamma} \le k$. We now check each possible step length $\gamma = 1, 2, \dots, \hat{\gamma}$ separately. To find a best possible choice for $\bar{\mathbf{y}}^{(1)}, \dots, \bar{\mathbf{y}}^{(N)}$, we solve the following N-fold IP:

$$\min\left\{ f(\mathbf{v}) : \begin{array}{c} \left(\begin{smallmatrix} C & D \\ B & A \end{smallmatrix}\right)^{(N)} (\mathbf{z}_0 + \mathbf{v}) = \mathbf{b}, \\ \mathbf{l}' \le (\mathbf{z}_0 + \mathbf{v}) \le \mathbf{u}', \\ \mathbf{v} = (\gamma \bar{\mathbf{x}}, \bar{\mathbf{y}}^{(1)}, \dots, \bar{\mathbf{y}}^{(N)}) \in \mathbb{Z}^{n_B + Nn_A} \end{array} \right\}.$$

Since the matrices A and D are fixed, each such N-fold IP is solvable in polynomial time, by Theorem 4.1.8.

If the N-fold IP is infeasible, there does not exist an augmenting vector using the particular choice of $\bar{\mathbf{x}}$ and γ. If it is feasible, let $\mathbf{v} = (\gamma\bar{\mathbf{x}}, \bar{\mathbf{y}}^{(1)}, \ldots, \bar{\mathbf{y}}^{(N)})$ be an optimal solution. Now if we have $f(\mathbf{v}) \geq 0$, then no augmenting vector can be constructed using this particular choice of $\bar{\mathbf{x}}$ and γ. If, on the other hand, we have $f(\mathbf{v}) < 0$, then $\mathbf{v}$ is a candidate for the Graver-best augmentation vector.

By iterating over all $\bar{\mathbf{x}}$ and all γ, we efficiently construct a Graver-best augmentation vector. □

Remark 4.3.10. A more precise complexity analysis is as follows.

(a) For the construction in the special case $\bar{\mathbf{x}} = \mathbf{0}$: In fact, using Graver-based dynamic programming for N-fold IPs as presented in Section 4.1.3, we can find in linear time $O(N)$ for any of the possible step lengths $\gamma = 1, 2, \ldots, k$, an augmenting vector $\gamma\mathbf{v}$ that is at least as good as the best Graver step $\gamma\mathbf{g}$ with $\mathbf{g} \in \mathcal{G}\left(\begin{smallmatrix}\cdot & D\\ \cdot & A\end{smallmatrix}\right)^{(N)}$. Checking all step lengths, we get a complexity of $O(kN)$.

(b) For the solution of the N-fold subproblem in the general case $\bar{\mathbf{x}} \neq \mathbf{0}$: This optimization, in turn, uses another Graver-best augmentation technique. In phase I, the possible step lengths are large, but the auxiliary objective functions are linear, and so the running time is $O(N^3 L)$ by Graver-based dynamic programming for N-fold IPs, where L denotes the binary encoding length of $\mathbf{l}'$, $\mathbf{u}'$, $\mathbf{z}_0$, $\bar{\mathbf{x}}$; see Section 4.1.3. In phase II, there are few possible step lengths, $\gamma = 1, 2, \ldots, k$, so we can try them all. Again by using Graver-based dynamic programming for N-fold IPs, we can find for a fixed γ in linear time $O(N)$ an augmenting vector $\gamma\mathbf{v}$ that is at least as good as the best Graver step $\gamma\mathbf{g}$ with $\mathbf{g} \in \mathcal{G}\left(\begin{smallmatrix}\cdot & D\\ \cdot & A\end{smallmatrix}\right)^{(N)}$. Checking all step lengths, we get a complexity of $O(kN)$. Using Theorem 3.4.1 and the bound on the number of augmentation steps in its proof, we conclude that the number of Graver-best augmentations is bounded by $O(N \log \hat{f})$. Thus the complexity of this subproblem is $O(N^2 k \log \hat{f} + N^3 L)$.

(c) The number of steps in the overall Graver-best augmentation algorithm for the restricted 4-block decomposable problem is again bounded by $O(N \log \hat{f})$.

Remark 4.3.11. Other augmentation techniques can be used to prove Theorem 4.3.9. For example, following [170, Section 2], we can reformulate a separable convex integer minimization problem with small bounds as a 0/1 linear integer minimization problem in a straightforward way. Then we can apply various speedup techniques, including the very simple bit-scaling speedup technique; see [304].

4.3.5 Graver proximity algorithm

Let us now put all the pieces together to state an algorithm that proves the main result of this section, Theorem 4.3.2. For each set of fixed matrices A, B, C, D and for any function $\epsilon(N)$ that is bounded polynomially in N, we consider the following algorithm.

ALGORITHM 4.1. Graver proximity algorithm.

input $N \in \mathbb{Z}_+$, bounds $\mathbf{l}, \mathbf{u} \in \mathbb{Z}^{n_B + N n_A}$, right-hand side $\mathbf{b} \in \mathbb{Z}^{d_C + N d_A}$, evaluation oracle for a separable convex function $f : \mathbb{R}^{n_B + N n_A} \to \mathbb{R}$, approximate continuous convex optimization oracle.

output an optimal solution $\mathbf{z}^*$ to $(\mathrm{IP})_{N,\mathbf{b},\mathbf{l},\mathbf{u},f}$ or INFEASIBLE or UNBOUNDED.

Let $n = n_B + N n_A$ denote the dimension of the problem.

Call the approximate continuous convex optimization oracle with $\epsilon = \epsilon(N)$ to find an approximate solution $\mathbf{r}_\epsilon \in \mathbb{Q}^{n_B + N n_A}$ to the continuous relaxation

$$\min\left\{ f(\mathbf{r}) : \begin{pmatrix} C & D \\ B & A \end{pmatrix}^{(N)} \mathbf{r} = \mathbf{b},\ \mathbf{l} \leq \mathbf{r} \leq \mathbf{u},\ \mathbf{r} \in \mathbb{R}^{n_B + N n_A} \right\}.$$

if oracle returns INFEASIBLE **then**
  return INFEASIBLE.
else if oracle returns UNBOUNDED **then**
  return UNBOUNDED.
else
  Compute an upper bound ℓ on the maximum ℓ_1-norm of the vectors in $\mathcal{G}\left(\begin{pmatrix} C & D \\ B & A \end{pmatrix}^{(N)}\right)$, using Proposition 4.3.4 or 4.3.5.
  Let $\mathbf{l}' = \max\{\mathbf{l}, \lfloor \mathbf{r}_\epsilon - (n \cdot \ell + \epsilon)\mathbf{1} \rfloor\}$ and $\mathbf{u}' = \min\{\mathbf{u}, \lceil \mathbf{r}_\epsilon + (n \cdot \ell + \epsilon)\mathbf{1} \rceil\}$.
  Let $k = \|\mathbf{u}' - \mathbf{l}'\|_\infty$.
  Using the algorithm of Section 4.3.2, find a feasible solution $\mathbf{z}_0$ for the restricted convex integer minimization problem

$$\min\left\{ f(\mathbf{z}) : \begin{pmatrix} C & D \\ B & A \end{pmatrix}^{(N)} \mathbf{z} = \mathbf{b},\ \mathbf{l}' \leq \mathbf{z} \leq \mathbf{u}',\ \mathbf{z} \in \mathbb{Z}^{n_B + N n_A} \right\}.$$

  Solve the problem to optimality using the algorithm of Theorem 4.3.9.

By analyzing this algorithm, we now prove the main theorem of this section.

***Proof of Theorem* 4.3.2.** We first show that Algorithm 4.1 is correct. If the continuous relaxation $(\mathrm{CP})_{N,\mathbf{b},\mathbf{l},\mathbf{u},f}$ is infeasible or unbounded, then so is the problem $(\mathrm{IP})_{N,\mathbf{b},\mathbf{l},\mathbf{u},f}$. In the following, assume that $(\mathrm{CP})_{N,\mathbf{b},\mathbf{l},\mathbf{u},f}$ has an optimal solution. Then there exists an optimal solution $\hat{\mathbf{r}}$ to $(\mathrm{CP})_{N,\mathbf{b},\mathbf{l},\mathbf{u},f}$ with $\|\hat{\mathbf{r}} - \mathbf{r}_\epsilon\|_\infty \leq \epsilon$. By Theorem 4.3.7, there exists an optimal solution $\mathbf{z}^*$ of the integer optimization problem $(\mathrm{IP})_{N,\mathbf{b},\mathbf{l},\mathbf{u},f}$ with $\|\hat{\mathbf{r}} - \mathbf{z}^*\|_\infty \leq n \cdot \ell$. By the triangle inequality, this solution then satisfies $\|\mathbf{z}^* - \mathbf{r}_\epsilon\|_\infty \leq n \cdot \ell + \epsilon$ and is therefore a feasible solution to the restricted IP with variable bounds $\mathbf{l}'$ and $\mathbf{u}'$. Thus it suffices to solve the restricted IP to optimality, which is done with the algorithm of Theorem 4.3.9.

The algorithm has the claimed complexity because

$$k \leq 2((n_B + N n_A) \cdot \ell + \epsilon)$$

is bounded polynomially in N by Proposition 4.3.4 or Proposition 4.3.5. The complexity then follows from Theorem 4.3.9. □

4.4 Notes and further references

This chapter is based on the papers [96, 157, 159, 161, 177, 289]. Berstein and Onn [49] gave a lower bound for the Graver complexity $g(A_{3\times M})$ for the $3 \times M$ transportation matrix. They showed an exponential lower bound $g(A_{3\times M}) \geq 17 \cdot 2^{M-3} - 7$. This bound was generalized to arbitrary $L \times M$ transportation matrices by [210].

Aschenbrenner and Hemmecke [16] extended the finiteness result of Theorem 4.2.5 from the two-stage to the multistage setting. Thus, as a direct consequence, we obtain a generalization of Theorem 4.2.7: Multistage stochastic integer linear programs can be solved in polynomial time provided the matrices from which the problem matrix is composed of are fixed.

4.5 Exercises

Exercise 4.5.1. In analogy to Example 4.1.5, study $N \times 4$, $N \times 5$, and $N \times 6$ tables.

(a) Which matrix A do we have to choose in order to define $A^{(N)}$ for each of the three cases?

(b) What is the maximum type of a vector appearing in the Graver bases of $A^{(N)}$ for each of the three cases as $N \to \infty$? In other words, what is the Graver complexity of A for each of the three cases?

(c) State a conjecture about the maximum type appearing in the Graver bases of $N \times m$ tables, where m is kept fixed and where $N \to \infty$.

Exercise 4.5.2. In analogy to Example 4.1.5, study $N \times 3 \times 2$ tables.

(a) Which matrix A do we have to choose in order to define $A^{(N)}$?

(b) What is the maximum type of a vector appearing in the Graver basis of $A^{(N)}$ as $N \to \infty$? In other words, what is the Graver complexity of A?

(c) State a conjecture about the maximum types of vectors appearing in the Graver bases of $N \times 3$ tables and of $N \times 3 \times 2$ tables, as $N \to \infty$.

(d) Prove your conjecture!

Exercise 4.5.3. In analogy to Example 4.1.5, study $N \times 3 \times 3$ tables.

(a) Which matrix A do we have to choose in order to define $A^{(N)}$?

(b) What is the maximum type of a vector appearing in the Graver basis of $A^{(N)}$ as $N \to \infty$? In other words, what is the Graver complexity of A?

(c) Compute the Graver bases of $N \times 3 \times 3$ tables for $N = 3,4,5,\ldots$. Up to which N can you compute the Graver basis on your computer within one hour? How does it compare to the Graver degree $g(A)$? What do you conclude?

Exercise 4.5.4. Explain why the bound (4.3) is not tight in general.

Part III

Generating Function Methods

Chapter 5

Introduction to Generating Functions

5.1 The geometric series as a generating function

We begin with a very simple example to introduce generating functions, which are an important and classical tool in combinatorics. We refer to [314] for a much broader introduction. In the next few chapters we will use them as data structures for optimization. This chapter tries to develop intuition before a more formal treatment.

Let us consider the set S of integers in the interval $P = [0,\dots,n]$; see Figure 5.1. We think of S as the feasible region of an integer optimization problem. We shall associate with the set S the polynomial

$$g(S;z) = z^0 + z^1 + \cdots + z^{n-1} + z^n; \tag{5.1}$$

that is, every integer $\alpha \in S$ corresponds to a monomial z^α with coefficient 1 in the polynomial $g(S;z)$. This polynomial is called the *generating function* of S (or of P). From the viewpoint of computational complexity, this generating function is of exponential size (in the bit length of n), just as an explicit list of all the integers 0, 1, …, $n-1$, n would be. However, we can observe that (5.1) is a finite geometric series, so there exists a simple summation formula that expresses (5.1) in a much more compact way:

$$g(S;z) = z^0 + z^1 + \cdots + z^{n-1} + z^n = \frac{1 - z^{n+1}}{1 - z}. \tag{5.2}$$

The "long" polynomial has a "short" representation as a rational function. The bit length of this new formula is *linear* in the bit length of n.

Counting. Suppose now someone presents to us a finite set S of integers as a generating function $g(S;z)$. Can we decide whether the set is nonempty? In fact, we can do something

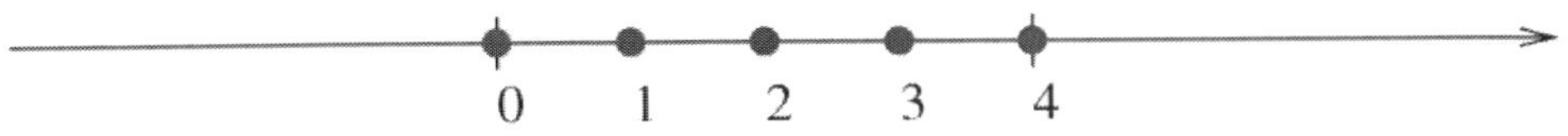

Figure 5.1. *A one-dimensional lattice-point set.*

much stronger—we can *count* the integers in the set S, simply by evaluating at $g(S;z)$ at $z = 1$. In our example we have

$$|S| = g(S;1) = 1^0 + 1^1 + \cdots + 1^{n-1} + 1^n = n + 1.$$

We can do the same on the shorter, rational-function formula. We need to be a bit careful, though: The point $z = 1$ is a singularity of the formula, but it is removable. In fact, we know that $g(S;z)$ is a polynomial, so it is regular everywhere; it has no poles. We just compute the limit using the Bernoulli–l'Hôpital rule:

$$|S| = \lim_{z\to 1} g(S;z) = \lim_{z\to 1} \frac{1 - z^{n+1}}{1 - z} = \lim_{z\to 1} \frac{-(n+1)z^n}{-1} = n + 1.$$

Note that we have avoided performing a polynomial division, which would have given us the long polynomial.

Can these simple observations be generalized and exploited to obtain an algorithmically efficient representation of lattice-point sets in arbitrary polyhedra? It turns out they can—Barvinok [30] pioneered a theory of "short" rational generating functions, which gives an efficient calculus for lattice-point sets in polyhedra for every fixed dimension. Later we will use this calculus to represent the feasible regions of integer optimization problems with linear constraints. Before presenting the general theory, though, we continue with our one-dimensional example (and later with a two-dimensional example) to investigate some of the features of the approach.

Rational functions and their Laurent expansions. We note that the summation formula (5.2) can also be written in a slightly different way:

$$g(S;z) = \frac{1}{1-z} - \frac{z^{n+1}}{1-z} = \frac{1}{1-z} + \frac{z^n}{1-z^{-1}}. \tag{5.3}$$

Each of the two summands on the right-hand side can be viewed as the summation formula of an infinite geometric series:

$$g_1(z) = \frac{1}{1-z} = z^0 + z^1 + z^2 + \cdots, \tag{5.4a}$$

$$g_2(z) = \frac{z^n}{1-z^{-1}} = z^n + z^{n-1} + z^{n-2} + \cdots. \tag{5.4b}$$

The two summands have a geometrical interpretation. If we view each geometric series as the generating function of an (infinite) lattice point set, we arrive at the picture shown in Figure 5.2. *Something in this calculation seems wrong—all integer points in the interval* $[0,n]$ *are covered twice, and also all integer points outside the interval are covered once.* Where is the mistake?

The theory of complex analysis establishes a rigorous way to study the connection between functions and their series expansions. For an introduction to this field from a practical point of view, we refer the reader to [164]. We have observed a phenomenon that is due to the *one-to-many correspondence* of rational functions to their series expansions, the *Laurent series*. When we consider Laurent series of the function $g_1(z)$ about $z = 0$, the pole $z = 1$ splits the complex plane into two domains of convergence (Figure 5.3): For $|z| < 1$, the power series

$$z^0 + z^1 + z^2 + \cdots \tag{5.5}$$

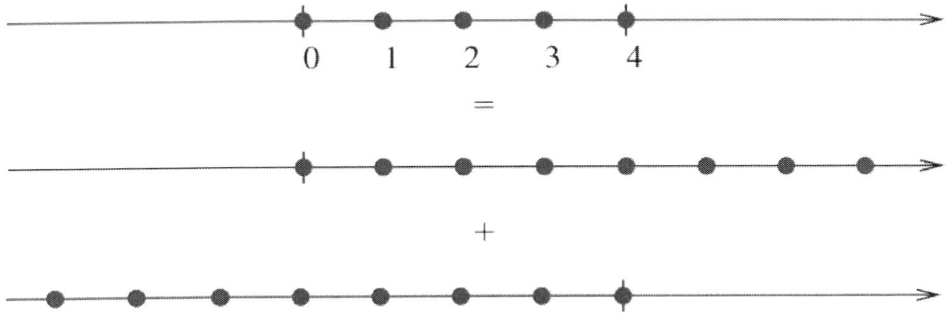

Figure 5.2. *One-dimensional identity (later seen to hold in general).*

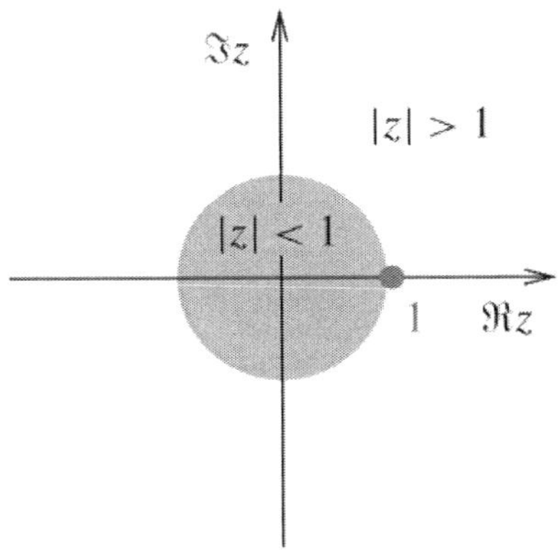

Figure 5.3. *The domains of convergence of the Laurent series.*

converges to $g_1(z)$. As a matter of fact, it converges absolutely and uniformly on every compact subset of the open circle $\{z \in \mathbb{C} : |z| < 1\}$. For $|z| > 1$, however, the series (5.5) diverges. On the other hand, the Laurent series

$$-z^{-1} - z^{-2} - z^{-3} - \cdots \tag{5.6}$$

converges (absolutely and uniformly on every compact subset) on the open circular ring $\{z \in \mathbb{C} : |z| > 1\}$ to the function $g_1(z)$, whereas it diverges for $|z| < 1$. The same holds for the second summand $g_2(z)$. Altogether, we have

$$g_1(z) = \begin{cases} z^0 + z^1 + z^2 + \cdots & \text{for } |z| < 1, \\ -z^{-1} - z^{-2} - z^{-3} - \cdots & \text{for } |z| > 1, \end{cases} \tag{5.7}$$

$$g_2(z) = \begin{cases} -z^{n+1} - z^{n+2} - z^{n+3} - \cdots & \text{for } |z| < 1, \\ z^n + z^{n-1} + z^{n-2} + \cdots & \text{for } |z| > 1. \end{cases} \tag{5.8}$$

We can now see that the "mistake" we observed in Formula (5.4) and Figure 5.2 is due to the fact that we had picked two Laurent series for the summands $g_1(z)$ and $g_2(z)$ that do not have a common domain of convergence. If for each summand we choose the series that

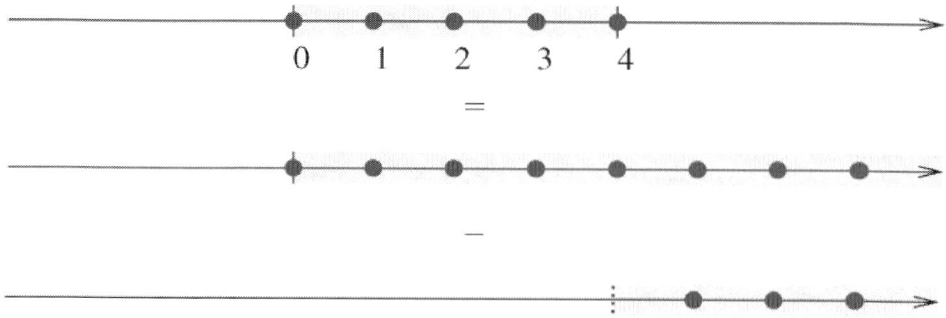

Figure 5.4. *Another one-dimensional identity.*

converge for $|z| < 1$, we obtain the more intuitive formula

$$g(S;z) = \left(z^0 + z^1 + z^2 + \cdots\right) - \left(z^{n+1} + z^{n+2} + z^{n+3} + \cdots\right), \tag{5.9}$$

which is illustrated in Figure 5.4. Nevertheless, it turns out that using Laurent series with disjoint domains of convergence is a powerful technique; we will meet the situation of Formula (5.4) and Figure 5.2 again in the multidimensional case as *Brion's theorem*.

5.2 Generating functions and objective functions

We have motivated our trying to find algorithmically efficient representations of feasible regions of integer optimization problems. But what about the objective functions? They come into play in two different ways.

Once more, we consider as an example the generating function of the set S of integer points of the interval $P = [0,4]$, as shown in Figure 5.1,

$$g(S;z) = z^0 + z^1 + z^2 + z^3 + z^4 = \frac{1}{1-z} - \frac{z^5}{1-z}. \tag{5.10}$$

Linear objective functions. As an example, let us consider the objective function $f(\alpha) = -3\alpha$. The idea is now to make a change of variables, $z = y\,t^{-3}$, where y and t are new variables. We obtain

$$g(S;yt^{-3}) = y^0t^0 + y^1t^{-3} + y^2t^{-6} + y^3t^{-9} + z^4t^{-12} = \frac{1}{1-yt^{-3}} - \frac{y^5t^{-15}}{1-yt^{-3}}. \tag{5.11}$$

By making the change of variables in the polynomial expression, we have obtained a Laurent polynomial in which each term is of the form $y^\alpha t^{f(\alpha)}$. On the other hand, we can make the same substitution in the rational function expression and obtain the right-hand expression in Equation (5.11). To maximize the function over the feasible region, then, is equivalent to finding the term with the largest exponent of the variable t. We will use this idea in Chapter 7.

Nonlinear (polynomial) objective functions. As an example, let us consider the objective function $f(\alpha) = \alpha^2$. The idea is now to use differentiation on the generating function.

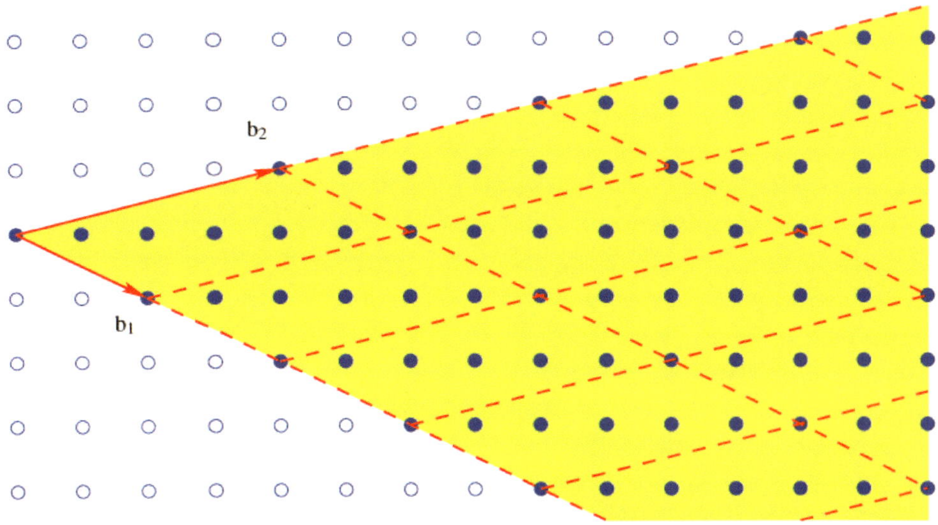

Figure 5.5. *Tiling a rational two-dimensional cone with copies of the fundamental parallelepiped.*

Let us apply the differential operator $z\frac{\mathrm{d}}{\mathrm{d}z}$ and obtain

$$\left(z\frac{\mathrm{d}}{\mathrm{d}z}\right) g(S;z) = 1z^1 + 2z^2 + 3z^3 + 4z^4 = \frac{1}{(1-z)^2} - \frac{-4z^5 + 5z^4}{(1-z)^2}. \tag{5.12}$$

Note that the right-hand side of (5.12) has been obtained by symbolically differentiating the right-hand side of (5.10). (As an exercise, the reader can verify by polynomial division that this rational function indeed is the polynomial shown in the middle of (5.12).)

Applying the same differential operator again, we obtain

$$\left(z\frac{\mathrm{d}}{\mathrm{d}z}\right)\left(z\frac{\mathrm{d}}{\mathrm{d}z}\right) g(S;z) = 1z^1 + 4z^2 + 9z^3 + 16z^4$$
$$= \frac{z+z^2}{(1-z)^3} - \frac{25z^5 - 39z^6 + 16z^7}{(1-z)^3}. \tag{5.13}$$

We have thus evaluated the objective function $f(\alpha)$ for $\alpha = 0, \ldots, 4$; the results appear as the coefficients of the respective monomials z^α.

To maximize the function over the feasible region, we "only" need to find the term that has the highest coefficient. In Chapter 8, we will show how to do that.

5.3 Generating functions in two dimensions

Let us consider a cone C spanned by the vectors $\mathbf{b}_1 = (\alpha, -1)$ and $\mathbf{b}_2 = (\beta, 1)$ (see Figure 5.5 for an example with $\alpha = 2$ and $\beta = 4$). We would like to write down a generating function for the integer points in this cone. We apparently need a generalization of the geometric series, of which we made use in the one-dimensional case. The key observation now is that by using copies of the half-open fundamental parallelepiped,

$$\Pi = \left\{ \lambda_1 \mathbf{b}_1 + \lambda_2 \mathbf{b}_2 : \lambda_1 \in [0,1), \lambda_2 \in [0,1) \right\}, \tag{5.14}$$

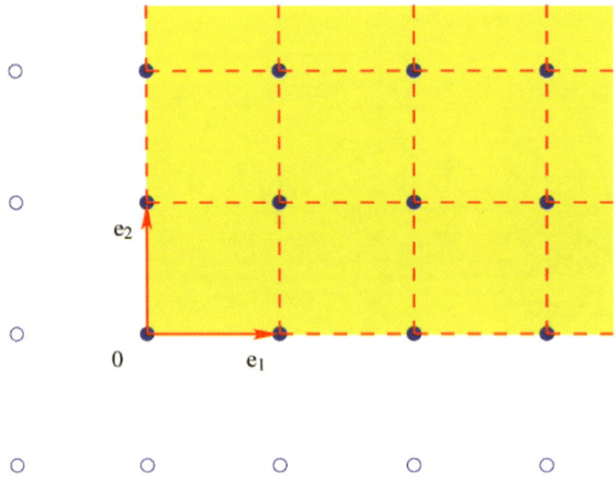

Figure 5.6. *The semigroup $S \subseteq \mathbb{Z}^2$ generated by $\mathbf{b}_1$ and $\mathbf{b}_2$ is a linear image of $\mathbb{Z}^2_+$.*

the cone can be tiled:

$$C = \bigcup_{\mathbf{s}\in S} (\mathbf{s} + \Pi) \quad \text{where} \quad S = \{\, \mu_1 \mathbf{b}_1 + \mu_2 \mathbf{b}_2 : (\mu_1, \mu_2) \in \mathbb{Z}^2_+ \,\} \tag{5.15}$$

(a disjoint union); cf. Corollary 2.3.13. Moreover, because we have chosen *integral* generators $\mathbf{b}_1, \mathbf{b}_2$ for our cone, the integer points are "the same" in each copy of the fundamental parallelepiped. Therefore, the integer points of C can also be tiled by copies of the integer points of Π:

$$C \cap \mathbb{Z}^2 = \bigcup_{\mathbf{s}\in S} (\mathbf{s} + (\Pi \cap \mathbb{Z}^2)). \tag{5.16}$$

We can also see $C \cap \mathbb{Z}^2$ as a finite disjoint union of copies of the set S, shifted by the integer points of the fundamental parallelepiped:

$$C \cap \mathbb{Z}^2 = \bigcup_{\mathbf{x}\in \Pi\cap\mathbb{Z}^2} (\mathbf{x} + S). \tag{5.17}$$

The benefit of this representation is that the set S is just the image of $\mathbb{Z}^2_+$ under the matrix $(\mathbf{b}_1, \mathbf{b}_2) \in \mathbb{Z}^{2\times 2}$; cf. Figures 5.5 and 5.6. Now $\mathbb{Z}^2_+$ is the direct product of $\mathbb{Z}_+$ with itself, whose generating function we already know—it is given by the geometric series:

$$g(\mathbb{Z}_+; z) = z^0 + z^1 + z^2 + z^3 + \cdots = \frac{1}{1-z}.$$

We thus obtain the generating function as a product,

$$g(\mathbb{Z}^2_+; z_1, z_2) = (z_1^0 + z_1^1 + z_1^2 + z_1^3 + \cdots)(z_2^0 + z_2^1 + z_2^2 + z_2^3 + \cdots) = \frac{1}{1-z_1} \cdot \frac{1}{1-z_2}.$$

Applying the linear transformation $(\mathbf{b}_1, \mathbf{b}_2) = \begin{pmatrix} \alpha & \beta \\ -1 & 1 \end{pmatrix}$, we obtain the generating function

$$g(S; z_1, z_2) = \frac{1}{(1 - z_1^{\alpha} z_2^{-1})(1 - z_1^{\beta} z_2^{1})}.$$

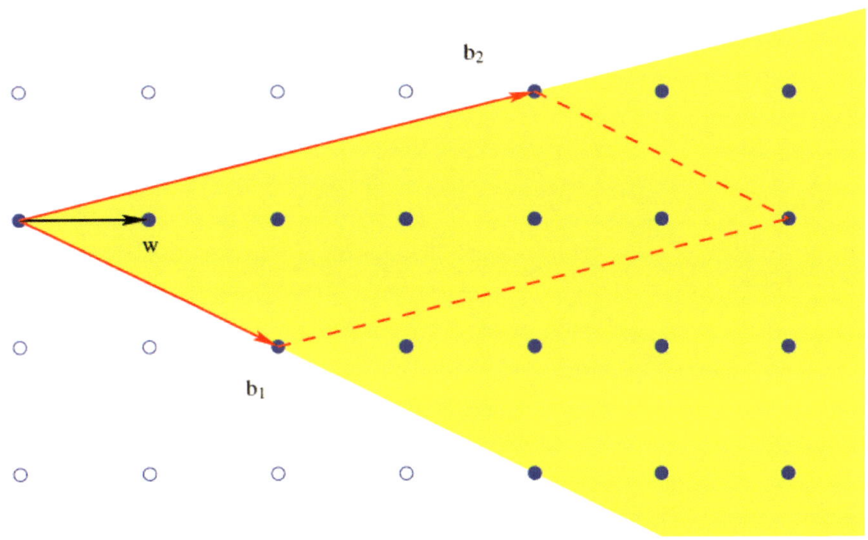

Figure 5.7. *A two-dimensional cone of index* 6 *with its fundamental parallelepiped. Using the interior vector* **w**, *a triangulation can be constructed.*

From the representation (5.17) it is now clear that

$$g(C;z_1,z_2) = \sum_{\mathbf{x}\in\Pi\cap\mathbb{Z}^2} z_1^{x_1} z_2^{x_2} g(S;z_1,z_2);$$

the multiplication with the monomial $z_1^{x_1} z_2^{x_2}$ corresponds to the shifting of the set S by the vector (x_1,x_2). In our example, it is easy to see that

$$\Pi\cap\mathbb{Z}^2 = \{(i,0) : i = 0,\dots,\alpha+\beta-1\} \tag{5.18}$$

(see Figure 5.7). We thus obtain the generating function

$$g(C;z_1,z_2) = \frac{z_1^0 + z_1^1 + \cdots + z_1^{\alpha+\beta-2} + z_1^{\alpha+\beta-1}}{(1-z_1^\alpha z_2^{-1})(1-z_1^\beta z_2^1)}.$$

Unfortunately, this formula has an exponential size since the numerator contains $\alpha+\beta$ summands. In our example, the numerator again is a finite geometric series, so we could use a short summation formula. However, this technique does not seem to be helpful in general because the structure of the integer points in the fundamental parallelepiped is usually more complicated than in our example.

Triangulations. A different idea to make the formula shorter is to break the cone into "smaller" cones, each of which has a shorter formula. We have observed that the length of the formula is essentially determined by the number of summands in the numerator—which correspond to the integer points in the fundamental parallelepiped. Thus, the right measure of size of a cone seems to be the number of integer points in the fundamental parallelepiped; this is called the *index* of the cone (cf. Theorem 2.3.19).

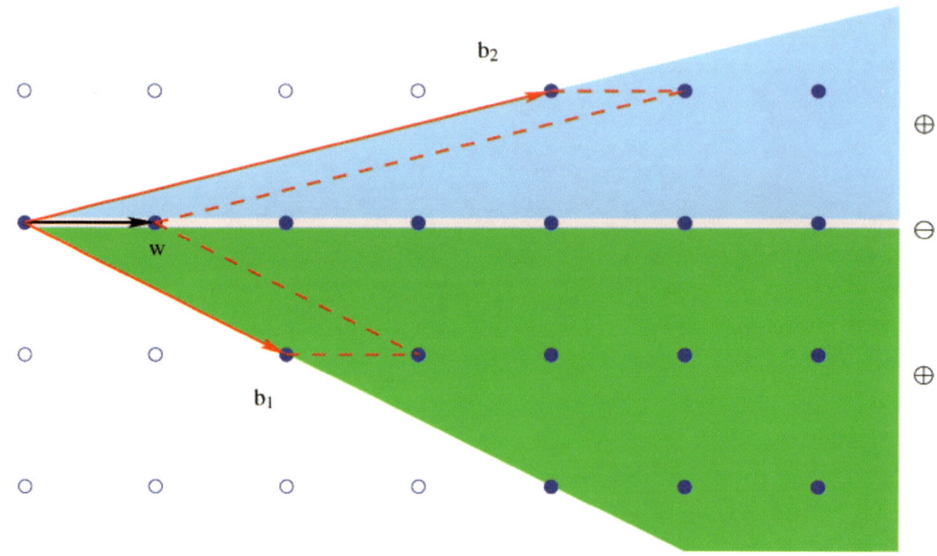

Figure 5.8. *A two-dimensional cone of index 6, triangulated into two unimodular cones. The integer points in the one-dimensional intersection would be counted twice, so we subtract them once (inclusion-exclusion principle).*

We show in our example that triangulations (which appeared briefly in Section 2.1) can be used to reduce the index of cones. Indeed, by using an interior vector $\mathbf{w} = (1,0)$, we can triangulate the cone into the cones $C_1 = \text{cone}\{\mathbf{b}_1, \mathbf{w}\}$ and $C_2 = \text{cone}\{\mathbf{w}, \mathbf{b}_2\}$. For each of the cones, the fundamental parallelepiped contains a single integer point—the origin; see Figure 5.8. Such cones are called *unimodular* cones. We can write down their generating functions:

$$\begin{aligned} g(C_1; z_1, z_2) &= \frac{1}{(1 - z_1^{\alpha} z_2^{-1})(1 - z_1)}, \\ g(C_2; z_1, z_2) &= \frac{1}{(1 - z_1^{\beta} z_2^{1})(1 - z_1)}. \end{aligned}$$

Note, however, that if we just added these two functions, all the integer points in the intersection $C_1 \cap C_2$ would be counted *twice*. However, the intersection $C_1 \cap C_2$ is just a one-dimensional cone, whose generating function we can easily write as

$$g(C_1 \cap C_2; z_1, z_2) = \frac{1}{1 - z_1}.$$

Thus, we can fix the overcounting using the inclusion-exclusion principle, writing

$$\begin{aligned} g(C; z_1, z_2) &= g(C_1; z_1, z_2) + g(C_2; z_1, z_2) - g(C_1 \cap C_2; z_1, z_2) \\ &= \frac{1}{(1 - z_1^{\alpha} z_2^{-1})(1 - z_1)} + \frac{1}{(1 - z_1^{\beta} z_2^{1})(1 - z_1)} - \frac{1}{1 - z_1}. \end{aligned} \tag{5.19}$$

We have now obtained a short formula.

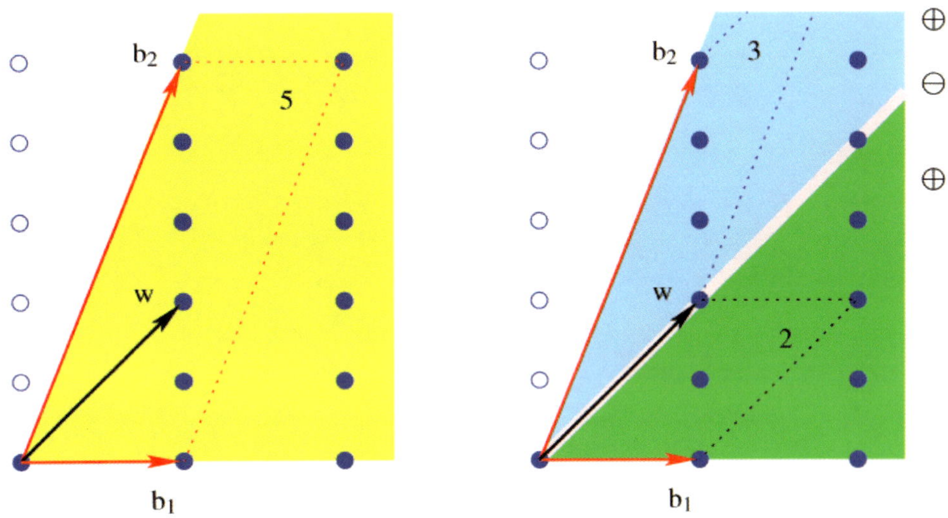

Figure 5.9. *A triangulation of the cone of index* 5 *generated by* $\mathbf{b}^1$ *and* $\mathbf{b}^2$ *into the two cones spanned by* $\{\mathbf{b}^1, \mathbf{w}\}$ *and* $\{\mathbf{b}^2, \mathbf{w}\}$*, having an index of* 2 *and* 3*, respectively. We have the inclusion-exclusion formula* $g(\text{cone}\{\mathbf{b}_1, \mathbf{b}_2\}; \mathbf{z}) = g(\text{cone}\{\mathbf{b}_1, \mathbf{w}\}; \mathbf{z}) + g(\text{cone}\{\mathbf{b}_2, \mathbf{w}\}; \mathbf{z}) - g(\text{cone}\{\mathbf{w}\}; \mathbf{z})$*; here the one-dimensional cone spanned by* $\mathbf{w}$ *needed to be subtracted.*

The bad news is that triangulations do not always help to reduce the size of the formula. Let us consider another two-dimensional example, a cone C' generated by $\mathbf{b}_1 = (1,0)$ and $\mathbf{b}_2 = (1,\alpha)$ (an example for $\alpha = 5$ is shown in Figure 5.9). The integer points in the fundamental parallelepiped are

$$\Pi' \cap \mathbb{Z}^2 = \{(0,0)\} \cup \{(1,i) : i = 1, \ldots, \alpha - 1\},$$

so again the rational generating function would have α summands in the numerator, and thus have exponential size. Unfortunately, every attempt to use triangulations to reduce the size of the formula fails in this example. The choice of an interior vector $\mathbf{w}$ in Figure 5.9, for instance, splits the cone of index 5 into two cones of index 2 and 3, respectively, and also a one-dimensional cone. Indeed, every possible triangulation of C' into unimodular cones contains at least α two-dimensional cones!

Signed decompositions. An important new idea by Barvinok [30] was to use so-called *signed decompositions* in addition to triangulations in order to reduce the index of a cone. In our example, we can choose the vector $\mathbf{w} = (0,1)$ *from the outside of the cone* to define cones $C_1 = \text{cone}\{\mathbf{b}_1, \mathbf{w}\}$ and $C_2 = \text{cone}\{\mathbf{w}, \mathbf{b}_2\}$; see Figure 5.10. Using these cones, we have the inclusion-exclusion formula

$$g(C'; z_1, z_2) = g(C_1; z_1, z_2) - g(C_2; z_1, z_2) + g(C_1 \cap C_2; z_1, z_2).$$

It turns out that both cones C_1 and C_2 are unimodular, with the only integer point in the fundamental parallelepiped being the origin for both cones. We obtain the rational generating

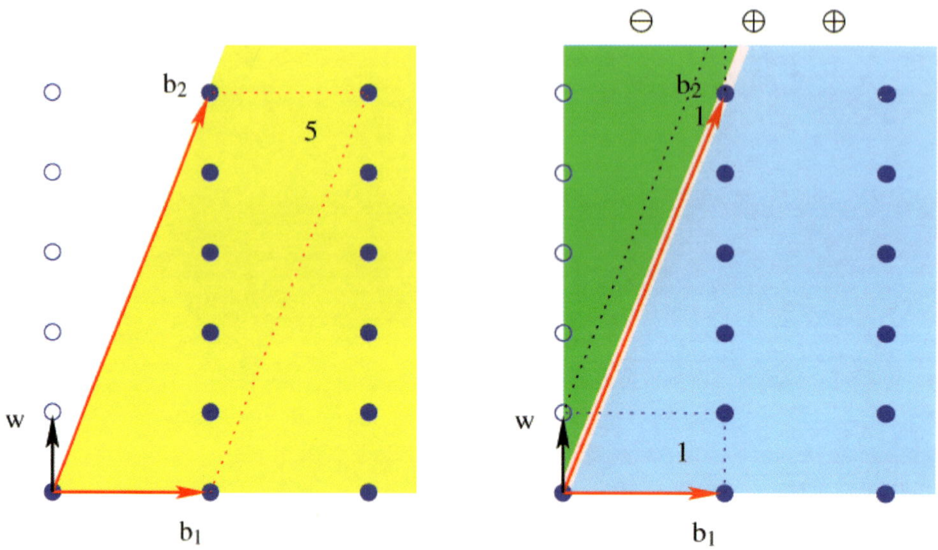

Figure 5.10. *A signed decomposition of the cone of index* 5 *generated by* $\mathbf{b}^1$ *and* $\mathbf{b}^2$ *into the two unimodular cones spanned by* $\{\mathbf{b}^1, \mathbf{w}\}$ *and* $\{\mathbf{b}^2, \mathbf{w}\}$*. We have the inclusion-exclusion formula* $g(\text{cone}\{\mathbf{b}_1, \mathbf{b}_2\}; \mathbf{z}) = g(\text{cone}\{\mathbf{b}_1, \mathbf{w}\}; \mathbf{z}) - g(\text{cone}\{\mathbf{b}_2, \mathbf{w}\}; \mathbf{z}) + g(\text{cone}\{\mathbf{w}\}; \mathbf{z})$.

functions

$$g(C_1; z_1, z_2) = \frac{1}{(1-z_1)(1-z_2)},$$
$$g(C_2; z_1, z_2) = \frac{1}{(1-z_1 z_2^{\alpha})(1-z_2)},$$
$$g(C_1 \cap C_2; z_1, z_2) = \frac{1}{1-z_1 z_2^{\alpha}};$$

hence the short formula,

$$g(C'; z_1, z_2) = \frac{1}{(1-z_1)(1-z_2)} - \frac{1}{(1-z_1^1 z_2^{\alpha})(1-z_2)} + \frac{1}{1-z_1 z_2^{\alpha}}.$$

In general, as we will see in the next chapters, we will not be able to obtain a decomposition into unimodular cones in one simple step; however, we will show that a signed decomposition can be constructed that provably reduces the index of cones very quickly.

We continue in the next chapter with a preparation for the general multidimensional theory, which is then developed in Chapter 7.

5.4 Notes and further references

It is often convenient to make a change of variables, $z = \mathrm{e}^{\xi}$ with $\xi \in \mathbb{C}$, in the generating function. In Formula (5.2), this leads to the *exponential sum*

$$g(S; \mathrm{e}^{\xi}) = \mathrm{e}^{0\xi} + \mathrm{e}^{1\xi} + \cdots + \mathrm{e}^{(n-1)\xi} + \mathrm{e}^{n\xi} = \frac{1}{1-\mathrm{e}^{\xi}} - \frac{\mathrm{e}^{(n+1)\xi}}{1-\mathrm{e}^{\xi}}. \tag{5.20}$$

The terms in the right-hand side of (5.20) are now meromorphic functions of ξ. This setting is convenient, particularly for a mixed-integer setting, in which both exponential sums and exponential integrals appear [20, 22]. We do not use it in this book, however. The relations of our generating functions and exponential sums to various classical discrete and continuous transforms, such as the Laplace transform, are discussed, for instance, in the monograph [217].

5.5 Exercises

Exercise 5.5.1. Suppose that $a < b$ are rational numbers. Let $P = [a,b]$. Write a formula for the generating function $g(P;z)$ as a rational function.

Exercise 5.5.2. Let P be an axis-parallel square with rational coordinates. Write a formula for $g(P;z_1,z_2)$ as a rational function.

Exercise 5.5.3. Verify the list of integer points in the fundamental parallelepiped given in (5.18) using the method of Lemma 2.3.16.

Exercise 5.5.4. Find the domain of convergence of the series corresponding to Figure 5.5.

Exercise 5.5.5. Let C be the cone generated by the vectors (α,β) and (γ,δ), where $\alpha,\beta,\gamma,\delta$ are random integers between -10 and 10. Determine the points in the fundamental parallelepiped of C. Then investigate possible triangulations and signed decompositions to reduce the length of the rational generating function.

Exercise 5.5.6. The cone in Figure 5.8 could be triangulated into two unimodular cones, so the sum of the indices of the subcones is $1+1=2$, which is smaller than 6, the index of the original cone. However, the same calculation for the cone in Figure 5.9 gives $2+3=5$, which equals the index of the original cone. Characterize the two-dimensional cones for which there is a triangulation such that the sum of the indices of the subcones is smaller than the index of the original cone.

Chapter 6

Decompositions of Indicator Functions of Polyhedra

In this chapter we study useful decompositions of polyhedra. They are unrelated to the decompositions with respect to the operation of taking Minkowski sums, which were studied in Section 1.4. Rather, they take place in the vector space generated by indicator functions of polyhedra. We ignore questions regarding lattice points for the moment; they will become relevant again in Chapter 7.

6.1 Indicator functions and inclusion-exclusion

Definition 6.1.1. The *indicator function* of a set $A \subseteq \mathbb{R}^n$ will be denoted by $[A]$, so it is the function $[A] : \mathbb{R}^n \to \mathbb{R}$ with $[A](\mathbf{x}) = 1$ if $\mathbf{x} \in A$ and 0 otherwise.

The indicator functions of sets of $\mathbb{R}^n$ span a vector space by pointwise addition (and scalar multiplication), which is also an *algebra* with respect to pointwise multiplication of functions. These operations are convenient because they represent the intersection of sets

$$[A] \cdot [B] = [A \cap B]$$

and the disjoint union

$$[A] + [B] = [A \cup B] \qquad \text{if } A \cap B = \emptyset.$$

Calculations with indicator functions often give short and elegant proofs. We illustrate this by giving a proof of the standard inclusion-exclusion identity that generalizes the formula

$$[A_1 \cup A_2] = [A_1] + [A_2] - [A_1 \cap A_2]$$

to several sets. Inclusion-exclusion is an important principle in combinatorics; again we refer to [314]. We will use this formula shortly.

Lemma 6.1.2 (inclusion-exclusion). *Let $A_1, \ldots, A_m \subseteq \mathbb{R}^n$. Let $M = \{1, \ldots, m\}$. Then*

$$[A_1 \cup \cdots \cup A_m] = \sum_{\emptyset \neq I \subseteq M} (-1)^{|I|-1} \left[\bigcap_{i \in I} A_i \right].$$

Proof. We write the "De Morgan formula"

$$\mathbb{R}^n \setminus (A_1 \cup \cdots \cup A_m) = (\mathbb{R}^n \setminus A_1) \cap \cdots \cap (\mathbb{R}^n \setminus A_m)$$

as

$$1 - [A_1 \cup \cdots \cup A_m] = (1 - [A_1]) \cdots (1 - [A_m]).$$

Multiplying out gives

$$1 - [A_1 \cup \cdots \cup A_m] = 1 + \sum_{\emptyset \neq I \subseteq M} (-1)^{|I|} \prod_{i \in I} [A_i],$$

which proves the result. □

6.2 Gram–Brianchon and Brion

We now apply this identity to obtain an interesting formula for the indicator function of a simplex in terms of the tangent cones of its faces.

Definition 6.2.1. Let $P \subseteq \mathbb{R}^n$ be a polyhedron. Let $\mathbf{x} \in P$. Then the *tangent cone* (*supporting cone*) of P at $\mathbf{x}$ is the shifted ("affine") polyhedral cone defined by

$$\operatorname{tcone}(P, \mathbf{x}) = \mathbf{x} + \operatorname{cone}(P - \mathbf{x}).$$

The tangent cone $\operatorname{tcone}(P, \mathbf{x})$ is the same as the cone of feasible directions of P at $\mathbf{x}$, with the apex shifted to $\mathbf{x}$. The inequality description of the tangent cones consists of the inequalities of P that are *active* (tight, satisfied with equality) at $\mathbf{x}$.

We will need the tangent cones of vertices and also, more generally, those of faces.

Definition 6.2.2. Let $F \subseteq P$ be a face. Then the *tangent cone* of F is defined as

$$\operatorname{tcone}(P, F) = \operatorname{tcone}(P, \mathbf{x}_F),$$

where $\mathbf{x}_F$ is any point in the relative interior of F.

Note that the tangent cone of a face F always contains the affine hull of the face, and so the tangent cones of vertices are the only ones that are pointed (cf. Section 1.4).

Theorem 6.2.3 (Brianchon [60], Gram [143]). *Let $P \subseteq \mathbb{R}^n$ be a polyhedron. Then*

$$[P] = \sum_{\substack{\emptyset \neq F \\ \text{face of } P}} (-1)^{\dim F} [\operatorname{tcone}(P, F)],$$

where the summation includes the face $F = P$.

We only prove it for the case of the standard simplex $\Delta = \operatorname{conv}\{\mathbf{e}_i : i = 1, \ldots, n\}$. The theorem holds, however, for arbitrary (also, unbounded) polyhedra. A short complete proof can be found in [309]; see also [224] and [32, Section VIII.4].

Proof. The simplex has the inequality description

$$\Delta = \left\{ \mathbf{x} \in \mathbb{R}^n : \mathbf{1}^\top \mathbf{x} = 1,\ x_i \geq 0 \text{ for } i = 1, \ldots, n \right\}.$$

Its affine hull is the hyperplane

$$A = \left\{ \mathbf{x} \in \mathbb{R}^n : \mathbf{1}^\top \mathbf{x} = 1 \right\}.$$

At a vertex $\mathbf{e}_i$, all inequalities "$x_j \geq 0$" for $j \neq i$ are tight, so the corresponding tangent cone has the description

$$\operatorname{tcone}(\Delta, \mathbf{e}_i) = \left\{ \mathbf{x} \in A : x_j \geq 0 \text{ for } j \neq i \right\}.$$

More generally, the simplex has precisely 2^n faces (including the empty set and Δ itself). They are indexed by the subsets $I \subseteq \{1, \ldots, n\}$,

$$F_I = \{ \mathbf{x} \in \Delta : x_j = 0 \text{ for } j \in I \} = \operatorname{conv}\{ \mathbf{e}_i : i \notin I \}.$$

Thus $\dim F_I = n - |I| - 1$. We have

$$\operatorname{tcone}(\Delta, F_I) = \{ \mathbf{x} \in A : x_j \geq 0 \text{ for } j \in I \}.$$

The facets (largest proper faces) of Δ are indexed by singletons $\{j\}$, and their tangent cones are affine half spaces of A,

$$\operatorname{tcone}(\Delta, F_{\{j\}}) = \{ \mathbf{x} \in A : x_j \geq 0 \} =: H_j.$$

Thus, in general, we can write each tangent cone as an intersection of these half spaces,

$$\operatorname{tcone}(\Delta, F_I) = \bigcap_{j \in I} H_j,$$

and of course

$$\Delta = \bigcap_{j=1}^{n} H_j.$$

On the other hand, the *union* of the H_j is the entire affine space A (because if the sum of the coordinates $\mathbf{1}^\top \mathbf{x}$ is positive, at least one coordinate is positive).

From our inclusion-exclusion lemma, we get

$$[A] = [H_1 \cup \cdots \cup H_n] = \sum_{\emptyset \neq I \subseteq [n]} (-1)^{|I|-1} \left[\bigcap_{i \in I} H_i \right].$$

Since $A = \operatorname{tcone}(\Delta, \Delta)$, we get

$$[\operatorname{tcone}(\Delta, \Delta)] = \sum_{\emptyset \neq I \subsetneq [n]} (-1)^{|I|-1} \left[\operatorname{tcone}(\Delta, F_I) \right] + (-1)^{|n|-1} [\Delta]$$

and, by rearranging, the desired identity. □

When we read this identity "modulo contributions of nonpointed polyhedra," it simplifies considerably because the only tangent cones that are pointed belong to the vertices; see Figure 6.1.

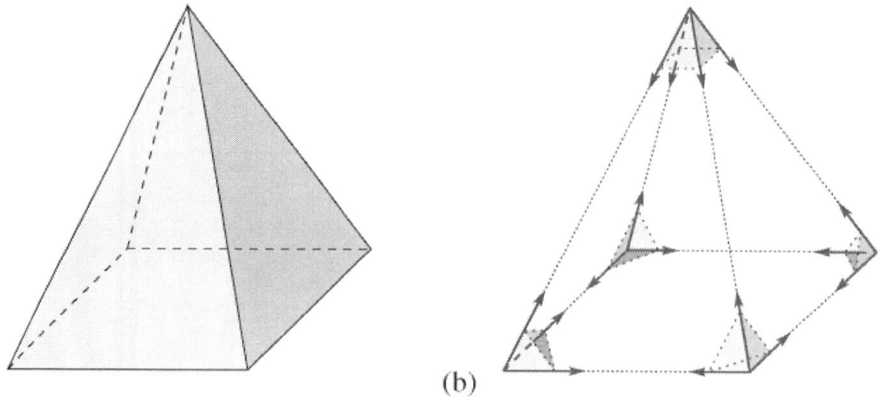

Figure 6.1. *Brion's theorem.*

Corollary 6.2.4 (Brion's theorem: Decomposition modulo nonpointed polyhedra). *Let $P \subseteq \mathbb{R}^n$ be a polyhedron. Then*

$$[P] \equiv \sum_{\mathbf{v} \text{ vertex of } P} [\operatorname{tcone}(P,\mathbf{v})] \qquad (\text{mod } \textit{indicator functions of nonpointed polyhedra}).$$

This is an important viewpoint, because we will later see that these contributions of nonpointed polyhedra are irrelevant for the purpose of computing rational generating functions. We will make use of the simplification of reading identities modulo contributions of nonpointed polyhedra to obtain simpler and faster algorithms.

6.3 Triangulations of cones and polytopes

Recall that an n-dimensional polyhedron P is called *simple* if every supporting cone $[\operatorname{tcone}(P,\mathbf{v})]$ is *simplicial*, i.e., generated by n linearly independent vectors. If P fails to be simple, there is a degenerate vertex $\mathbf{v}$, which has a supporting cone $[\operatorname{tcone}(P,\mathbf{v})]$ that is generated by more than n vectors. In this section we show how to construct a decomposition (triangulation) of such a cone into simplicial cones.

We start by defining triangulations first for polytopes (or point configurations); later we extend this notion to polyhedral cones (or vector configurations).

Triangulations of polytopes. Recall that a *simplex* is the convex hull of any set of affinely independent points. A *point configuration* is a finite set A of points in $\mathbb{R}^n$. We are particularly interested in the case where A is in *convex position*, when A consists of the vertices of a polytope Q. Our goal is to describe and compute decompositions of the convex hull of A using simplices whose vertices form a subset of A.

Definition 6.3.1. A *triangulation* of a point configuration A is a set Γ of simplices S_i of the same dimension as the affine hull of A with the following properties:

(i) The union of all the simplices equals $\operatorname{conv}(A)$.

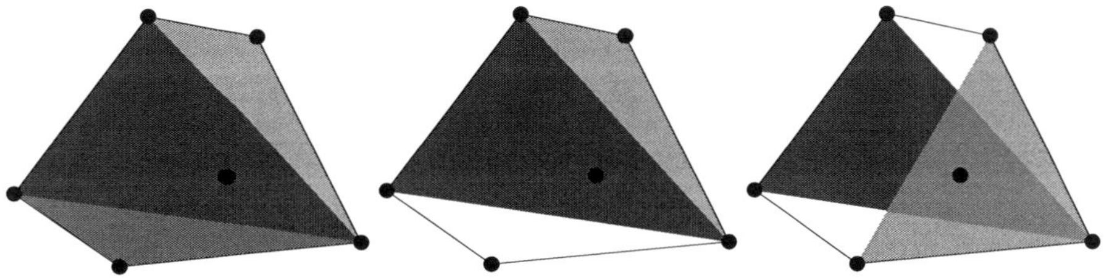

Figure 6.2. *Triangulations and nontriangulations. Out of these three pictures, only the left one gives a triangulation. In the middle picture, condition* (i) *of Definition* 6.3.1 *is violated. In the right picture, two full-dimensional simplices have an intersection which is another full-dimensional simplex, and thus not a face.* [102]

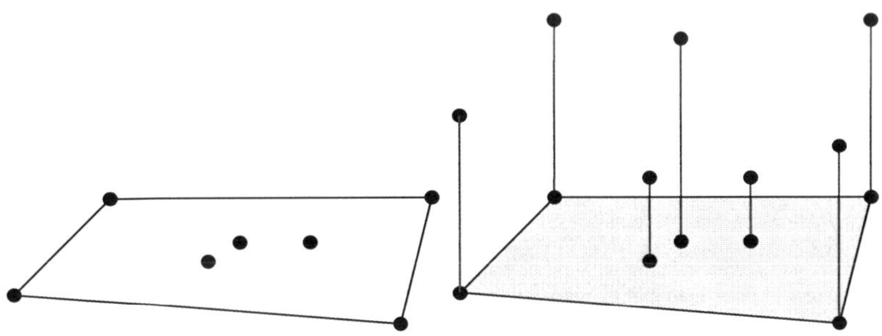

Figure 6.3. *A lifted configuration.*

(ii) Any pair of simplices intersects in a (possibly empty) common face.

(iii) The vertex set of each simplex is contained in A.

Figure 6.2 illustrates this definition. We remark that a triangulation does not need to use all (interior) points of A.

We note that one can generalize triangulations to the notion of a *polytopal subdivision*. This is a collection of polytopes that satisfies all of the same properties of triangulations, except that the polytopes are not required to be simplices.

We next introduce an effective way to compute triangulations.

ALGORITHM 6.1. Regular triangulation algorithm.

input a point configuration $A = \{\mathbf{a}_1, \dots, \mathbf{a}_d\} \subseteq \mathbb{R}^n$, a vector $\mathbf{h} = (h_1, \dots, h_d)^\top \in \mathbb{R}^d$ of *heights*.
output a polytopal subdivision Γ of $Q = \operatorname{conv}(A)$.
Consider the *lifted configuration*

$$\tilde{A} = \left\{ \begin{pmatrix} \mathbf{a}_1 \\ h_1 \end{pmatrix} \cdots \begin{pmatrix} \mathbf{a}_d \\ h_d \end{pmatrix} \right\} \subseteq \mathbb{R}^{n+1},$$

which is illustrated in Figure 6.3.
Compute the convex hull $\tilde{Q} = \operatorname{conv}(\tilde{A})$ of the lifted points.
Recover the *lower envelope* of $\operatorname{conv}(\tilde{A})$; see Figure 6.4. These are exactly those facets F of $\tilde{Q}$ that satisfy $\mathbf{x} - \lambda \mathbf{e}_{n+1} \notin \operatorname{conv}(\tilde{A})$ for each $\mathbf{x} \in F$ and $\lambda > 0$. Equivalently, F

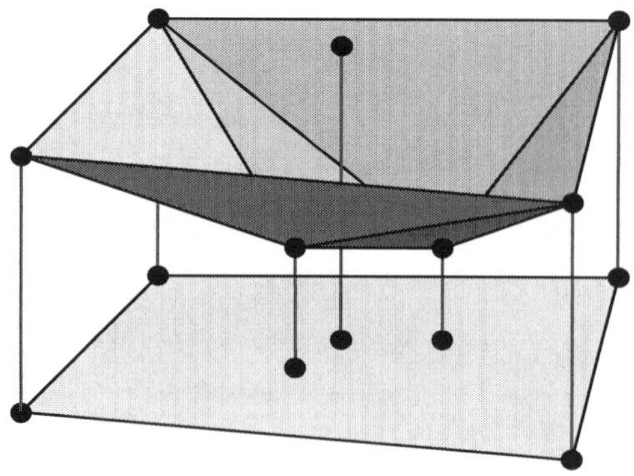

Figure 6.4. *Lower envelope of a lifted configuration.*

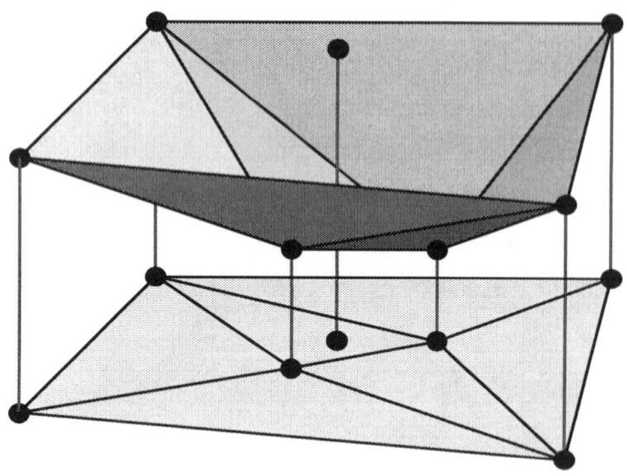

Figure 6.5. *Projecting down a lifted configuration.*

is a lower facet if $F = \{\mathbf{x} \in P : \mathbf{c}^\top \mathbf{x} = c_0\}$, where $\mathbf{c}^\top \mathbf{x} \le c_0$ is valid for $\operatorname{conv}(\tilde{A})$ and $c_{n+1} < 0$.

6: Project down to the faces of the lower envelope to $\mathbb{R}^n$. The projected faces form the desired subdivision Γ of Q; see Figure 6.5.

Definition 6.3.2. A subdivision Γ of a point configuration $A \subseteq \mathbb{R}^n$ is said to be *regular* or *convex* if it can be produced by Algorithm 6.1.

If the height vector $\mathbf{h}$ is sufficiently "generic," then it gives rise to a triangulation of A, and otherwise only to a subdivision of A. Note that the set of height vectors that give subdivisions, but not triangulations, is a measure zero set in $\mathbb{R}^n$, as it is the union of finitely many hyperplanes describing possible coplanarities or collinearities. Note that different lifting vectors $\mathbf{h}, \mathbf{h}'$ may still provide the same triangulation.

It is worth mentioning that several famous triangulation constructions, e.g., Delaunay triangulations, are in fact regular triangulations. What is more surprising perhaps is that, for some configurations A, not all triangulations are regular.

Example 6.3.3. Here is *the* classical example of six points laid out in two concentric equilateral triangles of parallel sides. For the convenience of having rational coordinates we

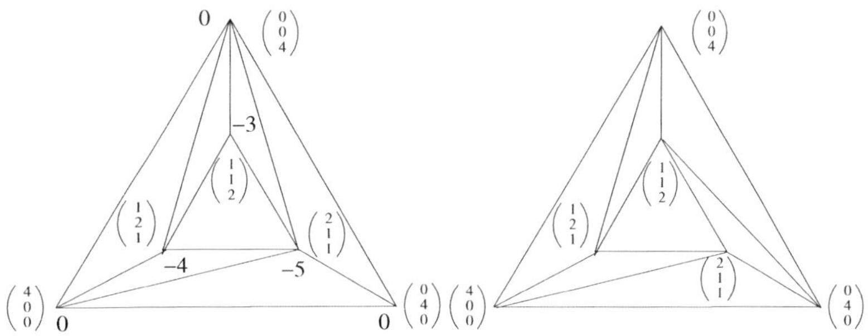

Figure 6.6. *A regular and a nonregular triangulation of a point configuration.*

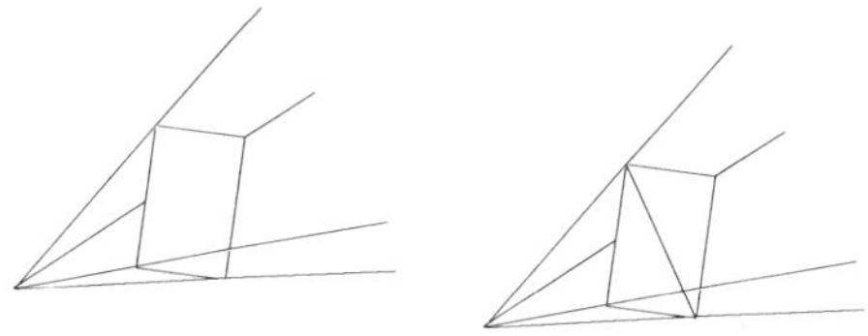

Figure 6.7. *Triangulation of pointed cones from triangulations of point configurations.*

show the six points not embedded in $\mathbb{R}^2$, but instead we set them inside the hyperplane $x_1 + x_2 + x_3 = 4$. The points are the columns of the matrix

$$A = \begin{pmatrix} 4 & 0 & 0 & 2 & 1 & 1 \\ 0 & 4 & 0 & 1 & 2 & 1 \\ 0 & 0 & 4 & 1 & 1 & 2 \end{pmatrix}.$$

Figure 6.6 shows one triangulation of the set that is regular (left figure). We give explicit height values that make it so, but the other triangulation (right figure), is not possible to obtain from lifting vectors. Why? We leave this as an exercise for the reader.

Triangulations of polyhedral cones. Let B be a *vector configuration*, i.e., a finite set B of vectors in $\mathbb{R}^n$. We assume that B spans a pointed cone $C = \text{cone}(B)$. We would like to decompose C in terms of *simplicial cones*, i.e., cones that are generated by linearly independent vectors. Triangulations of a pointed cone can be reduced to triangulations of point sets as follows. Since C is pointed, there exists some $\mathbf{c} \in \mathbb{R}^n$ such that $C \subseteq \{\mathbf{x} \in \mathbb{R}^n : \mathbf{c}^\top \mathbf{x} \geq 0\}$ and $C \cap \{\mathbf{x} \in \mathbb{R}^n : \mathbf{c}^\top \mathbf{x} = 0\} = \{\mathbf{0}\}$. Then $Q := C \cap \{\mathbf{x} \in \mathbb{R}^n : \mathbf{c}^\top \mathbf{x} = 1\}$ is a bounded section of C, and thus a polytope (see Figure 6.7). Let Γ_Q be a triangulation of Q, for instance one obtained by Algorithm 6.1. Every simplex $S_i \in \Gamma_Q$ then gives rise to a simplicial cone $C_i = \text{cone}(S_i)$. The collection Γ of these cones then is a triangulation of C, satisfying the following definition.

Definition 6.3.4. A *triangulation* of a vector configuration B is a set Γ of simplicial cones C_i of the same dimension as the linear hull B with the following properties:

(i) The union of all the simplicial cones equals $\text{cone}(B)$.

(ii) Any pair of cones intersects in a (possibly empty) common face.

(iii) Every ray of every simplicial cone has a representative in B.

Triangulations as identities of indicator functions. Now let C be a full-dimensional pointed polyhedral cone. Let us write a triangulation Γ of a cone C as

$$\Gamma = \{ C_i : i \in I_1 \},$$

where I_1 is a finite index set. By definition, we have $C = \bigcup_{i\in I_1} C_i$, but this is not a disjoint union. By the inclusion-exclusion principle (Lemma 6.1.2), we can write the following linear identity of indicator functions:

$$[C] = \left[\bigcup_{i\in I_1} C_i\right] = \sum_{\emptyset\neq J\subseteq I_1} (-1)^{|J|-1}\left[\bigcap_{j\in J} C_j\right]. \tag{6.1}$$

The terms $\bigcap_{j\in J} C_j$ with $|J| = 1$ are just the (full-dimensional) cones C_i in the triangulation. The terms $\bigcap_{j\in J} C_j$ with $|J| \geq 2$ are proper intersections of cones and, by virtue of property (ii), they are proper common faces of the cones C_i. In particular, they are lower-dimensional cones. The same face may arise from different index sets J. We now collect the corresponding terms $[\bigcap_{j\in J} C_j]$ as follows. Let I_2 be an index set for the different proper faces, and denote the faces by C_i for $i \in I_2$. Collecting the terms give rise to general integer coefficients ϵ_i, not just ± 1 as in (6.1). We obtain a linear identity of the form

$$[C] = \sum_{i\in I_1} \epsilon_i [C_i] + \sum_{i\in I_2} \epsilon_i [C_i]. \tag{6.2}$$

The same identity holds, of course, if we shift all cones by the same translation vector. For example, the triangulation of a tangent cone of a vertex $\mathbf{v}$, as it appears in Brion's theorem (Corollary 6.2.4), would be expressed by an identity of the form

$$[\operatorname{tcone}(P,\mathbf{v})] = \sum_{i\in I_1} \epsilon_i [\mathbf{v} + C_i] + \sum_{i\in I_2} \epsilon_i [\mathbf{v} + C_i]. \tag{6.3}$$

6.4 Avoiding inclusion-exclusion with half-open decompositions

In this section we show that identities of indicator functions of full-dimensional polyhedra modulo lower-dimensional polyhedra can be translated into "exact" identities of indicator functions of full-dimensional *half-open polyhedra*. Thus, in triangulations and other decompositions, it is possible to avoid any hint of inclusion-exclusion, or other computationally intensive standard techniques such as shelling. This improves the computational complexity of the methods.

Figure 6.8 shows an example of a triangulation, expressed as an identity of indicator functions of two-dimensional polyhedra modulo lower-dimensional polyhedra. Figure 6.9 illustrates the basic idea of making some of the polyhedra half-open and also shows an ad-hoc construction that works for the one-dimensional faces (rays), but not the zero-dimensional face (apex). Figure 6.10 shows a correct construction, which has been made according to the following theorem.

Theorem 6.4.1. *Let*

$$[P_0] = \sum_{i\in I_1} \epsilon_i [P_i] + \sum_{i\in I_2} \epsilon_i [P_i] \tag{6.4}$$

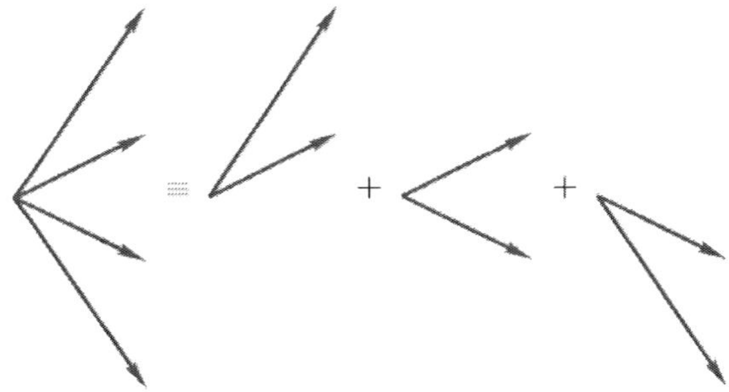

Figure 6.8. *An identity, valid modulo lower-dimensional cones, corresponding to a polyhedral subdivision of a cone.* [203]

be a (finite) linear identity of indicator functions of (closed) polyhedra $P_i \subseteq \mathbb{R}^n$, *where the polyhedra* P_0 *and* P_i *for* $i \in I_1$ *are full dimensional and the polyhedra* P_i *for* $i \in I_2$ *are lower dimensional, and where* $\epsilon_i \in \mathbb{Q}$. *Let each closed polyhedron be given as*

$$P_i = \left\{ \mathbf{x} : {\mathbf{b}^*_{i,j}}^\top \mathbf{x} \leq \beta_{i,j} \text{ for } j \in J_i \right\}. \tag{6.5}$$

Let $\mathbf{y} \in \mathbb{R}^n$ *be a vector such that*

$${\mathbf{b}^*_{0,j}}^\top \mathbf{y} < 0 \quad \forall j \in J_0, \tag{6.6}$$

$${\mathbf{b}^*_{i,j}}^\top \mathbf{y} \neq 0 \quad \forall i \in I_1, j \in J_i. \tag{6.7}$$

For $i \in I_1$, *we define the half-open polyhedron*

$$\begin{aligned}\tilde{P}_i = \Big\{ \mathbf{x} \in \mathbb{R}^n : {\mathbf{b}^*_{i,j}}^\top \mathbf{x} \leq \beta_{i,j} \text{ for } j \in J_i \text{ with } {\mathbf{b}^*_{i,j}}^\top \mathbf{y} < 0, \\ {\mathbf{b}^*_{i,j}}^\top \mathbf{x} < \beta_{i,j} \text{ for } j \in J_i \text{ with } {\mathbf{b}^*_{i,j}}^\top \mathbf{y} > 0 \Big\}.\end{aligned} \tag{6.8}$$

Then

$$[P_0] = \sum_{i \in I_1} \epsilon_i [\tilde{P}_i]. \tag{6.9}$$

The geometry of Theorem 6.4.1 is illustrated in Figures 6.8 and 6.10.

Proof. Without loss of generality, we can further assume that also

$${\mathbf{b}^*_{i,j}}^\top \mathbf{y} \neq 0 \quad \forall i \in I_2, j \in J_i. \tag{6.10}$$

We will show that (6.9) holds for an arbitrary $\bar{\mathbf{x}} \in \mathbb{R}^n$. Observe that, by (6.6), we have $\tilde{P}_0 = P_0$. Setting $\epsilon_0 = -1$ and $I_{0,1} = I_1 \cup \{0\}$, we now prove the simpler identity $\sum_{i \in I_{0,1}} \epsilon_i [\tilde{P}_i] = 0$.

To this end, fix an arbitrary $\bar{\mathbf{x}} \in \mathbb{R}^n$. We define $\mathbf{x}_\lambda = \bar{\mathbf{x}} + \lambda \mathbf{y}$ for $\lambda \in [0, +\infty)$. Consider the function

$$f \colon [0, +\infty) \ni \lambda \mapsto \left(\sum_{i \in I_{0,1}} \epsilon_i [\tilde{P}_i] \right) (\mathbf{x}_\lambda).$$

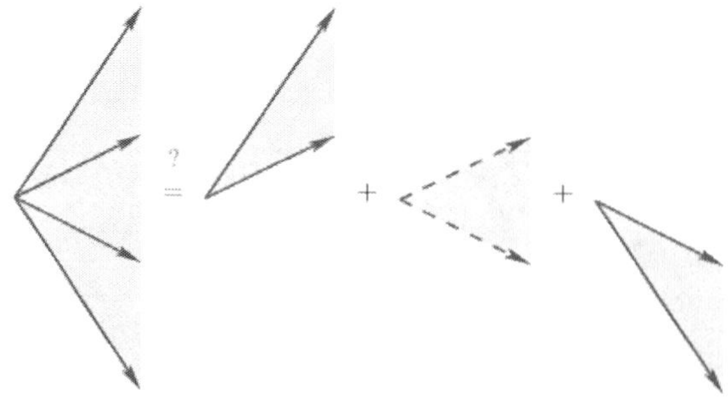

Figure 6.9. *The technique of half-open exact decomposition. The above ad hoc choice of strict inequalities (dashed lines) and weak inequalities (solid lines) appears to give an exact identity on first look. However, the apex of the cone is still counted twice.* [203]

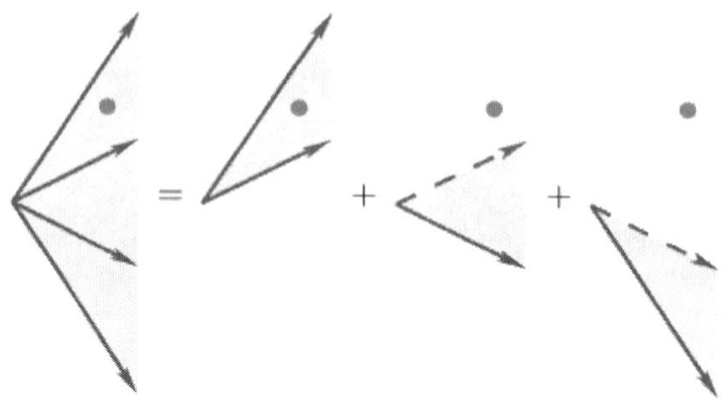

Figure 6.10. *The technique of half-open exact decomposition. The relative location of the vector* **y** *(represented by a dot) determines which defining inequalities are strict (dashed lines) and which are weak (solid lines).* [203]

We need to show that $f(0) = 0$. To this end, we first show that f is constant in a neighborhood of 0.

First, let $i \in I_{0,1}$ such that $\bar{\mathbf{x}} \in \tilde{P}_i$. For $j \in J_i$ with ${\mathbf{b}^*_{i,j}}^\top \mathbf{y} < 0$, we have ${\mathbf{b}^*_{i,j}}^\top \bar{\mathbf{x}} \leq \beta_{i,j}$, and thus ${\mathbf{b}^*_{i,j}}^\top \mathbf{x}_\lambda \leq \beta_{i,j}$. For $j \in J_i$ with ${\mathbf{b}^*_{i,j}}^\top \mathbf{y} > 0$, we have ${\mathbf{b}^*_{i,j}}^\top \bar{\mathbf{x}} < \beta_{i,j}$, and thus ${\mathbf{b}^*_{i,j}}^\top \mathbf{x}_\lambda < \beta_{i,j}$ for $\lambda > 0$ small enough. Hence, $\mathbf{x}_\lambda \in \tilde{P}_i$ for $\lambda > 0$ small enough.

Second, let $i \in I_{0,1}$ such that $\bar{\mathbf{x}} \notin \tilde{P}_i$. Then either there exists a $j \in J_i$ with ${\mathbf{b}^*_{i,j}}^\top \mathbf{y} < 0$ and ${\mathbf{b}^*_{i,j}}^\top \bar{\mathbf{x}} > \beta_{i,j}$. Then ${\mathbf{b}^*_{i,j}}^\top \mathbf{x}_\lambda > \beta_{i,j}$ for $\lambda > 0$ small enough. Otherwise, there exists a $j \in J_i$ with ${\mathbf{b}^*_{i,j}}^\top \mathbf{y} > 0$ and ${\mathbf{b}^*_{i,j}}^\top \bar{\mathbf{x}} \geq \beta_{i,j}$. Then ${\mathbf{b}^*_{i,j}}^\top \mathbf{x}_\lambda \geq \beta_{i,j}$. Hence, in either case, $\mathbf{x}_\lambda \notin \tilde{P}_i$ for $\lambda > 0$ small enough.

Next we show that f vanishes on some interval $(0, \lambda_0)$. We consider the function

$$g\colon [0, +\infty) \ni \lambda \mapsto \left(\sum_{i \in I_{0,1}} \epsilon_i [P_i] + \sum_{i \in I_2} \epsilon_i [P_i] \right) (\mathbf{x}_\lambda)$$

which is identically zero by equation (6.4). Since $[P_i](\mathbf{x}_\lambda)$ for $i \in I_2$ vanishes on all but finitely many λ, we have

$$g(\lambda) = \left(\sum_{i \in I_{0,1}} \epsilon_i [P_i] \right) (\mathbf{x}_\lambda)$$

for λ from some interval $(0, \lambda_1)$. Also, $[P_i](\mathbf{x}_\lambda) = [\tilde{P}_i](\mathbf{x}_\lambda)$ for some interval $(0, \lambda_2)$. Hence $f(\lambda) = g(\lambda) = 0$ for some interval $(0, \lambda_0)$.

Hence, since f is constant in a neighborhood of 0, it is also zero at $\lambda = 0$. Thus the identity (6.9) holds for $\bar{\mathbf{x}}$. □

Remark 6.4.2 (half-open triangulations). For example, let us consider the triangulation of the tangent cone of a vertex again, which is expressed by (6.3) as

$$[\operatorname{tcone}(P, \mathbf{v})] = \sum_{i \in I_1} \epsilon_i [\mathbf{v} + C_i] + \sum_{i \in I_2} \epsilon_i [\mathbf{v} + C_i]. \tag{6.11}$$

We have $\operatorname{tcone}(P, \mathbf{v}) = \mathbf{v} + \operatorname{cone}\{\mathbf{v}_1 - \mathbf{v}, \ldots, \mathbf{v}_k - \mathbf{v}\}$, where $\mathbf{v}_1, \ldots, \mathbf{v}_k$ are the vertices of P that are adjacent to $\mathbf{v}$. Now it is easy to choose a vector $\mathbf{y}$ as required by Theorem 6.4.1. For example, we can choose

$$\mathbf{y} = \sum_{i=1}^{k} (1 + \gamma^i)(\mathbf{v}_i - \mathbf{v})$$

for almost all $\gamma > 0$; there are only finitely many values of γ that hit a facet of one of the cones. By applying Theorem 6.4.1, we obtain the shorter identity

$$[\operatorname{tcone}(P, \mathbf{v})] = \sum_{i \in I_1} [\mathbf{v} + \tilde{C}_i].$$

6.5 Notes and further references

For more on triangulations (Section 6.3), we refer the reader to the monograph [102].

Section 6.4 on the half-open decomposition technique is based on Köppe and Verdoolaege [203], which contains a more detailed discussion on how to choose the vector $\mathbf{y}$, and also derives combinatorial decomposition rules from lexicographical choices of the vector.

6.6 Exercises

Exercise 6.6.1. Write the indicator function of a cube as a sum/subtraction of indicator functions of tangent cones of its faces. Which of these tangent cones contain a line?

Exercise 6.6.2. Prove that all triangulations of an n-gon are regular.

Exercise 6.6.3. Show that the triangulation of the point configuration of Example 6.3.3 given in Figure 6.6 (right) is not regular.

Exercise 6.6.4. List all triangulations of the point configuration in Example 6.3.3.

Chapter 7

Barvinok's Short Rational Generating Functions

7.1 Generating functions and the algorithm of Barvinok

In the following, let $P \subseteq \mathbb{R}^n$ be a rational polyhedron.

Generating functions as formal Laurent series. We first define the generating function of P as a *formal series*, that is, without any consideration of convergence properties. Since we do not wish to confine our polyhedron to the nonnegative orthant, the generating function will have monomials with negative exponents. Therefore a formal *power series* is not sufficient; we need a formal *Laurent series*.

We note that usually formal Laurent series are defined as series of the form

$$\sum_{(-M,\dots,-M)\leq\alpha\in\mathbb{Z}^n} c_\alpha \mathbf{z}^\alpha$$

for some integer M. Here (and in the following) we are using the *multiexponent* notation $\mathbf{z}^\alpha = \prod_{i=1}^n z_i^{\alpha_i}$. Since there is only a finite number of monomials with negative exponents, the multiplication of series is well-defined (because each coefficient of the product series is just a *finite* sum). Hence these "one-sided" infinite series form a ring, which is usually denoted by $\mathbb{Q}((z_1,\dots,z_n))$.

However, in order to deal with arbitrary polyhedra, we will be in need of "two-sided" formal Laurent series, in which an infinite number of monomials with negative exponents can appear. Note that the multiplication of such series is not defined in general, so the series only form a module $\mathbb{Z}[[z_1,\dots,z_n,z_1^{-1},\dots,z_n^{-1}]]$ (over the ring of integers $\mathbb{Z}$ or of Laurent polynomials $\mathbb{Z}[z_1,\dots,z_n,z_1^{-1},\dots,z_n^{-1}]$), but not a ring.

Definition 7.1.1. The *generating function* of $P\cap\mathbb{Z}^n$ is defined as the formal (two-sided) Laurent series

$$\tilde{g}(P;\mathbf{z}) = \sum_{\alpha\in P\cap\mathbb{Z}^n} \mathbf{z}^\alpha \in \mathbb{Z}[[z_1,\dots,z_n,z_1^{-1},\dots,z_n^{-1}]].$$

As we remarked in the introduction, the encoding of the set of lattice points of a polyhedron as a formal Laurent series does not give an immediate benefit in terms of complexity. We will get short formulas only when we can identify the Laurent series with certain rational functions.

The map from formal Laurent series to rational functions. If P is a bounded polyhedron (a polytope) or if P is unbounded, but does not contain any lattice points, then $\tilde{g}(P;\mathbf{z})$ is a *Laurent polynomial* (i.e., a *finite* sum of monomials with arbitrary—positive or negative—integer exponents). Clearly every such Laurent polynomial can be naturally identified with a rational function $g(P;\mathbf{z})$,

$$\mathbb{Z}[z_1,\dots,z_n,z_1^{-1},\dots,z_n^{-1}] \hookrightarrow \mathbb{Q}(z_1,\dots,z_n),$$
$$\mathbf{z}^{\alpha} \mapsto \mathbf{z}^{\alpha}.$$

Convergence comes into play whenever P is not bounded, since then $\tilde{g}(P;\mathbf{z})$ can be an infinite formal sum. We first consider the case of a *pointed* polyhedron P (i.e., P does not contain a straight line).

Theorem 7.1.2. *Let $P \subseteq \mathbb{R}^n$ be a pointed rational polyhedron. Then there exists a nonempty open subset $U \subseteq \mathbb{C}^n$ such that the series $\tilde{g}(P;\mathbf{z})$ converges absolutely and uniformly on every compact subset of U to a rational function $g(P;\mathbf{z}) \in \mathbb{Q}(z_1,\dots,z_n)$. The rational function $g(P;\mathbf{z})$ is independent of the choice of U.*

Remark 7.1.3. An arbitrary formal Laurent series $\tilde{g}(\mathbf{z}) \in \mathbb{Z}[[z_1,\dots,z_n,z_1^{-1},\dots,z_n^{-1}]]$, when it converges absolutely and uniformly on every compact subset of some nonempty open set U, usually defines a *meromorphic function* $g(\mathbf{z}) \in \mathbb{C}((z_1,\dots,z_n))$ on U. This is a much larger class of functions than rational functions. This already happens when we allow irrational polyhedra.

Proof of Theorem* 7.1.2 *(sketch). First consider the case of a *simplicial rational cone*, i.e., a cone generated by linearly independent *basis vectors* $\mathbf{b}_1,\dots,\mathbf{b}_n \in \mathbb{Z}^n$. These vectors are the representatives of the extreme rays.

By making the basis vectors the columns of some matrix $B \in \mathbb{Z}^{n\times n}$, we can write $K = \mathbf{v} + B\mathbb{R}^n_+$. We assume that the basis vectors are primitive vectors of the standard lattice $\mathbb{Z}^n$. Then the *index* of K is defined to be $\operatorname{ind}(K) = |\det(B)|$. Recall from Theorem 2.3.19 that this number can be interpreted as the cardinality of $\Pi \cap \mathbb{Z}^n$, where Π is the fundamental parallelepiped of K, i.e., the half-open parallelepiped

$$\Pi = \mathbf{v} + \left\{ \sum_{i=1}^n \lambda_i \mathbf{b}_i : 0 \le \lambda_i < 1 \right\}.$$

Then the generating function is, as illustrated in the two-dimensional case (Section 5.3), given by a geometric series. The region U of convergence is related to the polar cone K° of K (Section 1.3) and is always full dimensional because K is pointed.

For the case of a *pointed rational cone* K, we first compute a triangulation into simplicial cones. Using the technique of Section 6.4, we construct a set-theoretic partition of K into half-open simplicial cones. The corresponding series have domains of convergence that overlap in a full-dimensional set related to the polar cone of K.

Finally, for the case of a *pointed rational polyhedron* P, we form the homogenization of P (Section 1.4), i.e., we consider the cone $K := \{(\xi\mathbf{x},\xi) : \mathbf{x} \in P,\ \xi \ge 0\} \subset \mathbb{R}^{n+1}$. This is a pointed rational cone, to which we can associate the rational function $g(K;\mathbf{z},\zeta)$, where ζ corresponds to the extra dimension. Then

$$\left.\frac{\partial}{\partial \zeta} g(K;\mathbf{z},\zeta)\right|_{\zeta=0}$$

is the desired rational function for P. □

When P contains an integer point and also a straight line, there does not exist any point $\mathbf{z} \in \mathbb{C}^n$ where the series $\tilde{g}(P;\mathbf{z})$ converges absolutely.

Example 7.1.4. We consider the univariate two-sided infinite series

$$\tilde{g}(\mathbb{R};z) = \sum_{k=-\infty}^{+\infty} z^k \in \mathbb{Z}[[z,z^{-1}]], \tag{7.1}$$

which is the generating function of $\mathbb{Z}$. It is clear that this series does not converge absolutely for any $z \in \mathbb{C}$, since for each z, the positive or the negative half-series of magnitudes diverges:

$$\sum_{k=0}^{+\infty} |z|^k = +\infty \quad \text{for } |z| \geq 1,$$

$$\sum_{k=-\infty}^{0} |z|^k = +\infty \quad \text{for } |z| \leq 1.$$

Now let us consider the general case. Let $\mathbf{x} \in P \cap \mathbb{Z}^n$ and $\mathbf{t} \in \mathbb{Z}^n \setminus \{\mathbf{0}\}$ such that $\mathbf{x} + \mathbf{t}\mathbb{R} \subseteq P$. Then $\tilde{g}(P;\mathbf{z})$ contains the subseries

$$\sum_{k=-\infty}^{+\infty} \mathbf{z}^{\mathbf{x}+k\mathbf{t}},$$

which is equivalent (by a monomial substitution) to series (7.1) from the example. Thus there does not exist any point where the series converges absolutely, so we cannot use convergence to define a rational function $g(P;\mathbf{z})$.

Lawrence [225] and, independently, Pukhlikov and Khovanskii [281] showed how to assign a rational function to arbitrary polyhedra in a "consistent" (valuative) way.

Theorem 7.1.5 (Lawrence–Khovanskii–Pukhlikov). *There exists a* linear *map F* (valuation) *from the vector space spanned by the indicator functions of rational polyhedra in $\mathbb{R}^n$ to the space $\mathbb{Q}(z_1,\ldots,z_n)$ of rational functions such that:*

(i) *For a pointed rational polyhedron P, the function $F([P])(\mathbf{z})$ equals the function $g(P;\mathbf{z})$ defined by the above convergent series.*

(ii) *For any integer vector $\mathbf{t} \in \mathbb{Z}^n$, we have $F([\mathbf{t}+P]) = \mathbf{z}^{\mathbf{t}} F([P])$.*

(iii) *For any nonpointed rational polyhedron P, we have $F([P]) = 0$.*

Proof (sketch). For pointed rational polyhedra, define F using g.

Next we show the linearity of the map F for pointed rational polyhedra. Given a linear equation $\sum_i \alpha_i [P_i] = 0$, where the P_i are pointed, one needs to show that $\sum_i \alpha_i F[P_i] = 0$ holds. Note that the series associated with the P_i do not necessarily have a common domain of convergence, but using inclusion-exclusion one can break the formula down to obtain a common domain of convergence, so the definition of $g(P_i,\mathbf{z})$ using convergent series can be used.

Thus, F can be extended by linearity to all rational polyhedra. The translation property holds for pointed polyhedra and clearly extends by linearity. Finally, if P is nonpointed, it contains a rational line, so $[P] = [\mathbf{t}+P]$ for some $\mathbf{0} \neq \mathbf{t} \in \mathbb{Z}^n$, and thus $F([P]) = \mathbf{z}^{\mathbf{t}} F([P])$, and so $F([P]) = 0$ because $\mathbf{z}^{\mathbf{t}}$ is not a zero divisor. $\square$

Taking all together, we define the following.

Definition 7.1.6. The rational function $g(P;\mathbf{z}) = F([P])(\mathbf{z}) \in \mathbb{Q}(z_1,\dots,z_n)$ defined as above is called the *rational generating function* of $P \cap \mathbb{Z}^n$.

Generating function version of Brion's theorem. Because of Theorem 7.1.5 (iii), decompositions of polyhedra, modulo nonpointed polyhedra, give rise to decompositions of rational generating functions. From the decomposition of polyhedra, modulo nonpointed polyhedra, into vertex cones (Corollary 6.2.4), the following theorem follows.

Theorem 7.1.7. *Let P be a rational polyhedron and $V(P)$ be the set of vertices of P. Then*

$$g(P;\mathbf{z}) = \sum_{\mathbf{v}\in V(P)} g(\operatorname{tcone}(P,\mathbf{v});\mathbf{z}).$$

It needs to be remarked that Brion obtained this theorem with different techniques and did so before the results of Lawrence [225] and Pukhlikov and Khovanskii [281] appeared; see also Barvinok and Pommersheim [35]. See also [46] for an interesting discussion of this and related theorems.

We remark that in the case of a nonpointed polyhedron P, i.e., a polyhedron that has no vertices because it contains a straight line, both sides of the equation are zero.

Construction of Barvinok's signed decomposition. By Brion's theorem and triangulations, we have reduced the problem of computing rational generating functions to the base case of a simplicial cone K (spanned by linearly independent integer vectors $\mathbf{b}_1,\dots,\mathbf{b}_n$). The missing link for an efficient algorithm is a procedure to compute a *signed decomposition* of K into other simplicial cones with smaller index.

Theorem 7.1.8 (Barvinok's signed decomposition). *Let $K = \operatorname{cone}\{\mathbf{b}_1,\dots,\mathbf{b}_n\}$ be a simplicial cone in $\mathbb{R}^n$. Let $\mathbf{w}$ be any nonzero vector of $\mathbb{Z}^n$ such that the cone generated by $\mathbf{b}_1,\dots,\mathbf{b}_n,\mathbf{w}$ is pointed. For $i=1,\dots,n$, let K_i denote the cone spanned by the vectors $\mathbf{b}_1,\dots,\mathbf{b}_{i-1},\mathbf{w},\mathbf{b}_{i+1},\dots,\mathbf{b}_n$.*

Then there exist $\epsilon_i \in \{0,\pm 1\}$ such that

$$[K] \equiv \sum_{i=1}^{n} \epsilon_i [K_i] \quad (\text{mod } \textit{indicator functions of lower-dimensional polyhedra}),$$

where $\epsilon_i = 0$ if K_i is a lower-dimensional cone.

We leave the proof as an interesting exercise. In general, these cones form a signed decomposition of K (see Figure 5.10); if $\mathbf{w}$ lies inside K, then $\epsilon_i = 1$, and the cones form a triangulation of K (see Figure 5.9). We remark that if $\operatorname{cone}\{\mathbf{b}_1,\dots,\mathbf{b}_n,\mathbf{w}\}$ is not pointed, then we can just replace $\mathbf{w}$ by $-\mathbf{w}$ and arrive in a situation where we can apply Theorem 7.1.8.

Now the goal is to simultaneously reduce the index of the cones K_i by a specific choice of $\mathbf{w}$. The index of the cone K_i is the absolute value of the determinant of the matrix with columns $\mathbf{b}_1,\dots,\mathbf{b}_{i-1},\mathbf{w},\mathbf{b}_{i+1},\dots,\mathbf{b}_n$. We collect these determinants into a vector:

$$\boldsymbol{\lambda} := \begin{pmatrix} \det(\mathbf{w},\mathbf{b}_2,\dots,\mathbf{b}_{n-1},\mathbf{b}_n) \\ \vdots \\ \det(\mathbf{b}_1,\mathbf{b}_2,\dots,\mathbf{b}_{n-1},\mathbf{w}) \end{pmatrix}.$$

Now let $B^* = (B^{-1})^\top$, with column vectors $\mathbf{b}_1^*,\dots,\mathbf{b}_n^*$, be the dual (biorthonormal) basis, i.e., $\mathbf{b}_i^{*\top}\mathbf{b}_j = \delta_{ij}$. If we write $\mathbf{w} = \sum_{i=1}^n \alpha_i \mathbf{b}^i$, then $\alpha_i = \mathbf{b}_i^{*\top}\mathbf{w}$. Then we can calculate

$$\begin{aligned} \boldsymbol{\lambda} &= \begin{pmatrix} \det(\alpha_1\mathbf{b}_1,\mathbf{b}_2,\dots,\mathbf{b}_{n-1},\mathbf{b}_n) \\ \vdots \\ \det(\mathbf{b}_1,\mathbf{b}_2,\dots,\mathbf{b}_{n-1},\alpha_n\mathbf{b}_n) \end{pmatrix} \\ &= \det(B)\cdot \begin{pmatrix} \mathbf{b}_1^{*\top}\mathbf{w} \\ \vdots \\ \mathbf{b}_n^{*\top}\mathbf{w} \end{pmatrix} \\ &= \det(B)\cdot B^*\mathbf{w}. \end{aligned}$$

Since $\mathbf{w}$ can be chosen as an arbitrary nonzero vector in $\mathbb{Z}^n$, simultaneously reducing the index means to find the shortest nonzero vector $\boldsymbol{\lambda}$ of the lattice $\mathcal{L} = (\det(B)B^*)\mathbb{Z}^n$, with respect to the ℓ_∞-norm. (This construction is due to Dyer and Kannan [118].)

As discussed in Section 2.7, several algorithms for solving this shortest vector problem are available. In fixed dimension, they run in polynomial time. We note that for the complexity results proved below it actually suffices to construct an *approximate* shortest vector, such as the one provided by the first basis vector of an LLL-reduced lattice basis (Theorem 2.7.5). This is also the method employed in the practical computer implementations LattE integrale [103] and barvinok [330].

Estimates for the index descent. Now let $\mathbf{w}^*$ be a shortest vector. Using the bound from Theorem 2.4.4, derived from Minkowski's first theorem, we obtain

$$\|\mathbf{w}^*\|_\infty \le \left((\det B)^n(\det B)^{-1}\right)^{1/n} = (\det B)^{(n-1)/n}.$$

Thus the cones in the decomposition have

$$\log(\operatorname{ind}(K_i)) \le \tfrac{n-1}{n}\log(\operatorname{ind}(K));$$

that is, the logarithm of the index decreases geometrically in this construction.

The resulting full-dimensional cones are recursively processed, until *unimodular* cones (cones of index 1), or cones of index smaller than some chosen constant are obtained. For such low-index cones $\mathbf{v}+B\mathbb{R}^n_+$, the rational generating function can be written down explicitly as

$$\frac{\sum_{\mathbf{a}\in\Pi\cap\mathbb{Z}^n}\mathbf{z}^{\mathbf{a}}}{\prod_{j=1}^n(1-\mathbf{z}^{\mathbf{b}_j})}, \tag{7.2}$$

where Π is the fundamental parallelepiped of the cone. The lattice points in $\Pi\cap\mathbb{Z}^n$ can be effectively enumerated using the Smith normal form; see Lemma 2.3.16.

In practical implementations of Barvinok's algorithm, one observes that in the hierarchy of cone decompositions, the index of the decomposed cones quickly descends from large numbers to fairly low numbers. The "last mile," that is, decomposing many cones with fairly low index, creates a huge number of unimodular cones and thus is the bottleneck of the whole computation in many instances. In practice, for computing the number of lattice points, one stops decomposing cones that have an index smaller than about 1000. If the desired result is a particularly short formula for the generating function, the threshold should be somewhat smaller, about the dimension of the problem.

The recursive decomposition of cones defines a *decomposition tree*. Due to the fast descent of the indices in the signed decomposition procedure, the following estimate holds for its depth:

Lemma 7.1.9 (Barvinok [30]). *Let $B\mathbb{R}^n_+$ be a simplicial full-dimensional cone, whose basis is given by the columns of the matrix $B \in \mathbb{Z}^{n\times n}$. Let $D = |\det B|$. Then the depth of the decomposition tree is at most*

$$k(D) = \left\lfloor 1 + \frac{\log_2 \log_2 D}{\log_2 \frac{n}{n-1}} \right\rfloor. \tag{7.3}$$

Because at each decomposition step at most n cones are created and the depth of the tree is doubly logarithmic in the index of the input cone, one obtains a polynomiality result *in fixed dimension*, which was first proved by Barvinok [30]:

Theorem 7.1.10 (Barvinok [30]). *Let n be fixed. There exists a polynomial-time algorithm for computing the rational generating function of a polyhedron $P \subseteq \mathbb{R}^n$ given by rational inequalities.*

The above discussion contains algorithmic refinements upon Barvinok's original algorithm, which improve the theoretical and practical efficiency of the algorithm.

The overall Barvinok algorithm. We summarize Barvinok's algorithm below.

ALGORITHM 7.1. Barvinok's algorithm, primal half-open variant.

1: **input** a polyhedron $P \subset \mathbb{R}^n$ given by rational inequalities.
2: **output** the rational generating function for $P \cap \mathbb{Z}^n$ in the form

$$g(P;\mathbf{z}) = \sum_{i\in I} \epsilon_i \frac{\sum_{\mathbf{a}\in A_i} \mathbf{z}^{\mathbf{a}}}{\prod_{j=1}^n (1-\mathbf{z}^{\mathbf{b}_{ij}})}, \tag{7.4}$$

where $\epsilon_i \in \{\pm 1\}$, $\mathbf{b}_{ij} \in \mathbb{Z}^n$, and $A_i \subseteq \mathbb{Z}^n$ for $i \in I$ are finite sets.
3: Compute all vertices $\mathbf{v}_i$ and corresponding supporting cones C_i of P.
4: Triangulate C_i into simplicial cones C_{ij}.
5: Apply signed decomposition to the cones $\mathbf{v}_i + C_{ij}$ to obtain unimodular (or low-index) cones $\mathbf{v}_i + C_{ijl}$.
6: Replace all cones $\mathbf{v}_i + C_{ijl}$ by half-open variants $\mathbf{v}_i + \tilde{C}_{ijl}$ using Theorem 6.4.1.
7: Enumerate the integer points in the fundamental parallelepipeds of all resulting cones $\mathbf{v}_i + \tilde{C}_{ijl}$ to obtain the sets A_i (Exercise 7.8.2).
8: Write down the formula (7.4).

This algorithm and many of its variants are implemented and readily available in the state-of-the-art software packages `LattE integrale` [103] and `barvinok` [330].

7.2 Specialization of the generating function

We now compute the number of integer points $\#(P \cap \mathbb{Z}^n)$ from the multivariate rational generating function $g(P;\mathbf{z})$. This amounts to the problem of evaluating or *specializing* a rational generating function $g(P;\mathbf{z})$ at the point $\mathbf{z} = \mathbf{1}$. This is a pole of each of its summands but a regular point (removable singularity) of the function itself.

To this end, let the generating function of a polytope $P \subseteq \mathbb{R}^n$ be given in the form

$$g(P;\mathbf{z}) = \sum_{i \in I} \epsilon_i \frac{\mathbf{z}^{\mathbf{a}_i}}{\prod_{j=1}^{s_i}(1 - \mathbf{z}^{\mathbf{b}_{ij}})}, \tag{7.5}$$

where $\epsilon_i \in \{\pm 1\}$, $\mathbf{a}_i \in \mathbb{Z}^n$, and $\mathbf{b}_{ij} \in \mathbb{Z}^n \setminus \{\mathbf{0}\}$. Let $s = \max_{i \in I} s_i$ be the maximum number of binomials in the denominators. In general, if s is allowed to grow, more poles need to be considered for each summand, so the evaluation will need more computational effort.

Theorem 7.2.1 (polynomial-time specialization).

(a) *There exists an algorithm for computing the specialization of a rational function of the form*

$$g(P;\mathbf{z}) = \sum_{i \in I} \epsilon_i \frac{\mathbf{z}^{\mathbf{a}_i}}{\prod_{j=1}^{s_i}(1 - \mathbf{z}^{\mathbf{b}_{ij}})} \tag{7.6}$$

at its removable singularity $\mathbf{z} = \mathbf{1}$, which runs in time that is bounded polynomially by the encoding size of its data $\epsilon_i \in \mathbb{Q}$, $\mathbf{a}_i \in \mathbb{Z}^n$ for $i \in I$, and $\mathbf{b}_{ij} \in \mathbb{Z}^n$ for $i \in I$, $j = 1, \dots, s_i$, even when the dimension n and the numbers s_i of terms in the denominators are not fixed.

(b) *In particular, there exists a polynomial-time algorithm that, given data $\epsilon_i \in \mathbb{Q}$, $\mathbf{a}_i \in \mathbb{Z}^n$ for $i \in I$, and $\mathbf{b}_{ij} \in \mathbb{Z}^n$ for $i \in I$, $j = 1, \dots, s_i$ describing a rational function in the form* (7.6), *computes a vector $\boldsymbol{\lambda} \in \mathbb{Q}^n$ with $\boldsymbol{\lambda}^\top \mathbf{b}_{ij} \neq 0$ for all i, j and rational weights $w_{i,l}$ for $i \in I$ and $l = 0, \dots, s_i$. Then the number of integer points is given by*

$$\#(P \cap \mathbb{Z}^n) = \sum_{i \in I} \epsilon_i \sum_{l=0}^{s_i} w_{i,l} (\boldsymbol{\lambda}^\top \mathbf{a}_i)^l. \tag{7.7}$$

As a corollary of Theorem 7.1.10 and Theorem 7.2.1, we obtain Barvinok's celebrated result from [30].

Corollary 7.2.2 (Barvinok's theorem). *Let n be fixed. There exists a polynomial-time algorithm for computing the number $\#(P \cap \mathbb{Z}^n)$ of integer points of a polyhedron $P \subseteq \mathbb{R}^n$ given by rational inequalities.*

A much more general theorem on monomial specialization appears in [39]. Here Barvinok and Woods show that a general change of variables can be done efficiently using only rational functions and avoiding difficulties with singularities:

Theorem 7.2.3 (substitution lemma, Theorem 2.6 in [39]). *Let us fix s, the number of binomials present in the denominator of a rational function. Given a rational function $g(\mathbf{z})$, $\mathbf{z} \in \mathbb{C}^n$, of the form*

$$g(\mathbf{z}) = \sum_{i \in I} \epsilon_i \frac{\mathbf{z}^{\mathbf{u}_i}}{\prod_{j=1}^{s} (1 - \mathbf{z}^{\mathbf{v}_{ij}})},$$

where $\mathbf{u}_i, \mathbf{v}_{ij}$ are integral n-dimensional vectors, ϵ_i are rational numbers, and a monomial map $\phi\colon \mathbb{C}^k \to \mathbb{C}^n$ given by the variable change $z_i = \mathbf{y}^{\mathbf{t}_i}$ (with $\mathbf{t}_i \in \mathbb{Z}^k$) whose image does not lie entirely in the set of poles of $g(\mathbf{z})$, then there exists a polynomial-time algorithm which computes the function $g(\phi(\mathbf{y}))$ of $\mathbf{y} \in \mathbb{C}^k$ as a sum of rational functions of the same shape as $g(\mathbf{z})$, with at most s binomials in the denominator.

Here the "set of poles" of $g(\mathbf{z})$ does *not* refer to the poles of the individual basic rational terms, but only those of the function $g(\mathbf{z})$. For instance, if $g(\mathbf{z})$ is the generating function of a polytope in $\mathbb{R}^n_+$, then $g(\mathbf{z})$ is a polynomial and thus regular at every point. Thus, in this case, the lemma can be applied unconditionally.

Substitution and integer projection. This change of variables corresponds to an integer projection $\boldsymbol{\alpha} \mapsto T\boldsymbol{\alpha} \in \mathbb{Z}^k$, where $T \in \mathbb{Z}^{k \times n}$ is the matrix with columns $\mathbf{t}_1, \dots, \mathbf{t}_n$. Indeed, consider a lattice point $\boldsymbol{\alpha} \in S \subseteq \mathbb{Z}^n$, which is represented in $g(S;\mathbf{z})$ as the monomial $\mathbf{z}^{\boldsymbol{\alpha}}$. The monomial substitution gives

$$\mathbf{z}^{\boldsymbol{\alpha}}\big|_{\mathbf{z}=\phi(\mathbf{y})} = \prod_{i=1}^{n} z_i^{\alpha_i}\big|_{z_i=\mathbf{y}^{\mathbf{t}_i}} = \prod_{i=1}^{n} \mathbf{y}^{\alpha_i \mathbf{t}_i} = \mathbf{y}^{\sum_{i=1}^{n} \alpha_i \mathbf{t}_i} = \mathbf{y}^{T\boldsymbol{\alpha}},$$

which represents the lattice point $T\boldsymbol{\alpha} \in \mathbb{Z}^k$.

Let us denote by $T(S)$ the image $\{ T\boldsymbol{\alpha} : \boldsymbol{\alpha} \in S \}$. If we apply the monomial substitution to $g(S;\mathbf{z}) = \sum_{\boldsymbol{\alpha} \in S} \mathbf{z}^{\boldsymbol{\alpha}}$, then we obtain a function

$$g(S;\phi(\mathbf{y})) = \sum_{\boldsymbol{\beta} \in T(S)} k_{\boldsymbol{\beta}} \mathbf{y}^{\boldsymbol{\beta}}, \tag{7.8}$$

where each monomial $\mathbf{y}^{\boldsymbol{\beta}}$ is weighted with a coefficient

$$k_{\boldsymbol{\beta}} = |\{ \boldsymbol{\alpha} \in S : T\boldsymbol{\alpha} = \boldsymbol{\beta} \}|,$$

which is the cardinality of the fiber of $\boldsymbol{\beta}$. Thus we observe:

Corollary 7.2.4 (one-to-one projection). *If the projection $\boldsymbol{\alpha} \mapsto T\boldsymbol{\alpha}$ is one-to-one (injective) from S, then all $k_{\boldsymbol{\beta}} = 1$ for all $\boldsymbol{\beta} \in T(S)$, and $g(S;\phi(\mathbf{y}))$ coincides with the generating function of the projection, $g(T(S);\mathbf{y})$.*

Otherwise, we shall call $\hat{g}(T(S);\mathbf{y}) = g(S;\phi(\mathbf{y}))$ a *positively weighted generating function* of $T(S)$.

Remark 7.2.5. Many (but not all) operations with generating functions that we are interested in also work with positively weighted generating functions. Of course, we systematically overcount the cardinality of $T(S)$ if we evaluate $\hat{g}(T(S);\mathbf{1})$ instead of $g(T(S);\mathbf{1})$.

Since all coefficients k_β are positive, there is no cancellation, however, and therefore we can still correctly detect whether $T(S)$ is empty or not. For cases when a positively weighted generating function is not good enough, a different, much more computationally expensive algorithm can be used to compute the generating function $g(T(S);\mathbf{y})$ (Section 7.6).

Remark 7.2.6. The specialization $g(S;\mathbf{1})$ can be seen as the special case where T is the zero matrix in $\mathbb{Z}^{0\times n}$. Then there is only one fiber, for $\boldsymbol{\beta}=\mathbf{0}$, which has cardinality $k_{\mathbf{0}}=|S|$.

Proof of the specialization theorem. The remainder of this section contains a detailed proof of Theorem 7.2.1, which directly leads to an efficient implementation. We use standard techniques of computational complex analysis; cf. [164].

We first define Todd polynomials. We will prove that they can be efficiently evaluated in rational arithmetic.

Definition 7.2.7. We consider the function

$$H(x,\xi_1,\dots,\xi_s)=\prod_{i=1}^{s}\frac{x\xi_i}{1-\exp\{-x\xi_i\}},$$

a function that is analytic in a neighborhood of $\mathbf{0}$. The m-th (s-variate) *Todd polynomial* is the coefficient of x^m in the Taylor expansion

$$H(x,\xi_1,\dots,\xi_s)=\sum_{m=0}^{\infty}\operatorname{td}_m(\xi_1,\dots,\xi_s)x^m.$$

We remark that, when the numbers s and m are allowed to vary, the Todd polynomials have an exponential number of monomials.

Theorem 7.2.8. *The Todd polynomial* $\operatorname{td}_m(\xi_1,\dots,\xi_s)$ *can be evaluated for given rational data* $\xi_1,\dots,\xi_s$ *in time that is polynomial in* s, m, *and the encoding length of* $\xi_1,\dots,\xi_s$.

The proof makes use of the following lemma.

Lemma 7.2.9. *The function* $h(x)=x/(1-\mathrm{e}^{-x})$ *is a function that is analytic in a neighborhood of* 0. *Its Taylor series about* $x=0$ *is of the form*

$$h(x)=\sum_{n=0}^{\infty}b_nx^n \quad \textit{where} \quad b_n=\frac{1}{n!(n+1)!}c_n \tag{7.9}$$

with integer coefficients c_n *that have a bit length of* $\mathrm{O}(n^2\log n)$. *The coefficients* c_n *can be computed from the recursion*

$$\begin{aligned} c_0&=1,\\ c_n&=\sum_{j=1}^{n}(-1)^{j+1}\binom{n+1}{j+1}\frac{n!}{(n-j+1)!}c_{n-j} \quad \textit{for } n=1,2,\dots. \end{aligned} \tag{7.10}$$

Proof. The reciprocal function $1/h(x) = (1-e^{-x})/x$ has the Taylor series

$$1/h(x) = \sum_{i=0}^{\infty} a_n x^n \quad \text{with} \quad a_n = \frac{(-1)^n}{(n+1)!}.$$

Using the identity $1/h(x) \cdot h(x) = \left(\sum_{n=0}^{\infty} a_n x^n\right)\left(\sum_{n=0}^{\infty} b_n x^n\right) = 1$, we obtain the recursion

$$\begin{aligned} b_0 &= \tfrac{1}{a_0} = 1, \\ b_n &= -(a_1 b_{n-1} + a_2 b_{n-2} + \cdots + a_n b_0) \quad \text{for } n = 1,2,\ldots. \end{aligned} \tag{7.11}$$

We prove (7.9) inductively. Clearly $b_0 = c_0 = 1$. For $n = 1,2,\ldots$, we have

$$\begin{aligned} c_n &= n!(n+1)!\,b_n \\ &= -n!(n+1)!(a_1 b_{n-1} + a_2 b_{n-2} + \cdots + a_n b_0) \\ &= n!(n+1)! \sum_{j=1}^{n} \frac{(-1)^{j+1}}{(j+1)!} \cdot \frac{1}{(n-j)!(n-j+1)!} c_{n-j} \\ &= \sum_{j=1}^{n} (-1)^{j+1} \frac{(n+1)!}{(j+1)!(n-j)!} \cdot \frac{n!}{(n-j+1)!} c_{n-j}. \end{aligned}$$

Thus we obtain the recursion formula (7.10), which also shows that all c_n are integers. A rough estimate shows that

$$|c_n| \le n(n+1)!n!\,|c_{n-1}| \le ((n+1)!)^2\,|c_{n-1}|,$$

and thus $|c_n| \le ((n+1)!)^{2n}$, so c_n has a binary encoding length of $O(n^2 \log n)$. □

***Proof of Theorem* 7.2.8.** By definition, we have

$$H(x,\xi_1,\ldots,\xi_s) = \sum_{m=0}^{\infty} \mathrm{td}_m(\xi_1,\ldots,\xi_s) x^m = \prod_{j=1}^{s} h(x\xi_j).$$

From Lemma 7.2.9 we have

$$h(x\xi_j) = \sum_{n=0}^{m} \beta_{j,n} x^n + o(x^m) \quad \text{where} \quad \beta_{j,n} = \frac{\xi_j^n}{n!(n+1)!} c_n \tag{7.12}$$

with integers c_n given by the recursion (7.10). Thus we can evaluate $\mathrm{td}_m(\xi_1,\ldots,\xi_s)$ by summing over all the possible compositions $n_1 + \cdots + n_s = m$ of the order m from the orders n_j of the factors:

$$\mathrm{td}_m(\xi_1,\ldots,\xi_s) = \sum_{\substack{(n_1,\ldots,n_s)\in\mathbb{Z}_+^s \\ n_1+\cdots+n_s=m}} \beta_{1,n_1} \cdots \beta_{s,n_s}. \tag{7.13}$$

We remark that the length of the above sum is equal to the number of weak compositions of m into s nonnegative parts,

$$\begin{aligned}C'_s(m) &= \binom{m+s-1}{s-1}\\ &= \frac{(m+s-1)(m+s-2)\dots(m+s-(s-1))}{(s-1)(s-2)\dots 2\cdot 1}\\ &= \Omega\left(\left(1+\tfrac{m}{s-1}\right)^s\right),\end{aligned}$$

which is exponential in s (whenever $m \geq s$). Thus we cannot evaluate the formula (7.13) efficiently when s is allowed to grow.

However, we show that we can evaluate $\mathrm{td}_m(\xi_1,\dots,\xi_s)$ more efficiently. To this end, we multiply up the s truncated Taylor series (7.12), one factor at a time, truncating after order m. Let us denote

$$\begin{aligned}H_1(x) &= h(x\xi_1),\\ H_2(x) &= H_1(x)\cdot h(x\xi_2),\\ &\vdots\\ H_s(x) &= H_{s-1}(x)\cdot h(x\xi_s) = H(x,\xi_1,\dots,\xi_s).\end{aligned}$$

Each multiplication can be implemented in $\mathrm{O}(m^2)$ elementary rational operations. We finally show that all numbers appearing in the calculations have polynomial encoding size. Let Ξ be the largest binary encoding size of any of the rational numbers $\xi_1,\dots,\xi_s$. Then every $\beta_{j,n}$ given by (7.12) has a binary encoding size $\mathrm{O}(\Xi n^5\log^3 n)$. Let $H_j(x)$ have the truncated Taylor series $\sum_{n=0}^m \alpha_{j,n}x^n + o(x^m)$ and let A_j denote the largest binary encoding length of any $\alpha_{j,n}$ for $n \leq m$. Then

$$H_{j+1}(x) = \sum_{n=0}^{m} \alpha_{j+1,n}x^n + o(x^m) \quad \text{with} \quad \alpha_{j+1,n} = \sum_{l=0}^{n} \alpha_{j,l}\beta_{j,n-l}.$$

Thus the binary encoding size of $\alpha_{j+1,n}$ (for $n \leq m$) is bounded by $A_j + \mathrm{O}(\Xi m^5 \log^3 m)$. Thus, after s multiplication steps, the encoding size of the coefficients is bounded by $\mathrm{O}(s\,\Xi m^5\log^3 m)$, a polynomial quantity. □

***Proof of Theorem* 7.2.1.** *Parts (a) and (b).* We first construct a rational vector $\boldsymbol{\lambda} \in \mathbb{Z}^n$ such that $\boldsymbol{\lambda}^\top \mathbf{b}_{ij} \neq 0$ for all i,j. One such construction is to consider the *moment curve* $\boldsymbol{\lambda}(\xi) = (1,\xi,\xi^2,\dots,\xi^{n-1}) \in \mathbb{R}^n$. For each exponent vector $\mathbf{b}_{ij}$ occurring in a denominator of (7.4), the function $f_{ij}\colon \xi \mapsto \boldsymbol{\lambda}(\xi)^\top \mathbf{b}_{ij}$ is a polynomial function of degree at most $n-1$. Since $\mathbf{b}_{ij} \neq \mathbf{0}$, the function f_{ij} is not identically zero. Hence f_{ij} has at most $n-1$ zeros. By evaluating all functions f_{ij} for $i \in I$ and $j = 1,\dots,s_i$ at $M = (n-1)s|I|+1$ different values for ξ, for instance at the integers $\xi = 0,\dots,M$, we can find one $\xi = \bar{\xi}$ that is not a zero of any f_{ij}. Clearly this search can be implemented in polynomial time, even when the dimension n and the number s of terms in the denominators are not fixed. We set $\boldsymbol{\lambda} = \boldsymbol{\lambda}(\bar{\xi})$.

For $\tau > 0$, let us consider the points $\mathbf{z}_\tau = \mathbf{e}^{\tau\boldsymbol{\lambda}} = (\exp\{\tau\lambda_1\},\dots,\exp\{\tau\lambda_n\})$. We have

$$\mathbf{z}_\tau^{\mathbf{b}_{ij}} = \prod_{l=1}^{n} \exp\{\tau\lambda_l b_{ijl}\} = \exp\{\tau\boldsymbol{\lambda}^\top \mathbf{b}_{ij}\};$$

since $\boldsymbol{\lambda}^\top \mathbf{b}_{ij} \neq 0$ for all i,j, all the denominators $1-\mathbf{z}_\tau^{\mathbf{b}_{ij}}$ are nonzero. Hence for every $\tau > 0$, the point $\mathbf{z}_\tau$ is a regular point not only of $g(\mathbf{z})$ but also of the individual summands of (7.4). We have

$$\begin{aligned}
g(\mathbf{1}) &= \lim_{\tau\to 0^+} \sum_{i\in I} \epsilon_i \frac{\mathbf{z}_\tau^{\mathbf{a}_i}}{\prod_{j=1}^{s_i}(1-\mathbf{z}_\tau^{\mathbf{b}_{ij}})} \\
&= \lim_{\tau\to 0^+} \sum_{i\in I} \epsilon_i \frac{\exp\{\tau\boldsymbol{\lambda}^\top \mathbf{a}_i\}}{\prod_{j=1}^{s_i}(1-\exp\{\tau\boldsymbol{\lambda}^\top \mathbf{b}_{ij}\})} \\
&= \lim_{\tau\to 0^+} \sum_{i\in I} \epsilon_i \tau^{-s_i} \exp\{\tau\boldsymbol{\lambda}^\top \mathbf{a}_i\} \prod_{j=1}^{s_i} \frac{\tau}{1-\exp\{\tau\boldsymbol{\lambda}^\top \mathbf{b}_{ij}\}} \\
&= \lim_{\tau\to 0^+} \sum_{i\in I} \epsilon_i \tau^{-s_i} \exp\{\tau\boldsymbol{\lambda}^\top \mathbf{a}_i\} \prod_{j=1}^{s_i} \frac{-1}{\boldsymbol{\lambda}^\top \mathbf{b}_{ij}} h(-\tau\boldsymbol{\lambda}^\top \mathbf{b}_{ij}) \\
&= \lim_{\tau\to 0^+} \sum_{i\in I} \epsilon_i \frac{(-1)^{s_i}}{\prod_{j=1}^{s_i}\boldsymbol{\lambda}^\top \mathbf{b}_{ij}} \tau^{-s_i} \exp\{\tau\boldsymbol{\lambda}^\top \mathbf{a}_i\} H(\tau, -\boldsymbol{\lambda}^\top \mathbf{b}_{i1}, \dots, -\boldsymbol{\lambda}^\top \mathbf{b}_{is_i}),
\end{aligned}$$

where $H(x,\xi_1,\dots,\xi_{s_i})$ is the function from Definition 7.2.7. We will compute the limit by computing the constant term of the Laurent expansion of each summand about $\tau = 0$. Now the function $\tau \mapsto \exp\{\tau\boldsymbol{\lambda}^\top \mathbf{a}_i\}$ is holomorphic and has the Taylor series

$$\exp\{\tau\boldsymbol{\lambda}^\top \mathbf{a}_i\} = \sum_{l=0}^{s_i} \alpha_{i,l}\tau^l + o(\tau^{s_i}) \quad \text{where} \quad \alpha_{i,l} = \frac{(\boldsymbol{\lambda}^\top \mathbf{a}_i)^l}{l!}, \tag{7.14}$$

and $H(\tau,\xi_1,\dots,\xi_{s_i})$ has the Taylor series

$$H(\tau,\xi_1,\dots,\xi_s) = \sum_{m=0}^{s_i} \operatorname{td}_m(\xi_1,\dots,\xi_s)\tau^m + o(\tau^{s_i}).$$

Because of the factor τ^{-s_i}, which gives rise to a pole of order s_i in the summand, we can compute the constant term of the Laurent expansion by summing over all the possible compositions $s_i = l + (s_i - l)$ of the order s_i:

$$g(\mathbf{1}) = \sum_{i\in I} \epsilon_i \frac{(-1)^{s_i}}{\prod_{j=1}^{s_i}\boldsymbol{\lambda}^\top \mathbf{b}_{ij}} \sum_{l=0}^{s_i} \frac{(\boldsymbol{\lambda}^\top \mathbf{a}_i)^l}{l!} \operatorname{td}_{s_i-l}(-\boldsymbol{\lambda}^\top \mathbf{b}_{i1},\dots,-\boldsymbol{\lambda}^\top \mathbf{b}_{is_i}). \tag{7.15}$$

We use the notation

$$w_{i,l} = (-1)^{s_i} \frac{\operatorname{td}_{s_i-l}(-\langle\boldsymbol{\lambda},\mathbf{b}_{i,1}\rangle,\dots,-\langle\boldsymbol{\lambda},\mathbf{b}_{i,s_i}\rangle)}{l!\cdot\langle\boldsymbol{\lambda},\mathbf{b}_{i,1}\rangle\cdots\langle\boldsymbol{\lambda},\mathbf{b}_{i,s_i}\rangle} \quad \text{for } i \in I \text{ and } l = 0,\dots,s_i;$$

these rational numbers can be computed in polynomial time using Theorem 7.2.8. We now obtain the formula of the claim,

$$g(\mathbf{1}) = \sum_{i\in I} \epsilon_i \sum_{l=0}^{s_i} w_{i,l}(\boldsymbol{\lambda}^\top \mathbf{a}_i)^l. \qquad \square$$

7.3 Explicit enumeration of lattice points

In this section we show how short rational generating functions can be used to efficiently list all the integer points in a polytope. Since the number of integer points is exponential in general, there cannot exist a polynomial-time enumeration algorithm, even if the dimension is fixed. In order to analyze the running time of an enumeration algorithm, we must turn to *output-sensitive complexity analysis*.

Various notions of output-sensitive efficiency have appeared in the literature; we follow the discussion of [183]. Let $W \subseteq \mathbb{Z}^p$ be a finite set to be enumerated. An enumeration algorithm is said to run in *polynomial total time* if its running time is bounded by a polynomial in the encoding size of the input and the output. A stronger notion is that of *incremental polynomial time*: Such an algorithm receives a list of solutions $\mathbf{w}_1, \dots, \mathbf{w}_N \in W$ as an additional input. In polynomial time, it outputs one solution $\mathbf{w} \in W \setminus \{\mathbf{w}_1, \dots, \mathbf{w}_N\}$ or asserts that there are no more solutions. An even stronger notion is that of a *polynomial-delay* algorithm, which takes only polynomial time (in the encoding size of the input) before the first solution is output, between successive outputs of solutions, and after the last solution is output to the termination of the algorithm. Since the algorithm could take exponential time to output all solutions, it could also build exponential-size data structures in the course of the enumeration. This observation gives rise to an even stronger notion of efficiency, a *polynomial-space polynomial-delay* enumeration algorithm.

We prove the existence of a polynomial-space polynomial-delay prescribed-order enumeration algorithm in a general setting, where the set W to be enumerated is given as the projection of the set of integer points of a polytope.

Theorem 7.3.1. *Let the dimension n be fixed. Let $P \subseteq \mathbb{R}^n$ be a polytope, given by rational inequalities. Let $V = P \cap \mathbb{Z}^n$ and*

$$W = \{\mathbf{w} \in \mathbb{Z}^p : \exists \mathbf{t} \in \mathbb{Z}^{n-p} \text{ such that } (\mathbf{t}, \mathbf{w}) \in V\}$$

denote the projection of V onto the last p components.

There exists a polynomial-space polynomial-delay enumeration algorithm for the points in the projection W.

We prove the theorem by analyzing the following simple recursive algorithm that is based on the iterative bisection of intervals.

Algorithm 7.2. Output-sensitive enumeration with generating functions.

input an oracle to determine if the set $W \cap \{\mathbf{w} : \mathbf{l}' \leq \mathbf{w} \leq \mathbf{u}'\}$ for given $\mathbf{l}', \mathbf{u}'$ is nonempty; lower and upper bound vectors $\mathbf{l}, \mathbf{u} \in \mathbb{Z}^p$.
output all vectors $\mathbf{w}$ in W with $\mathbf{l} \leq \mathbf{w} \leq \mathbf{u}$, sorted in lexicographic order.
Call the oracle to find if the set $W \cap \{\mathbf{w} : \mathbf{l} \leq \mathbf{w} \leq \mathbf{u}\}$ is empty.
if it is empty **then**
    Do nothing.
else if $\mathbf{l} = \mathbf{u}$ **then**
    Output $\mathbf{w} = \mathbf{l} = \mathbf{u}$.
else
    Let j be the smallest index with $l_j \neq u_j$.
    Bisect the integer interval $\{l_j, \dots, u_j\}$ evenly into $\{l_j, \dots, m_j\}$ and $\{m_j + 1, \dots, u_j\}$, where $m_j = \lfloor \frac{l_j + u_j}{2} \rfloor$.

11: Invoke the algorithm recursively on the first part, then on the second part, using the corresponding lower and upper bound vectors.

***Proof of Theorem* 7.3.1.** We first compute appropriate lower and upper bound vectors $\mathbf{l}, \mathbf{u}$ using linear optimization to start Algorithm 7.2.

The oracle, which is called in line 3 of the algorithm to determine whether

$$W \cap \{\mathbf{w} : \mathbf{l} \leq \mathbf{w} \leq \mathbf{u}\} = \emptyset, \tag{7.16}$$

can be implemented as follows. The set is nonempty if and only if the polytope

$$P' := \{(\mathbf{t}, \mathbf{w}) \in \mathbb{R}^n : (\mathbf{t}, \mathbf{w}) \in P,\ \mathbf{l} \leq \mathbf{w} \leq \mathbf{u}\}$$

contains a lattice point. Using Barvinok's algorithm, compute a rational generating function $g(P';\mathbf{z})$ in polynomial time. Then the specialization $g(P';\mathbf{1})$ can be computed in polynomial time. It gives the number of lattice points in P'. In particular, we can decide, whether P' contains a lattice point.

It is clear that the algorithm outputs the elements of W in lexicographic order. We next show that the algorithm is a polynomial-space polynomial-delay enumeration algorithm. The subproblem in line 3 only depends on the input data as stated in the theorem and on the vectors $\mathbf{l}$ and $\mathbf{u}$, whose encoding length only decreases in recursive invocations. Therefore each of the subproblems can be solved in polynomial time (thus also in polynomial space).

The recursion of the algorithm corresponds to a binary tree whose nodes are labeled by the bound vectors $\mathbf{l}$ and $\mathbf{u}$. There are two types of leaves in the tree, one corresponding to the "empty-box" situation (7.16) in line 3, and one corresponding to the "solution-output" situation in line 7. Inner nodes of the tree correspond to the recursive invocation of the algorithm in line 11. It is clear that the depth of the recursion is $\mathrm{O}(p \log(pMN))$, because the integer intervals are bisected evenly. Thus the stack space of the algorithm is polynomially bounded. Since the algorithm does not maintain any global data structures, the whole algorithm uses polynomial space only.

Let $\mathbf{w}_i \in W$ be an arbitrary solution and let $\mathbf{w}_{i+1}$ be its direct successor in the lexicographic order. We shall show that the algorithm only spends polynomial time between the output of $\mathbf{w}_i$ and the output of $\mathbf{w}_{i+1}$. The key property of the recursion tree of the algorithm is the following:

> Every inner node is the root of a subtree that contains at least one solution-output leaf. (7.17)

The reason for that property is the test for situation (7.16) in line 3 of the algorithm. Therefore, the algorithm can visit only $\mathrm{O}(p \log(pMN))$ inner nodes and empty-box leaves between the solution-output leaves for $\mathbf{w}_i$ and $\mathbf{w}_{i+1}$. For the same reason, also the time before the first solution is output and the time after the last solution is output are polynomially bounded. □

7.4 Integer linear optimization with Barvinok's algorithm

In this chapter we have introduced the details of Barvinok's algorithm and the encoding of the lattice points in a convex rational polytope in terms of rational generating functions. Our purpose is now to show how this encoding can be useful to solve *integer linear optimization problems* of the form

$$\begin{aligned} \max \quad & \mathbf{c}^{\top}\mathbf{x} \\ \text{subject to} \quad & A\mathbf{x} \leq \mathbf{b},\ \mathbf{x} \geq \mathbf{0},\ \mathbf{x} \in \mathbb{Z}^n, \end{aligned} \tag{7.18}$$

where the input data are an $m \times n$ integral matrix A, an integral m-vector $\mathbf{b}$, and an integral n-vector $\mathbf{c}$.

7.4.1 The binary search and bisection algorithm

We begin with the most straightforward integer programming algorithm. It is an immediate consequence of the fact that we can count the lattice points of a polytope in polynomial time. As discussed in Section 2.8 in the context of Lenstra's algorithm, one can turn *any* feasibility oracle into an algorithm that finds the optimal value of problem (7.18) using binary search. Since we even have a counting oracle, not just a feasibility oracle, available, we are able to do at least as well.

Below we make the algorithm explicit. By counting the number of lattice points in P that satisfy $\mathbf{c}^{\top}\mathbf{x} \geq M$, we can narrow the range for the maximum value of $\mathbf{c}^{\top}\mathbf{x}$, then we iteratively look for the largest integer M where the count is nonzero:

ALGORITHM 7.3. Binary search for the optimal value.

input $A \in \mathbb{Z}^{m\times n}$, $\mathbf{b} \in \mathbb{Z}^m$, $\mathbf{c} \in \mathbb{Z}^n$.
output the optimal value $M = \max\{\mathbf{c}^{\top}\mathbf{x} : A\mathbf{x} \leq \mathbf{b},\ \mathbf{x} \geq \mathbf{0},\ \mathbf{x} \in \mathbb{Z}^n\}$.
Let $P = \{\mathbf{x} : A\mathbf{x} \leq \mathbf{b},\ \mathbf{x} \geq \mathbf{0}\}$.
Using the linear programming relaxation of problem (7.18), compute

$$M \leftarrow \lceil \max\{\mathbf{c}^{\top}\mathbf{x} : A\mathbf{x} \leq \mathbf{b},\ \mathbf{x} \geq \mathbf{0}\}\rceil,$$

$$m \leftarrow \lfloor \min\{\mathbf{c}^{\top}\mathbf{x} : A\mathbf{x} \leq \mathbf{b},\ \mathbf{x} \geq \mathbf{0}\}\rfloor.$$

Thus $[m, M]$ is the initial range for the binary search.
while $M > m$ **do**
  $\alpha \leftarrow \lceil \frac{M+m}{2} \rceil$.
  Using Barvinok's algorithm compute $q_\alpha \leftarrow |P \cap \{\mathbf{x} : \mathbf{c}^{\top}\mathbf{x} \geq \alpha\} \cap \mathbb{Z}^n|$.
  if $q_\alpha > 0$ **then**
    $m \leftarrow \alpha$.
  else
    $M \leftarrow \alpha - 1$.
return M.

Once the optimal function value M is found, we can run our enumeration procedure (Algorithm 7.2) based on iterated bisection on the polyhedron

$$P \cap \{\mathbf{x} : \mathbf{c}^{\top}\mathbf{x} = M\}$$

to list the feasible integer solutions that attain the optimal function value. By simply stopping when the first integer solution has been output, we obtain an algorithm that runs in polynomial time in fixed dimension. We have thus given a proof of the following theorem:

Theorem 7.4.1. *Let n be fixed. There exists a polynomial-time algorithm to solve the integer linear optimization problem* (7.18).

Of course, in Section 2.8 we saw that this theorem was proved first by H.W. Lenstra, Jr. using his famous branching-on-hyperplanes algorithm [231]. It is worth remarking that the original version of Barvinok's counting algorithm in [30] relied on Lenstra's algorithm as a subroutine. However with the modification by Dyer and Kannan [118], as presented in this chapter, this dependence can be avoided in favor of a shortest-vector computation. Therefore, as pointed out in [35], the binary search and bisection search algorithm using Barvinok's algorithm gives a new proof of the polynomiality of integer linear optimization in fixed dimension, independent of Lenstra's algorithm.

In the following sections we explain heuristic methods proposed in the literature that avoid binary search and the repeated calls to Barvinok's algorithm. These methods, though exponential time even in fixed dimension in the worst case, could be faster in practice for some concrete problems. They can thus complement the exact method introduced above.

7.4.2 Monomial substitutions with the objective function

As outlined already in Section 5.2, we can include the linear objective function $\mathbf{c}^\top\mathbf{x}$ in the generating function by making a change of variables, a monomial substitution. We now present this idea in more detail.

Barvinok's algorithm computes the function

$$g(P;\mathbf{z}) = \sum_{\alpha\in P\cap\mathbb{Z}^n} \mathbf{z}^{\alpha} \tag{7.19}$$

as the short rational function expression

$$g(P;\mathbf{z}) = \sum_{i\in I} \epsilon_i \frac{\mathbf{z}^{\mathbf{a}_i}}{\prod_{j=1}^{n}(1-\mathbf{z}^{\mathbf{b}_{ij}})}. \tag{7.20}$$

Now, for the given (integral) objective function vector $\mathbf{c}$, we make the substitutions $z_k = y_k t^{c_k}$, for $k = 1,\dots,n$. These substitutions into (7.20) yield a sum of *multivariate* rational functions in the vector variable $\mathbf{y} \in \mathbb{C}^n$ and scalar variable $t \in \mathbb{C}$:

$$g(P;\mathbf{y},t) = \sum_{i\in I} \epsilon_i \frac{\mathbf{y}^{\mathbf{a}_i} t^{\mathbf{c}^\top\mathbf{a}_i}}{\prod_{j=1}^{n}(1-\mathbf{y}^{\mathbf{b}_{ij}} t^{\mathbf{c}^\top\mathbf{b}_{ij}})}. \tag{7.21}$$

(This substitution can always be done in this direct and elementary way; we do not need to invoke Theorem 7.2.3.) On the other hand, the same substitution in (7.19) gives the following sum of monomials,

$$g(P;\mathbf{y},t) = \sum_{\alpha\in P\cap\mathbb{Z}^n} \mathbf{y}^{\alpha} t^{\mathbf{c}^\top\alpha} = \sum_{s=-\infty}^{M}\left(\sum_{\substack{\alpha\in P\cap\mathbb{Z}^n\\ \mathbf{c}^\top\alpha=s}} \mathbf{y}^{\alpha}\right) t^{s}. \tag{7.22}$$

Both equations, (7.22) and (7.21), represent the same function $g(P;\mathbf{y},t)$. Thus, if we compute a Laurent series of (7.21) that shares a region of convergence with the series in (7.22), then the corresponding coefficients of both series must be equal. In particular, because P is a polytope, the series in (7.22) converges almost everywhere. Thus if we compute a Laurent series of (7.21) that has any nonempty region of convergence, then the corresponding coefficients of both series must be equal. Barvinok's algorithm provides us with the right-hand side of (7.21). We need to obtain the coefficient of highest degree in t from the expanded equation (7.22).

In a similar way, we can also a produce a generating function in a single variable that only keeps track of the objective function values. To this end, we use the substitution $z_k = t^{c_k}$ with a new scalar variable $t \in \mathbb{C}$.

If $\mathbf{c}^\top \mathbf{b}_{ij} \neq 0$ for all i, j, then the substitution does not create any zero factors in the denominators of the right-hand side of (7.20), and we can make the substitution in an elementary way. We obtain

$$g(P;t) = \sum_{i \in I} \epsilon_i \frac{t^{\mathbf{c}^\top \mathbf{a}_i}}{\prod_{j=1}^n (1 - t^{\mathbf{c}^\top \mathbf{b}_{ij}})}. \tag{7.23}$$

Otherwise, if $\mathbf{c}^\top \mathbf{b}_{ij} = 0$ somewhere, then this means that the substitution creates a zero factor in a denominator. Thus we need to use Theorem 7.2.3 to make this substitution. We obtain a univariate rational function in t of the form

$$g(P;t) = \sum_{i \in I'} \epsilon'_i \frac{t^{\alpha_i}}{\prod_{j=1}^n (1 - t^{\beta_{ij}})}. \tag{7.24}$$

On the other hand, observe that if we make the substitution directly in (7.19), we have that the multivariate monomial $\mathbf{z}^{\boldsymbol{\alpha}}$ becomes $t^{\mathbf{c}^\top \boldsymbol{\alpha}}$. We obtain the relation

$$g(P;t) = \sum_{\boldsymbol{\alpha} \in P \cap \mathbb{Z}^n} t^{\mathbf{c}^\top \boldsymbol{\alpha}} = k_M t^M + k_{M-1} t^{M-1} + k_{M-2} t^{M-2} + \cdots, \tag{7.25}$$

where M is the optimal value of our integer program and where k_s is the number of feasible integer solutions with objective function value s. Relating to the discussion in section 7.2, k_s is the cardinality of the fiber of s in the projection $\boldsymbol{\alpha} \mapsto \mathbf{c}^\top \boldsymbol{\alpha}$.

Thus, after the monomial substitution, the IP maximum value equals the highest degree of the univariate Laurent polynomial $g(P;t)$. If we have a way to compute the highest degree of this Laurent polynomial, then we have solved problem (7.18).

Example 7.4.2. For the polygon shown in Figure 7.1, we find the rational generating function

$$g(P;\mathbf{z}) = \frac{1}{(1-z_1)(1-z_2)} + \frac{z_1^{50}}{(1-z_1^{-1})(1-z_2)} + \frac{z_2^{100}}{(1-z_1^{-1})(1-z_2)} + \frac{z_1^{50} z_2^{50}}{(1-z_1^{-1})(1-z_1^{-1} z_2)}.$$

If we substitute $z_1 = t^{100}$ and $z_2 = t^{90}$, then we have

$$g(P;t) = t^{9500} + \text{lower degree terms in } t.$$

7.4.3 Digging: Opportunistic series expansion

Next we present a technique that can be seen as an extension of a heuristic proposed by Lasserre in [214]. The technique was introduced in [89, 92] as the *digging algorithm*. This algorithm "digs" for the coefficient of the highest exponent of t that "survives" the cancellation of terms in (7.21), by doing a controlled Laurent series expansion. This is a worst-case exponential-time technique, but the series expansion can be implemented very efficiently in practice; see [105] for a discussion on a cache-efficient implementation. When only a few terms need to be produced, it could be quite fast for some problems.

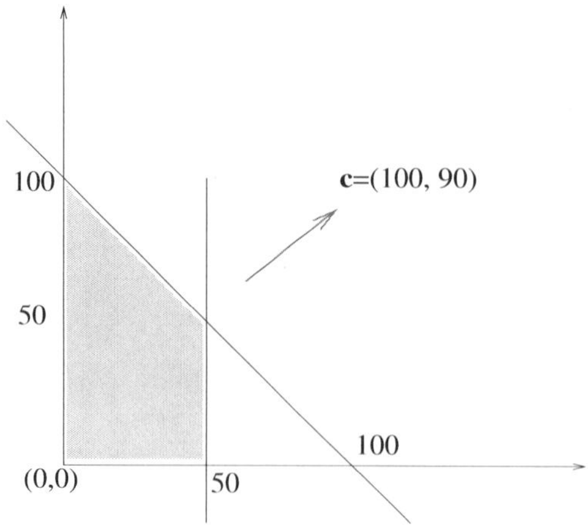

Figure 7.1. *Integer linear optimization problem of Example* 7.4.2

For simplicity our explanation assumes the genericity condition $\mathbf{c}^\top \mathbf{b}_{ij} \neq 0$ for all $\mathbf{b}_{ij}$, which can be achieved by a suitable perturbation of the objective function vector $\mathbf{c}$.

Thus we compute a Laurent series for (7.21) using the following procedure. For any $\mathbf{b}_{ij}$ with $\mathbf{c}^\top \mathbf{b}_{ij} > 0$, we first apply the identity

$$\frac{1}{1-\mathbf{y}^{\mathbf{b}_{ij}} t^{\mathbf{c}^\top \mathbf{b}_{ij}}} = -\frac{\mathbf{y}^{-\mathbf{b}_{ij}} t^{-\mathbf{c}^\top \mathbf{b}_{ij}}}{1-\mathbf{y}^{-\mathbf{b}_{ij}} t^{-\mathbf{c}^\top \mathbf{b}_{ij}}}, \tag{7.26}$$

which makes the exponent of t in the denominator nonpositive. After this operation, we still have a rational function of the form (7.21), but now $\mathbf{c}^\top \mathbf{b}_{ij} < 0$ holds for all i, j. Then, for each of the basic rational functions in the sum of equation (7.21), we compute a Laurent series of the form

$$\epsilon_i \, \mathbf{y}^{\mathbf{a}_i} t^{\mathbf{c}^\top \mathbf{a}_i} \prod_{j=1}^{n} \left(1 + \mathbf{y}^{\mathbf{b}_{ij}} t^{\mathbf{c}^\top \mathbf{b}_{ij}} + \left(\mathbf{y}^{\mathbf{b}_{ij}} t^{\mathbf{c}^\top \mathbf{b}_{ij}}\right)^2 + \left(\mathbf{y}^{\mathbf{b}_{ij}} t^{\mathbf{c}^\top \mathbf{b}_{ij}}\right)^3 + \cdots \right). \tag{7.27}$$

Multiply out each such product of series and add the resulting series. This yields precisely the Laurent series in (7.22).

ALGORITHM 7.4. Digging algorithm.

1: **input** $A \in \mathbb{Z}^{m\times n}$, $\mathbf{b} \in \mathbb{Z}^m$, $\mathbf{c} \in \mathbb{Z}^n$.
2: **output** the optimal value and an optimal solution of

$$\max\{\mathbf{c}^\top \mathbf{x} : A\mathbf{x} \leq \mathbf{b},\ \mathbf{x} \geq \mathbf{0},\ \mathbf{x} \in \mathbb{Z}^n\}.$$

3: Using Barvinok's algorithm (Algorithm 7.1), compute the rational generating function of the form (7.21). Use the identity (7.26) as necessary to enforce that all $\mathbf{b}_{ij}$ in (7.21) satisfy $\mathbf{c}^\top \mathbf{b}_{ij} < 0$.
4: Via the expansion formula (7.27), multiply out the factors and add the terms, grouping together those of the same degree in t. Thus we find (7.22) by calculating the terms' coefficients. Proceed in decreasing order with respect to the degree of t. (This can be

done because, for each series appearing in the expansion formulas (7.27), all $\mathbf{c}^\top \mathbf{b}_{ij}$ are negative, so that the terms of the series are given in decreasing order with respect to the degree of t.)

5: Continue calculating the terms of the expansion (7.22), in decreasing order with respect to the degree of t, until a degree M of t is found such that for some $\boldsymbol{\alpha} \in \mathbb{Z}^n$, the coefficient of $\mathbf{y}^{\boldsymbol{\alpha}} t^M$ is nonzero in the expansion (7.22).

6: **return** M as the optimal value of the integer program and $\boldsymbol{\alpha}$ as an optimal solution.

Unfortunately, although the necessary rational functions can be computed in polynomial time when the dimension is assumed to be fixed, the digging algorithm may have to compute an exponential number of coefficients before finding one that does not vanish.

By stopping Algorithm 7.4 after a polynomial number of terms have been expanded, the exponential-time exact algorithm turns into a polynomial-time algorithm that computes an upper bound for the optimal value. By stopping already at the first term, one recovers the heuristic proposed by Lasserre in [214].

7.5 Boolean operations on generating functions

Barvinok and Woods [39] further developed a set of powerful manipulation rules for using these short rational functions in Boolean constructions on various sets of lattice points.

Theorem 7.5.1 (intersection lemma; Theorem 3.6 in [39]). *Let ℓ be a fixed integer. Let S_1, S_2 be finite subsets of $\mathbb{Z}^n$. Let $g(S_1;\mathbf{z})$ and $g(S_2;\mathbf{z})$ be their generating functions, given as short rational functions with at most ℓ binomials in each denominator. Then there exists a polynomial-time algorithm which computes*

$$g(S_1 \cap S_2;\mathbf{z}) = \sum_{i\in I} \epsilon_i \frac{\mathbf{z}^{\mathbf{a}_i}}{(1-\mathbf{z}^{\mathbf{b}_{i1}})\cdots(1-\mathbf{z}^{\mathbf{b}_{is}})}$$

with $s \le 2\ell$, where the ϵ_i are rational numbers, $\mathbf{a}_i, \mathbf{b}_{ij}$ are nonzero integer vectors, and I is a polynomial-size index set.

Proof (sketch). The essential step in the intersection algorithm is the use of the *Hadamard product* [39, Definition 3.2]. The Hadamard product is a bilinear operation on series expansions of rational functions, denoted by $\star$. The Hadamard product $\mathbf{z}^{\alpha_1} \star \mathbf{z}^{\alpha_2}$ of two monomials $\mathbf{z}^{\alpha_1}, \mathbf{z}^{\alpha_2}$ is zero unless $\boldsymbol{\alpha}_1 = \boldsymbol{\alpha}_2$, in which case it is $\mathbf{z}^{\alpha_1} = \mathbf{z}^{\alpha_2}$. The computation is carried out for pairs of basic rational summands. For each pair, an auxiliary lattice point problem in dimension 2ℓ is created, for which a generating function is computed using Barvinok's algorithm (Algorithm 7.1). This is followed by a monomial specialization (Theorem 7.2.3). □

We remark that the algorithm also works if the sets are given by positively weighted generating functions (Section 7.2). By the bilinearity of the Hadamard product, we have $k_\alpha \mathbf{z}^\alpha \star k'_\alpha \mathbf{z}^\alpha = k_\alpha k'_\alpha \mathbf{z}^\alpha$, i.e., the coefficients multiply, and so we obtain another positively weighted generating function as the result.

Using Theorem 7.5.1, we can extend the theorem on efficient explicit enumeration (Theorem 7.3.1) from a lattice point set of the form $P \cap \mathbb{Z}^n$ to *arbitrary* lattice point sets given by a short rational generating function (or a positively weighted generating function). This will become relevant in Chapter 9.

Theorem 7.5.2. *Let the dimension n and the maximum number ℓ of binomials in the denominator of a rational generating function be fixed.*

Let $V \subseteq \mathbb{Z}^n$ be a bounded set of lattice points with $V \subseteq [-M,M]^n$, given only by the bound $M \in \mathbb{Z}_+$ and its rational generating function encoding $g(V;\mathbf{z})$ with at most ℓ binomials in each denominator. Let

$$W = \{\mathbf{w} \in \mathbb{Z}^p : \exists \mathbf{t} \in \mathbb{Z}^{n-p} \text{ such that } (\mathbf{t},\mathbf{w}) \in V\}$$

denote the projection of V onto the last p components.

There exists a polynomial-space polynomial-delay enumeration algorithm for the points in the projection W, which outputs the points of W in lexicographic order.

Proof. We use the same algorithm (Algorithm 7.2), but this time the oracle for determining whether $W \cap \{\mathbf{w} : \mathbf{l} \leq \mathbf{w} \leq \mathbf{u}\} \neq \emptyset$ is implemented using the intersection lemma (Theorem 7.5.1) as follows. Consider the polytope

$$Q_{\mathbf{l},\mathbf{u}} = [-M,M]^{n-p} \times \{\mathbf{w} \in \mathbb{R}^p : \mathbf{l} \leq \mathbf{w} \leq \mathbf{u}\} \subseteq \mathbb{R}^n, \tag{7.28}$$

a parallelepiped in $\mathbb{R}^n$. Since W is the projection of V and since $V \subseteq [-M,M]^n$, we have (7.16) if and only if $V \cap Q_{\mathbf{l},\mathbf{u}} = \emptyset$. The rational generating function $g(Q_{\mathbf{l},\mathbf{u}};\mathbf{z})$ can be computed in polynomial time. By using the intersection lemma, we can compute the rational generating function $g(V \cap Q_{\mathbf{l},\mathbf{u}};\mathbf{z})$ in polynomial time. The specialization $g(V \cap Q_{\mathbf{l},\mathbf{u}};\mathbf{z}=\mathbf{1})$ can also be computed in polynomial time. It gives the number of lattice points in $V \cap Q_{\mathbf{l},\mathbf{u}}$; in particular, we can decide whether $V \cap Q_{\mathbf{l},\mathbf{u}} = \emptyset$. □

The following theorem was proved by Barvinok and Woods.

Theorem 7.5.3 (Boolean operations lemma; Corollary 3.7 in [39]). *Let m and ℓ be fixed integers. Let $S_1, S_2, \ldots, S_m$ be finite subsets of $\mathbb{Z}^n$. Let $g(S_i;\mathbf{z})$ for $i = 1,\ldots,m$ be their generating functions, given as short rational functions with at most ℓ binomials in each denominator. Let a set $S \subseteq \mathbb{Z}^n$ be defined as a Boolean combination of $S_1,\ldots,S_m$ (i.e., using any of the operations $\cup$, $\cap$, $\setminus$). Then there exists a polynomial-time algorithm, which computes*

$$g(S;\mathbf{z}) = \sum_{i \in I} \epsilon_i \frac{\mathbf{z}^{\mathbf{a}_i}}{(1-\mathbf{z}^{\mathbf{b}_{i1}})\ldots(1-\mathbf{z}^{\mathbf{b}_{is}})},$$

where $s = s(\ell,m)$ is a constant, the ϵ_i are rational numbers, $\mathbf{a}_i, \mathbf{b}_{ij}$ are nonzero integer vectors, and I is a polynomial-size index set.

Proof (sketch). It follows by iterating the construction of Theorem 7.5.1 by noting

$$g(A \cup B;\mathbf{z}) = g(A;\mathbf{z}) + g(B;\mathbf{z}) - g(A \cap B;\mathbf{z})$$

and

$$g(A \setminus B;\mathbf{z}) = g(A;\mathbf{z}) - g(A \cap B;\mathbf{z}).$$ □

We remark that for positively weighted generating functions (Section 7.2), the operations $\cap$ (via the Hadamard product) and $\cup$ (via the sum of generating functions) are available, but $\setminus$ is *not* available because there is no control over the coefficients of the monomials.

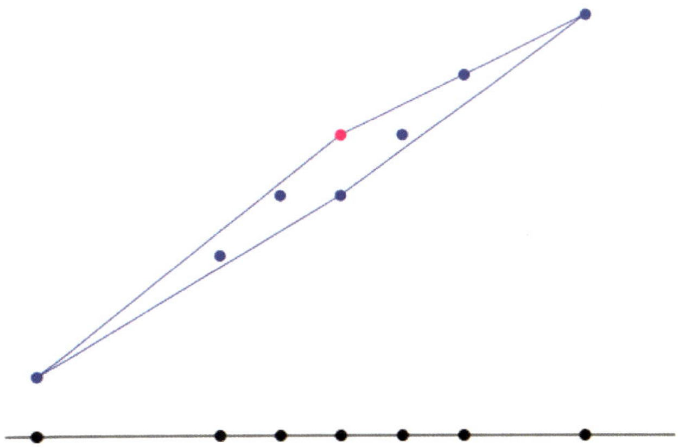

Figure 7.2. *Projection of a lattice point set, codimension* 1 *case. The resulting lattice point set on the horizontal axis (bottom) has gaps. All nonempty fibers in this example have cardinality* 1, *except for the middle one, which has cardinality* 2. [204]

7.6 Integer projections

Another key subroutine introduced by Barvinok and Woods is the following *projection theorem*.

Theorem 7.6.1 (projection theorem; Theorem 1.7 in [39]). *Assume the dimension n is a fixed constant. Consider a rational polytope $P \subset \mathbb{R}^n$ and a matrix $T \in \mathbb{Z}^{k \times n}$, which induce a linear map $\mathbb{Z}^n \to \mathbb{Z}^k$. There is a polynomial-time algorithm which computes a short rational generating function $g\big(T(P \cap \mathbb{Z}^n); \mathbf{y}\big)$, $\mathbf{y} \in \mathbb{C}^k$.*

In contrast to the computation of a positively weighted generating function discussed in Section 7.2, every element $\boldsymbol{\beta}$ of the integer projection $T(P \cap \mathbb{Z}^n)$ gives rise to a monomial $\mathbf{y}^{\boldsymbol{\beta}}$ with coefficient 1 in the generating function $g\big(T(P \cap \mathbb{Z}^n); \mathbf{y}\big)$. With the resulting generating function, we can determine the cardinality $|T(P \cap \mathbb{Z}^n)|$. Later this and other Boolean operations as in Theorem 7.5.3 will be important in the application in Chapter 9.

We also emphasize that the integer projection $T(P \cap \mathbb{Z}^n)$ is *not* the same as $T(P) \cap \mathbb{Z}^k$. In fact, as Figure 7.2 illustrates, $T(P \cap \mathbb{Z}^n)$ may have *gaps*. In this case, it cannot be the set of integer points of any polyhedron.

It is important to note that, in contrast to Theorem 7.5.3, the input of the algorithm of Theorem 7.6.1 is a polytope and not a generating function. It is unknown whether, in fixed dimension, there exists a polynomial-time algorithm that computes the projection of an arbitrary lattice point set given by its rational generating function.

The Barvinok–Woods projection theorem is a very powerful tool. Let us give one example of its algorithmic implications; many more can be found in [39].

Example 7.6.2 (enumeration of Hilbert basis elements). Theorem 7.5.2 implies the existence of a polynomial-space polynomial-delay prescribed-order enumeration algorithm for Hilbert bases (cf. section 2.6) of rational polyhedral cones in fixed dimension.

Indeed, fix the dimension n and let $C = \operatorname{cone}\{\mathbf{b}_1, \dots, \mathbf{b}_k\} \subseteq \mathbb{R}^n$ be a pointed rational polyhedral cone, and denote by $H \subseteq C \cap \mathbb{Z}^n$ the inclusion-minimal Hilbert basis. For *simplicial* cones C (where $\mathbf{b}_1, \dots, \mathbf{b}_k$ are linearly independent), Barvinok and Woods [39] proved that one can compute the rational generating function $g(H; \mathbf{z})$ (having a constant

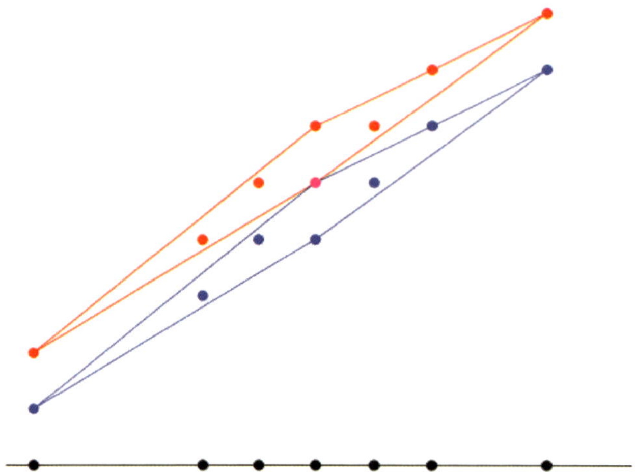

Figure 7.3. *Projection of a lattice point set, codimension* 1 *case. We remove the extra element in the fiber by taking the set difference with a shifted copy of* $P \cap \mathbb{Z}^n$ *in* (7.29). [204]

number of binomials in the denominators) of the Hilbert basis of $C \cap \mathbb{Z}^n$ using the projection theorem. The same technique works for nonsimplicial pointed cones (Exercise 7.8.5). Now Theorem 7.5.2 gives a polynomial-space polynomial-delay prescribed-order enumeration algorithm.

Proof of the Barvinok–Woods projection theorem. The remainder of this section contains a proof sketch of Theorem 7.6.1.

Let $m = n - k$. After making a unimodular change of variables, we may assume that the projection $V = T(P \cap \mathbb{Z}^n)$ can be expressed in the form

$$V = \{\, \mathbf{v} \in \mathbb{Z}^k : \exists \mathbf{w} \in \mathbb{Z}^m : (\mathbf{v}, \mathbf{w}) \in P \,\}.$$

If there is only one existentially quantified variable w ($m = 1$, so $k + 1 = n$), then computing the generating function $g(V; \mathbf{y})$ works as follows. We shift P by 1 in the w direction and subtract this shifted copy from the original,

$$V'' = (P \cap \mathbb{Z}^n) \setminus \big(\mathbf{e}_{k+1} + (P \cap \mathbb{Z}^n)\big). \tag{7.29}$$

In the difference V'' there will be *exactly* one value of w for each $\mathbf{v}$ for which there was *at least* one value of w in P. Thus the projection from V'' onto V is one-to-one. This is illustrated in Figure 7.3. All of this can then be implemented on the level of rational generating functions, using Boolean operations (Theorem 7.5.3) and monomial substitution (Theorem 7.2.3).

If there is more than one existentially quantified variable ($m > 1$), then we can in principle apply this shifting and subtracting technique recursively. A complication is that, after applying the technique in one direction, the resulting set, say S, in general does not correspond to the integer points in some polytope, but can contain gaps. In the example of Figure 7.3, consider the case $k = 0$, so $m = 2$ and we want to project down to a single point. After projecting onto the horizontal direction, the biggest gap is 3, so we need to compute

$$S \setminus (\mathbf{e}_{k+1} + S) \setminus (2\mathbf{e}_{k+1} + S) \setminus (3\mathbf{e}_{k+1} + S).$$

In general, there is no bound on the widths of the gaps we may encounter. However, as a consequence of flatness theory for lattice-point-free convex bodies (Theorem 2.8.2), it is possible to *construct* directions of small gaps for the polytopes

$$P_{\mathbf{v}} = \{\, \mathbf{w} \in \mathbb{R}^m : (\mathbf{v}, \mathbf{w}) \in P \,\},$$

which describe the fibers $P_{\mathbf{v}} \cap \mathbb{Z}^m$ in a parametric way.

The following "small-gaps theorem" is a version of Theorem 4.3 in [39]. It is a corollary of Theorem 2.8.2.

Theorem 7.6.3. *Let $\kappa \geq 1$ and let $\mathbf{d} \in \mathbb{Z}^m$ be a κ-approximative lattice width direction, i.e.,*

$$\operatorname{width}_{\mathbf{d}}(P_{\mathbf{v}}) \leq \kappa \cdot \operatorname{width}(P_{\mathbf{v}}).$$

Then the image $Y = \{\, \mathbf{d}^\top \mathbf{x} : \mathbf{x} \in P_{\mathbf{v}} \cap \mathbb{Z}^m \,\} \subset \mathbb{Z}$ does not have gaps larger than $\kappa \cdot f^*(m)$, i.e., for any numbers $y_1, y_2 \in Y$ with $y_2 - y_1 > \kappa \cdot f^*(m)$, there exists a number $y \in Y$ with $y_1 < y < y_2$.

Here $f^*(m)$ is the flatness constant from Theorem 2.8.2. Kannan [189] proved the following theorem:

Theorem 7.6.4 (Kannan). *Let the total dimension $n = k + m$ be fixed. Then there exists a polynomial-time algorithm for the following problem. Given as input, in binary encoding, inequalities describing a rational polytope $P \subset \mathbb{R}^{k+m}$, output, in binary encoding,*

- *inequality systems describing half-open rational polyhedra $\tilde{Q}_1, \dots, \tilde{Q}_M \subset \mathbb{R}^k$ that form a set-theoretic partition of the projection Q of P onto the first k coordinates,*
- *integer vectors $\mathbf{d}_1, \dots, \mathbf{d}_M \in \mathbb{Z}^m$,*

such that $\mathbf{d}_i$ is a 2-approximative lattice width direction for every polytope $P_{\mathbf{v}}$ when $\mathbf{v} \in \tilde{Q}_i$.

Eisenbrand and Shmonin [123] improved this result from 2-approximative to exact lattice widths ($\kappa = 1$). Our proof sketch follows their method.

Proof. In the defining equation (2.7) for the width of the polytope along any given direction $\mathbf{d}$, it is clear that the minimum and maximum are attained at (different) vertices of $P_{\mathbf{v}}$. The idea of the algorithm is then to consider all pairs of basic solutions that correspond to vertices $\mathbf{w}(\mathbf{v})$ of $P_{\mathbf{v}}$, to compute all candidate integer directions for a given pair of such vertices, and then to compute the minimum width over all candidate integer directions found. For any given basic solution $\mathbf{w}(\mathbf{v})$, the (rational) directions for which this vertex is minimal form the inner normal cone C^* of the vertex. Suppose we have a pair of basic solutions, $\mathbf{w}_1(\mathbf{v})$ attaining the minimum and $\mathbf{w}_2(\mathbf{v})$ attaining the maximum; thus the set of (rational) directions for this pair of vertices is the "truncated cone" $C_{1,2} = \left(C_1^* \cap -C_2^*\right) \setminus \{\mathbf{0}\}$. A sufficient set of candidate *integer* directions are the vertices of the integer hull (Section 2.9) of $C_{1,2}$; all other integer directions in $C_{1,2}$ are dominated. It now follows from [74] that this set of candidates is of polynomial size (because the dimension is fixed); it can be computed with Hartmann's algorithm [149]. After computing all candidates $\mathbf{d}_i$, we construct the chambers where each of the widths is minimal, i.e.,

$$Q_i = \{\, \mathbf{v} \in Q : \forall j : \operatorname{width}_{\mathbf{d}_i}(P_{\mathbf{v}}) \leq \operatorname{width}_{\mathbf{d}_j}(P_{\mathbf{v}}) \,\}.$$

Many of the Q_i will be empty or of lower dimension. Note that some of the Q_i will have nonempty intersection (along common faces). To obtain a set-theoretic *partition* of Q, we use the technique of Section 6.4. (This is a minor improvement upon Eisenbrand–Shmonin's lexicographic technique.) We obtain a set-theoretic partition into half-open *full-dimensional* chambers $\tilde{Q}_i$, discarding all lower-dimensional chambers. □

Now let $P_i = P \cap (\tilde{Q}_i \times \mathbb{Q}^m)$ and let $g(V_i;\mathbf{y})$ be the generating function of the set

$$V_i = \{\, \mathbf{v} \in \mathbb{Z}^k : \exists \mathbf{w} \in \mathbb{Z}^m : (\mathbf{v},\mathbf{w}) \in P_i \,\}.$$

Then clearly, $g(V;\mathbf{y}) = \sum_i g(V_i;\mathbf{y})$. From now on, we will consider a particular P_i with corresponding κ-approximative lattice width direction $\mathbf{d}_i$ and drop the i subscript. We are thus given a polyhedron P such that a κ-approximative lattice width direction of $P_\mathbf{v}$ is $\mathbf{d}$. Without loss of generality, $\mathbf{d}$ is primitive, so we can extend the row vector $\mathbf{d}^\top$ to a unimodular matrix, which we use to transform the $\mathbf{w}$ variables in the polyhedron P; see Figure 7.4. Using the notations $\mathbf{w}'$ and P', we have

$$V = \{\, \mathbf{v} : \exists \mathbf{w}' \in \mathbb{Z}^m : (\mathbf{v},\mathbf{w}') \in P' \,\},$$

i.e., we have changed the values of the existentially quantified variables, but we have not changed the set V. Now consider the set

$$V' = \{\, (\mathbf{v}, w_1') : \exists w_2',\dots,w_m' \in \mathbb{Z} : (\mathbf{v},\mathbf{w}') \in P' \,\}.$$

This set has only $m-1$ existentially quantified variables, so we apply the projection algorithm recursively and obtain the generating function $g(V';\mathbf{y},z)$ for $V' \subseteq \mathbb{R}^{k+1}$. By construction, the gaps in the final coordinate of V' are small ($\leq \kappa \cdot f^*(m)$).

We now compute the generating function $g(V;\mathbf{y})$ for V. Define the set

$$V'' = V' \setminus (\mathbf{e}_{k+1} + V') \setminus (2\mathbf{e}_{k+1} + V') \setminus \cdots \setminus (\lfloor \kappa f^*(m) \rfloor \mathbf{e}_{k+1} + V'),$$

where only the smallest value of w_1' is retained. The generating function of V'' can be computed by Theorem 7.5.3 in polynomial time because $\lfloor \kappa f^*(m) \rfloor$ is fixed. Now the projection from V'' onto V is actually one-to-one, and we have $g(V;\mathbf{y}) = g(V'';\mathbf{y},1)$, which can be computed using Theorem 7.2.3 in polynomial time. This concludes the description of the algorithm and the proof of Theorem 7.6.1. □

We will make use of Theorem 7.6.1 in Chapter 9.

7.7 Notes and further references

Barvinok's algorithm first appeared in [30]. Later expositions can be found in [34], [35], and [50]. The original algorithm by Barvinok [30] used decompositions into closed cones, inclusion-exclusion to handle the intersection of proper faces, and Lenstra's algorithm to obtain the decomposition vector for constructing the signed decomposition. The use of a shortest vector algorithm for this purpose appeared in Dyer and Kannan [118]. A dual variant of the algorithm appeared in Barvinok and Pommersheim [35]; the "duality trick" (decomposing the polars of cones), attributed to [61], allowed us to remove the use of inclusion-exclusion to handle the intersection of proper faces. This variant was implemented in `LattE` [90, 91].

The practical benefit of stopping the decomposition before reaching unimodular cones was explored in [200]. A primal algorithm that avoids inclusion-exclusion using "irrational" [45] decompositions appeared in [200]. These variants were implemented in `LattE`

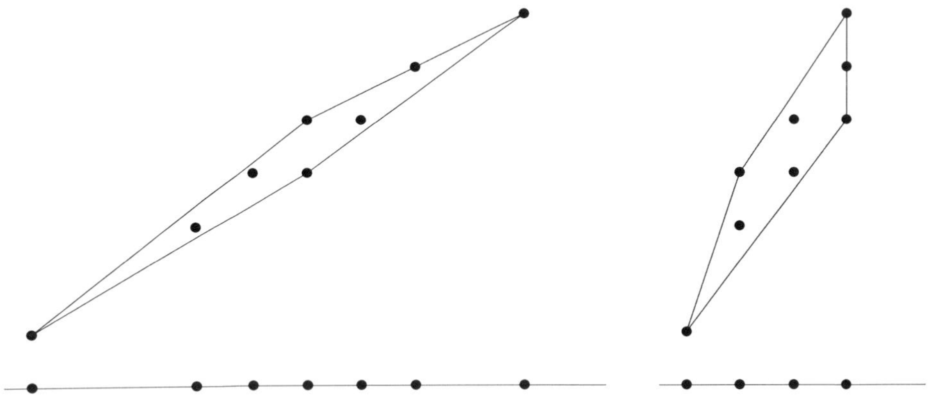

Figure 7.4. *A polytope and its integer projections; the same after unimodular transformation.* [204]

macchiato [201]. The variant using half-open decompositions is a refinement of this technique, which appeared first in Köppe and Verdoolaege [203] and was implemented in barvinok [330]. Similar decompositions in the primal space, replacing the use of half-open cones by closed cones modulo nonpointed polyhedra, were described earlier by Brion and Vergne [62].

Since stopped decomposition works significantly better with a primal algorithm than a dual algorithm, as observed in [200], the current method of choice is to use a primal algorithm (irrational or half-open) with stopped decomposition, as presented in this chapter.

Instead of using the vertex cones of a polytope via Brion's theorem, one can also do the computations using the cone over the polytope as in our proof sketch of the Lawrence–Khovanskii–Pukhlikov theorem (Theorem 7.1.5). This is known as the "homogenized" variant, which was first described in [88] and implemented in LattE.

The homogenized variant, combined with half-open primal decompositions, also gives a way to introduce Barvinok's algorithm in an elementary, algebraic way that only uses formal Laurent series, sidestepping the use of convergent series and rational functions and the "mysterious" identities like Brion's that are associated with them.

For some polytopes, it is more efficient to compute triangulations in the dual space. Using dual triangulations and primal signed decompositions is known as the "primal" variant, whereas using primal triangulations and primal signed decompositions is known as the "all-primal" algorithm.

Parametric versions of Barvinok's algorithm (where the right-hand side vector or the vertices of the polytope appear as parameters) have been studied in [203, 332].

The evaluation (specialization) problem of Section 7.2 has been considered in various degrees of generality in the literature. In the original paper by Barvinok [30, Lemma 4.3], the dimension n is fixed, and each summand has exactly $s_i = n$ binomials in the denominator. The same restriction can be found in the survey by Barvinok and Pommersheim [35]. In the more general algorithmic theory of monomial substitutions developed by Barvinok and Woods [39], Woods [336], there is no assumption on the dimension n, but the number s of binomials in each of the denominators is fixed. The same restriction appears in the paper by Verdoolaege and Woods [331, Lemma 2.15]. In a recent paper, Barvinok [33, section 5] gives a polynomial-time algorithm for the specialization problem for rational functions of the form

$$g(\mathbf{z}) = \sum_{i\in I} \epsilon_i \frac{\mathbf{z}^{\mathbf{a}_i}}{\prod_{j=1}^{s}(1-\mathbf{z}^{\mathbf{b}_{ij}})^{\gamma_{ij}}}, \tag{7.30}$$

where the dimension n is fixed, the number s of different binomials in each denominator equals n, but the multiplicity γ_{ij} is varying. Our exposition essentially follows the method of Barvinok [30, Lemma 4.3] and Barvinok [33, Section 5]. A refined analysis in De Loera et al. [97], on which Section 7.2 is based, showed that the method can be implemented in a way such that we obtain a polynomial-time algorithm even for the case of a general Formula (7.4), when the dimension and the number of binomials are allowed to grow.

The results on output-sensitive polynomial-time enumeration of lattice point sets (Theorem 7.3.1) appeared first in [98]. The idea is similar to the one used in [268]. Theorem 7.5.2 is a stronger result than what can be obtained by the repeated application of the monomial-extraction technique of Lemma 7 from [88], which accumulates exponential space and thus only gives an incremental polynomial-time enumeration algorithm.

Section 7.4 is based on the computational studies in [89, 92]. A variant of the digging algorithm, single-cone digging, is related to Gomory's group relaxation [140]. This connection is also studied in [214]; see also [217]. The connection to Gomory's group relaxation is also a topic in [176], where the authors use the Barvinok–Woods projection theorem and irreducible decompositions of monomial ideals to compute the gap between parametric families of integer linear programs and their linear programming relaxations.

Section 7.6 is based on the paper [204], which describes the implementation of the Barvinok–Woods integer projection algorithm and computational experiments.

7.8 Exercises

Exercise 7.8.1. Prove Theorem 7.1.8. (Hint: Let $d \le n$ and consider the vector configuration $\mathbf{e}^1,\dots,\mathbf{e}^d,-(\mathbf{e}^1+\cdots+\mathbf{e}^d)$. Then consider the triangulation of the space $\mathbb{R}^d$ into the cones spanned by d of these $d+1$ vectors.)

Exercise 7.8.2. For the Barvinok algorithm with half-open decomposition, we need to be able to write down the generating function of half-open simplicial cones $\tilde{C}$ of small index. Define the notion of a half-open fundamental parallelepiped Π of a half-open simplicial cone $\tilde{C}$, and prove a version of Lemma 2.3.16 to effectively enumerate its lattice points.

Exercise 7.8.3. Implement an explicit lattice point enumeration procedure using the technique of Section 7.3. You can use `LattE` or `barvinok` as a subroutine. For extra credit, combine the technique with other techniques, such as branching on hyperplanes corresponding to thin directions, or binary decision diagrams.

Exercise 7.8.4. (following an idea by Pak [268].) As in Section 7.3, suppose that a lattice point set W in fixed dimension is given by a rational generating function and bounds on the variables. Show that you can uniformly sample over W with a polynomial-time algorithm.

Exercise 7.8.5. (following Barvinok and Woods [39].) As in Example 7.6.2, let $C = \operatorname{cone}\{\mathbf{b}_1,\dots,\mathbf{b}_k\} \subseteq \mathbb{R}^n$ be a pointed rational polyhedral cone, and denote by $H \subseteq C\cap\mathbb{Z}^n$ the inclusion-minimal Hilbert basis. Show that one can compute the rational generating function $g(H;\mathbf{z})$ of H using the projection theorem (Theorem 7.6.1) in polynomial time if the dimension is fixed.

Exercise 7.8.6. Improve the complexity of the binary search algorithm (Section 7.4.1) by studying the counting problem $|P \cap \{\mathbf{x} : \mathbf{c}^\top \mathbf{x} \geq \alpha\} \cap \mathbb{Z}^n|$ in a parametric way. Compare with the complexity of Eisenbrand's algorithm [121].

Exercise 7.8.7. Prove Theorem 7.6.3.

Chapter 8

Global Mixed-Integer Polynomial Optimization via Summation

Here we consider the problem

$$\begin{array}{rl}\max & f(x_1,\dots,x_n)\\ \text{subject to} & A\mathbf{x} \le \mathbf{b},\\ & \mathbf{x} = (x_1,\dots,x_n) \in \mathbb{R}^{n_1} \times \mathbb{Z}^{n_2},\end{array} \tag{8.1}$$

where $n = n_1 + n_2$, A is a rational matrix, $\mathbf{b}$ is a rational vector, and f is a polynomial function of maximum total degree D with rational coefficients. We are interested in general polynomial objective functions f *without any convexity assumptions*.

It turns out that optimizing polynomials of degree 4 over problems with two integer variables ($n_1 = 0$, $n_2 = 2$) is already a hard problem; see [202]. Thus, even when we fix the dimension, we cannot get a polynomial-time algorithm for solving the optimization problem. The best we can hope for, even when the number of both the continuous and the integer variables is fixed, is an approximation result.

8.1 Approximation algorithms and schemes

Definition 8.1.1. An algorithm $\mathcal{A}$ is said to *efficiently approximate* an optimization problem if, for every value of the input parameter $\epsilon > 0$, it returns a rational vector $\mathbf{x}$ (not necessarily feasible) with $\|\mathbf{x} - \mathbf{x}^*\| \le \epsilon$, where $\mathbf{x}^*$ is an optimal solution, and the running time of $\mathcal{A}$ is polynomial in the input encoding of the instance and in $\log 1/\epsilon$.

The polynomial dependence of the running time in $\log 1/\epsilon$, as defined above, is a very strong requirement. For many problems, efficient approximation algorithms of this type do not exist, unless P = NP. The following, weaker notions of approximation are useful; here though, it is common to ask for the approximations to be *feasible solutions*.

Definition 8.1.2. (a) An algorithm $\mathcal{A}$ is an ϵ*-approximation algorithm* for a maximization problem with optimal value $f_{\max}$ if $\mathcal{A}$ runs in polynomial time in the encoding length and returns a feasible solution with value $f_{\mathcal{A}}$, such that

$$f_{\mathcal{A}} \ge (1-\epsilon)\cdot f_{\max}. \tag{8.2}$$

(b) A family of algorithms $\mathcal{A}_\epsilon$ is a *polynomial-time approximation scheme (PTAS)* if for every error parameter $\epsilon > 0$, $\mathcal{A}_\epsilon$ is an ϵ-approximation algorithm and its running time is polynomial in the encoding length for every fixed ϵ.

(c) A family $\{\mathcal{A}_\epsilon\}_\epsilon$ of ϵ-approximation algorithms is a *fully polynomial-time approximation scheme (FPTAS)* if the running time of $\mathcal{A}_\epsilon$ is polynomial in the encoding length and $1/\epsilon$.

These notions of approximation are popular in the domain of combinatorial optimization. It is clear that they are only useful when the function f (or at least the maximal value $f_{\max}$) is nonnegative. In Section 8.3 we introduce such an FPTAS for the pure integer, nonnegative case. Then we show an extension to mixed-integer optimization by discretization in Section 8.4, which is the following result:

Theorem 8.1.3 (fully polynomial-time approximation scheme). *Let the dimension $n = n_1 + n_2$ be fixed. There exists a fully polynomial-time approximation scheme (FPTAS) for the maximization problem* (8.1) *for all polynomial functions $f(x_1,\dots,x_n)$ with rational coefficients that are nonnegative on the feasible region .*

For polynomial or general nonlinear optimization problems, various authors [48, 83, 328] have proposed to use a different notion of approximation, where we compare the approximation error to the *range* of the objective function on the feasible region,

$$\left|f_{\mathcal{A}} - f_{\max}\right| \le \epsilon \left|f_{\max} - f_{\min}\right|. \tag{8.3}$$

(Here $f_{\min}$ denotes the minimal value of the function on the feasible region.) It enables us to study objective functions that are not restricted to be nonnegative on the feasible region. In addition, this notion of approximation is invariant under shifting of the objective function by a constant, and under exchanging minimization and maximization. On the other hand, it is not useful for optimization problems that have an infinite range. We remark that, when the objective function can take negative values on the feasible region, (8.3) is weaker than (8.2). We will call approximation algorithms and schemes with respect to this notion of approximation *weak* .

Remark 8.1.4. This terminology is not consistent in the literature; [82], for instance, uses the notion (8.3) without an additional attribute and instead reserves the word *weak* for approximation algorithms and schemes that give a guarantee on the absolute error:

$$\left|f_{\mathcal{A}} - f_{\max}\right| \le \epsilon. \tag{8.4}$$

Indeed, we prove:

Theorem 8.1.5 (fully polynomial-time weak-approximation scheme). *Let the dimension $n = n_1 + n_2$ be fixed. Let f be an arbitrary polynomial function with rational coefficients and maximum total degree D, and let $P \subset \mathbb{R}^n$ be a rational convex polytope.*

(a) *In time that is bounded polynomially by the input size and D, it is possible to decide whether f is constant on $P \cap \left(\mathbb{R}^{n_1} \times \mathbb{Z}^{n_2}\right)$.*

(b) *In time that is bounded polynomially in the input size, D, and $\frac{1}{\epsilon}$ it is possible to compute a solution $\mathbf{x}_\epsilon \in P \cap \left(\mathbb{R}^{n_1} \times \mathbb{Z}^{n_2}\right)$ with*

$$\left|f(\mathbf{x}_\epsilon) - f(\mathbf{x}_{\max})\right| \le \epsilon \left|f(\mathbf{x}_{\max}) - f(\mathbf{x}_{\min})\right|,$$

where $\mathbf{x}_{\max}$ denotes a maximizer and $\mathbf{x}_{\min}$ denotes a minimizer of the objective function on the feasible region.

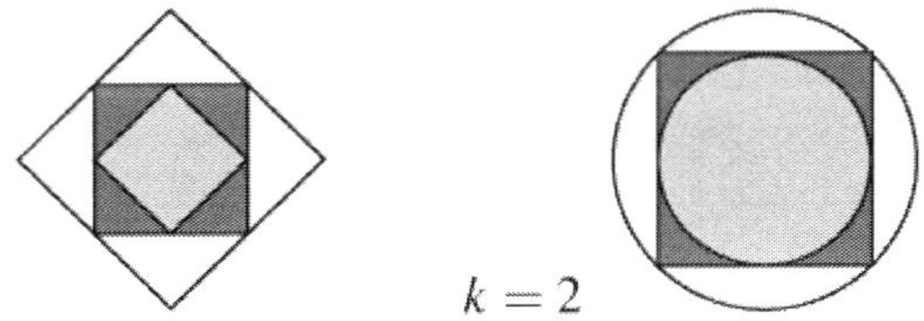

Figure 8.1. *Approximation properties of ℓ_k-norms.* [202]

8.2 The summation method

The summation method for optimization is the idea to use an elementary limit relation such as

$$\max\{s_1,\dots,s_N\} = \lim_{k\to\infty} \sqrt[k]{s_1^k+\cdots+s_N^k}, \tag{8.5}$$

which holds for any finite set $S=\{s_1,\dots,s_N\}$ of nonnegative real numbers. This relation can be viewed as an approximation result for ℓ_k-norms. Now if P is a polytope and f is an objective function nonnegative on $P\cap\mathbb{Z}^n$, let $\mathbf{x}^1,\dots,\mathbf{x}^N$ denote all the feasible integer solutions in $P\cap\mathbb{Z}^n$ and collect their objective function values $s_i=f(\mathbf{x}^i)$ in a vector $\mathbf{s}\in\mathbb{Q}^N$. Then, comparing the unit balls of the ℓ_k-norm and the ℓ_∞-norm (Figure 8.1), we get the relation

$$L_k := N^{-1/k}\|\mathbf{s}\|_k \le \|\mathbf{s}\|_\infty \le \|\mathbf{s}\|_k =: U_k.$$

Remark 8.2.1. These estimates are independent of the function f. Different estimates that make use of the properties of f, and that are suitable also for the continuous case, can be obtained from Hölder's inequality; see, for instance, [21].

Thus, for obtaining a good approximation of the maximum, it suffices to solve a summation problem of the polynomial function $h=f^k$ on $P\cap\mathbb{Z}^n$ for a value of k that is large enough. Indeed, for $k=\lceil(1+1/\epsilon)\log N\rceil$, we obtain $U_k-L_k\le\epsilon f(\mathbf{x}_{\max})$. On the other hand, this choice of k is polynomial in the input size (because $1/\epsilon$ is presented in the unary encoding scheme in the input, and $\log N$ is bounded by a polynomial in the binary encoding size of the polytope P). Hence, when the dimension n is fixed, we can expand the polynomial function f^k as a list of monomials in polynomial time.

Below we show how to solve the summation problem for the pure integer case.

8.3 FPTAS for maximizing nonnegative polynomials over integer points of polytopes

Here we prove Theorem 8.1.3 for the pure integer case ($n_1=0$). The proof uses the summation method. To solve the summation problem, we use the technique of short rational generating functions, as discussed in Chapter 7. The key is now to use differential operators that bring the objective function into the generating function. A first example of this technique appeared already in Section 5.2. There we considered the generating function $g(S;z)$ of the set S of integer points of the interval $P=[0,4]$. Applying a differential operator, we obtained

$$\left(z\frac{\mathrm{d}}{\mathrm{d}z}\right)\left(z\frac{\mathrm{d}}{\mathrm{d}z}\right)g(S;z) = 1z^1+4z^2+9z^3+16z^4$$
$$= \frac{z+z^2}{(1-z)^3} - \frac{25z^5-39z^6+16z^7}{(1-z)^3}.$$

We have thus evaluated the monomial function $f(\alpha)=\alpha^2$ for $\alpha = 0,\dots,4$; the results appear as the coefficients of the respective monomials. Substituting $z=1$ yields the desired sum

$$\left(z\frac{\mathrm{d}}{\mathrm{d}z}\right)\left(z\frac{\mathrm{d}}{\mathrm{d}z}\right)g(S;z)\bigg|_{z=1} = 1+4+9+16=30.$$

The idea now is to evaluate this sum instead by computing the limit of the rational function for $z\to 1$,

$$\sum_{\alpha=0}^{4}\alpha^2 = \lim_{z\to 1}\left[\frac{z+z^2}{(1-z)^3} - \frac{25z^5-39z^6+16z^7}{(1-z)^3}\right];$$

again this evaluation problem can be solved using residue techniques.

Such differential operators can be constructed and efficiently applied in general.

Lemma 8.3.1. *Let $g(P;\mathbf{z})$ be the rational generating function of the lattice points of P. Let f be a polynomial in $\mathbb{Z}[x_1,\dots,x_n]$ of maximum total degree D. We can compute, in time that is bounded polynomially by D and the size of the input data, a rational generating function $g(P,f;\mathbf{z})$ that represents $\sum_{\boldsymbol{\alpha}\in P\cap\mathbb{Z}^n} f(\boldsymbol{\alpha})\mathbf{z}^{\boldsymbol{\alpha}}$.*

Proof (sketch). In general, we apply the differential operator

$$D_f = f\left(z_1\frac{\partial}{\partial z_1},\dots,z_n\frac{\partial}{\partial z_n}\right).$$

Consider first the case $f(\mathbf{z}) = z_r$. Consider the action of the differential operator $z_r\frac{\partial}{\partial z_r}$ in the rational generating function $g(P;\mathbf{z})$. On the one hand, for the generating function,

$$z_r\frac{\partial}{\partial z_r}g(P;\mathbf{z}) = \sum_{\boldsymbol{\alpha}\in P\cap\mathbb{Z}^n} z_r\frac{\partial}{\partial z_r}\mathbf{z}^{\boldsymbol{\alpha}} = \sum_{\boldsymbol{\alpha}\in P\cap\mathbb{Z}^n}\alpha_r\mathbf{z}^{\boldsymbol{\alpha}}.$$

On the other hand, by linearity of the operator, we have that in terms of rational functions,

$$z_r\frac{\partial}{\partial z_r}g(P;\mathbf{z}) = \sum_{i\in I}\epsilon_i z_r\frac{\partial}{\partial z_r}\frac{\mathbf{z}^{\mathbf{a}_i}}{\prod_{j=1}^{n}(1-\mathbf{z}^{\mathbf{b}_{ij}})}.$$

Thus it is enough to prove that the summands of the expression above can be written in terms of rational functions computable in polynomial time. The quotient rule for derivatives says that

$$\frac{\partial}{\partial z_r}\frac{\mathbf{z}^{\mathbf{a}_i}}{\prod_{j=1}^{n}(1-\mathbf{z}^{\mathbf{b}_{ij}})} = \frac{(\frac{\partial\mathbf{z}^{\mathbf{a}_i}}{\partial z_r})\prod_{j=1}^{n}(1-\mathbf{z}^{\mathbf{b}_{ij}}) - \mathbf{z}^{\mathbf{a}_i}(\frac{\partial}{\partial z_r}\prod_{j=1}^{n}(1-\mathbf{z}^{\mathbf{b}_{ij}}))}{\prod_{j=1}^{n}(1-\mathbf{z}^{\mathbf{b}_{ij}})^2}.$$

We can expand the numerator as a sum of no more than 2^n monomials. This is a constant number because n, the number of variables, is assumed to be a constant.

For the case of f being any monomial, repeat this construction, remembering to cancel powers in the denominator after the repeated application of the quotient rule (this is important). The general case of a polynomial f of many monomial terms follows by linearity. □

Now we present the algorithm to obtain bounds U_k, L_k that approach the optimum.

ALGORITHM 8.1. Summation method for optimizing polynomials.

1: **input** a rational convex polytope $P \subset \mathbb{R}^n$, a polynomial objective function

$$f \in \mathbb{Z}[x_1,\ldots,x_n]$$

of maximum total degree D in sparse or dense encoding.

2: **output** an increasing sequence of lower bounds L_k, and a decreasing sequence of upper bounds U_k approaching the maximal function value f^* of f over all lattice points of P.

3: If f is known to be nonnegative in all points of P, then go directly to line 4. Otherwise, solving $2n$ linear programs over P, we find lower and upper integer bounds for each of the variables $x_1,\ldots,x_n$. Let M be the maximum of the absolute values of these $2n$ numbers. Thus $|x_i| \le M$ for all i. Let C be the maximum of the absolute values of all coefficients, and r be the number of monomials of $f(x)$. Then

$$L := -rCM^D \le f(x) \le rCM^D =: U,$$

as we can bound the absolute value of each monomial of $f(x)$ by CM^D. Replace f by $\bar{f}(x) = f(x) - L \le U - L$, a nonnegative polynomial over P, and keep track of this shift of the objective function.

4: Via Barvinok's algorithm (Section 7.1), compute a short rational function expression for the generating function $g(P;\mathbf{z}) = \sum_{\alpha \in P\cap\mathbb{Z}^n} \mathbf{z}^\alpha$. From $g(P;\mathbf{z})$ compute the number $|P \cap \mathbb{Z}^n| = g(P;\mathbf{1})$ of lattice points in P in polynomial time using the algorithm of Section 7.2.

5: From the rational function representation of the generating function

$$g(P;\mathbf{z}) = \sum_{\alpha\in P\cap\mathbb{Z}^n} \mathbf{z}^\alpha,$$

compute the rational function representation of

$$g(P,f^k;\mathbf{z}) = \sum_{\alpha\in P\cap\mathbb{Z}^n} f^k(\alpha)\mathbf{z}^\alpha$$

in polynomial time by application of Lemma 8.3.1.

6: Define

$$L_k := \sqrt[k]{\frac{g(P,f^k;\mathbf{1})}{g(P,f^0;\mathbf{1})}} \quad \text{and} \quad U_k := \sqrt[k]{g(P,f^k;\mathbf{1})}.$$

When $\lfloor U_k \rfloor - \lceil L_k \rceil < 1$ stop and return $\lceil L_k \rceil = \lfloor U_k \rfloor$ as the optimal value.

Lemma 8.3.2. *Algorithm* (8.1) *is correct.*

Proof. Using the fact that the arithmetic mean of a finite set of nonnegative values is at most as big as the maximum value, which in turn is at most as big as the sum of all values, we obtain the sequences of lower and upper bounds, L_k and U_k, for the maximum:

$$L_k = \sqrt[k]{\frac{\sum_{\alpha \in P\cap\mathbb{Z}^n} f(\alpha)^k}{|P\cap\mathbb{Z}^n|}} \leq \max\{f(\alpha) : \alpha \in P\cap\mathbb{Z}^n\} \leq \sqrt[k]{\sum_{\alpha \in P\cap\mathbb{Z}^n} f(\alpha)^k} = U_k.$$

Note that as $s \to \infty$, L_k and U_k approach this maximum value monotonically (from below and above, respectively). Trivially, if the difference between (rounded) upper and lower bounds becomes strictly less than 1, we have determined the value

$$\max\{f(\mathbf{x}) : \mathbf{x} \in P\cap\mathbb{Z}^n\} = \lceil L_k \rceil.$$

Thus the algorithm terminates with the correct answer. □

The main theorem will follow from the next lemma.

Lemma 8.3.3. *Let f be a polynomial with integer coefficients and maximum total degree D. When the dimension n is fixed, the following holds:*

(i) *The bounds L_k, U_k can be computed in time that is bounded polynomially by k, the input size of P and f, and the total degree D. The bounds satisfy the following inequality:*

$$U_k - L_k \leq f^* \cdot \left(\sqrt[k]{|P\cap\mathbb{Z}^n|} - 1\right).$$

(ii) *In addition, when f is nonnegative over P (i.e., $f(\mathbf{x}) \geq 0$ for all $\mathbf{x} \in P$), for $k = (1+1/\epsilon)\log(|P\cap\mathbb{Z}^n|)$, L_k is a $(1-\epsilon)$-approximation to the optimal value f^* and it can be computed in time that is bounded polynomially by the input size, the total degree D, and $1/\epsilon$. Similarly, U_k gives a $(1+\epsilon)$-approximation to f^*.*

(iii) *Moreover, with the same complexity, one can also find a feasible lattice point that approximates an optimal solution with similar quality.*

Proof. *Part* (i) . From Lemma 8.3.1 on fixed dimension n, we can compute $g(P, f; \mathbf{z}) = \sum_{\alpha \in P\cap\mathbb{Z}^n} f(\alpha)\mathbf{z}^\alpha$ as a rational function in time that is bounded polynomially by D, the total degree of f, and the input size of P. Thus, because f^k has a total degree of Dk and the encoding length for the coefficients of f^k is bounded by $k\log(kC)$ (with C being the largest coefficient in f), we can also compute $g(P, f^k; \mathbf{z}) = \sum_{\alpha \in P\cap\mathbb{Z}^n} f^k(\alpha)\mathbf{z}^\alpha$ in time that is polynomially bounded by k, the total degree D, and the input size of P. Note that using residue techniques (Section 7.2), we can evaluate $g(P, f^k; \mathbf{1})$ in polynomial time. Finally observe:

$$\begin{aligned} U_k - L_k &= \sqrt[k]{\sum_{\alpha \in P\cap\mathbb{Z}^n} f^k(\alpha)} - \sqrt[k]{\frac{\sum_{\alpha \in P\cap\mathbb{Z}^n} f^k(\alpha)}{|P\cap\mathbb{Z}^n|}} \\ &= \sqrt[k]{\frac{\sum_{\alpha \in P\cap\mathbb{Z}^n} f^k(\alpha)}{|P\cap\mathbb{Z}^n|}} \left(\sqrt[k]{|P\cap\mathbb{Z}^n|} - 1\right) \\ &= L_k \left(\sqrt[k]{|P\cap\mathbb{Z}^n|} - 1\right) \leq f^* \left(\sqrt[k]{|P\cap\mathbb{Z}^n|} - 1\right). \end{aligned}$$

Part (ii). Note that if $\left(\sqrt[k]{|P \cap \mathbb{Z}^n|} - 1\right) \leq \epsilon$ then L_k is indeed a $(1-\epsilon)$-approximation because

$$f^* \leq U_k = L_k + (U_k - L_k) \leq L_k + f^* \left(\sqrt[k]{|P \cap \mathbb{Z}^n|} - 1\right) \leq L_k + f^* \epsilon.$$

Observe that

$$\phi(\epsilon) := \frac{1 + 1/\epsilon}{1/\log(1+\epsilon)}$$

is an increasing function for $\epsilon < 1$ and $\lim_{\epsilon \to 0} \phi(\epsilon) = 1$, thus $\phi(\epsilon) \geq 1$ for $0 < \epsilon \leq 1$. Hence, for all

$$k \geq \log(|P \cap \mathbb{Z}^n|) + \log(|P \cap \mathbb{Z}^n|)/\epsilon \geq \log(|P \cap \mathbb{Z}^n|)/\log(1+\epsilon),$$

we have indeed $\sqrt[k]{|P \cap \mathbb{Z}^n|} - 1 \leq \epsilon$. Finally, from Lemma 8.3.1, the calculation of L_k for

$$k \geq \log(|P \cap \mathbb{Z}^n|) + \log(|P \cap \mathbb{Z}^n|)/\epsilon$$

would require a number of steps that is bounded polynomially in the input size and $1/\epsilon$. A very similar argument can be written for U_k but we leave it as an exercise.

Part (iii). It remains to show that not only can we approximate the optimal value f^*, but we can also efficiently find a lattice point $\boldsymbol{\alpha}$ with $f(\boldsymbol{\alpha})$ giving that quality approximation of f^*. Let $k \geq (1 + 1/\epsilon)\log(|P \cap \mathbb{Z}^n|)$; thus, by the above discussion, L_k is an $(1-\epsilon)$-approximation to f^*. Let $Q_0 := [-M, M]^n$ denote the box computed in step 1 of the algorithm such that $P \subseteq Q_0$. By bisecting Q_0, we obtain two boxes Q_1' and Q_1''. By applying the algorithm separately to the polyhedra $P \cap Q_1'$ and $P \cap Q_1''$, we compute lower bounds L_k' and L_k'' for the optimization problems restricted to Q_1' and Q_1'', respectively. Because L_k^k is the arithmetic mean of $f^k(\boldsymbol{\alpha})$ for $\boldsymbol{\alpha} \in P \cap \mathbb{Z}^n$, clearly

$$\min\{L_k', L_k''\} \leq L_k \leq \max\{L_k', L_k''\}.$$

Without loss of generality, let $L_k' \geq L_k''$. We now apply the bisection procedure iteratively on Q_k'. After $n \log M$ bisection steps, we obtain a box Q_k' that contains a single lattice point $\boldsymbol{\alpha} \in P \cap Q_k' \cap Z^n$, which has an objective value $f(\boldsymbol{\alpha}) = L_k' \geq L_k \geq (1-\epsilon) f^*$. □

We remark that if we need to apply the construction of step 1 of the algorithm because f takes negative values on P, then we can only obtain an $(1-\epsilon)$-approximation (and $(1+\epsilon)$-approximation, respectively) for the modified function $\bar{f}$ in polynomial time, but not the original function f. We also emphasize that, although our algorithm requires the computation of $\sum_{\boldsymbol{\alpha} \in P \cap \mathbb{Z}^n} f^q(\boldsymbol{\alpha})$ for different powers of f, these numbers are obtained without explicitly listing all lattice points (a hard task), nor do we assume any knowledge of the individual values $f(\boldsymbol{\alpha})$. We can access the power means $\sum_{\boldsymbol{\alpha} \in P \cap \mathbb{Z}^n} f^q(\boldsymbol{\alpha})$ indirectly via rational generating functions.

Here are two small examples.

Example 8.3.4 (monomial optimization over a quadrilateral). We consider the problem of maximizing the value of the monomial $x^3 y$ over the lattice points of the quadrilateral

$$\{(x,y) : 3991 \leq 3996x - 4y \leq 3993,\ 1/2 \leq x \leq 5/2\}.$$

It contains only 2 lattice points. The sum of rational functions encoding the lattice points is

$$\frac{x^2 y^{1000}}{\left(1-(xy^{999})^{-1}\right)\left(1-y^{-1}\right)} + \frac{xy}{\left(1-xy^{999}\right)\left(1-y^{-1}\right)} + \frac{xy}{\left(1-xy^{999}\right)(1-y)} + \frac{x^2 y^{1000}}{\left(1-(xy^{999})^{-1}\right)(1-y)}.$$

In the first iteration, $L_1 = 4000.50$ while $U_1 = 8001$. After thirty iterations, we see $L_{30} = 7817.279750$ while $U_{30} = 8000$, the true optimal value.

Example 8.3.5 (problem nvs04 from MINLPLIB). From a well-known library of test examples (see `http://www.gamsworld.org/minlp/`) comes a somewhat more complicated example, the problem given by

$$\begin{aligned} \min \quad & 100\left(\frac{1}{2}+i_2-\left(\frac{3}{5}+i_1\right)^2\right)^2+\left(\frac{2}{5}-i_1\right)^2 \\ \text{subject to} \quad & i_1, i_2 \in [0,200]\cap\mathbb{Z}. \end{aligned} \tag{8.6}$$

Its optimal solution as given in MINLPLIB is $i_1 = 1$, $i_2 = 2$ with an objective value of 0.72. Clearly, to apply our algorithm literally, the objective function needs to be multiplied by a factor of 100 to obtain an integer valued polynomial.

Using the bounds on i_1 and i_2 we obtain an upper bound of $165 \cdot 10^9$ for the objective function, which allows us to convert the problem into an equivalent maximization problem, where all feasible points have a nonnegative objective value. The new optimal objective value is 164999999999.28. Expanding the new objective function and translating it into a differential operator yields

$$\begin{aligned} &\frac{4124999999947}{25}\mathrm{Id} - 28z_2\frac{\partial}{\partial z_2} + \frac{172}{5}z_1\frac{\partial}{\partial z_1} - 117\left(z_1\frac{\partial}{\partial z_1}\right)^{(2)} - 100\left(z_2\frac{\partial}{\partial z_2}\right)^{(2)} \\ &\quad + 240\left(z_2\frac{\partial}{\partial z_2}\right)\left(z_1\frac{\partial}{\partial z_1}\right) + 200\left(z_2\frac{\partial}{\partial z_2}\right)\left(z_1\frac{\partial}{\partial z_1}\right)^{(2)} \\ &\quad - 240\left(z_1\frac{\partial}{\partial z_1}\right)^{(3)} - 100\left(z_1\frac{\partial}{\partial z_1}\right)^{(4)}. \end{aligned}$$

The short generating function can be written as

$$g(z_1,z_2) = \left(\frac{1}{1-z_1} - \frac{z_1^{201}}{1-z_1}\right)\left(\frac{1}{1-z_2} - \frac{z_2^{201}}{1-z_2}\right).$$

In this example, the number of lattice points is $|P\cap\mathbb{Z}^2| = 40401$. The first bounds are

$$L_1 = 139463892042.292155534, U_1 = 28032242300500.723262442.$$

After 30 iterations the bounds become

$$L_{30} = 164999998845.993553019 \text{ and } U_{30} = 165000000475.892451381.$$

The same proof as outlined in this section yields the following more general theorem on optimization over finite lattice point sets given by a rational generating function. This generality is important because it allows optimization over lattice point sets obtained by Boolean operations (Section 7.5) and projections (Section 7.6).

Theorem 8.3.6 (FPTAS for maximizing nonnegative polynomials over finite lattice point sets). *For all fixed integers n (dimension) and s (maximum number of binomials in the denominator), there exists an algorithm with running time that is bounded polynomially by the encoding size of the problem and $\frac{1}{\epsilon}$ for the following problem.*

Input: *Let $S \subseteq \mathbb{Z}^n$ be a finite set, given by a rational generating function in the form*

$$g(S;\mathbf{z}) = \sum_{i\in I} \epsilon_i \frac{\mathbf{z}^{\mathbf{a}_i}}{(1-\mathbf{z}^{\mathbf{b}_{i1}})\dots(1-\mathbf{z}^{\mathbf{b}_{is_i}})},$$

where the numbers s_i of binomials in the denominators are at most s. Furthermore, let two vectors $\mathbf{l}$, $\mathbf{u} \in \mathbb{Z}^n$ be given such that S is contained in the box $\{\mathbf{x} : \mathbf{l} \leq \mathbf{x} \leq \mathbf{u}\}$.

Let $f \in \mathbb{Q}[x_1,\dots,x_n]$ be a polynomial with rational coefficients that is nonnegative on S, given by a list of its monomials, whose coefficients are encoded in binary and whose exponents are encoded in unary.

Finally, let $\epsilon \in \mathbb{Q}$.

Output: *Compute a point $\mathbf{x}_\epsilon \in S$ that satisfies*

$$f(\mathbf{x}_\epsilon) \geq (1-\epsilon)f^* \quad \text{where} \quad f^* = \max_{\mathbf{x}\in S} f(\mathbf{x}).$$

We remark that the algorithm also works if a positively weighted generating function is given (Section 7.2).

8.4 Extension to mixed-integer optimization via discretization

We now extend the algorithm to mixed-integer problems. Our approach is to use grid refinement in order to approximate the mixed-integer optimal value via auxiliary pure integer problems. One of the difficulties in constructing approximations is the fact that not every sequence of grids whose widths converge to zero leads to a convergent sequence of optimal solutions of grid optimization problems. This difficulty is addressed in Section 8.4.1. In Section 8.4.2 we develop techniques for bounding differences of polynomial function values. Section 8.4.3 contains the proof of the main theorem.

It is convenient in this section to denote the continuous variables by $\mathbf{x} \in \mathbb{R}^{n_1}$ and the integer variables by $\mathbf{z} \in \mathbb{Z}^{n_2}$. We then write the objective function as $f(\mathbf{x},\mathbf{z})$ and the constraint system as $A\mathbf{x} + B\mathbf{z} \leq \mathbf{b}$.

8.4.1 Grid approximation results

An important step in the development of an FPTAS for the mixed-integer optimization problem is the reduction of the mixed-integer problem (8.1) to an auxiliary optimization problem over a lattice $\frac{1}{m}\mathbb{Z}^{n_1} \times \mathbb{Z}^{n_2}$. To this end, we consider the *grid problem* with grid size m,

$$\begin{aligned} \max \quad & f(x_1,\dots,x_{n_1},z_1,\dots,z_{n_2}) \\ \text{subject to} \quad & A\mathbf{x} + B\mathbf{z} \leq \mathbf{b}, \\ & x_i \in \tfrac{1}{m}\mathbb{Z} \qquad \text{for } i = 1,\dots,n_1, \\ & z_i \in \mathbb{Z} \qquad \text{for } i = 1,\dots,n_2. \end{aligned} \tag{8.7}$$

We can solve this problem approximately using the integer FPTAS (Lemma 8.3.3):

Corollary 8.4.1. *For fixed dimension $n = n_1 + n_2$ there exists an algorithm with running time that is bounded polynomially by $\log m$, the encoding length of f and of P,* *the maximum total degree D of f, and $\frac{1}{\epsilon}$ for computing a feasible solution $(\mathbf{x}_\epsilon^m, \mathbf{z}_\epsilon^m) \in P \cap \left(\frac{1}{m}\mathbb{Z}^{n_1} \times \mathbb{Z}^{n_2}\right)$ to the grid problem (8.7) with an objective function f that is nonnegative on the feasible region, with*

$$f(\mathbf{x}_\epsilon^m, \mathbf{z}_\epsilon^m) \geq (1-\epsilon) f(\mathbf{x}^m, \mathbf{z}^m), \tag{8.8}$$

where $(\mathbf{x}^m, \mathbf{z}^m) \in P \cap \left(\frac{1}{m}\mathbb{Z}^{n_1} \times \mathbb{Z}^{n_2}\right)$ is an optimal solution to (8.7).

Proof. We apply Lemma 8.3.3 to the pure integer optimization problem:

$$\begin{aligned} \max \quad & \tilde{f}(\tilde{\mathbf{x}}, \mathbf{z}) \\ \text{subject to} \quad & A\tilde{\mathbf{x}} + mB\mathbf{z} \leq m\mathbf{b}, \\ & \tilde{x}_i \in \mathbb{Z} \qquad \text{for } i = 1, \ldots, n_1, \\ & z_i \in \mathbb{Z} \qquad \text{for } i = 1, \ldots, n_2, \end{aligned} \tag{8.9}$$

where $\tilde{f}(\tilde{\mathbf{x}}, \mathbf{z}) := m^D f(\frac{1}{m}\tilde{\mathbf{x}}, \mathbf{z})$ is a polynomial function with integer coefficients. Clearly the binary encoding length of the coefficients of $\tilde{f}$ increases by at most $\lceil D \log m \rceil$, compared to the coefficients of f. Likewise, the encoding length of the coefficients of mB and $m\mathbf{b}$ increases by at most $\lceil \log m \rceil$. By Theorem 8.3.6, there exists an algorithm with running time that is bounded polynomially by the encoding length of $\tilde{f}$ and of $A\mathbf{x} + mB\mathbf{z} \leq m\mathbf{b}$, the maximum total degree D, and $\frac{1}{\epsilon}$ for computing a feasible solution $(\mathbf{x}_\epsilon^m, \mathbf{z}_\epsilon^m) \in P \cap \left(\frac{1}{m}\mathbb{Z}^{n_1} \times \mathbb{Z}^{n_2}\right)$ such that $\tilde{f}(\mathbf{x}_\epsilon^m, \mathbf{z}_\epsilon^m) \geq (1-\epsilon)\tilde{f}(\mathbf{x}^m, \mathbf{z}^m)$, which implies the estimate (8.8). □

One might be tempted to think that for large-enough choice of m, we immediately obtain an approximation to the mixed-integer optimum with arbitrary precision. However, this is not true, as the following example demonstrates.

Example 8.4.2. Consider the mixed-integer optimization problem

$$\begin{aligned} \max \quad & 2z - x \\ \text{subject to} \quad & z \leq 2x, \\ & z \leq 2(1-x), \\ & x \in \mathbb{R}_{\geq 0},\ z \in \{0, 1\}, \end{aligned} \tag{8.10}$$

whose feasible region consists of the point $(\frac{1}{2}, 1)$ and the segment $\{(x, 0) : x \in [0, 1]\}$. The unique optimal solution to (8.10) is $x = \frac{1}{2}$, $z = 1$. Now consider the sequence of grid approximations of (8.10) where $x \in \frac{1}{m}\mathbb{Z}_{\geq 0}$. For even m, the unique optimal solution to the grid approximation is $x = \frac{1}{2}$, $z = 1$. However, for odd m, the unique optimal solution is $x = 0$, $z = 0$. Thus the full sequence of the optimal solutions to the grid approximations does not converge since it has two limit points; see Figure 8.2.

Even though taking the limit does not work, taking the upper limit does. More strongly, we can prove that it is possible to construct, in polynomial time, a subsequence of finer and finer grids that contain a certain lattice point that is arbitrarily close to the mixed-integer optimum $(\mathbf{x}^*, \mathbf{z}^*)$. We denote this lattice point by $(\lfloor \mathbf{x}^* \rceil_\delta, \mathbf{z}^*)$ to suggest that the coordinates of $\mathbf{x}^*$ have been "rounded" to obtain a nearby lattice point.

This is the central statement of this section and a basic building block of the approximation result.

Theorem 8.4.3 (grid approximation). *Let n_1 be fixed. Let $P = \{(\mathbf{x}, \mathbf{z}) \in \mathbb{R}^{n_1+n_2} : A\mathbf{x} + B\mathbf{z} \leq \mathbf{b}\}$, where $A \in \mathbb{Z}^{p\times n_1}$, $B \in \mathbb{Z}^{p\times n_2}$. Let $M \in \mathbb{R}$ be given such that $P \subseteq \{(\mathbf{x}, \mathbf{z}) \in$*

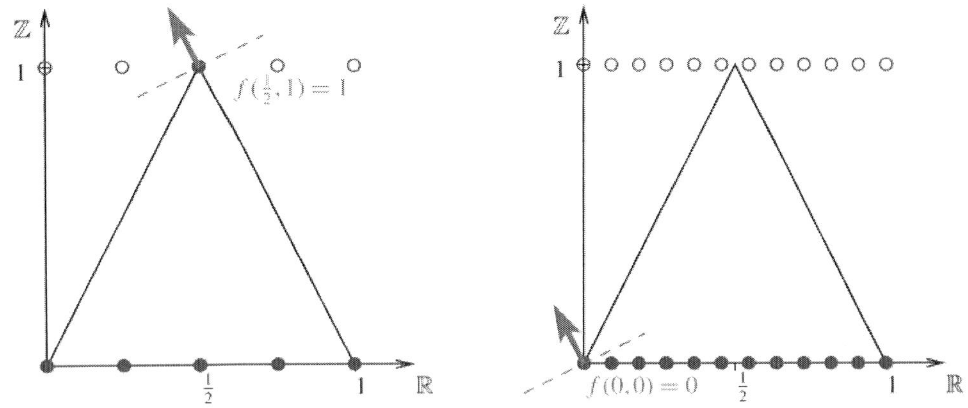

Figure 8.2. *A sequence of optimal solutions to grid problems with two limit points, for even m and for odd m.* [94]

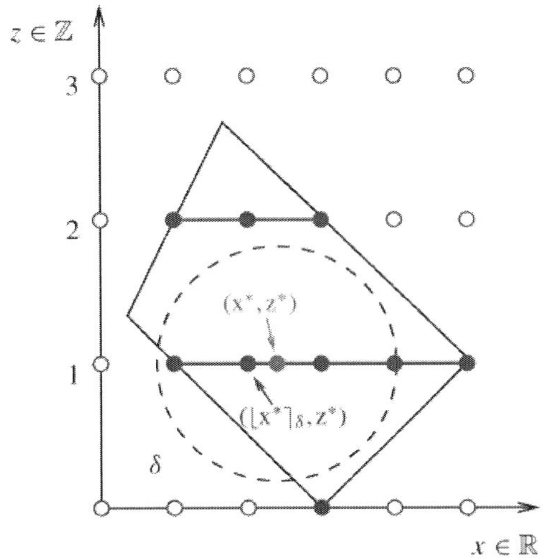

Figure 8.3. *The principle of grid approximation. Since we can refine the grid only in the direction of the continuous variables, we need to construct an approximating grid point* $(\mathbf{x}, \mathbf{z}^*)$ *in the same integral slice as the target point* $(\mathbf{x}^*, \mathbf{z}^*)$. [94]

$\mathbb{R}^{n_1+n_2} : |x_i| \leq M$ *for* $i = 1, \ldots, n_1\}$. *There exists a polynomial-time algorithm to compute a number* Δ *such that for every* $(\mathbf{x}^*, \mathbf{z}^*) \in P \cap (\mathbb{R}^{n_1} \times \mathbb{Z}^{n_2})$ *and* $\delta > 0$ *the following property holds:*

> *Every lattice* $\frac{1}{m}\mathbb{Z}^{n_1}$ *for* $m = k\Delta$ *and* $k \geq \frac{2}{\delta} n_1 M$ *contains a lattice point* $\lfloor \mathbf{x}^* \rceil_\delta$ *such that* $(\lfloor \mathbf{x}^* \rceil_\delta, \mathbf{z}^*) \in P \cap \left(\frac{1}{m}\mathbb{Z}^{n_1} \times \mathbb{Z}^{n_2}\right)$ *and* $\left\| \lfloor \mathbf{x}^* \rceil_\delta - \mathbf{x}^* \right\|_\infty \leq \delta$.

The geometry of Theorem 8.4.3 is illustrated in Figure 8.3. The rounding method is provided by the next two lemmas; Theorem 8.4.3 follows directly from them.

Lemma 8.4.4 (integral scaling lemma). *Let* $P = \{(\mathbf{x},\mathbf{z}) \in \mathbb{R}^{n_1+n_2} : A\mathbf{x} + B\mathbf{z} \leq \mathbf{b}\}$, *where* $A \in \mathbb{Z}^{p\times n_1}$, $B \in \mathbb{Z}^{p\times n_2}$. *For fixed* n_1, *there exists a polynomial-time algorithm to compute a number* $\Delta \in \mathbb{Z}_{>0}$ *such that for every* $\mathbf{z} \in \mathbb{Z}^{n_2}$ *the polytope*

$$\Delta P_{\mathbf{z}} = \left\{ \Delta \mathbf{x} : (\mathbf{x},\mathbf{z}) \in P \right\}$$

is integral, i.e., all vertices have integer coordinates. In particular, the number Δ *has an encoding length that is bounded by a polynomial in the encoding length of* P.

Proof. Because the dimension n_1 is fixed, there exist only polynomially many simplex bases of the inequality system $A\mathbf{x} \leq \mathbf{b} - B\mathbf{z}$, and they can be enumerated in polynomial time. The determinant of each simplex basis can be computed in polynomial time. Then Δ can be chosen as the least common multiple of all these determinants. □

Lemma 8.4.5. *Let* $Q \subset \mathbb{R}^n$ *be an integral polytope. Let* $M \in \mathbb{R}$ *be such that* $Q \subseteq \{\mathbf{x} \in \mathbb{R}^n : |x_i| \leq M \text{ for } i = 1,\ldots,n\}$. *Let* $\mathbf{x}^* \in Q$ *and let* $\delta > 0$. *Then every lattice* $\frac{1}{k}\mathbb{Z}^n$ *for* $k \geq \frac{2}{\delta}nM$ *contains a lattice point* $\mathbf{x} \in Q \cap \frac{1}{k}\mathbb{Z}^n$ *with* $\|\mathbf{x} - \mathbf{x}^*\|_\infty \leq \delta$.

Proof. By Carathéodory's theorem (Theorem 1.1.7), there exist $n+1$ vertices $\mathbf{x}^0,\ldots,\mathbf{x}^n \in \mathbb{Z}^n$ of Q and convex multipliers $\lambda_0,\ldots,\lambda_n$ such that $\mathbf{x}^* = \sum_{i=0}^n \lambda_i \mathbf{x}^i$. Let $\lambda_i' := \frac{1}{k}\lfloor k\lambda_i \rfloor \geq 0$ for $i = 1,\ldots,n$ and $\lambda_0' := 1 - \sum_{i=1}^n \lambda_i' \geq 0$. Moreover, we conclude $\lambda_i - \lambda_i' \leq \frac{1}{k}$ for $i = 1,\ldots,n$ and $\lambda_0' - \lambda_0 = \sum_{i=1}^n (\lambda_i - \lambda_i') \leq n\frac{1}{k}$. Then $\mathbf{x} := \sum_{i=0}^n \lambda_i' \mathbf{x}^i \in Q \cap \frac{1}{k}\mathbb{Z}^n$, and we have

$$\|\mathbf{x} - \mathbf{x}^*\|_\infty \leq \sum_{i=0}^n |\lambda_i' - \lambda_i| \cdot \|\mathbf{x}^i\|_\infty \leq 2n\frac{1}{k}M \leq \delta,$$

which proves the lemma. □

8.4.2 Bounding techniques

Using the results of Section 8.4.1 we are now able to approximate the mixed-integer optimal point by a point of a suitably fine lattice. The question arises as to how we can use the geometric distance of these two points to estimate the difference in objective function values. We prove Lemma 8.4.6 that provides us with a local Lipschitz constant for the polynomial to be maximized.

Lemma 8.4.6 (local Lipschitz constant). *Let* f *be a polynomial in* n *variables with maximum total degree* D. *Let* C *denote the largest absolute value of a coefficient of* f. *Then there exists a Lipschitz constant* L *such that* $|f(\mathbf{x}) - f(\mathbf{y})| \leq L\|\mathbf{x} - \mathbf{y}\|_\infty$ *for all* $|x_i|, |y_i| \leq M$. *The constant* L *is* $\mathrm{O}(D^{n+1}CM^D)$.

Proof. Let $f(\mathbf{x}) = \sum_{\boldsymbol{\alpha}\in\mathcal{D}} c_{\boldsymbol{\alpha}} \mathbf{x}^{\boldsymbol{\alpha}}$, where $\mathcal{D} \subseteq \mathbb{Z}^n_{\geq 0}$ is the set of exponent vectors of monomials appearing in f. Let $r = |\mathcal{D}|$ be the number of monomials of f. Then we have

$$|f(\mathbf{x}) - f(\mathbf{y})| \leq \sum_{\boldsymbol{\alpha}\neq\mathbf{0}} |c_{\boldsymbol{\alpha}}| \cdot |\mathbf{x}^{\boldsymbol{\alpha}} - \mathbf{y}^{\boldsymbol{\alpha}}|.$$

We estimate all summands separately. Let $\boldsymbol{\alpha} \neq \mathbf{0}$ be an exponent vector with $N := \sum_{i=1}^n \alpha_i \leq D$. Let

$$\boldsymbol{\alpha} = \boldsymbol{\alpha}^0 \geq \boldsymbol{\alpha}^1 \geq \cdots \geq \boldsymbol{\alpha}^N = \mathbf{0}$$

be a decreasing chain of exponent vectors with $\boldsymbol{\alpha}^{i-1} - \boldsymbol{\alpha}^i = \mathbf{e}^{j_i}$ for $i = 1,\dots,N$. Let $\boldsymbol{\beta}^i := \boldsymbol{\alpha} - \boldsymbol{\alpha}^i$ for $i = 0,\dots,N$. Then $\mathbf{x}^{\boldsymbol{\alpha}} - \mathbf{y}^{\boldsymbol{\alpha}}$ can be expressed as the "telescope sum"

$$\begin{aligned}\mathbf{x}^{\boldsymbol{\alpha}} - \mathbf{y}^{\boldsymbol{\alpha}} &= \mathbf{x}^{\boldsymbol{\alpha}^0}\mathbf{y}^{\boldsymbol{\beta}^0} - \mathbf{x}^{\boldsymbol{\alpha}^1}\mathbf{y}^{\boldsymbol{\beta}^1} + \mathbf{x}^{\boldsymbol{\alpha}^1}\mathbf{y}^{\boldsymbol{\beta}^1} - \mathbf{x}^{\boldsymbol{\alpha}^2}\mathbf{y}^{\boldsymbol{\beta}^2} + - \cdots - \mathbf{x}^{\boldsymbol{\alpha}^N}\mathbf{y}^{\boldsymbol{\beta}^N}\\ &= \sum_{i=1}^{N}\left(\mathbf{x}^{\boldsymbol{\alpha}^{i-1}}\mathbf{y}^{\boldsymbol{\beta}^{i-1}} - \mathbf{x}^{\boldsymbol{\alpha}^i}\mathbf{y}^{\boldsymbol{\beta}^i}\right)\\ &= \sum_{i=1}^{N}\left((x_{j_i} - y_{j_i})\mathbf{x}^{\boldsymbol{\alpha}^i}\mathbf{y}^{\boldsymbol{\beta}^{i-1}}\right).\end{aligned}$$

Since $\left|\mathbf{x}^{\boldsymbol{\alpha}^i}\mathbf{y}^{\boldsymbol{\beta}^{i-1}}\right| \le M^{N-1}$ and $N \le D$, we obtain

$$|\mathbf{x}^{\boldsymbol{\alpha}} - \mathbf{y}^{\boldsymbol{\alpha}}| \le D \cdot \|\mathbf{x} - \mathbf{y}\|_\infty \cdot M^{N-1},$$

and thus

$$|f(\mathbf{x}) - f(\mathbf{y})| \le CrDM^{D-1}\|\mathbf{x} - \mathbf{y}\|_\infty.$$

Let $L := CrDM^{D-1}$. Now, since $r = \mathrm{O}(D^n)$, we have $L = \mathrm{O}(D^{n+1}CM^D)$. □

Moreover, in order to obtain an FPTAS, we need to compare differences of function values to the maximum function value. To do this, we need to deal with the special case of polynomials that are constant on the feasible region; then, of course, every feasible solution is optimal. For nonconstant polynomials, we can prove a lower bound on the maximum function value. The technique is to bound the difference of the minimum and the maximum function value on the mixed-integer set from below. If the polynomial is nonconstant, this implies, for a nonnegative polynomial, a lower bound on the maximum function value. We will need a simple fact about the roots of multivariate polynomials.

Lemma 8.4.7. *Let $f \in \mathbb{Q}[x_1,\dots,x_n]$ be a polynomial and let D be the largest power of any variable that appears in f. Then $f = 0$ if and only if f vanishes on the set $\{0,\dots,D\}^n$.*

Proof. This is a simple consequence of the fundamental theorem of algebra. See, for instance, [78, Chapter 1, §1, Exercise 6 b]. □

Lemma 8.4.8. *Let $f \in \mathbb{Q}[x_1,\dots,x_n]$ be a polynomial with maximum total degree D. Let $Q \subset \mathbb{R}^n$ be an integral polytope of dimension $n' \le n$. Let $k \ge Dn'$. Then f is constant on Q if and only if f is constant on $Q \cap \frac{1}{k}\mathbb{Z}^n$.*

Proof. Let $\mathbf{x}^0 \in Q \cap \mathbb{Z}^n$ be an arbitrary vertex of Q. There exist vertices $\mathbf{x}^1,\dots,\mathbf{x}^{n'} \in Q \cap \mathbb{Z}^n$ such that the vectors $\mathbf{x}^1 - \mathbf{x}^0,\dots,\mathbf{x}^{n'} - \mathbf{x}^0 \in \mathbb{Z}^n$ are linearly independent. By convexity, Q contains the parallelepiped

$$S := \left\{\mathbf{x}^0 + \textstyle\sum_{i=1}^{n'} \lambda_i(\mathbf{x}^i - \mathbf{x}^0) : \lambda_i \in [0, \frac{1}{n'}] \text{ for } i = 1,\dots,n'\right\}.$$

We consider the set

$$S_k = \tfrac{1}{k}\mathbb{Z}^n \cap S \supseteq \left\{\mathbf{x}^0 + \textstyle\sum_{i=1}^{n'} \frac{n_i}{k}(\mathbf{x}^i - \mathbf{x}^0) : n_i \in \{0,1,\dots,D\} \text{ for } i = 1,\dots,n'\right\};$$

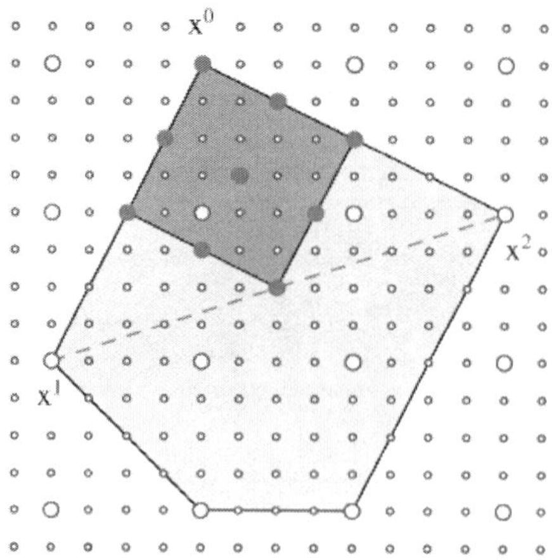

Figure 8.4. *The geometry of Lemma 8.4.8. For a polynomial with a maximum total degree of 2, we construct a refinement $\frac{1}{k}\mathbb{Z}^n$ (small circles) of the standard lattice (large circles) such that $P \cap \frac{1}{k}\mathbb{Z}^n$ contains an affine image of the set $\{0,1,2\}^n$ (large dots).* [94]

see Figure 8.4. Now if there exists a $c \in \mathbb{R}$ with $f(\mathbf{x}) = c$ for all $\mathbf{x} \in Q \cap \frac{1}{k}\mathbb{Z}^n$, then all the points in S_k are roots of the polynomial $f - c$, which has only maximum total degree D. By Lemma 8.4.7 (after an affine transformation), $f - c$ is zero on the affine hull of S_k; hence f is constant on the polytope Q. □

Theorem 8.4.9. *Let $f \in \mathbb{Z}[x_1,\dots,x_{n_1},z_1,\dots,z_{n_2}]$. Let P be a rational convex polytope, and let Δ be the number from Lemma 8.4.4. Let $m = k\Delta$ with $k \geq D n_1$, $k \in \mathbb{Z}$. Then f is constant on the feasible region $P \cap (\mathbb{R}^{n_1} \times \mathbb{Z}^{n_2})$ if and only if f is constant on $P \cap (\frac{1}{m}\mathbb{Z}^{n_1} \times \mathbb{Z}^{n_2})$. If f is not constant, then*

$$\left| f(\mathbf{x}_{\max}, \mathbf{z}_{\max}) - f(\mathbf{x}_{\min}, \mathbf{z}_{\min}) \right| \geq m^{-D}, \tag{8.11}$$

where $(\mathbf{x}_{\max}, \mathbf{z}_{\max})$ is an optimal solution to the maximization problem over the feasible region $P \cap (\mathbb{R}^{n_1} \times \mathbb{Z}^{n_2})$ and $(\mathbf{x}_{\min}, \mathbf{z}_{\min})$ is an optimal solution to the minimization problem.

Proof. Let f be constant on $P \cap (\frac{1}{m}\mathbb{Z}^{n_1} \times \mathbb{Z}^{n_2})$. For fixed integer part $\mathbf{z} \in \mathbb{Z}^{n_2}$, we consider the polytope $\Delta P_{\mathbf{z}} = \{\Delta \mathbf{x} : (\mathbf{x}, \mathbf{z}) \in P\}$, which is a slice of P scaled to become an integral polytope. By applying Lemma 8.4.8 with $k = (D+1)n$ on every polytope $\Delta P_{\mathbf{z}}$, we obtain that f is constant on every slice $P_{\mathbf{z}}$. Because f is also constant on the set $P \cap (\frac{1}{m}\mathbb{Z}^{n_1} \times \mathbb{Z}^{n_2})$, which contains a point of every nonempty slice $P_{\mathbf{z}}$, it follows that f is constant on P.

If f is not constant, there exist $(\mathbf{x}^1, \mathbf{z}^1)$, $(\mathbf{x}^2, \mathbf{z}^2) \in P \cap (\frac{1}{m}\mathbb{Z}^{n_1} \times \mathbb{Z}^{n_2})$ with $f(\mathbf{x}^1, \mathbf{z}^1) \neq f(\mathbf{x}^2, \mathbf{z}^2)$. By the integrality of all coefficients of f, we obtain the estimate

$$|f(\mathbf{x}^1, \mathbf{z}^1) - f(\mathbf{x}^2, \mathbf{z}^2)| \geq m^{-D}.$$

Because $(\mathbf{x}^1, \mathbf{z}^1)$, $(\mathbf{x}^2, \mathbf{z}^2)$ are both feasible solutions to the maximization problem and the minimization problem, this implies (8.11). □

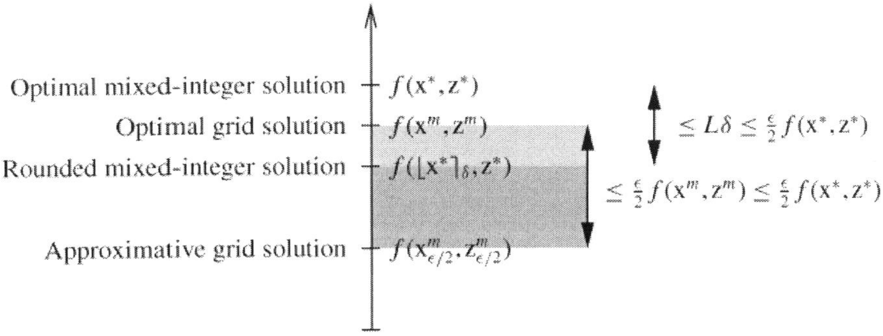

Figure 8.5. *Estimates in the proof of Theorem* 8.1.3*(a).* [94]

8.4.3 Proof of Theorem 8.1.3

Now we are in the position to prove the main result.

***Proof of Theorem* 8.1.3.** Let $(\mathbf{x}^*, \mathbf{z}^*)$ denote an optimal solution to the mixed-integer problem (8.1). Let $\epsilon > 0$. We show that, in time polynomial in the input length, the maximum total degree, and $\frac{1}{\epsilon}$, we can compute a point $(\mathbf{x}, \mathbf{z})$ that satisfies the constraints such that

$$|f(\mathbf{x},\mathbf{z}) - f(\mathbf{x}^*,\mathbf{z}^*)| \le \epsilon f(\mathbf{x}^*,\mathbf{z}^*). \tag{8.12}$$

We prove this by establishing several estimates, which are illustrated in Figure 8.5.

First we note that we can restrict ourselves to the case of polynomials with integer coefficients simply by multiplying f with the least common multiple of all denominators of the coefficients. We next establish a lower bound on $f(\mathbf{x}^*,\mathbf{z}^*)$. To this end, let Δ be the integer from Lemma 8.4.4, which can be computed in polynomial time. By Theorem 8.4.9 with $m = Dn_1\Delta$, either f is constant on the feasible region, or

$$f(\mathbf{x}^*,\mathbf{z}^*) \ge (Dn_1\Delta)^{-D}, \tag{8.13}$$

where D is the maximum total degree of f. Now let

$$\delta := \frac{\epsilon}{2(Dn_1\Delta)^D L(C,D,M)} \tag{8.14}$$

and let us choose the grid size

$$m := \Delta \left\lceil \frac{4}{\epsilon}(Dn_1\Delta)^D L(C,D,M) n_1 M \right\rceil, \tag{8.15}$$

where $L(C,D,M)$ is the Lipschitz constant from Lemma 8.4.6. Then we have $m \ge \Delta\frac{2}{\delta} n_1 M$, so by Theorem 8.4.3, there is a point $(\lfloor \mathbf{x}^*\rceil_\delta, \mathbf{z}^*) \in P \cap \left(\frac{1}{m}\mathbb{Z}^{n_1} \times \mathbb{Z}^{n_2}\right)$ with $\left\| \lfloor \mathbf{x}^*\rceil_\delta - \mathbf{x}^* \right\|_\infty \le \delta$. Let $(\mathbf{x}^m, \mathbf{z}^m)$ denote an optimal solution to the grid problem (8.7). Because $(\lfloor \mathbf{x}^*\rceil_\delta, \mathbf{z}^*)$ is a feasible solution to the grid problem (8.7), we have

$$f(\lfloor \mathbf{x}^*\rceil_\delta, \mathbf{z}^*) \le f(\mathbf{x}^m,\mathbf{z}^m) \le f(\mathbf{x}^*,\mathbf{z}^*). \tag{8.16}$$

Now we can estimate

$$\begin{aligned}\left|f(\mathbf{x}^*,\mathbf{z}^*)-f(\mathbf{x}^m,\mathbf{z}^m)\right| &\le \left|f(\mathbf{x}^*,\mathbf{z}^*)-f(\lfloor\mathbf{x}^*\rceil_\delta,\mathbf{z}^*)\right|\\ &\le L(C,D,M)\left\|\mathbf{x}^*-\lfloor\mathbf{x}^*\rceil_\delta\right\|_\infty\\ &\le L(C,D,M)\,\delta\\ &=\frac{\epsilon}{2}(Dn_1\Delta)^{-D}\\ &\le \frac{\epsilon}{2}f(\mathbf{x}^*,\mathbf{z}^*),\end{aligned} \tag{8.17}$$

where the last estimate is given by (8.13) in the case that f is not constant on the feasible region. On the other hand, if f is constant, the estimate (8.17) holds trivially.

By Corollary 8.4.1 we can compute a point $(\mathbf{x}^m_{\epsilon/2},\mathbf{z}^m_{\epsilon/2})\in P\cap\left(\frac{1}{m}\mathbb{Z}^{n_1}\times\mathbb{Z}^{n_2}\right)$ such that

$$(1-\tfrac{\epsilon}{2})f(\mathbf{x}^m,\mathbf{z}^m)\le f(\mathbf{x}^m_{\epsilon/2},\mathbf{z}^m_{\epsilon/2})\le f(\mathbf{x}^m,\mathbf{z}^m) \tag{8.18}$$

in time that is bounded polynomially by $\log m$, the encoding length of f and P, the maximum total degree D, and $1/\epsilon$. Here $\log m$ is bounded by a polynomial in $\log M$, D, and $\log C$, so we can compute $(\mathbf{x}^m_{\epsilon/2},\mathbf{z}^m_{\epsilon/2})$ in time that is bounded polynomially by the input size, the maximum total degree D, and $1/\epsilon$. Now, using (8.17) and (8.18), we can estimate

$$\begin{aligned}f(\mathbf{x}^*,\mathbf{z}^*)-f(\mathbf{x}^m_{\epsilon/2},\mathbf{z}^m_{\epsilon/2}) &\le f(\mathbf{x}^*,\mathbf{z}^*)-(1-\tfrac{\epsilon}{2})f(\mathbf{x}^m,\mathbf{z}^m)\\ &=\tfrac{\epsilon}{2}f(\mathbf{x}^*,\mathbf{z}^*)+(1-\tfrac{\epsilon}{2})\big(f(\mathbf{x}^*,\mathbf{z}^*)-f(\mathbf{x}^m,\mathbf{z}^m)\big)\\ &\le \tfrac{\epsilon}{2}f(\mathbf{x}^*,\mathbf{z}^*)+\tfrac{\epsilon}{2}f(\mathbf{x}^*,\mathbf{z}^*)\\ &=\epsilon f(\mathbf{x}^*,\mathbf{z}^*).\end{aligned}$$

Hence $f(\mathbf{x}^m_{\epsilon/2},\mathbf{z}^m_{\epsilon/2})\ge(1-\epsilon)f(\mathbf{x}^*,\mathbf{z}^*)$. □

8.5 Extension to objective functions of arbitrary range

In this section we drop the requirement of the polynomial objective function being positive over the feasible region and prove Theorem 8.1.5.

The approximation algorithms for the integer case (Lemma 8.3.3) and the mixed-integer case (Theorem 8.1.3) only work for polynomial objective functions that are non-negative on the feasible region. In order to apply them to an arbitrary polynomial objective function f, we need to add a constant term to f that is large enough. As indicated in Section 8.3, we can use linear programming techniques to obtain a bound M on the variables and then estimate

$$f(\mathbf{x})\ge -rCM^D =: L_0,$$

where C is the largest absolute value of a coefficient, r is the number of monomials of f, and D is the maximum total degree. However, the range $\left|f(\mathbf{x}_{\max},\mathbf{z}_{\max})-f(\mathbf{x}_{\min},\mathbf{z}_{\min})\right|$ can be exponentially small compared to L_0, so in order to obtain an approximation $(\mathbf{x}_\epsilon,\mathbf{z}_\epsilon)$ satisfying

$$\left|f(\mathbf{x}_\epsilon,\mathbf{z}_\epsilon)-f(\mathbf{x}_{\max},\mathbf{z}_{\max})\right|\le\epsilon\left|f(\mathbf{x}_{\max},\mathbf{z}_{\max})-f(\mathbf{x}_{\min},\mathbf{z}_{\min})\right|, \tag{8.19}$$

we would need an $(1-\epsilon')$-approximation to the problem of maximizing $g(\mathbf{x},\mathbf{z}):=f(\mathbf{x},\mathbf{z})-L_0$ with an exponentially small value of ϵ'.

To address this difficulty, we will first apply an algorithm which will compute an approximation $[L_i, U_i]$ of the range $[f(\mathbf{x}_{\min}, \mathbf{z}_{\min}), f(\mathbf{x}_{\max}, \mathbf{z}_{\max})]$ with constant quality. To this end, we first prove a simple corollary of Theorem 8.1.3.

Corollary 8.5.1 (computation of upper bounds for mixed-integer problems). *Let the dimension $n = n_1 + n_2$ be fixed. Let $P \subseteq \mathbb{R}^n$ be a rational convex polytope. Let $f \in \mathbb{Z}[x_1, \dots, x_{n_1}, z_1, \dots, z_{n_2}]$ be a polynomial function with integer coefficients and maximum total degree D that is nonnegative on $P \cap \left(\mathbb{R}^{n_1} \times \mathbb{Z}^{n_2}\right)$. Let $\delta > 0$. There exists an algorithm with running time that is bounded polynomially by the input size, D, and $\frac{1}{\delta}$ for computing an upper bound u such that*

$$f(\mathbf{x}_{\max}, \mathbf{z}_{\max}) \le u \le (1+\delta) f(\mathbf{x}_{\max}, \mathbf{z}_{\max}), \tag{8.20}$$

where $(\mathbf{x}_{\max}, \mathbf{z}_{\max})$ is an optimal solution to the maximization problem of f over $P \cap \left(\mathbb{R}^{n_1} \times \mathbb{Z}^{n_2}\right)$.

Proof. Let $\epsilon = \frac{\delta}{1+\delta}$. By Theorem 8.1.3, we can compute, in time that is bounded polynomially by the input size, D, and $\frac{1}{\epsilon} = 1 + \frac{1}{\delta}$, a solution $(\mathbf{x}_\epsilon, \mathbf{z}_\epsilon)$ with

$$\left| f(\mathbf{x}_{\max}, \mathbf{z}_{\max}) - f(\mathbf{x}_\epsilon, \mathbf{z}_\epsilon) \right| \le \epsilon f(\mathbf{x}_{\max}, \mathbf{z}_{\max}). \tag{8.21}$$

Let $u := \frac{1}{1-\epsilon} f(\mathbf{x}_\epsilon, \mathbf{z}_\epsilon) = (1+\delta) f(\mathbf{x}_\epsilon, \mathbf{z}_\epsilon)$. Then

$$f(\mathbf{x}_{\max}, \mathbf{z}_{\max}) \le \frac{1}{1-\epsilon} f(\mathbf{x}_\epsilon, \mathbf{z}_\epsilon) = u \tag{8.22}$$

and

$$\begin{aligned} (1+\delta) f(\mathbf{x}_{\max}, \mathbf{z}_{\max}) &\ge (1+\delta) f(\mathbf{x}_\epsilon, \mathbf{z}_\epsilon) \\ &= (1+\delta)(1-\epsilon) u \\ &= (1+\delta)\left(1 - \frac{\delta}{1+\delta}\right) u = u. \end{aligned} \tag{8.23}$$

This proves the estimate (8.20). □

ALGORITHM 8.2. Approximation of the range of the objective function.

1: **input** mixed-integer polynomial optimization problem (8.1), a number $0 < \delta < 1$.
2: **output** sequences $\{L_i\}$, $\{U_i\}$ of lower and upper bounds of f over the feasible region $P \cap \left(\mathbb{R}^{n_1} \times \mathbb{Z}^{n_2}\right)$ such that

$$L_i \le f(\mathbf{x}_{\min}, \mathbf{z}_{\min}) \le f(\mathbf{x}_{\max}, \mathbf{z}_{\max}) \le U_i \tag{8.24}$$

and

$$\lim_{i \to \infty} |U_i - L_i| = c\big(f(\mathbf{x}_{\max}, \mathbf{z}_{\max}) - f(\mathbf{x}_{\min}, \mathbf{z}_{\min})\big), \tag{8.25}$$

where c depends only on the choice of δ.

3: By solving $2n$ linear programs over P, we find lower and upper integer bounds for each of the variables $x_1,\ldots,x_{n_1},z_1,\ldots,z_{n_2}$. Let M be the maximum of the absolute values of these $2n$ numbers. Thus $|x_i|,|z_i|\le M$ for all i. Let C be the maximum of the absolute values of all coefficients, and r be the number of monomials of $f(x)$. Then

$$L_0 := -rCM^D \le f(\mathbf{x},\mathbf{z}) \le rCM^D =: U_0,$$

as we can bound the absolute value of each monomial of $f(x)$ by CM^D.

4: $i \leftarrow 0$.

5: Using the algorithm of Corollary 8.5.1, compute an upper bound u for the problem

$$\begin{array}{rl} \max & g(\mathbf{x},\mathbf{z}) := f(\mathbf{x},\mathbf{z}) - L_i \\ \text{subject to} & (\mathbf{x},\mathbf{z}) \in P \cap (\mathbb{R}^{n_1}\times\mathbb{Z}^{n_2}) \end{array}$$

that gives a $(1+\delta)$-approximation to the optimal value. Let $U_{i+1} := L_i + u$.

6: Likewise, compute an upper bound u for the problem

$$\begin{array}{rl} \max & h(\mathbf{x},\mathbf{z}) := U_i - f(\mathbf{x},\mathbf{z}) \\ \text{subject to} & (\mathbf{x},\mathbf{z}) \in P \cap (\mathbb{R}^{n_1}\times\mathbb{Z}^{n_2}) \end{array}$$

that gives a $(1+\delta)$-approximation to the optimal value. Let $L_{i+1} := U_i - u$.

7: $i := i+1$.

8: Go to 5.

Lemma 8.5.2. *Algorithm 8.2 is correct. For fixed $0<\delta<1$, it computes the bounds L_n, U_n satisfying equations* (8.24) *and* (8.25) *in time that is bounded polynomially by the input size and n.*

Proof. We have

$$U_i - L_{i+1} \le (1+\delta)\big(U_i - f(\mathbf{x}_{\min},\mathbf{z}_{\min})\big) \tag{8.26}$$

and

$$U_{i+1} - L_i \le (1+\delta)\big(f(\mathbf{x}_{\max},\mathbf{z}_{\max}) - L_i\big). \tag{8.27}$$

This implies

$$U_{i+1} - L_{i+1} \le \delta(U_i - L_i) + (1+\delta)\big(f(\mathbf{x}_{\max},\mathbf{z}_{\max}) - f(\mathbf{x}_{\min},\mathbf{z}_{\min})\big).$$

Therefore

$$\begin{aligned} U_n - L_n &\le \delta^n(U_0 - L_0) + (1+\delta)\left(\sum_{i=0}^{n-2}\delta^i\right)\big(f(\mathbf{x}_{\max},\mathbf{z}_{\max}) - f(\mathbf{x}_{\min},\mathbf{z}_{\min})\big) \\ &= \delta^n(U_0 - L_0) + (1+\delta)\frac{1-\delta^{n-1}}{1-\delta}\big(f(\mathbf{x}_{\max},\mathbf{z}_{\max}) - f(\mathbf{x}_{\min},\mathbf{z}_{\min})\big) \\ &\to \frac{1+\delta}{1-\delta}\big(f(\mathbf{x}_{\max},\mathbf{z}_{\max}) - f(\mathbf{x}_{\min},\mathbf{z}_{\min})\big) \quad (n\to\infty). \end{aligned}$$

The bound on the running time requires a careful analysis. Because in each step the result u (a rational number) of the bounding procedure (Corollary 8.5.1) becomes part of the input in the next iteration, the encoding length of the input could grow exponentially after only polynomially many steps. However, we will show that the encoding length only grows very slowly.

First we need to remark that the auxiliary objective functions g and h have integer coefficients except for the constant term, which may be rational. It turns out that the estimates in the proof of Theorem 8.1.3 (in particular, the local Lipschitz constant L and the lower bound on the optimal value) are independent from the constant term of the objective function. Therefore, the *same* approximating grid $\frac{1}{m}\mathbb{Z}^{n_1} \times \mathbb{Z}^{n_2}$ can be chosen in all iterations of Algorithm 8.2; the number m only depends on δ, the polytope P, the maximum total degree D, and the coefficients of f with the exception of the constant term.

The construction in the proof of Corollary 8.5.1 obtains the upper bound u by multiplying the approximation $f(\mathbf{x}_\epsilon, \mathbf{z}_\epsilon)$ by $(1+\delta)$. Therefore we have

$$\begin{aligned} U_{i+1} &= L_i + u \\ &= L_i + (1+\delta)\big(f(\mathbf{x}_\epsilon, \mathbf{z}_\epsilon) - L_i\big) \\ &= -\delta L_i + (1+\delta) f(\mathbf{x}_\epsilon, \mathbf{z}_\epsilon). \end{aligned} \tag{8.28}$$

Because the solution $(\mathbf{x}_\epsilon, \mathbf{z}_\epsilon)$ lies in the grid $\frac{1}{m}\mathbb{Z}^{n_1} \times \mathbb{Z}^{n_2}$, the value $f(\mathbf{x}_\epsilon, \mathbf{z}_\epsilon)$ is an integer multiple of m^{-D}. This implies that, because $L_0 \leq f(\mathbf{x}_\epsilon, \mathbf{z}_\epsilon) \leq U_0$, the encoding length of the rational number $f(\mathbf{x}_\epsilon, \mathbf{z}_\epsilon)$ is bounded by a polynomial in the input size of f and P. Therefore the encoding length U_{i+1} (and likewise L_{i+1}) only increases by an additive term that is bounded by a polynomial in the input size of f and P. □

We are now in the position to prove Theorem 8.1.5.

***Proof of Theorem* 8.1.5.** Clearly we can restrict ourselves to polynomials with integer coefficients. Let $m = (D+1)n_1\Delta$, where Δ is the number from Theorem 8.4.3. We apply Algorithm 8.2 using $0 < \delta < 1$ arbitrary to compute bounds U_n and L_n for

$$n = \left\lceil -\log_\delta\big(2m^D(U_0 - L_0)\big)\right\rceil.$$

Because n is bounded by a polynomial in the input size and the maximum total degree D, this can be done in polynomial time. Now, by the proof of Lemma 8.5.2, we have

$$\begin{aligned} U_n - L_n &\leq \delta^n(U_0 - L_0) + (1+\delta)\frac{1-\delta^{n-1}}{1-\delta}\big(f(\mathbf{x}_{\max}, \mathbf{z}_{\max}) - f(\mathbf{x}_{\min}, \mathbf{z}_{\min})\big) \\ &\leq \frac{1}{2}m^{-D} + \frac{1+\delta}{1-\delta}\big(f(\mathbf{x}_{\max}, \mathbf{z}_{\max}) - f(\mathbf{x}_{\min}, \mathbf{z}_{\min})\big). \end{aligned} \tag{8.29}$$

If f is constant on $P \cap (\mathbb{R}^{n_1} \times \mathbb{Z}^{n_2})$, it is constant on $P \cap (\frac{1}{m}\mathbb{Z}^{n_1} \times \mathbb{Z}^{n_2})$, then $U_n - L_n \leq \frac{1}{2}m^{-D}$. Otherwise, by Theorem 8.4.9, we have $U_n - L_n \geq f(\mathbf{x}_{\max}, \mathbf{z}_{\max}) - f(\mathbf{x}_{\min}, \mathbf{z}_{\min}) \geq m^{-D}$. This settles part (a).

For part (b), if f is constant on $P \cap (\mathbb{R}^{n_1} \times \mathbb{Z}^{n_2})$, we return an arbitrary solution as an optimal solution. Otherwise, we can estimate further:

$$U_n - L_n \leq \left(\frac{1}{2} + \frac{1+\delta}{1-\delta}\right)\big(f(\mathbf{x}_{\max}, \mathbf{z}_{\max}) - f(\mathbf{x}_{\min}, \mathbf{z}_{\min})\big). \tag{8.30}$$

Now we apply the algorithm of Theorem 8.1.3 to the maximization problem of the polynomial function $f' := f - L_n$, which is nonnegative over the feasible region $P \cap$

$\left(\mathbb{R}^{n_1} \times \mathbb{Z}^{n_2}\right)$. We compute a point $(\mathbf{x}_{\epsilon'}, \mathbf{z}_{\epsilon'})$ where $\epsilon' = \epsilon\left(\frac{1}{2} + \frac{1+\delta}{1-\delta}\right)^{-1}$ such that

$$\left| f'(\mathbf{x}_{\epsilon'}, \mathbf{z}_{\epsilon'}) - f'(\mathbf{x}_{\max}, \mathbf{z}_{\max}) \right| \leq \epsilon' f'(\mathbf{x}_{\max}, \mathbf{z}_{\max}).$$

Then we obtain the estimate

$$\begin{aligned} \left| f(\mathbf{x}_{\epsilon'}, \mathbf{z}_{\epsilon'}) - f(\mathbf{x}_{\max}, \mathbf{z}_{\max}) \right| &\leq \epsilon' \big(f(\mathbf{x}_{\max}, \mathbf{z}_{\max}) - L_n \big) \\ &\leq \epsilon' \big(U_n - L_n \big) \\ &\leq \epsilon' \left(\frac{1}{2} + \frac{1+\delta}{1-\delta} \right) \big(f(\mathbf{x}_{\max}, \mathbf{z}_{\max}) - f(\mathbf{x}_{\min}, \mathbf{z}_{\min}) \big) \\ &= \epsilon \big(f(\mathbf{x}_{\max}, \mathbf{z}_{\max}) - f(\mathbf{x}_{\min}, \mathbf{z}_{\max}) \big), \end{aligned}$$

which proves part (b). □

8.6 Notes and further references

The summation method for optimization was introduced by Barvinok in [29], who investigated both linear and polynomial objective functions. In the case of linear objective functions $f(\mathbf{x}) = \mathbf{c}^\top \mathbf{x}$, it is convenient to study the summation of $e^{\mathbf{c}^\top \mathbf{x}}$ over the feasible region. This case is also the central theme in Lasserre's paper [215] and his monograph [217].

The FPTAS for the pure integer case (Section 8.3) was introduced in [93]. The important Lemma 8.3.1 is due to Barvinok [33]. The mixed-integer case (Section 8.4) was first published in [94]. The extension to polynomials of arbitrary range appeared in [95]. This chapter is based on [93, 95].

A continuous version of the summation problem, using exponential integrals and the decomposition of polynomials into powers of linear forms, is explored in Baldoni et al. [21]; see also De Loera et al. [105]. This can serve as the basis of a summation method for global (nonconvex) polynomial optimization.

For the mixed-integer case, as an alternative to the discretization technique (Section 8.4), mixed-integer summation techniques can be used, where one computes discrete sums of integrals along lattice slices. The algorithmic theory of the corresponding mixed-integer ("intermediate") generating functions is developed in [20, 22], based on earlier work by Barvinok [33].

8.7 Exercises

Exercise 8.7.1. Implement a branch and bound algorithm for integer polynomial optimization using the bounds developed in this chapter. Then compare its performance with that of traditional global optimization tools. Try various choices of k.

Exercise 8.7.2. As a variation of Exercise 8.7.1, form a simple relaxation of the feasible region of the node problem, such as a Gomory corner polyhedron, truncated by a hyperplane to make it bounded.

Exercise 8.7.3. Model the problem of finding a nontrivial factor of a given integer as a nonlinear integer optimization problem of the form (8.1). Then explain why the algorithm of this chapter does *not* imply a polynomial-time algorithm for factoring.

Exercise 8.7.4 (continuation of Exercise 7.8.4). Suppose that a lattice point set $W \subseteq \mathbb{Z}^n$ in fixed dimension n is given by a rational generating function and bounds on the variables. In addition, a density function $\rho\colon \mathbb{Z}^n \to \mathbb{Q}_+$ is given as a polynomial, whose coefficients are encoded in the binary encoding scheme and whose exponents are encoded in the unary encoding scheme. Let the probability measure $\bar{\rho}$ on W be defined by $\bar{\rho}(\{\mathbf{w}\}) = \rho(\mathbf{w}) / \sum_{\mathbf{v} \in W} \rho(\mathbf{v})$. Show that you can sample from $\bar{\rho}$ with a polynomial-time algorithm.

Chapter 9

Multicriteria Integer Linear Optimization via Integer Projection

In this chapter, we give algorithmic solutions to fundamental questions related to multicriteria integer linear programs, when the dimensions of the strategy space and of the outcome space are considered fixed constants. For general information about multicriteria problems we refer the reader to [127, 290].

9.1 Introduction

Let $A = (a_{ij})$ be an integral $m \times n$ matrix and $\mathbf{b} \in \mathbb{Z}^m$ such that the convex polyhedron $P = \{\mathbf{u} \in \mathbb{R}^n : A\mathbf{u} \leq \mathbf{b}\}$ is bounded. Given k linear functionals $f_1, f_2, \ldots, f_k \in \mathbb{Z}^n$, we consider the *multicriterion integer linear programming problem*

$$\begin{aligned} \text{Pareto-min} \quad & \big(f_1(\mathbf{u}), f_2(\mathbf{u}), \ldots, f_k(\mathbf{u})\big) \\ \text{subject to} \quad & A\mathbf{u} \leq \mathbf{b}, \\ & \mathbf{u} \in \mathbb{Z}^n, \end{aligned} \tag{9.1}$$

where Pareto-min is defined as the problem of finding all Pareto optima and a corresponding Pareto strategy; see Figure 9.1.

For a lattice point $\mathbf{u}$ the vector $\mathbf{f}(\mathbf{u}) = \big(f_1(\mathbf{u}), \ldots, f_k(\mathbf{u})\big)$ is called an *outcome vector*. Such an outcome vector is a *Pareto optimum* for the above problem if and only if there is no other point $\tilde{\mathbf{u}}$ in the feasible set such that $f_i(\tilde{\mathbf{u}}) \leq f_i(\mathbf{u})$ for all i and $f_j(\tilde{\mathbf{u}}) < f_j(\mathbf{u})$ for at least one index j; see Figure 9.2.

The corresponding feasible point $\mathbf{u}$ is called a *Pareto strategy*. Thus a feasible vector is a Pareto strategy if no feasible vector can decrease some criterion without causing a simultaneous increase in at least one other criterion; see Figure 9.1.

In general multiobjective problems, the number of Pareto optimal solutions may be infinite, but in our situation the number of Pareto optima and strategies is finite. There are several well-known techniques to generate Pareto optima. Some popular methods used to solve such problems include weighting the objectives or using a so-called global criterion approach (see [119]). In particularly nice situations, such as multicriteria *linear* programs [181], one knows a way to generate all Pareto optima, but most techniques reach only some of the Pareto optima.

Here we study the sets of *all* Pareto optima and strategies of a multicriterion integer linear program using rational generating functions. The set of Pareto points can be

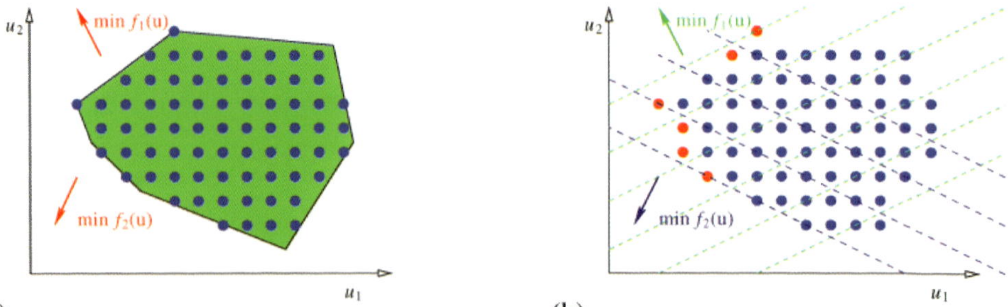

(a) (b)

Figure 9.1. *Strategy space. Red filled circles indicate the Pareto strategies.*

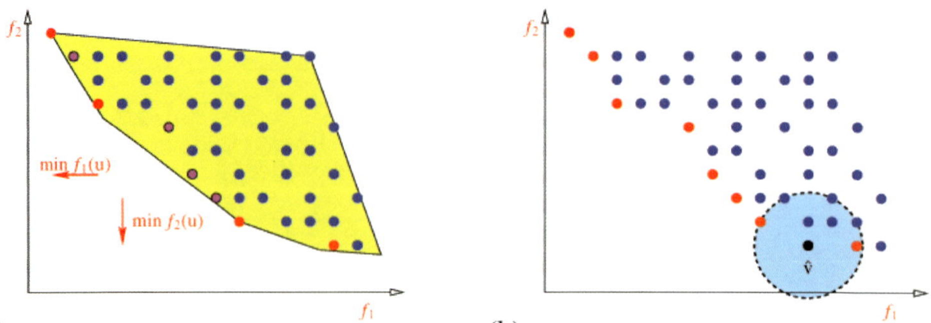

(a) (b)

Figure 9.2. *Outcome space.* (a) *Blue filled circles indicate the outcomes of feasible strategies; note that this set is a projection of a lattice point set and may have holes. Red filled circles indicate supported Pareto outcomes; magenta filled circles indicate non-supported Pareto outcomes.* (b) *Global criteria.*

described as the sum of monomials

$$\sum \left\{ \mathbf{z}^{\mathbf{v}} : \ \mathbf{u} \in P \cap \mathbb{Z}^n \text{ and } \mathbf{v} = \mathbf{f}(\mathbf{u}) \in \mathbb{Z}^k \text{ is a Pareto optimum} \right\}. \tag{9.2}$$

Under the assumption that the number of variables is fixed, we can compute in polynomial time a short expression for the huge polynomial above, thus *all* its Pareto optima can in fact be counted exactly. The same can be done for the corresponding Pareto strategies when written in the form

$$\sum \left\{ \mathbf{x}^{\mathbf{u}} : \ \mathbf{u} \in P \cap \mathbb{Z}^n \text{ and } \mathbf{f}(\mathbf{u}) \text{ is a Pareto optimum} \right\}. \tag{9.3}$$

Theorem 9.1.1. *Let $A \in \mathbb{Z}^{m \times n}$, a vector $\mathbf{b} \in \mathbb{Z}^m$, and linear functions $f_1, \ldots, f_k \in \mathbb{Z}^n$ be given. There are algorithms to perform the following tasks:*

(i) *Compute the generating function* (9.2) *of all the Pareto optima as a sum of rational functions. In particular, we can count how many Pareto optima are there. If we assume k and n are fixed, the algorithm runs in time polynomial in the size of the input data.*

(ii) *Compute the generating function* (9.3) *of all the Pareto strategies as a sum of rational functions. In particular, we can count how many Pareto strategies are there in P. If* *we assume k and n are fixed, the algorithm runs in time that is bounded polynomially by the size of the input data.*

(iii) *Generate the full sequence of Pareto optima ordered lexicographically. If we assume k and n are fixed, the algorithm runs in polynomial time on the input size and the number of Pareto optima. (More strongly, there exists a polynomial-space polynomial-delay prescribed-order enumeration algorithm.)*

In contrast it is known that for nonfixed dimension it is #P-hard to count Pareto optima and NP-hard to find them [126, 307]. The proof of Theorem 9.1.1 parts (i) and (ii) will be given in Section 9.2. Again it is based on the theory of rational generating functions. Part (iii) of Theorem 9.1.1 will be proved in Section 9.3.

For a user that knows some or all of the Pareto optima or strategies, a goal is to select the "best" member of the family. One is interested in selecting a Pareto optimum that realizes the "best" compromise between the individual objective functions. The quality of the compromise is often measured by the distance of a Pareto optimum $\mathbf{v}$ from a user-defined comparison point $\hat{\mathbf{v}}$. For example, often users take as a good comparison point the so-called *ideal point* $\mathbf{v}^{\text{ideal}} \in \mathbb{Z}^k$ of the multicriterion problem, which is defined as

$$v_i^{\text{ideal}} = \min\{\, f_i(\mathbf{u}) : \mathbf{u} \in P \cap \mathbb{Z}^n \,\}.$$

The criteria of comparison with the point $\hat{\mathbf{v}}$ are quite diverse, but some popular ones include computing the minimum over the possible sums of absolute differences of the individual objective functions, evaluated at the different Pareto strategies, from the comparison point $\hat{\mathbf{v}}$, i.e.,

$$f(\mathbf{u}) = |f_1(\mathbf{u}) - \hat{v}_1| + \cdots + |f_k(\mathbf{u}) - \hat{v}_k|, \tag{9.4a}$$

or the maximum of the absolute differences,

$$f(\mathbf{u}) = \max\bigl\{|f_1(\mathbf{u}) - \hat{v}_1|, \ldots, |f_k(\mathbf{u}) - \hat{v}_k|\bigr\}, \tag{9.4b}$$

over all Pareto optima $(f_1(\mathbf{u}), \ldots, f_k(\mathbf{u}))$. Another popular criterion, sometimes called the *global criterion*, is to minimize the sum of relative distances of the individual objectives from their known minimal values, i.e.,

$$f(\mathbf{u}) = \frac{f_1(\mathbf{u}) - v_1^{\text{ideal}}}{|v_1^{\text{ideal}}|} + \cdots + \frac{f_k(\mathbf{u}) - v_k^{\text{ideal}}}{|v_k^{\text{ideal}}|}. \tag{9.4c}$$

We stress that if we take any one of these functions as an objective function of an integer program, the optimal solution will be a non-Pareto solution of the multicriterion problem (9.1) in general; see Figure 9.2. In contrast, we show here that by encoding Pareto optima and strategies as a rational function we avoid this problem, since we evaluate the objective functions directly on the space of Pareto optima.

All of the above criteria (9.4) measure the distance from a prescribed point with respect to a *polyhedral norm*. In Section 9.4, we prove:

Theorem 9.1.2. *Let the dimension n and the number k of objective functions be fixed. Let a multicriterion integer linear program* (9.1) *be given. Let a polyhedral norm* $\|\cdot\|_Q$ *be given by the vertex or inequality description of its unit ball* $Q \subseteq \mathbb{R}^k$. *Finally, let a prescribed point* $\hat{\mathbf{v}} \in \mathbb{Z}^k$ *be given.*

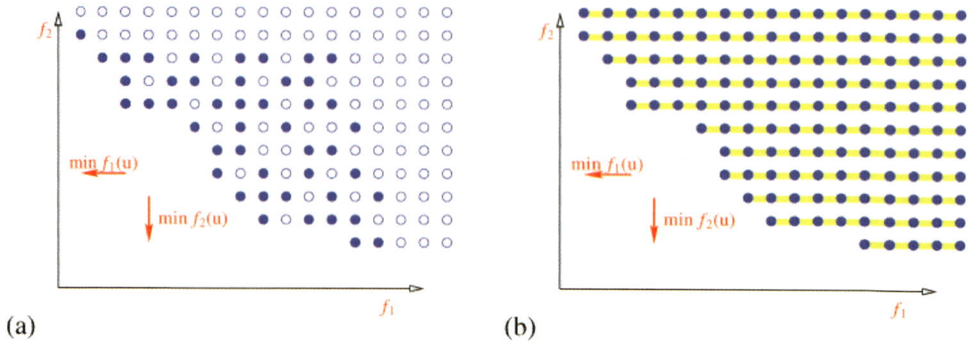

Figure 9.3. *Outcome space.* (a) *Feasible outcomes.* (b) *Their discrete epigraph.*

(i) *There exists a polynomial-time algorithm to find a Pareto optimum* $\mathbf{v}$ *of* (9.1) *that minimizes the distance* $\|\mathbf{v}-\hat{\mathbf{v}}\|_Q$ *from the prescribed point.*

(ii) *There exists a polynomial-space polynomial-delay enumeration algorithm for enumerating the Pareto optima of* (9.1) *in the order of increasing distances from the prescribed point* $\hat{\mathbf{v}}$.

Often users are actually interested in finding a Pareto optimum that minimizes the *Euclidean* distance from a prescribed comparison point $\hat{\mathbf{v}}$,

$$f(\mathbf{u}) = \sqrt{|f_1(\mathbf{u}) - \hat{v}_1|^2 + \cdots + |f_k(\mathbf{u}) - \hat{v}_k|^2}. \tag{9.5}$$

In Section 9.4 we prove the following theorem, which gives a very strong approximation result.

Theorem 9.1.3. *Let the dimension n and the number k of objective functions be fixed. There exists a fully polynomial-time approximation scheme for the problem of minimizing the Euclidean distance of a Pareto optimum of* (9.1) *from a prescribed comparison point* $\hat{\mathbf{v}} \in \mathbb{Z}^k$.

We actually prove this theorem in a somewhat more general setting, using an arbitrary norm whose unit ball is representable by a homogeneous polynomial inequality.

9.2 The rational generating function encoding of all Pareto optima

One has to be careful when using the Barvinok–Woods theory, especially the projection theorem (Theorem 7.6.1), that the sets in question are finite. The proof of Theorem 9.1.1 will require us to project and intersect sets of lattice points represented by rational functions. Fortunately, in our setting it is possible to restrict our attention to finite sets.

***Proof of Theorem* 9.1.1, *parts* (i) *and* (ii).** The proof of part (i) has three steps:

Step 1. For $i = 1,\dots,k$ let $\bar{v}_i \in \mathbb{Z}$ be an upper bound of polynomial encoding size for the value of f_i over P. Such a bound exists because of the boundedness of P, and it can be

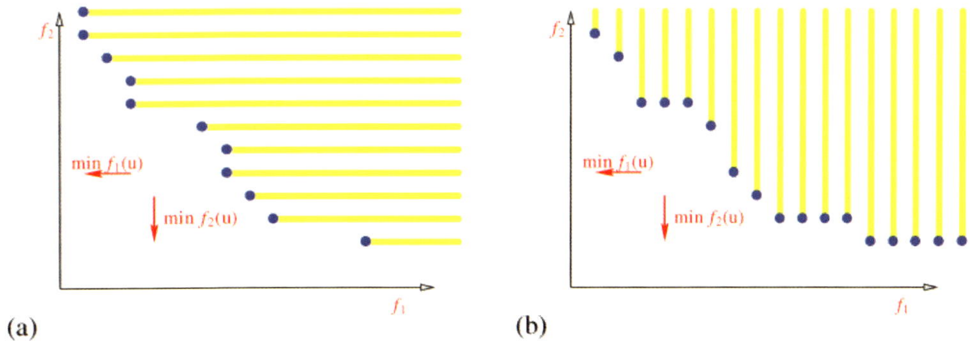

Figure 9.4. *Outcome space.* (a) *After erasing horizontally dominated solutions.* (b) *After erasing vertically dominated solutions.*

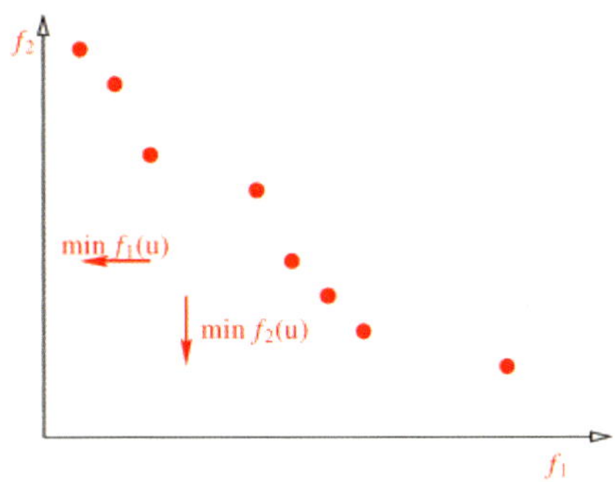

Figure 9.5. *Outcome space, after erasing all dominated solutions.*

computed in polynomial time by linear programming. We will denote the vector of upper bounds by $\bar{\mathbf{v}} \in \mathbb{Z}^k$. We consider the *truncated multiepigraph* of the objective functions $f_1, \ldots, f_k$ over the linear relaxation of the feasible region P,

$$P^{\geq}_{f_1,\ldots,f_k} = \big\{ (\mathbf{u}, \mathbf{v}) \in \mathbb{R}^n \times \mathbb{R}^k : \mathbf{u} \in P, \\ \bar{v}_i \geq v_i \geq f_i(\mathbf{u}) \text{ for } i = 1, \ldots, k \big\}, \tag{9.6}$$

which is a rational convex polytope in $\mathbb{R}^n \times \mathbb{R}^k$ (see Figure 9.3). Let $V^{\geq} \subseteq \mathbb{Z}^k$ denote the integer projection of $P^{\geq}_{f_1,\ldots,f_k}$ on the $\mathbf{v}$ variables, i.e., the set

$$V^{\geq} = \big\{ \mathbf{v} \in \mathbb{Z}^k : \exists \mathbf{u} \in \mathbb{Z}^n \text{ with } (\mathbf{u}, \mathbf{v}) \in P^{\geq}_{f_1,\ldots,f_k} \cap (\mathbb{Z}^n \times \mathbb{Z}^k) \big\}. \tag{9.7}$$

Clearly, the vectors in $V^{\geq}$ are all integer vectors in the outcome space which are weakly dominated by some outcome vector $\big(f_1(\mathbf{u}), f_2(\mathbf{u}), \ldots, f_k(\mathbf{u})\big)$ for a feasible solution $\mathbf{u}$ in $P \cap \mathbb{Z}^n$; however, we have truncated away all outcome vectors which weakly dominate the computed bound $\bar{\mathbf{v}}$. Let us consider the generating function of $V^{\geq}$, the multivariate polynomial

$$g(V^{\geq}; \mathbf{z}) = \sum \big\{ \mathbf{z}^{\mathbf{v}} : \mathbf{v} \in V^{\geq} \big\}.$$

In the terminology of polynomial ideals, the monomials in $g(V^{\geq}; \mathbf{z})$ form a truncated ideal generated by the Pareto optima. By the projection theorem (Theorem 7.6.1), we can compute $g(V^{\geq};\mathbf{z})$ in the form of a polynomial-size rational function in polynomial time.

Step 2. Let $V^{\text{Pareto}} \subseteq \mathbb{Z}^k$ denote the set of Pareto optima. Clearly we have

$$V^{\text{Pareto}} = \left(V^{\geq} \setminus (\mathbf{e}_1 + V^{\geq})\right) \cap \cdots \cap \left(V^{\geq} \setminus (\mathbf{e}_k + V^{\geq})\right),$$

where $\mathbf{e}_i \in \mathbb{Z}^k$ denotes the i-th unit vector and

$$\mathbf{e}_i + V^{\geq} = \{\mathbf{e}_i + \mathbf{v} : \mathbf{v} \in V^{\geq}\}.$$

This construction is illustrated in Figures 9.4 and 9.5. The generating function $g(V^{\text{Pareto}};\mathbf{z})$ can be computed by the Boolean operations lemma (Theorem 7.5.3) in polynomial time from $g(V^{\geq};\mathbf{z})$ as

$$\begin{aligned} g(V^{\text{Pareto}};\mathbf{z}) = \big(&g(V^{\geq};\mathbf{z}) - g(V^{\geq};\mathbf{z}) \star z_1 g(V^{\geq};\mathbf{z})\big) \\ &\star \cdots \star \big(g(V^{\geq};\mathbf{z}) - g(V^{\geq};\mathbf{z}) \star z_k g(V^{\geq};\mathbf{z})\big), \end{aligned} \tag{9.8}$$

where $\star$ denotes taking the Hadamard product (Section 7.5) of the rational functions.

Step 3. To obtain the number of Pareto optima, we finally compute the specialization $g(V^{\text{Pareto}};\mathbf{z}{=}\mathbf{1})$ using Theorem 7.2.1.

Proof of part (ii). Now we recover the Pareto *strategies* that gave rise to the Pareto optima, i.e., we compute a generating function for the set

$$U^{\text{Pareto}} = \left\{\mathbf{u} \in \mathbb{Z}^n : \mathbf{u} \in P \cap \mathbb{Z}^n \text{ and } \mathbf{f}(\mathbf{u}) \text{ is a Pareto optimum}\right\}.$$

To this end, we first compute the generating function for the set

$$S^{\text{Pareto}} = \left\{(\mathbf{u},\mathbf{v}) \in \mathbb{Z}^n \times \mathbb{Z}^k : \mathbf{v} \text{ is a Pareto point with Pareto strategy } \mathbf{u}\right\}.$$

For this purpose, we consider the multigraph of the objective functions $f_1,\dots,f_k$ over P,

$$\begin{aligned} P^{=}_{f_1,\dots,f_k} = \big\{(\mathbf{u},\mathbf{v}) \in \mathbb{R}^n \times \mathbb{R}^k : \mathbf{u} \in P, \\ v_i = f_i(\mathbf{u}) \text{ for } i = 1,\dots,k\big\}. \end{aligned} \tag{9.9}$$

Using Barvinok's theorem, we can compute in polynomial time the generating function for the integer points in P,

$$g(P;\mathbf{x}) = \sum\{\mathbf{x}^{\mathbf{u}} : \mathbf{u} \in P \cap \mathbb{Z}^n\},$$

and also, using the monomial substitution $x_j \to x_j z_1^{f_1(\mathbf{e}_j)} \cdots z_k^{f_k(\mathbf{e}_j)}$ for all j, the generating function is transformed into

$$g(P^{=}_{f_1,\dots,f_k};\mathbf{x},\mathbf{z}) = \sum\{\mathbf{x}^{\mathbf{u}}\mathbf{z}^{\mathbf{v}} : (\mathbf{u},\mathbf{v}) \in P^{=}_{f_1,\dots,f_k} \cap (\mathbb{Z}^n \times \mathbb{Z}^k)\},$$

where the variables $\mathbf{x}$ carry on the monomial exponents the information of the $\mathbf{u}$-coordinates of $P^{=}_{f_1,\dots,f_k}$ and the $\mathbf{z}$ variables of the generating function carry the $\mathbf{v}$-coordinates of lattice points in $P^{=}_{f_1,\dots,f_k}$. Now

$$g(S^{\text{Pareto}};\mathbf{x},\mathbf{z}) = \big(g(P;\mathbf{x})\, g(V^{\text{Pareto}};\mathbf{z})\big) \star g(P^{=}_{f_1,\dots,f_k};\mathbf{x},\mathbf{z}), \tag{9.10}$$

which can be computed in polynomial time for fixed dimension by the theorems outlined earlier in this section. Finally, to obtain the generating function $g(U^{\text{Pareto}};\mathbf{x})$ of the Pareto strategies, we need to compute the projection of S^{Pareto} into the space of the strategy variables $\mathbf{u}$. Since the projection is one-to-one, it suffices to compute the specialization

$$g(U^{\text{Pareto}};\mathbf{x}) = g(S^{\text{Pareto}};\mathbf{x},\mathbf{z}{=}\mathbf{1}),$$

which can be done in polynomial time using Theorem 7.2.3. □

9.3 Efficiently listing all Pareto optima

The Pareto optimum that corresponds to the "best" compromise between the individual objective functions is often chosen in an *interactive mode*, where a visualization of the Pareto optima is presented to the user, who then chooses a Pareto optimum. Since the outcome space is frequently too large of a dimension for visualization, an important task is to list (explicitly enumerate) the elements of the *projection* of the Pareto set into some lower-dimensional linear space.

It is clear that the set of Pareto optima (and thus also any projection) is of exponential size in general, ruling out the existence of a polynomial-time enumeration algorithm. Thus we turn to output-sensitive complexity analysis as in Section 7.3.

Indeed, from Theorem 7.5.2 we deduce the following immediate corollary, which is a stronger formulation of Theorem 9.1.1(iii).

Corollary 9.3.1. *Let n and k be fixed integers. There exist polynomial-space polynomial-delay enumeration algorithms to enumerate the set of Pareto optima of the multicriterion integer linear program* (9.1)*, the set of Pareto strategies, or arbitrary projections thereof in lexicographic order.*

9.4 Selecting a Pareto optimum using polyhedral global criteria

Now that we know that all Pareto optima of a multicriteria integer linear programs can be encoded in a rational generating function, and that they can be listed efficiently on the output size, we can aim to apply selection criteria stated by a user. The advantage of our setup is that, when we optimize a global objective function, it guarantees to return a Pareto optimum, because we evaluate the global criterion only on the Pareto optima. Let us start with the simplest global criterion which generalizes the use of the ℓ_1-norm distance function:

Theorem 9.4.1. *Let the dimension k and the maximum number ℓ of binomials in the denominator be fixed.*

Let $V \subseteq \mathbb{Z}^k$ be a bounded set of lattice points with $V \subseteq [-M,M]^{n+k}$, given only by the bound $M \in \mathbb{Z}_+$ and its rational generating function encoding $g(V;\mathbf{z})$ with at most ℓ binomials in the denominators.

Let $Q \subseteq \mathbb{R}^k$ be a rational convex central-symmetric polytope with $\mathbf{0} \in \operatorname{int}(Q)$, given by its vertex or inequality description. Let the polyhedral norm $\|\cdot\|_Q$ be defined using the Minkowski functional

$$\|\mathbf{y}\|_Q = \inf\{\lambda \geq 0 : \mathbf{y} \in \lambda Q\}. \tag{9.11}$$

Finally, let a prescribed point $\hat{\mathbf{v}} \in \mathbb{Z}^k$ be given.

(i) *There exists a polynomial-time algorithm to find a point $\mathbf{v} \in V$ that minimizes the distance $d_Q(\mathbf{v},\hat{\mathbf{v}}) = \|\mathbf{v}-\hat{\mathbf{v}}\|_Q$ from the prescribed point.*

(ii) *There exists a polynomial-space polynomial-delay enumeration algorithm for enumerating the points of V in the order of increasing distances d_Q from the prescribed point $\hat{\mathbf{v}}$.*

Theorem 9.1.2, as stated in the introduction to this chapter, is an immediate corollary of this theorem.

Proof. Since the dimension k is fixed, we can compute an inequality description of

$$Q = \{\mathbf{y} \in \mathbb{R}^k : A\mathbf{y} \leq \mathbf{b}\}$$

with $A \in \mathbb{Z}^{m\times k}$ and $\mathbf{b} \in \mathbb{Z}^k$ in polynomial time, if Q is not already given by an inequality description. Let $\mathbf{v} \in V$ be arbitrary; then

$$\begin{aligned} d_Q(\hat{\mathbf{v}}, \mathbf{v}) &= \|\mathbf{v} - \hat{\mathbf{v}}\|_Q \\ &= \inf\big\{\lambda \geq 0 : \mathbf{v} - \hat{\mathbf{v}} \in \lambda Q\big\} \\ &= \min\big\{\lambda \geq 0 : \lambda\mathbf{b} \geq A(\mathbf{v} - \hat{\mathbf{v}})\big\}. \end{aligned}$$

Thus there exists an index $i \in \{1, \dots, m\}$ such that

$$d_Q(\hat{\mathbf{v}}, \mathbf{v}) = \frac{(A\mathbf{v})_i - (A\hat{\mathbf{v}})_i}{b_i};$$

so $d_Q(\hat{\mathbf{v}}, \mathbf{v})$ is an integer multiple of $1/b_i$. Hence for every $\mathbf{v} \in V$, we have that

$$d_Q(\hat{\mathbf{v}}, \mathbf{v}) \in \frac{1}{\operatorname{lcm}(b_1, \dots, b_m)} \mathbb{Z}_+, \tag{9.12}$$

where $\operatorname{lcm}(b_1, \dots, b_m)$ clearly is a number of polynomial encoding size. On the other hand, every $\mathbf{v} \in V$ certainly satisfies

$$d_Q(\hat{\mathbf{v}}, \mathbf{v}) \leq ka\big(M + \max\{|\hat{v}_1|, \dots, |\hat{v}_k|\}\big) \tag{9.13}$$

where a is the largest number in A, which is also a bound of polynomial encoding size.

Using Barvinok's algorithm, we can compute the rational generating function $g(\hat{\mathbf{v}} + \lambda Q; \mathbf{z})$ for any rational λ of polynomial enoding size in polynomial time. We can also compute the rational generating function $g(V \cap (\hat{\mathbf{v}} + \lambda Q); \mathbf{z})$ using the intersection lemma. By computing the specialization $g(V \cap (\hat{\mathbf{v}} + \lambda Q); \mathbf{z} = \mathbf{1})$, we can compute the number of points in $V \cap (\hat{\mathbf{v}} + \lambda Q)$, and thus we can decide whether this set is empty or not.

Hence we can employ binary search for the smallest $\lambda \geq 0$ such that $V \cap (\hat{\mathbf{v}} + \lambda Q)$ is nonempty. Because of (9.12) and (9.13), it runs in polynomial time. By using the recursive bisection algorithm of Theorem 7.5.2, it is then possible to construct one Pareto optimum in $V \cap (\hat{\mathbf{v}} + \lambda Q)$ for part (i), or to construct a sequence of Pareto optima in the desired order for part (ii). □

9.5 Selecting a Pareto optimum using a nonpolyhedral global criterion

Now we consider a global criterion using a distance function corresponding to a nonpolyhedral norm like the Euclidean norm $\|\cdot\|_2$ (or any other ℓ_p-norm for $1 < p < \infty$). We are able to prove fully polynomial-time approximation scheme (FPTAS), in a somewhat more general setting. In contrast to the definitions in Section 8.1, we consider an FPTAS for a *minimization* problem. We define it as follows.

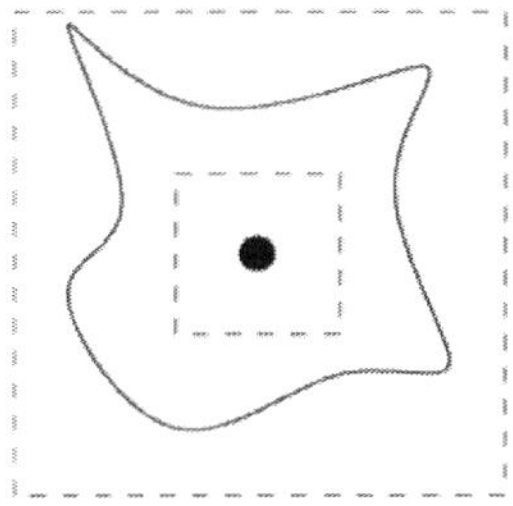

Figure 9.6. *A set defining a pseudonorm with the inscribed and circumscribed cubes αB_∞ and βB_∞ (dashed lines).* [98]

Definition 9.5.1 (FPTAS). A *fully polynomial-time approximation scheme (FPTAS)* for the minimization problem $\min\{\, f(\mathbf{v}) : \mathbf{v} \in V \,\}$ is a family $\{\, \mathcal{A}_\epsilon : \epsilon \in \mathbb{Q}, \epsilon > 0 \}$ of approximation algorithms $\mathcal{A}_\epsilon$, each of which returns an *ϵ-approximation*, i.e., a solution $\mathbf{v}_\epsilon \in V$ with

$$f(\mathbf{v}_\epsilon) \leq (1+\epsilon) f^* \quad \text{where} \quad f^* = \min_{\mathbf{v}\in V} f(\mathbf{v}), \tag{9.14}$$

such that the algorithms $\mathcal{A}_\epsilon$ run in time bounded polynomially by the input size and $\frac{1}{\epsilon}$.

Remark 9.5.2. An FPTAS is based on the notion of ϵ-approximation, which gives an approximation guarantee relative to the value f^* of an optimal solution. Since the approximation quality of a solution changes when the objective function is changed by an additive constant, it is nontrivial to convert an FPTAS for a maximization problem to an FPTAS for a minimization problem.

We shall present an FPTAS for the problem of minimizing the distance of a Pareto optimum from a prescribed outcome vector $\hat{\mathbf{v}} \in \mathbb{Z}^k$. We consider distances $d(\hat{\mathbf{v}}, \cdot)$ induced by a pseudonorm $\|\cdot\|_Q$ via

$$d(\hat{\mathbf{v}}, \mathbf{v}) = \|\mathbf{v} - \hat{\mathbf{v}}\|_Q. \tag{9.15a}$$

To this end, let $Q \subseteq \mathbb{R}^k$ be a compact basic semialgebraic set with $\mathbf{0} \in \operatorname{int}(Q)$, which is described by one polynomial inequality,

$$Q = \left\{ \mathbf{y} \in \mathbb{R}^k : q(\mathbf{y}) \leq 1 \right\}, \tag{9.15b}$$

where $q \in \mathbb{Q}[y_1, \ldots, y_k]$ is a homogeneous polynomial of (even) degree D. The pseudonorm $\|\cdot\|_Q$ is now defined using the Minkowski functional

$$\|\mathbf{y}\|_Q = \inf\left\{ \lambda \geq 0 : \mathbf{y} \in \lambda Q \right\}. \tag{9.15c}$$

Note that we do not make any assumptions of the convexity of Q, which would make $\|\cdot\|_Q$ a norm. Since Q is compact and $\mathbf{0} \in \operatorname{int}(Q)$, there exist positive rational numbers (norm equivalence constants) α, β with

$$\alpha B_\infty \subseteq Q \subseteq \beta B_\infty \quad \text{where} \quad B_\infty = \left\{ \mathbf{y} \in \mathbb{R}^k : \|\mathbf{y}\|_\infty \leq 1 \right\}; \tag{9.16}$$

see Figure 9.6.

Now we can formulate our main theorem, which has Theorem 9.1.3, which we stated in the introduction to this chapter, as an immediate corollary.

Theorem 9.5.3. *Let the dimension n and the number k of objective functions be fixed. Moreover, let a degree D and two rational numbers $0 < \alpha \leq \beta$ be fixed. Then there exists a fully polynomial-time approximation scheme for the problem of minimizing the distance $d_Q(\hat{\mathbf{v}}, \mathbf{v})$, defined via* (9.15) *by a homogeneous polynomial $q \in \mathbb{Q}[y_1, \ldots, y_k]$ of degree D satisfying* (9.16)*, whose coefficients are encoded in binary and whose exponent vectors are encoded in unary, of a Pareto optimum of* (9.1) *from a prescribed outcome vector $\hat{\mathbf{v}} \in \mathbb{Z}^k$.*

The proof is based on the FPTAS for polynomial maximization over lattice point sets (Theorem 8.3.6).

***Proof of Theorem* 9.5.3.** Using Theorem 9.1.1, we first compute the rational generating function $g(V^{\text{Pareto}}; \mathbf{z})$ of the Pareto optima. With binary search using the intersection lemma with generating functions of cubes as in Section 9.3, we can find the smallest nonnegative integer γ such that

$$(\hat{\mathbf{v}} + \gamma B_\infty) \cap V^{\text{Pareto}} \neq \emptyset. \tag{9.17}$$

If $\gamma = 0$, then the prescribed outcome vector $\hat{\mathbf{v}}$ itself is a Pareto optimum, so it is the optimal solution to the problem.

Otherwise, let $\mathbf{v}_0$ be an arbitrary outcome vector in $(\hat{\mathbf{v}} + \gamma B_\infty) \cap V^{\text{Pareto}}$. Then

$$\begin{aligned} \gamma \geq \|\mathbf{v}_0 - \hat{\mathbf{v}}\|_\infty &= \inf\{\lambda : \mathbf{v}_0 - \hat{\mathbf{v}} \in \lambda B_\infty\} \\ &\geq \inf\{\lambda : \mathbf{v}_0 - \hat{\mathbf{v}} \in \lambda \tfrac{1}{\alpha} Q\} = \alpha \|\mathbf{v}_0 - \hat{\mathbf{v}}\|_Q, \end{aligned}$$

thus $\|\mathbf{v}_0 - \hat{\mathbf{v}}\|_Q \leq \gamma/\alpha$. Let $\delta = \beta\gamma/\alpha$. Then, for every $\mathbf{v}_1 \in \mathbb{R}^k$ with $\|\mathbf{v}_1 - \hat{\mathbf{v}}\|_\infty \geq \delta$ we have

$$\begin{aligned} \delta \leq \|\mathbf{v}_1 - \hat{\mathbf{v}}\|_\infty &= \inf\{\lambda : \mathbf{v}_1 - \hat{\mathbf{v}} \in \lambda B_\infty\} \\ &\leq \inf\{\lambda : \mathbf{v}_1 - \hat{\mathbf{v}} \in \lambda \tfrac{1}{\beta} Q\} = \beta \|\mathbf{v}_1 - \hat{\mathbf{v}}\|_Q, \end{aligned}$$

thus

$$\|\mathbf{v}_1 - \hat{\mathbf{v}}\|_Q \geq \frac{\delta}{\beta} = \frac{\gamma}{\alpha} \geq \|\mathbf{v}_0 - \hat{\mathbf{v}}\|_Q.$$

Therefore, a Pareto optimum $\mathbf{v}^* \in V^{\text{Pareto}}$ minimizing the distance d_Q from the prescribed outcome vector $\hat{\mathbf{v}}$ is contained in the cube $\hat{\mathbf{v}} + \delta B_\infty$. Moreover, for all points $\mathbf{v} \in \hat{\mathbf{v}} + \delta B_\infty$ we have

$$\|\mathbf{v}_0 - \hat{\mathbf{v}}\|_Q \leq \frac{\delta}{\alpha} = \frac{\beta\gamma}{\alpha^2}.$$

We define a function f by

$$f(\mathbf{v}) = \left(\frac{\beta\gamma}{\alpha^2}\right)^D - \|\mathbf{v} - \hat{\mathbf{v}}\|_Q^D, \tag{9.18}$$

which is nonnegative over the cube $\hat{\mathbf{v}} + \delta B_\infty$. Since q is a homogeneous polynomial of degree D, we obtain

$$f(\mathbf{v}) = \left(\frac{\beta\gamma}{\alpha^2}\right)^D - q(\mathbf{v} - \hat{\mathbf{v}}), \tag{9.19}$$

so f is a polynomial.

We next compute the rational generating function

$$g\big(V^{\text{Pareto}} \cap (\hat{\mathbf{v}} + \delta B_\infty); \mathbf{z}\big)$$

from $g(V^{\text{Pareto}}; \mathbf{z})$ using the intersection lemma (Theorem 7.5.1). Let $\epsilon' > 0$ be a rational number, which we will determine later. By Theorem 8.3.6, we compute a solution $\mathbf{v}_{\epsilon'} \in V^{\text{Pareto}}$ with

$$f(\mathbf{v}_{\epsilon'}) \geq (1-\epsilon') f(\mathbf{v}^*),$$

or, equivalently,

$$f(\mathbf{v}^*) - f(\mathbf{v}_{\epsilon'}) \leq \epsilon' f(\mathbf{v}^*).$$

Thus,

$$\begin{aligned}
[d_Q(\hat{\mathbf{v}}, \mathbf{v}_{\epsilon'})]^D - [d_Q(\hat{\mathbf{v}}, \mathbf{v}^*)]^D &= \|\mathbf{v}_{\epsilon'} - \hat{\mathbf{v}}\|_Q^D - \|\mathbf{v}^* - \hat{\mathbf{v}}\|_Q^D \\
&= f(\mathbf{v}^*) - f(\mathbf{v}_{\epsilon'}) \\
&\leq \epsilon' f(\mathbf{v}^*) \\
&= \epsilon' \left(\left(\frac{\beta\gamma}{\alpha^2}\right)^D - \|\mathbf{v}^* - \hat{\mathbf{v}}\|_Q^D \right).
\end{aligned}$$

Since γ is the smallest integer satisfying (9.17) and also $\|\mathbf{v}^* - \hat{\mathbf{v}}\|_\infty$ is an integer, we have

$$\gamma \leq \|\mathbf{v}^* - \hat{\mathbf{v}}\|_\infty \leq \beta \|\mathbf{v}^* - \hat{\mathbf{v}}\|_Q.$$

Thus,

$$[d_Q(\hat{\mathbf{v}}, \mathbf{v}_{\epsilon'})]^D - [d_Q(\hat{\mathbf{v}}, \mathbf{v}^*)]^D \leq \epsilon' \left[\left(\frac{\beta}{\alpha}\right)^{2D} - 1 \right] \|\mathbf{v}^* - \hat{\mathbf{v}}\|_Q^D.$$

An elementary calculation yields

$$d_Q(\hat{\mathbf{v}}, \mathbf{v}_{\epsilon'}) - d_Q(\hat{\mathbf{v}}, \mathbf{v}^*) \leq \frac{\epsilon'}{D} \left[\left(\frac{\beta}{\alpha}\right)^{2D} - 1 \right] d_Q(\hat{\mathbf{v}}, \mathbf{v}^*).$$

Thus we can choose

$$\epsilon' = \epsilon D \left[\left(\frac{\beta}{\alpha}\right)^{2D} - 1 \right]^{-1} \tag{9.20}$$

to get the desired estimate. Since α, β, and D are fixed constants, we have $\epsilon' = \Theta(\epsilon)$. Thus the computation of $\mathbf{v}_{\epsilon'} \in V^{\text{Pareto}}$ by Theorem 8.3.6 runs in time bounded polynomially in the input encoding size and $\frac{1}{\epsilon}$. □

Remark 9.5.4. It is straightforward to extend this result to also include the ℓ_p norms for *odd* integers p by solving the approximation problem separately for all of the $2^k = \mathrm{O}(1)$ shifted orthants $\hat{\mathbf{v}} + \mathbb{O}_\sigma = \{\mathbf{v} : \sigma_i(v_i - \hat{v}_i) \geq 0\}$, where $\boldsymbol{\sigma} \in \{\pm 1\}^k$. On each of the orthants, the ℓ_p-norm has a representation by a polynomial as required by Theorem 9.5.3.

9.6 Notes and further references

This chapter is based on De Loera et al. [98].

A simplified version of the construction of the rational generating function in Section 9.2 was given by Blanco and Puerto [56]. At the same time, it removes the need to fix the number of criteria in advance.

The Barvinok–Woods projection theorem (Theorem 7.6.1) is very powerful and can be used for many other applications. An application to the computation of pure-strategy Nash equilibria in a class of strategic games called integer programming games appears in Köppe, Ryan, and Queyranne [205]; an extension to certain symmetric games appears in Ryan, Jiang, and Leyton-Brown [287].

9.7 Exercises

Exercise 9.7.1. Implement the algorithm of Theorem 9.1.2 or its refinement by Blanco and Puerto [56]. You can use the implementation of the Barvinok–Woods projection theorem from the library `barvinok` [330] as a subroutine. Test it on a number of examples.

Exercise 9.7.2. Develop an algorithm for bilevel integer linear programming using the Barvinok–Woods projection theorem.

Part IV

Gröbner Basis Methods

Chapter 10

Computations with Polynomials

The key purpose of this chapter is to introduce the reader to a few useful points of "algorithmic algebraic computation" (for a deeper introduction to these algebraic notions we recommend the books [58, 78, 79, 241]). Given a system of polynomial equations and inequalities, we wish to know how to solve them or decide when no solution exists; or, we may also wish to find a solution maximizing a certain objective. In Part V of the book we will see how many combinatorial optimization problems can be modeled by systems of polynomial equations.

10.1 Introduction

Consider the following system of polynomial equations:

$$\begin{cases} 3x^2+11y^3-7=0, \\ x^3+7y^5-2=0 \end{cases}$$

We can ask a series of questions about it:

- How many complex/real/rational/integer solutions are there?
- How many solutions lie within the triangle $x+y \geq 0, x \leq 1, y \leq 1$?

In what follows, $\mathbb{K}$ denotes a field and when the distinction is necessary we denote its algebraic closure by $\overline{\mathbb{K}}$. Moreover, let $\mathbb{K}[x_1,\dots,x_n]$ denote the ring of polynomials in n indeterminates with coefficients over $\mathbb{K}$, and denote the monomials in this ring by $\mathbf{x}^{\boldsymbol{\alpha}} := x_1^{\alpha_1}x_2^{\alpha_2}\cdots x_n^{\alpha_n}$ for $\boldsymbol{\alpha} \in \mathbb{Z}_+^n$. The degree of $\mathbf{x}^{\boldsymbol{\alpha}}$ is $\deg(\mathbf{x}^{\boldsymbol{\alpha}}) := |\boldsymbol{\alpha}| := \sum_{i=1}^n \alpha_i$. The degree of a polynomial $f = \sum_{\boldsymbol{\alpha}\in\mathbb{Z}_+^n} f_{\boldsymbol{\alpha}}\mathbf{x}^{\boldsymbol{\alpha}}$, written $\deg(f)$, is the maximum degree of $\mathbf{x}^{\boldsymbol{\alpha}}$, where $f_{\boldsymbol{\alpha}} \neq 0$ for $\boldsymbol{\alpha} \in \mathbb{Z}_+^n$. Given a set of polynomials $F \subset \mathbb{K}[x_1,\dots,x_n]$, we write $\deg(F)$ for the maximum degree of the polynomials in the set F.

Given a field $\mathbb{K}$ (e.g., $\mathbb{Q},\mathbb{R},\mathbb{C}$, or finite fields), we consider the n-dimensional affine space over $\mathbb{K}$, $\mathbb{K}^n = \{(a_1,\dots,a_n) : a_i \in \mathbb{K}\}$. Then $\bar{a} \in \mathbb{K}^n$ is a *solution*, or *root*, of a polynomial $f(x_1,\dots,x_n)$ if $f(a_1,\dots,a_n) = 0$. For any field $\mathbb{K}$, a polynomial can also be seen as a function $f\colon \mathbb{K}^n \to \mathbb{K}$ using evaluation.

Proposition 10.1.1. *If $\mathbb{K}$ is an infinite field, then $f = g$ in $\mathbb{K}[x_1,\dots,x_n]$ if and only if f,g are the same function.*

Proof. ($\Rightarrow$) Clearly, if f, g are the same polynomial, that implies that they are the same function.

($\Leftarrow$) If they are the same function, then $h = f - g$ is the constant zero function. It remains to show that h is the zero polynomial. We proceed by induction on the number of variables.

(a) If $n = 1$, it is easy to see that a nonzero polynomial of degree m has no more than m distinct roots, therefore it must be the zero polynomial because it has infinitely many roots.

(b) Assume that our claim is true for $n-1$ variables. By collecting powers of x_n, we can rewrite h as:

$$h = \sum_{i=1}^{N} l_i(x_1, \ldots, x_{n-1}) x_n^i.$$

This univariate polynomial (in x_n^i) has infinitely many roots and thus has to be the zero polynomial, that is, $l_i(a_1, \ldots, a_{n-1}) = 0$ for all $(a_1, \ldots, a_{n-1})$. By induction hypothesis for $n-1$ indeterminates, this implies that all coefficients of h are zero. □

First off, let us take a look at finding the roots of polynomials in a *single variable*, that is, we wish to solve $f := a_n x^n + a_{n-1} x^{n-1} + \cdots + a_0 = 0$. In this case we have a fairly good understanding on how to solve them, and it will serve as a good inspiration for what is to come. Nevertheless, solving univariate polynomial equations is not totally trivial. For example, while there are general formulas that can help find roots of a given function (such as the quadratic formula for polynomials of degree two), Galois theory indicates there does not exist such a formula for general polynomials of degree five or larger. One important point that we will later see is that solving multivariate systems of equations has a nice process of elimination of variables until we reduce the finding of solutions to a single variable problem; thus, it is worth understanding this now.

10.2 Univariate polynomials

A polynomial in one variable will always be described as $f = \sum_{i=1}^{n} a_i x^i$, where a_i is a coefficient in a field $\mathbb{K}$ (say $\mathbb{Q}$, $\mathbb{R}$, or $\mathbb{C}$). Each summand $a_i x^i$ is called a *term*. We say that x^i is of *degree i*. One key property that will be exploited again and again is that the terms of any polynomial can be ordered by the degrees of monomials. A *root* of f is a number c such that $f(c) = 0$. The *leading term* $\mathrm{LT}(f)$ is the term in f of highest degree. This important notion can be generalized to higher dimensions.

10.2.1 Division algorithm and gcd

A first important procedure for univariate polynomials that will be generalized below is the well-known division algorithm for polynomials, which divides one polynomial by another with remainder.

Lemma 10.2.1 (division algorithm). *Let $\mathbb{K}$ be a field and let g be a nonzero polynomial in $\mathbb{K}[x]$. Then every polynomial $f \in \mathbb{K}[x]$ can be written (uniquely) as $f = qg + r$, where $q, r \in \mathbb{K}[x]$ and either $r = 0$ or $\deg(r) < \deg(g)$.*

Proof. We give an explicit algorithmic way to find q and r.

ALGORITHM 10.1. Division algorithm.

input polynomials f, g.
output polynomials q, r.
Set $q \leftarrow 0, r \leftarrow f$.
while $r \neq 0, \mathrm{LT}(g) \mid \mathrm{LT}(r)$ **do**
  $q \leftarrow q + \mathrm{LT}(r)/\mathrm{LT}(g)$.
  $r \leftarrow r - (\mathrm{LT}(r)/\mathrm{LT}(g))g$.
return (q, r).

We leave the simple correctness and uniqueness proofs as an exercise to the reader. □

It is important to note that successive values of $\deg(r)$ decrease, so this sequence cannot be infinite because any strictly decreasing sequence of positive integers has to stop. The division algorithm is very useful, as we will see in the following corollaries and it serves to motivate other important multivariate calculations.

Suppose f is a polynomial and $g = (x - c)$ for some $c \in \mathbb{K}$. Then, by the division algorithm, $f = q(x - c) + r$ which implies $f(c) = r(c)$. In particular, $f(c) = 0$ implies $r(c) = 0$. That is, if c is a root of f, then we must have $r = 0$ and hence $(x - c)$ must divide f.

Corollary 10.2.2. *When c is a constant and f is divided by $(x - c)$ the remainder equals $f(c)$. (The remainder equals zero if and only if c is a root.)*

Corollary 10.2.3. *If $\mathbb{K}$ is a field and $f \in \mathbb{K}[x]$, then f has at most* $\deg(f)$ *roots in $\mathbb{K}$.*

The division algorithm is also useful for the computation of the *greatest common divisor* of univariate polynomials.

Definition 10.2.4. Given polynomials $f_1, \ldots, f_k \in \mathbb{K}[x]$, their greatest common divisor $\gcd(f_1, f_2, \ldots, f_k)$ is a polynomial $h \in \mathbb{K}[x]$ such that

1. $h \mid f_i$ for all $i = 1, \ldots, k$,
2. if $p \mid f_i$ for all $i = 1, \ldots, k$, then $p \mid h$.

Both of these conditions define a polynomial from $\mathbb{K}[x]$ that is unique up to a constant factor from $\mathbb{K}$.

Theorem 10.2.5. *Given two polynomials $f \neq 0, g \neq 0$, there exist $A, B \in \mathbb{K}[x]$ such that $\gcd(f, g) = Af + Bg$.*

Proof. Let $X_{f,g} = \{Af + Bg : A, B \in \mathbb{K}[x]\}$. Pick $h \in X_{f,g}$ nonzero and of smallest possible degree. We claim that $h = \gcd(f, g)$. It is enough to show they divide each other.

1. First, $h = Af + Bg = Ap_1 \gcd(f, g) + Bp_2 \gcd(f, g) = (Ap_1 + Bp_2)(\gcd(f, g))$, and thus, $\gcd(f, g) \mid h$.
2. Conversely, suppose (by contradiction) that $h \nmid f$ (and similarly $h \nmid g$). As also $f \in X_{f,g}$, we must have $\deg(f) \geq \deg(h)$. Hence we can write f as $f = qh + r$, where

$\deg(r) < \deg(h)$. (Note that $r \not\equiv 0$, as this would mean $h \mid f$). But now we have $r = f - qh = f - (Af + Bg) = (1-A)f + Bg$ and thus $r \in X_{f,g}$ with $\deg(r) < \deg(h)$, a contradiction to the minimality of $\deg(h)$. □

How do we compute the gcd of two polynomials? The following algorithm is a modification of the Euclidean algorithm that will compute not just the gcd of two polynomials f, g, but also explicitly finds the linear combination $\gcd(f,g) = sf + tg = \gcd(f,g)$ (and even more information). In what follows $\mathrm{LC}(f)$ denotes the leading coefficient of the polynomial f, that is, the coefficient of the leading term $\mathrm{LT}(f)$, and we define $\mathrm{LC}(0) = 1$.

ALGORITHM 10.2. Extended Euclidean algorithm.

input $f, g \in \mathbb{K}[x]$.
output integer l, polynomials $p_i, r_i, s_i, t_i \in \mathbb{K}[x]$ for $0 \le i \le l+1$, and polynomials $q_i \in \mathbb{K}[x]$ for $1 \le i \le l$, such that $s_i f + t_i g = r_i$, and in particular, $s_l f + t_l g = r_l = \gcd(f,g)$. The q_i are the quotients of the r_i.
Set $p_0 \leftarrow \mathrm{LC}(f)$; $p_1 \leftarrow \mathrm{LC}(g)$; $r_0 \leftarrow f/p_0$; $r_1 \leftarrow g/p_1$.
Set $s_0 \leftarrow 1/p_0$; $t_0 \leftarrow 0$; $s_1 \leftarrow 0$; $t_1 \leftarrow 1/p_1$.
Set $i \leftarrow 1$ (counter).
while $r_i \ne 0$ **do**
    $q_i \leftarrow r_{i-1}$ quotient r_i.
    $p_{i+1} \leftarrow \mathrm{LC}(r_{i-1} - q_i r_i)$.
    $r_{i+1} \leftarrow (r_{i-1} - q_i r_i)/p_{i+1}$.
    $s_{i+1} \leftarrow (s_{i-1} - q_i s_i)/p_{i+1}$.
    $t_{i+1} \leftarrow (t_{i-1} - q_i t_i)/p_{i+1}$.
    $i \leftarrow i+1$.
$l \leftarrow i-1$.
return $(l, p_i, r_i, s_i, t_i, q_i)$.

Our goal now is to prove the correctness of the extended Euclidean algorithm. We use matrices to represent intermediate evaluations. This simplifies the arguments enormously.

We let

$$R_0 = \begin{pmatrix} s_0 & t_0 \\ s_1 & t_1 \end{pmatrix}, \quad Q_i = \begin{pmatrix} 0 & 1 \\ p_{i+1}^{-1} & -q_i p_{i+1}^{-1} \end{pmatrix}.$$

Here, $s_0 = 1/p_0$, $t_0 = 0$, $s_1 = 0$, $t_1 = 1/p_1$, where $p_0 = LC(f)$, $p_1 = LC(g)$, so we can now write

$$R_0 = \begin{pmatrix} \frac{1}{p_0} & 0 \\ 0 & \frac{1}{p_1} \end{pmatrix}.$$

We also let $r_0 = f/p_0$, $r_1 = g/p_1$.

We define $R_i = Q_i Q_{i-1} \cdots Q_2 Q_1 R_0$, so that $R_i = Q_i R_{i-1}$. Then we can state the following lemma.

Lemma 10.2.6. *For $0 \le i \le l$, we have*

1. $R_i \begin{pmatrix} f \\ g \end{pmatrix} = \begin{pmatrix} r_i \\ r_{i+1} \end{pmatrix}$.

2. $R_i = \begin{pmatrix} s_i & t_i \\ s_{i+1} & t_{i+1} \end{pmatrix}$.

3. $\gcd(f,g) = \gcd(r_i, r_{i+1}) = r_l$.

4. $s_i f + t_i g = r_i$.

Proof. 1. We prove this claim by induction on i. Clearly, for $i = 0$, we are done by our previous definition. Suppose it is true for $i-1$. Take

$$R_i \begin{pmatrix} f \\ g \end{pmatrix} = Q_i R_{i-1} \begin{pmatrix} f \\ g \end{pmatrix} = Q_i \begin{pmatrix} r_{i-1} \\ r_i \end{pmatrix} = \begin{pmatrix} 0 & 1 \\ p_{i+1}^{-1} & -q_i p_{i+1}^{-1} \end{pmatrix} \begin{pmatrix} r_{i-1} \\ r_i \end{pmatrix}$$
$$= \begin{pmatrix} r_i \\ \frac{r_{i-1}}{p_{i+1}} - q_i \frac{r_i}{p_{i+1}} \end{pmatrix} = \begin{pmatrix} r_i \\ \frac{1}{p_{i+1}}(r_{i-1} - q_i r_i) \end{pmatrix} = \begin{pmatrix} r_i \\ r_{i+1} \end{pmatrix}.$$

2. Again, one can prove it by induction on i. It is true for $i = 0$, and assume it is true for $i-1$.

$$R_i = Q_i R_{i-1} = \begin{pmatrix} 0 & 1 \\ p_{i+1}^{-1} & -q_i p_{i+1}^{-1} \end{pmatrix} \begin{pmatrix} s_{i-1} & t_{i-1} \\ s_i & t_i \end{pmatrix}$$
$$= \begin{pmatrix} s_i & t_i \\ \frac{s_{i-1} - q_i s_i}{p_{i+1}} & \frac{t_{i-1} - q_i t_i}{p_{i+1}} \end{pmatrix} = \begin{pmatrix} s_i & t_i \\ s_{i+1} & t_{i+1} \end{pmatrix}.$$

3. We have

$$R_l \begin{pmatrix} f \\ g \end{pmatrix} = Q_l Q_{l-1} \cdots Q_{i+1} R_i \begin{pmatrix} f \\ g \end{pmatrix} = Q_l Q_{l-1} \cdots Q_{i+1} \begin{pmatrix} r_i \\ r_{i+1} \end{pmatrix}.$$

On the other hand, from part (1), we have

$$R_l \begin{pmatrix} f \\ g \end{pmatrix} = \begin{pmatrix} r_l \\ 0 \end{pmatrix}.$$

Thus, r_l is a linear combination of r_i, r_{i+1} for all i, and therefore $\gcd(r_i, r_{i+1}) \mid r_l$. Now $\det(Q_i) = -\frac{1}{p_{i+1}} \neq 0$ and consequently, Q_i is invertible. Therefore we can write

$$\begin{pmatrix} r_i \\ r_{i+1} \end{pmatrix} = Q_{i+1}^{-1} Q_i^{-1} \cdots Q_l^{-1} \begin{pmatrix} r_l \\ 0 \end{pmatrix}.$$

Hence, r_i and r_{i+1} are divisible by r_l and we conclude that $\gcd(r_i, r_{i+1}) = r_l$.

4. From part (1), we get $R_i \begin{pmatrix} f \\ g \end{pmatrix} = \begin{pmatrix} r_i \\ r_{i+1} \end{pmatrix}$ and part (2) states $R_i = \begin{pmatrix} s_i & t_i \\ s_{i+1} & t_{i+1} \end{pmatrix}$. Therefore, we have that $s_i f + t_i g = r_i$. (This again proves that $\gcd(f,g)$ is a linear combination of f and g.) □

We end with a useful application of the gcd. We say for a polynomial p, that α is a root of multiplicity $k > 1$ if $p(\alpha) = 0$ and $p = (x-\alpha)^k h$ for some $h \in \mathbb{K}[x]$. But how can we tell whether we have roots with multiplicity?

Lemma 10.2.7. *The set of roots of p with multiplicity is the same as the roots of the polynomial* $\gcd(p, p')$.

Proof. Suppose $p = (x-\alpha)^k h$, where $k > 1$ and $(x-\alpha)$ does not divide h.

$$\begin{aligned} p' &= (x-\alpha)^k h' + k(x-\alpha)^{k-1} h \\ &= (x-\alpha)^{k-1} \cdot [\text{polynomial not divisible by } (x-\alpha)]. \end{aligned}$$

Hence we have $(x-\alpha)^{k-1} \mid \gcd(p, p')$. □

10.2.2 Real roots of univariate polynomials

For applications in optimization, we will be interested mostly in the real solutions. How can we find those solutions? We will now review a few basic facts about the real roots of a univariate polynomial. We wish to know how many real roots there are and we want to find ways to isolate them. Say, for example, $p = ax^{203} - bx^{100} + c$ for $a, b, c > 0$. How many real roots does p have? Can you give a bound? As we will see below, p has no more than two positive real roots and one negative real root.

Lemma 10.2.8 (Rolle's theorem). *Between two consecutive real roots a, b of f there is an odd number of real roots of f'. Here, a root of multiplicity k is counted k times.*

Corollary 10.2.9. *Between two consecutive real roots of the derivative f', there is at most one real root of f.*

Proof. Suppose on the contrary that there are two roots a, b of f in $[\alpha, \beta]$; then, by Rolle's theorem, there exists at least one root γ of f' in $(a, b) \subset [\alpha, \beta]$, a contradiction. □

Proof of Rolle's theorem. Let a and b be consecutive roots of f and let $f = (x-a)^r \times (x-b)^s Q$ for $a < b$. Assume further that Q is not divisible by $(x-a)$ or $(x-b)$. Note that $f' = r(x-a)^{r-1}(x-b)^s Q + (x-a)^r (x-b)^{s-1} Q + (x-a)^r (x-b)^s Q'$. Consider the expression

$$g = \frac{(x-a)(x-b)f'}{f}.$$

- Claim 1: g is a continuous function inside (a, b). This follows, since there is no root of f in this interval.
- Claim 2: $g(a) < 0$ and $g(b) > 0$. This holds, since $g = r(x-b) + s(x-a) + (x-a)(x-b)\frac{Q'}{Q}$ and also $g(a) = r(a-b) < 0$ and $g(b) = s(b-a) > 0$.

By continuity, g has an odd number of roots inside (a, b). But note that $(x-a)$, $(x-b)$, and f remain nonzero and of the same sign inside (a, b) (as they have no roots in this interval). Consequently, f' must vanish an odd number of times. □

Example 10.2.10. Let us consider the polynomial $f = 3x^5 - 25x^3 + 60x - 20$. Then $\frac{1}{15} f'(x) = x^4 - 5x^2 + 4 = (x^2-1)(x^2-4)$, and thus, the roots of f' are ± 1, ± 2. Now we compute

$$f(-\infty) = -\infty, \; f(-2) = -36, \; f(-1) = -58, \; f(1) = 18, \; f(2) = -4, \; f(\infty) = \infty.$$

Consequently, there exists a single real root inside each of the intervals $(-1, 1)$, $(1, 2)$, and $(2, +\infty)$.

Definition 10.2.11. For a polynomial p, two consecutive terms (ignoring those terms with zero coefficients) present a *sign variation* if their coefficients have different signs.

Theorem 10.2.12 (Descartes' rule of signs). *The number P of positive real roots of a polynomial f with real coefficients is bounded above by the number V of sign variations on the sequence of coefficients of f. More strongly, $P \le V$ and $P = V - 2k$ for some integer k (which means $V - P$ is even). Here, roots are counted with multiplicity.*

Example 10.2.13. We see $f = 3x^5 + 0x^4 - 25x^3 + x^2 + 60x - 20$ has three sign variations, from x^4 to x^3, from x^3 to x^2, and from x^1 to x^0. This predicts no more than three positive real roots.

Consider also $f = x^6 - 3x^2 + x + 1$. p has either two positive real roots or none.

Now consider $f = x^4 + 3x^3 + x - 1 = 0$. What are its roots? By Descartes' rule of signs, it has one positive real root (which cannot be a multiple root). How about negative real roots? $f(-x) = x^4 - 3x^3 - x - 1$ has one sign variation, which implies that f has one negative real root (no multiplicity), since the number of negative real roots f is equal to the number of positive roots of $f(-x)$.

Lemma 10.2.14. *If f is a polynomial with real coefficients and $[a,b]$ is an interval with $f(a) \neq 0, f(b) \neq 0$, then the number of real roots of f, counting multiplicities, in (a,b), is even when $f(a), f(b)$ have the same sign and odd when $f(a), f(b)$ have different signs.*

Proof. Rewrite f as

$$f = (x - r_1) \cdots (x - r_k)(x - s_1) \cdots (x - s_m)(x - t_1) \cdots (x - t_n) g,$$

where

$r_1, \ldots, r_k$ are roots of f in (a,b),

$s_1, \ldots, s_m$ are roots of f less than a,

$t_1, \ldots, t_n$ are roots of f greater than b,

g is positive when evaluated at any real number.

Therefore, we have

$$\frac{f(a)}{f(b)} = \frac{\prod(a - r_i)}{\prod(b - r_i)} \cdot \underbrace{\frac{\prod(a - s_i)}{\prod(b - s_i)}}_{\ge 0} \cdot \underbrace{\frac{\prod(a - t_i)}{\prod(b - t_i)}}_{\ge 0} \underbrace{\frac{g(a)}{g(b)}}_{\ge 0},$$

and thus

$$\frac{f(a)}{f(b)} = (-1)^{\#\text{ of roots in } (a,b)},$$

proving the claim. □

Now we will use these lemmas to prove the theorem.

Proof of Descartes' rule of signs. First we prove that $V - P$ is even: Note that $V =$ (number of sign variations) is even if the leading coefficient and constant term have the same sign; and V is odd if the leading coefficient and constant term have different signs. By Lemma 10.2.14 applied to the interval $(0, \infty)$, $P =$ (number of roots) is even in the first situation and odd in the second. Thus $V - P$ is even, as claimed.

Second, we prove $P \le V$. We prove this by induction on deg(f).

1. If $\deg(f) = 1$, then either f has no positive root, that is $P = 0 \le V$, or f has a positive root and therefore the two coefficients of f must have different signs, that is $P = V = 1$.

2. Now assume it is true for polynomials of degree $n-1$. Take f of degree n. Consider f', the derivative of f.

$$\begin{aligned} P - 1 &\le \#\text{ of roots of } f'(x) \\ &\le \#\text{ of sign variations of } f'(x) \\ &\le \#\text{ sign variations of } f(x) = V. \end{aligned}$$

Therefore $P - 1 \le V$, so $P \le V + 1$, but $P - V$ is even, thus $P \le V$. □

Descartes' rule of signs gives only a bound by which we can do more precise computations to obtain the exact number using *Sturm sequences*. Suppose we wish to know the real roots of f exactly (within an error interval). We will determine the number of real roots of f inside $[a,b]$.

Definition 10.2.15. Let $p_0 = f$, $p_1 = f'$, $p_i = -\text{remainder}(p_{i-2}, p_{i-1})$ and let p_m be the last nonzero polynomial in the sequence. The sequence $(p_0, p_1, \ldots, p_m)$ is called a (canonical) *Sturm sequence* of f.

Example 10.2.16. Let $f = x^3 + x + 1$. Then the Sturm sequence is

$$\left(x^3 + x + 1, 3x^2 + 1, -\tfrac{2}{3}x - 1, -\tfrac{31}{4}\right).$$

Theorem 10.2.17 (Sturm, 1829). *Let f be a polynomial without multiple roots. If $a < b$ and $f(a) \neq 0, f(b) \neq 0$, then the number of distinct real roots of f inside $[a,b]$ is the number of sign variations in the sequence $[p_0(a), p_1(a), \ldots, p_m(a)]$ minus the number of sign variations in $[p_0(b), p_1(b), \ldots, p_m(b)]$.*

Note that zeros do not count as sign variations!

Example 10.2.18. Taking the same sequence as in Example 10.2.16, let us find the distinct real roots of $f = x^3 + x + 1$. Descartes' rule says there are no roots in $(0,\infty)$ and there is one root in $(-\infty,0)$. How many real roots are there in the interval $[-1,0]$? Substituting $x = -1$ into the Sturm sequence of f yields

$$\left(-1, 4, -\tfrac{1}{3}, -\tfrac{31}{4}\right),$$

which has two sign variations. Now substituting $x = 0$ into the Sturm sequence of f yields

$$\left(1, 1, -1, -\tfrac{31}{4}\right),$$

which has one sign variation. Thus, we have one real root in $[-1,0]$.

One can iterate the above procedure by dividing the interval in half to locate each root. This procedure is fairly fast!

10.3 Systems of multivariate polynomial equations

So far, we have only looked at polynomials in one variable. From here on, we are interested in the simultaneous solutions of

$$
\begin{aligned}
f_1(x_1,x_2,\ldots,x_n) &= 0,\\
f_2(x_1,x_2,\ldots,x_n) &= 0,\\
&\vdots\\
f_k(x_1,x_2,\ldots,x_n) &= 0.
\end{aligned}
$$

We will ask questions about the number of solutions or which of the solutions are real.

Let us first make an analogy with a familiar situation. In linear algebra, to solve a system of linear equations, we use Gaussian elimination (row reduction). This allows us to find a new "triangular" system of linear equations with the same solutions, but one that is easy to recover from back-substitution. The row reduction operations are simple elimination steps (pivots). Our goal now is to extend this procedure to apply it to systems of *nonlinear polynomials*. We will need the notion of *ideal* to do this.

Similarly, when solving systems of multivariate linear equations, linear subspaces and affine subspaces play an important role. This time for multivariate polynomials we have the analogous notion of *variety*.

Definition 10.3.1. Let $F = \{f_1, f_2, \ldots, f_k\} \subset \mathbb{K}[x_1,\ldots,x_n]$ be a set of polynomials. A vector $\bar{\mathbf{x}} \in \overline{\mathbb{K}}^n$ is a *solution of the system* $f_1 = f_2 = \cdots = f_k = 0$ (or $F = 0$ for short) if $f_i(\bar{\mathbf{x}}) = 0$ for all $i = 1,\ldots,k$. The *variety* $V(f_1, f_2, \ldots, f_k)$ (or $V(F)$ for short) is the set of all solutions in $\overline{\mathbb{K}}^n$ to $F = 0$.

Example 10.3.2. Let $\mathbb{K} = \mathbb{C}$ and let $f_1 = x_1^2$, $f_2 = x_2^2$, and $f_3 = x_1x_2$. Then, $V(f_1, f_2, f_3) = \{(0,0)\}$, since the only solution that satisfies $f_1 = 0$, $f_2 = 0$, and $f_3 = 0$ simultaneously is the vector $(0,0)$. However, $V(f_1, f_3) = \{(0,c) : c \in \mathbb{C}\}$. This can be seen as follows: $f_1 = 0$ implies $x_1 = 0$, but then x_2 can take any value and still $f_3(x_1,x_2) = x_1x_2 = 0$.

As another example, let $g_1 = x_1^2 + x_2^2 - 1$, $g_2 = x_1x_2 - x_2^3 - 2$, and $g_3 = x_1 - x_2 - 11$. It is not hard to see that the line $g_3 = 0$ does not intersect the circle $g_1 = 0$, that is, $V(g_1, g_3) = \emptyset$. Since $V(g_1, g_2, g_3) \subseteq V(g_1, g_3)$, we know that $V(g_1, g_2, g_3) = \emptyset$.

Proposition 10.3.3. *Suppose* $V = V(f_1, \ldots, f_k)$ *and* $W = V(g_1, \ldots, g_s)$. *Then* $V \cap W = V(f_1, \ldots, f_k, g_1, \ldots, g_s)$. *Similarly,* $V \cup W = V(f_i g_j, 1 \le i \le k, 1 \le j \le s)$.

Now, we need to take a look at two important questions:

1. When is the variety $V(F)$ empty? That, is, when do the polynomials not share a common solution?

2. When is $V(F)$ finite? We will take a look at this later on when we begin to talk about Gröbner bases.

How can we answer these questions? We must first learn about polynomial ideals.

Definition 10.3.4. We say that a set of polynomials $I \subseteq \mathbb{K}[x_1,\ldots,x_n]$ forms an *ideal* if it satisfies the following properties:

1. $0 \in I$.

2. If $f, g \in I$, then $f+g \in I$.

3. If $f \in I$, $g \in \mathbb{K}[x_1,\dots,x_n]$, then $fg \in I$.

If $F \subseteq \mathbb{K}[x_1,\dots,x_n]$, then

$$\langle F \rangle := \left\{ f = \sum_{i=1}^{k} h_i f_i : f_i \in F, h_i \in \mathbb{K}[x_1,\dots,x_n] \forall i = 1,\dots,k \right\} \subseteq \mathbb{K}[x_1,\dots,x_n]$$

is the (polynomial) ideal generated by F in $\mathbb{K}[x_1,\dots,x_n]$.

Example 10.3.5. As an example, let $f_1 = x_1^2 + x_2^2 - 1$ and $f_2 = x_1 - x_2$. Then, with f_1 and f_2, any polynomial combination of f_1 and f_2 lies in the ideal generated by f_1 and f_2, for example:

$$\begin{aligned} f_1 + (x_1 + x_2) f_2 &= (x_1^2 + x_2^2 - 1) + x_1^2 - x_2^2 \\ &= 2x_1^2 - 1. \end{aligned}$$

What this identity states is that $2x_1^2 - 1 \in \langle x_1^2 + x_2^2 - 1, x_1 - x_2 \rangle$.

Suppose $S \subseteq \overline{\mathbb{K}}^n$. Let $I(S) = \{ f \in \mathbb{K}[x_1,\dots,x_n] : f(\mathbf{s}) = 0 \text{ for all } \mathbf{s} \in S \}$. One may check this is an ideal by checking the properties from Definition 10.3.4. This is called the *vanishing ideal* of S. To better understand this, let us take a look at an example in one indeterminate.

Example 10.3.6. Let $S = \{2,3,5\}$. Then for any $f \in I(S)$, 2, 3, and 5 are roots of f. Thus, any $f \in I(S)$ can be written as

$$f = (x-2)(x-3)(x-5)h$$

for some polynomial $h \in \mathbb{K}[x]$. Therefore, the ideal $I(S)$ is the ideal consisting of all multiples of $(x-2)(x-3)(x-5)$, that is, $I(S) = \langle (x-2)(x-3)(x-5) \rangle$.

From this, we can derive the following lemma. Its proof is left as an exercise.

Lemma 10.3.7. *Let $S = V(f_1, f_2, \dots, f_k)$. Then $\langle f_1, f_2, \dots, f_k \rangle \subseteq I(S)$.*

There is also a relation in the opposite direction. Its proof is left as an exercise.

Lemma 10.3.8. *If $I(F) \subseteq I(G)$, then $V(G) \subseteq V(F)$.*

Some very natural questions arise. First of all, how does one decide whether a polynomial f belongs to a given polynomial ideal $\langle f_1, f_2, \dots, f_k \rangle$? This is the so-called *ideal membership test* for polynomial ideals. For linear systems, this question is easy to solve using matrix operations. Similarly, again making the analogy to linear algebra, we may ask whether polynomial ideals have a basis like vector spaces? Although there is no analogue to the concept of linear independence, we will find some special generating sets do exist for every ideal.

10.4 Monomial orders and the multivariate division algorithm

We would like to solve the following problems:

- Given $f, f_1, \ldots, f_k \in \mathbb{K}[x_1, \ldots, x_n]$, decide whether $f \in \langle f_1, \ldots, f_k \rangle$.
- Given $f_1, \ldots, f_k$, find $V(f_1, \ldots, f_k)$.

There are two special case for which the answer should be known to the reader:

1. $f, f_1, f_2, \ldots, f_k$ are linear polynomials. To decide ideal membership for linear polynomials, it is in fact sufficient to decide whether f is a *linear* combination of $f_1, \ldots, f_k$. This linear system of equations can be solved using Gaussian elimination.

2. $f, f_1, f_2, \ldots, f_k$ are univariate polynomials. Then the ideal membership problem can be answered with the computation of $\gcd(f_1, f_2, \ldots, f_k)$: we have $f \in \langle f_1, f_2, \ldots, f_k \rangle$ if and only if $\gcd(f_1, f_2, \ldots, f_k) \mid f$. (We leave the proof as an exercise.) Thus, finding the variety of a system of univariate equations is as hard as finding all roots of one single univariate polynomial, the gcd of the input polynomials. For example, given $\langle x^7 + x^3 - 6x, x^3 + 2x \rangle$, we compute $\gcd(x^7 + x^3 - 6x, x^3 + 2x) = x$ and see that the only common root is 0, that is, $V(x^7 + x^3 - 6x, x^3 + 2x) = \{0\}$.

Note that both Gaussian elimination and the gcd computation need an ordering of variables or more generally of terms. Our goal now is to generalize what we have learned in the previous examples to solve the ideal membership problem and to find the variety for polynomials $\mathbb{K}[x_1, \ldots, x_n]$ of arbitrary degree. First we must establish an ordering of all possible terms $\mathbf{x}^{\alpha} := x_1^{\alpha_1} x_2^{\alpha_2} \ldots x_n^{\alpha_n}$ in n indeterminates, a so-called *term order*. (The vector $\alpha = (\alpha_1, \alpha_2, \ldots, \alpha_n) \in \mathbb{Z}_+^n$ is called the *exponent vector*.)

Definition 10.4.1. A *monomial order* (or *term order*) $\succ$ in $\mathbb{K}[x_1, \ldots, x_n]$ is an ordering relation on the monomials that satisfies the following conditions:

1. $\succ$ is a total order: for any pair of monomials $\mathbf{x}^{\alpha}$ and $\mathbf{x}^{\beta}$ we have $\mathbf{x}^{\alpha} \succ \mathbf{x}^{\beta}$ or $\mathbf{x}^{\beta} \succ \mathbf{x}^{\alpha}$, and if both hold, we have $\boldsymbol{\alpha} = \boldsymbol{\beta}$.

2. If $\mathbf{x}^{\alpha} \succ \mathbf{x}^{\beta}$, then $\mathbf{x}^{\gamma}\mathbf{x}^{\alpha} \succ \mathbf{x}^{\gamma}\mathbf{x}^{\beta}$.

3. The ordering $\succ$ is a *well-ordering*: Every family of monomials in $\mathbb{K}[x_1, \ldots, x_n]$ has a smallest element with respect to $\succ$.

It should be noted that in our situation, the third condition is equivalent to saying that $\mathbf{x}^{\alpha} \succ \mathbf{x}^{\mathbf{0}}$ for all $\boldsymbol{\alpha} \in \mathbb{Z}_+^n$. (Exercise!) In fact, we have another nice lemma. The easy proof is left as an exercise.

Lemma 10.4.2. *An ordering $\succ$ of monomials (equivalently, of exponent vectors) is a* well-ordering *if and only if any strictly decreasing sequence of monomials terminates: $x^{\alpha_1} \succ x^{\alpha_2} \succ \cdots \succ x^{\alpha_k} = x^{\alpha_{k+1}}$.*

Examples of monomial orders are:

1. *Lexicographic order (or dictionary order):* We say that $\mathbf{x}^{\alpha} \succ_{\text{lex}} \mathbf{x}^{\beta}$ if the leftmost nonzero entry of $\boldsymbol{\alpha} - \boldsymbol{\beta}$ is positive.

For example, $x_1^4 \succ_{\text{lex}} x_2^{10000}$ because $(4,0,0)-(0,10000,0)=(4,-10000,0)$, and the leftmost entry is positive. Similarly $x_1^4 x_3^{11} \succ_{\text{lex}} x_1^3 x_2^{11} x_3^{11}$, because the leftmost entry of $(4,0,11)-(3,11,11)=(1,-11,0)$ is positive.

2. *Graded lexicographic order:* We say that $\mathbf{x}^{\boldsymbol{\alpha}} \succ_{\text{deglex}} \mathbf{x}^{\boldsymbol{\beta}}$ if $\mathbf{1}^\top\boldsymbol{\alpha} > \mathbf{1}^\top\boldsymbol{\beta}$ (that is, the *total degree* of $\mathbf{x}^{\boldsymbol{\alpha}}$ is greater than the total degree of $\mathbf{x}^{\boldsymbol{\beta}}$), or if $\mathbf{1}^\top\boldsymbol{\alpha} = \mathbf{1}^\top\boldsymbol{\beta}$ and the tie is broken by the lexicographic order $\mathbf{x}^{\boldsymbol{\alpha}} \succ_{\text{lex}} \mathbf{x}^{\boldsymbol{\beta}}$.

   For example, $x_2^{10000} \succ_{\text{deglex}} x_1^4$ because the total degree of the first monomial is greater. Similarly $x_3^{100} \succ_{\text{deglex}} x_1^4 x_2 x_3^3 \succ_{\text{deglex}} x_1^5 x_2 x_3 > x_1^6$.

3. *Graded reverse lexicographic order:* We say that $\mathbf{x}^{\boldsymbol{\alpha}} \succ_{\text{degrevlex}} \mathbf{x}^{\boldsymbol{\beta}}$ if $\mathbf{1}^\top\boldsymbol{\alpha} > \mathbf{1}^\top\boldsymbol{\beta}$, or if $\mathbf{1}^\top\boldsymbol{\alpha} = \mathbf{1}^\top\boldsymbol{\beta}$ and the rightmost nonzero entry of $\boldsymbol{\alpha}-\boldsymbol{\beta}$ is negative.

   For example, $x_1^4 x_2^7 z \succ_{\text{degrevlex}} x_1^4 x_2^2 z^3$ and $x_1 x_2^5 x_3^2 \succ_{\text{degrevlex}} x_1^4 x_2 x_3^3$.

Instead of writing $\mathbf{x}^{\boldsymbol{\alpha}} \succ \mathbf{x}^{\boldsymbol{\beta}}$, we often simply write $\boldsymbol{\alpha} \succ \boldsymbol{\beta}$, since we can understand $\succ$ as an order on $\mathbb{Z}_+^n$, the space of exponent vectors. Thus the exponent vectors are used for establishing the order of monomials. Exponent vectors are lattice points, and many authors have described term orders using discrete geometry. See [317] and the references therein.

It will be important to be able specify any possible monomial order. This can be achieved by using the *matrix term order*.

Definition 10.4.3. Let W be an $r\times n$ matrix with nonnegative integer coordinates, of rank r. We define the monomial order $\succ_W$ in $\mathbb{K}[x_1,\dots,x_n]$ by saying that $\mathbf{x}^{\boldsymbol{\alpha}} \succ_W \mathbf{x}^{\boldsymbol{\beta}}$ if and only if $W\boldsymbol{\alpha} \succ_{\text{lex}} W\boldsymbol{\beta}$.

Theorem 10.4.4. *Every term order can be obtained as a matrix term order.*

Example 10.4.5. If $w_1 = \mathbf{1}$, then we have an ordering that is graded with respect to the total degree. The graded lexicographic ordering $\succ_{\text{deglex}}$ is then given by the matrix

$$W = \begin{pmatrix} 1 & 1 & \cdots & 1 & 1 \\ 1 & 0 & \cdots & 0 & 0 \\ 0 & 1 & \cdots & 0 & 0 \\ & & \ddots & & \\ 0 & 0 & \cdots & 1 & 0 \end{pmatrix}.$$

Let us now turn to the multivariate division algorithm. Recall that when $f,g\in\mathbb{K}[x]$, we can always write $f = qg + r$ with $q,r\in\mathbb{K}[x]$ such that $\deg(r) < \deg(g)$ or $r = 0$. Now we wish to extend this to many polynomials and many indeterminates. First, let us extend some notation from the univariate to the multivariate situation.

Definition 10.4.6. We use the notation $\mathrm{LT}_\succ(g)$ to denote the leading term (largest term) of g with respect to the monomial order $\succ$. When there is no possibility for confusion, we remove the subscript $\succ$ and write simply $\mathrm{LT}(g)$. Analogously, we define $\mathrm{LC}_\succ(g)$ and $\mathrm{LM}_\succ(g)$ as the leading coefficient (coefficient of the leading term) and the leading monomial (leading term without coefficient).

Theorem 10.4.7. *Given polynomials $f, g_1,\dots,g_k \in \mathbb{K}[x_1,\dots,x_n]$ and a monomial order $\succ$, there exist $q_1,\dots,q_k\in\mathbb{K}[x_1,\dots,x_n]$, and a remainder $r\in\mathbb{K}[x_1,\dots,x_n]$ such that*

$$f = g_1q_1 + \cdots + g_kq_k + r,$$

and for the remainder polynomial r none of its monomials is divisible by any of the leading monomials of $g_1,\dots,g_k$.

We find the polynomials q_i and r by using a *multivariate division algorithm* which extends both the univariate division algorithm and Gaussian elimination. Here are two examples to illustrate the procedure.

Example 10.4.8. Let us divide $f=xy^2+1$ by $g_1=xy+1, g_2=y+1$ using lexicographic ordering with $x \succ_{\text{lex}} y$.

$$
\begin{array}{r|l}
 & q_1 = y \\
 & q_2 = -1 \\
\hline
xy+1 & xy^2+1 \\
y+1 & -(xy^2+y) \\
\hline
 & 1-y \\
 & -(-1-y) \\
\hline
 & 2
\end{array}
$$

Thus, we see that $r=2$.

Example 10.4.9. Now we divide $f=xy^2+1$ by $g_2=xy+1, g_1=y+1$ using lexicographic ordering with $x \succ_{\text{lex}} y$.

$$
\begin{array}{r|l}
 & q_1 = 0 \\
 & q_2 = xy-x \\
\hline
y+1 & xy^2+1 \\
xy+1 & -(xy^2+xy) \\
\hline
 & -xy+1 \\
 & -(-xy-x) \\
\hline
 & x+1
\end{array}
$$

Thus, we see that $r=x+1$.

Below is the pseudocode for the multivariate division algorithm.

ALGORITHM 10.3. Multivariate division algorithm.

input nonzero polynomials $f,g_1,\dots,g_k \in \mathbb{K}[x_1,\dots,x_n]$, monomial order $\succ$.
output polynomials $q_1,\dots,q_k,r$ that satisfy the conditions of Theorem 10.4.7.
Set $q_1 \leftarrow q_2 \leftarrow \cdots \leftarrow q_k \leftarrow r \leftarrow 0$.
Set $p \leftarrow f$.
while $p \neq 0$ **do**
    $i \leftarrow 1$.
    division_occurred $\leftarrow$ false.
    while $i \leq k$ and division_occurred $=$ false **do**
        if $\mathrm{LT}(g_i) \mid \mathrm{LT}(p)$ **then**
            $q_i \leftarrow q_i + \dfrac{\mathrm{LT}(p)}{\mathrm{LT}(q_i)}$.
            $p \leftarrow p - \dfrac{\mathrm{LT}(p)}{\mathrm{LT}(q_i)} \cdot g_i$.

```
            division_occurred ← true.
        else
            i ← i + 1.
    if division_occurred = false then
        r ← r + LT(p).
        p ← p − LT(p).
return (q_1, ..., q_k, r).
```

Proposition 10.4.10. *The multivariate division algorithm terminates and is correct.*

Proof. Let us first show termination of the algorithm. The code contains two while loops. $\mathrm{LT}(p)$ decreases in each iteration of the outer loop because either $\mathrm{LT}(p)$ is canceled from p or $\mathrm{LT}(p)$ is added to the remainder r. In either case the leading monomial $\mathrm{LM}(p)$ decreases (with respect to $\succ$) at each iteration. However, as $\succ$ is a term order, every strictly decreasing sequence of monomials must terminate. Thus, the algorithm must reach $p = 0$ after finitely many steps.

To show correctness, we claim that $f = q_1 g_1 + \cdots + q_k g_k + r$ is satisfied at every iteration of the algorithm. We know the code is correct before the outer while loop of the algorithm starts because at that moment $f = 0 + \cdots + 0 + p$ and $p = f$ from the initialization. Next we have two cases:

Case 1: There exists some g_i such that $\mathrm{LT}(g_i) \mid \mathrm{LT}(p)$. Then we have to show that

$$f = q_1^{\mathrm{new}} g_1 + \cdots + q_i^{\mathrm{new}} g_i + \cdots + q_s^{\mathrm{new}} g_s + r^{\mathrm{new}} + p^{\mathrm{new}}. \tag{10.1}$$

Clearly, $r^{\mathrm{new}} = r^{\mathrm{old}}$, $q_j^{\mathrm{new}} = q_j^{\mathrm{old}}$ for $j \neq i$, and

$$q_i^{\mathrm{new}} g_i = (q_i^{\mathrm{old}} + \mathrm{LT}(p^{\mathrm{old}})/\mathrm{LT}(g_i)) \cdot g_i = q_i^{\mathrm{old}} g_i + (\mathrm{LT}(p^{\mathrm{old}})/\mathrm{LT}(g_i)) \cdot g_i.$$

Moreover, we have $p^{\mathrm{new}} = p^{\mathrm{old}} - (\mathrm{LT}(p^{\mathrm{old}})/\mathrm{LT}(g_i)) \cdot g_i$ and therefore we get $q_i^{\mathrm{new}} g_i + p^{\mathrm{new}} = q_i^{\mathrm{old}} g_i + p^{\mathrm{old}}$. Substituting this into the right-hand side of (10.1) we obtain f, as claimed.

Case 2: No division occurred, that is, $\mathrm{LT}(g_i) \nmid \mathrm{LT}(p)$ for all i. Thus, we have $q_i^{\mathrm{new}} = q_i^{\mathrm{old}}$ for all i, $r^{\mathrm{new}} = r^{\mathrm{old}} + \mathrm{LT}(p^{\mathrm{old}})$, and $p^{\mathrm{new}} = p^{\mathrm{old}} - \mathrm{LT}(p^{\mathrm{old}})$. We conclude that $r^{\mathrm{new}} + p^{\mathrm{new}} = r^{\mathrm{old}} + p^{\mathrm{old}}$. Substituting this into the right-hand side of (10.1) we again obtain f, as claimed. □

Clearly if the remainder of the division algorithm is zero, then the polynomial being divided must belong to the ideal generated by the g_is, but is the converse true? No, it is possible to obtain a nonzero remainder for f even if $f \in \langle g_1, \ldots, g_s \rangle$.

Example 10.4.11. Let $f = xy^2 - x$ and divide it by $\{xy+1, y^2-1\}$. Then $q_1 = y$ with $r = -x - y$. Or, if we divide f by $\{y^2-1, xy-1\}$, then $q_1 = x$ with $r = 0$.

We need better generators for an ideal to avoid such an unpleasant situation. But first we will reduce the problem to easier ideals.

Definition 10.4.12. An ideal $I \subseteq \mathbb{K}[x_1, \ldots, x_n]$ is a *monomial ideal* if there is a subset A (not necessarily finite) of monomials such that $I = \langle \mathbf{x}^{\alpha} : \alpha \in A \rangle$.

Monomial ideals are special because they have a special membership test, demonstrated in the following lemma.

Lemma 10.4.13. *Let $I = \langle \mathbf{x}^\alpha : \alpha \in A \rangle$ be a monomial ideal. A monomial $\mathbf{x}^\beta$ belongs to I if and only if $\mathbf{x}^\beta$ is divisible by some $\mathbf{x}^\alpha, \alpha \in A$.*

Proof. ($\Rightarrow$) If $\mathbf{x}^\beta \in I$, where $\mathbf{x}^\beta = \sum h_i \mathbf{x}^{\alpha(i)}$, then every (noncanceling) term is divisible by some $\mathbf{x}^{\alpha(i)}$. Thus $\mathbf{x}^\beta$ is divisible by some $\mathbf{x}^\alpha$, as desired.

($\Leftarrow$) The converse is trivial. □

Let us consider the following example.

Example 10.4.14. Let us decide whether $f = x^4 - z + y$ belongs to $I = \langle x^3y, x^2z, x^2 \rangle$. x^4 is divisible by x^2 while z is not divisible by any monomial generator of I. Thus f is not in I.

Lemma 10.4.15. *Let $I \subseteq \mathbb{K}[x_1, \dots, x_n]$ be a monomial ideal and f be a polynomial. Then the following are equivalent:*

1. *$f \in I$.*
2. *Each term of f is in I.*
3. *f is a $\mathbb{K}$-linear combination of monomials in I.*

Corollary 10.4.16. *Two monomial ideals are equal if and only if they contain the same monomials.*

The following is just a restatement of the set version of the Gordan–Dickson lemma (Lemma 2.5.2) in algebraic terms.

Lemma 10.4.17 (Dickson's lemma). *Let $I = \langle \mathbf{x}^\alpha : \alpha \in A \rangle$ be a monomial ideal. Every monomial ideal is finitely generated; more precisely, there exists a finite set $B \subseteq A$ such that $I = \langle \mathbf{x}^\alpha : \alpha \in A \rangle = \langle \mathbf{x}^\alpha : \alpha \in B \rangle$.*

Definition 10.4.18. For any ideal $I = \langle f_1, f_2, \dots, f_n \rangle \subseteq \mathbb{K}[x_1, \dots, x_n]$, denoted by $\mathrm{LT}_\succ(I)$ the set of (infinitely many) *leading monomials* of the polynomials in the ideal I, that is, $\mathrm{LT}_\succ(I) = \{\mathrm{LT}_\succ(f) : f \in I\}$. As always, we remove the subscript $\succ$ when there is no confusion as to which monomial order we are using.

Lemma 10.4.19. *Let I be an ideal in $\mathbb{K}[x_1, \dots, x_n]$ and let $G \subseteq I$ be a finite subset such that $\langle \mathrm{LT}(G) \rangle = \langle \mathrm{LT}(I) \rangle$. Then $\langle G \rangle = I$.*

Proof. Clearly, $\langle G \rangle \subseteq I$. To show the other inclusion, choose any $f \in I$ and apply the division algorithm to divide f by the elements in G. That implies that

$$f = q_1 g_1 + q_2 g_2 + \cdots + q_s g_s + r.$$

If $r \neq 0$, then $r \in I$ implies that $\mathrm{LT}(r) \in \langle \mathrm{LT}(I) \rangle = \langle \mathrm{LT}(G) \rangle$. Therefore, $\mathrm{LT}(r)$ is divisible by $\mathrm{LT}(g)$ for some $g \in G$, in contradiction to r being the remainder by division by G. Consequently, we must have $r = 0$ and thus $f \in \langle G \rangle$, as claimed. □

Corollary 10.4.20 (Hilbert's basis theorem). *Every ideal $I \subseteq \mathbb{K}[x_1,\dots,x_n]$ is finitely generated.*

Proof. Let I be an ideal. We associate it to $\langle \mathrm{LT}_{\succ}(f) : f \in I \rangle$ for some monomial order $\succ$. By Lemma 10.4.17, there exists a finite set $g_1, g_2, \dots, g_s \in I$ such that

$$\langle \mathrm{LT}_{\succ}(g_1), \dots, \mathrm{LT}_{\succ}(g_s)\rangle = \langle\, \mathrm{LT}_{\succ}(f) : f \in I \rangle.$$

By Lemma 10.4.19, we have $I = \langle g_1, g_2, \dots, g_s\rangle$, that is, I is finitely generated. □

The following two corollaries are immediate consequences of Hilbert's basis theorem.

Corollary 10.4.21. *Every ascending chain of ideals $I_1 \subseteq I_2 \subseteq \cdots$ eventually stabilizes, that is, $I_n = I_{n+1} = \cdots$ holds after some index n.*

Corollary 10.4.22. *For every ideal $I \subseteq \mathbb{K}[x_1,\dots,x_n]$, there exist finitely many polynomials $f_1,\dots,f_s \in I$ such that $V(I) = V(f_1,\dots,f_s)$.*

10.5 Gröbner bases and Buchberger's algorithm

Finally, we are in the position to identify special generating sets of an ideal that allow us to decide ideal membership by the multivariate division algorithm.

Definition 10.5.1. Given a monomial order $\succ$, a *Gröbner basis* of an ideal $I \subseteq \mathbb{K}[x_1,\dots,x_n]$ with respect to $\succ$ is a finite set of polynomials $\{\, f_1,\dots,f_n \,\} \subseteq I$ with the property that $\langle \mathrm{LT}_{\succ}(I)\rangle = \langle \mathrm{LT}_{\succ}(f_1),\dots,\mathrm{LT}_{\succ}(f_n)\rangle$.

Example 10.5.2. Let $I = \langle x^3 - 2xy, x^2y - 2y^2 + x\rangle$ consider the monomial order $\succ_{\mathrm{deglex}}$. Is $G = \{g_1, g_2\} := \{x^3 - 2xy, x^2y - 2y^2 + x\}$ a Gröbner basis of I with respect to $\succ_{\mathrm{deglex}}$? Let us look at the polynomial

$$x^2 = -yg_1 + xg_2 = -y(x^3 - 2xy) + x(x^2y - 2y^2 + x).$$

Thus, $x^2 \in \mathrm{LT}_{\succ_{\mathrm{deglex}}}(I)$ but $x^2 \notin \langle \mathrm{LT}_{\succ_{\mathrm{deglex}}}(g_1), \mathrm{LT}_{\succ_{\mathrm{deglex}}}(g_2)\rangle = \langle x^3 - 2xy, x^2y\rangle$. This implies that $\{g_1, g_2\}$ is a basis for I, but it is not a Gröbner basis for I with respect to $\succ_{\mathrm{deglex}}$. However, if we add $g_3 = x^2$, $g_4 = 2xy$, and $g_5 = -2y^2 + x$ (all belonging to I) to G, then $G = \{g_1, g_2, g_3, g_4, g_5\}$ is a Gröbner basis of I with respect to $\succ_{\mathrm{deglex}}$.

Proposition 10.5.3. *Let $\{\, f_1,\dots,f_k \,\}$ be a Gröbner basis for the ideal $I = \langle f_1,\dots,f_k\rangle$ with respect to the monomial order $\succ$. Let g be any polynomial; then there exists a unique polynomial r such that if we apply the division algorithm, we get:*

(a) $g = \alpha_1 f_1 + \alpha_2 f_2 + \cdots + \alpha_k f_k + r$,

(b) No term of r is divisible by $\mathrm{LT}_{\succ}(f_i)$.

Be aware that although the remainder r is unique, the α_i's are not!

Proof. We can find $\alpha_1, \alpha_2, \dots, \alpha_k$ and r by the division algorithm. To show that r is unique, assume on the contrary that we may obtain two different remainders r_1, r_2 when performing the division algorithm with different ordering of the dividing polynomials. Being remainders upon division by $\{ f_1,\dots,f_k \}$, none of the terms of r_1 or r_2 are divisible by any of the monomials $\mathrm{LT}_{\succ}(f_1),\dots,\mathrm{LT}_{\succ}(f_k)$. However, we have:

$$\begin{aligned} r_1 &= g-(\alpha_1 f_1+\alpha_2 f_2+\cdots+\alpha_k f_k),\\ r_2 &= g-(\beta_1 f_1+\beta_2 f_2+\cdots+\beta_k f_k),\\ r_1-r_2 &= (\beta_1-\alpha_1)f_1+(\beta_2-\alpha_2)f_2+\cdots+(\beta_k-\alpha_k)f_k. \end{aligned}$$

This implies that $r_1-r_2\in I$. Because $\{ f_1,\dots,f_k \}$ is a Gröbner basis of I with respect to $\succ$, we conclude that either $r_1-r_2=0$ or $\mathrm{LT}_{\succ}(r_1-r_2)$ is divisible by $\mathrm{LT}_{\succ}(f_i)$ for some i. As we have shown above that the latter is impossible, we conclude that $r_1-r_2=0$, and thus the remainder upon division by $\{ f_1,\dots,f_k \}$ is unique. □

Theorem 10.5.4. *Let $G=\{ f_1,\dots,f_k \}$ be a Gröbner basis for an ideal $I\subseteq\mathbb{K}[x_1,\dots,x_n]$ and $f\in\mathbb{K}[x_1,\dots,x_n]$. Then $f\in I$ if and only if the remainder of f upon division by $f_1,\dots,f_k$ is zero.*

Proof. ($\Leftarrow$) If the remainder is zero, f must be in I, since we may write it as a combination of elements of the G, by the division algorithm.

($\Rightarrow$) If $f\in I$, then f can be written as a combination of elements of G. But by the uniqueness of the remainder, dividing f by G always yields a zero as remainder. □

This means, that once we know a Gröbner basis of the ideal I with respect to any monomial ordering $\succ$, then we can decide ideal membership using the division algorithm. But how can we check whether a given set $\{ f_1,\dots,f_s \}$ is a Gröbner basis of $\langle f_1,\dots,f_s\rangle$ with respect to $\succ$? And, if it is not a Gröbner basis, how can we compute one?

First, let us observe the following nice corollary to Theorem 10.5.

Corollary 10.5.5. *For an ideal I, $G\subseteq I$ is a Gröbner basis of I with respect to $\succ$ if and only if* remainder$(f,G)=0$ *for all $f\in I$.*

Unfortunately, this corollary does not provide us with a practical tool to decide whether G is a Gröbner basis, since we have to test for infinitely many polynomials f whether remainder$(f,G)=0$. In 1965, Bruno Buchberger proved that only finitely many "critical polynomials" $f\in I$, so-called *S-polynomials*, have to be tested.

Definition 10.5.6. Given two polynomials f,g and a monomial order $\succ$, the *S-polynomial* of f and g is

$$S(f,g)=\left(\frac{\mathrm{lcm}(LM_{>}(f),\mathrm{LM}_{\succ}(g))}{\mathrm{LT}_{\succ}(f)}\right)f-\left(\frac{\mathrm{lcm}(\mathrm{LM}_{\succ}(f),\mathrm{LM}_{\succ}(g))}{\mathrm{LT}_{\succ}(g)}\right)g.$$

Example 10.5.7. For example, if $f=x^3y^2-x^2y^3$ and $g=3x^4y+y^2$, let us use the graded lexicographic order with x bigger than y, and let us compute $S(f,g)$. We know

$\text{lcm}(x^3y^2, x^4y) = x^4y^2$. Thus, we get

$$\begin{aligned} S(f,g) &= \left(\frac{x^4y^2}{x^3y^2}\right)(x^3y^2 - x^2y^3) - \left(\frac{x^4y^2}{3x^4y}\right)(3x^4y + y^2) \\ &= x(x^3y^2 - x^2y^3) - \left(\frac{y}{3}\right)(3x^4y + y^2) \\ &= x^4y^2 - x^3y^3 - x^4y^2 - \frac{y^3}{3} \\ &= -x^3y^3 - \frac{y^3}{3}. \end{aligned}$$

Now we can state and prove Buchberger's theorem.

Theorem 10.5.8 (Buchberger's theorem). *A finite set of polynomials $G = \{g_1, g_2, \ldots, g_s\}$ is a Gröbner basis for the ideal $\langle G \rangle$ with respect to a given monomial order $\succ$ if and only if the remainder of any S-polynomial $S(g_i, g_j)$ by division with G is zero for all pairs (i, j). (This is called* Buchberger's S-pair criterion.*)*

Example 10.5.9. Take $G = \{y - x^2, z - x^3\}$. We claim that G is a Gröbner basis for the ideal $\langle G \rangle$ with respect to lexicographic order with $y \succ_{\text{lex}} z \succ_{\text{lex}} x$. First, we compute the S-polynomial

$$S(y - x^2, z - x^3) = \frac{yz}{y}(y - x^2) - \frac{yz}{z}(z - x^3) = yx^3 - zx^2,$$

and then we divide by G:

$$\begin{array}{r|l} & q_1 = x^3 \\ & q_2 = -x^2 \\ \hline y - x^2 & yx^3 - zx^2 \\ z - x^3 & yx^3 - x^5 \\ \hline & x^5 - zx^2 \\ & x^5 - zx^2 \\ \hline & 0 \end{array}$$

So we have shown that indeed G is a Gröbner basis!

We will later prove this theorem for the special case of lattice ideals. For a proof of this general statement, see, for example, [78].

Interestingly, Theorem 10.5.8 can be easily turned into an algorithm: given $I = \langle F \rangle = \langle f_1, \ldots, f_s \rangle$ and a monomial order $\succ$, compute the remainder of all S-polynomials by division with F. If all remainders are zero, we are done. F is a desired Gröbner basis. If there are nonzero remainders, we put them into F and repeat the process until all remainders turn out to be zero. Then we have found a Gröbner basis $\langle g_1, \ldots, g_t \rangle$. Unfortunately, it is possible that $t \gg s$. Moreover, the degrees of the polynomials g_i may grow very large. We introduce Buchberger's algorithm below and consider its shortcomings afterwards.

ALGORITHM 10.4. Buchberger's algorithm.

1: **input** $F = \{f_1, \ldots, f_s\} \in \mathbb{K}[x_1, \ldots, x_n]$, monomial order $\succ$.
2: **output** $G = \{g_1, \ldots, g_t\}$, a Gröbner basis for $\langle F \rangle$ with respect to $\succ$.

$G \leftarrow F$.
$C \leftarrow \{(f_i, f_j) : f_i, f_j \in G, f_i \neq f_j\}$.
while $C \neq \emptyset$ **do**
  Choose $(f_i, f_j) \in C$.
  $C \leftarrow C \setminus (f_i, f_j)$.
  $h \leftarrow$ remainder($S(f_i, f_j), G$).
  if $h \neq 0$ **then**
    $C \leftarrow C \cup \{(f_i, h) : f_i \in G\}$.
    $G \leftarrow G \cup h$.
return G.

Example 10.5.10. Consider $G = \{f_1, f_2, f_3\}$, where $f_1 = y^2 + yx + x^2$, $f_2 = y + x$, and $f_3 = y$. Moreover, let us use $\succ_{\text{lex}}$ where $y \succ_{\text{lex}} x$. Then we have $C = \{(f_1, f_2), (f_1, f_3), (f_2, f_3)\}$.

- First pass through the while loop:

$$\begin{aligned} S(f_1, f_2) &= \frac{y^2}{y^2} f_1 - \frac{y^2}{y} f_2 = (y^2 + yx + x^2) - y(y + x) = x^2, \\ f_4 &:= \text{remainder}(S(f_1, f_2, G)) = x^2, \\ G &:= \{f_1, f_2, f_3, f_4\}, \\ C &:= \{(f_1, f_3), (f_2, f_3), (f_1, f_4), (f_2, f_4), (f_3, f_4)\}. \end{aligned}$$

- Second pass through the while loop:

$$S(f_1, f_3) = \frac{y^2}{y^2} f_1 - \frac{y^2}{y} f_3 = y^2 + xy + x^2 - y^2 = xy + x^2,$$

$$\begin{array}{r|l} & q_2 = x \\ \hline f_1 = y^2 + yx + x^2 & xy + x^2 \\ f_2 = y + x & xy + x^2 = x f_2 \\ \cline{2-2} f_3 = y & \quad 0 \\ f_4 = x^2 & \end{array} ,$$

$$\text{remainder}(S(f_1, f_3), G) = 0.$$

- Third pass through the while loop:

$$\begin{aligned} S(f_2, f_3) &= \frac{y}{y}(y + x) - \frac{y}{y} y = y + x - y = x, \\ f_5 &:= \text{remainder}(S(f_2, f_3), G) = x, \\ G &:= \{f_1, \ldots, f_5\}, \\ C &:= \{(f_1, f_4), (f_2, f_4), (f_3, f_4), (f_1, f_5), (f_2, f_5), (f_3, f_5), (f_4, f_5)\}. \end{aligned}$$

Since now $x, y \in G$, all subsequent remainders are zero, and thus the algorithm terminates and returns $G = \{f_1, \ldots, f_5\}$ as a desired Gröbner basis.

By Buchberger's S-pair criterion in Theorem 10.5.8, we know that upon termination, Buchberger's algorithm indeed returns a correct answer. To see that it terminates, observe that whenever $h :=$ remainder$(S(f_i, f_j, G)) \neq 0$, then $\mathrm{LT}(h) \notin \langle \mathrm{LT}(G) \rangle$. Consequently, we have $\langle \mathrm{LT}(G) \rangle \subsetneq \langle \mathrm{LT}(G \cup \{h\}) \rangle$ whenever a new element h has to be added to G. Hence, if the algorithm did not terminate, the ideal $\langle \mathrm{LT}(G) \rangle$ would be part of an infinite ascending chain of ideals, which is impossible by Corollary 10.4.21.

Note that Buchberger's algorithm often returns an unnecessarily large set of polynomials. Moreover, it may return different Gröbner bases for the same ideal and the same monomial order $\succ$. In the following, we reduce the size of the Gröbner basis and indeed end up with a unique smallest Gröbner basis of the given ideal with respect to $\succ$.

Definition 10.5.11. A *reduced Gröbner basis* for a polynomial ideal I is a Gröbner basis G such that for all $p \in G$ the leading coefficient is 1 and no monomial of p lies in $\langle \mathrm{LT}(G \setminus \{p\}) \rangle$.

Proposition 10.5.12. *Given an ideal I and a monomial order $\succ$, then I has a unique reduced Gröbner basis with respect to $\succ$. Two ideals are the same if and only if they have the same reduced Gröbner basis with respect to $\succ$.*

Gröbner bases have many applications. Let us conclude by listing some of them.

Ideal membership

Problem. Let $I = \langle f_1, \dots, f_k \rangle \subseteq \mathbb{K}[x_1, \dots, x_n]$ be given. Decide for $h \in \mathbb{K}[x_1, \dots, x_n]$ whether or not $h \in I$.

Solution. Compute any Gröbner basis G for I. Then $h \in I$ if and only if remainder$(h, G) = 0$.

Elimination of variables

Many applications rely on the fact that Gröbner bases allow us to eliminate variables and to compute the so called *elimination ideals*.

Definition 10.5.13. Given an ideal $I = \langle f_1, f_2, \dots, f_s \rangle \subseteq \mathbb{K}[x_1, \dots, x_n]$, the j-th elimination ideal of I is defined as

$$I_j = I \cap K[x_{j+1}, x_{j+2}, \dots, x_n].$$

Theorem 10.5.14 (elimination theorem). *Let $I \subseteq \mathbb{K}[x_1, \dots, x_n]$ be an ideal and let G be a Gröbner basis of I with respect to lexicographic order with $x_1 \succ_{lex} x_2 \succ_{lex} \cdots \succ_{lex} x_n$. Then*

$$G_j = G \cap \mathbb{K}[x_{j+1}, x_{j+2}, \dots, x_n]$$

is a Gröbner basis for the j-th elimination ideal.

Proof. Relabeling the elements of G if necessary, we can assume the first r elements $g_1, g_2, \dots, g_r$ are in G_j. To check that G_j is a Gröbner basis, we have to show that remainder$(S(g_i, g_l), G) = 0$ for $1 \leq i, l \leq r$.

1. Observe that $S(g_i, g_l)$ lies in $\mathbb{K}[x_{j+1}, \dots, x_n]$.

2. Since we used a lexicographic order with $x_1 \succ_{\text{lex}} \cdots \succ_{\text{lex}} x_n$, the leading terms of $g_{r+1}, \ldots, g_m$ must involve other variables from $\{x_1, \ldots, x_j\}$. Thus, when we apply the division algorithm to compute remainder($S(g_i, g_l), G$) with $1 \leq i, l \leq r$, the polynomials $g_{r+1}, \ldots, g_m$ will not be used.

3. Thus, remainder($S(g_i, g_l), G_j$) $= 0$. □

Ideal intersection

Problem. Let two ideals $I = \langle f_1, \ldots, f_r \rangle, J = \langle g_1, \ldots, g_s \rangle \subseteq \mathbb{K}[x_1, \ldots, x_n]$ be given. Compute $I \cap J$.

Solution. Using an additional indeterminate z, we compute

$$I \cap J = \langle z f_1, \ldots, z f_r, (1-z) g_1, \ldots, (1-z) g_s \rangle \cap \mathbb{K}[x_1, \ldots, x_n]$$

using a lexicographic ordering with z biggest (which thus eliminates z).

Solvability of polynomial equations

Problem. Given $f_1, \ldots, f_m \in \mathbb{K}[x_1, \ldots, x_n]$, does there exist some $\mathbf{a} \in \overline{\mathbb{K}}^n$ with $f_1(\mathbf{a}) = \cdots = f_m(\mathbf{a}) = 0$?

Solution. Compute the reduced Gröbner basis G of $\langle f_1, \ldots, f_m \rangle$ with respect to any term order. Then Hilbert's Nullstellensatz tells us that the above system has no solution if and only if $G = \{1\}$.

Example 10.5.15. Consider the following system of equations over $\mathbb{C}$.

$$\begin{aligned} x^2 + y^2 + z^2 &= 1, \\ x^2 + y^2 - y &= 0, \\ x - z &= 0. \end{aligned}$$

Using lexicographic order, we obtain $g_1 = x - z$, $g_2 = -y + 2z^2$, $g_3 = z^4 + \frac{z^2}{2} - \frac{1}{4}$ as the three elements in the reduced Gröbner basis of I. From g_3 we obtain $z = \pm\frac{1}{2}\sqrt{\pm\sqrt{5} - 1}$. For each possible value of z, we can now find the roots y of g_2, and so on.

Clearly if one is solving a system of equations using Gröbner bases via lexicographic computation one can select the real roots of the system starting from the real roots of a univariate polynomial. We have seen how to do this in earlier sections. Readers that still remember Descartes' rule of signs may ask: Can one also write a bound on the number of isolated real solutions of a system of n real polynomial equations in n variables that does not depend on the degrees of the given equations? Somehow the bound, one hopes, will only depend on the monomials present in the system. Indeed, such a bound exists although it is rather large and probably not tight. See the book [196].

Theorem 10.5.16 (Khovanskii). *For a system of n real polynomial equations in n variables, involving k distinct monomials in total, the number of isolated roots in the positive orthant $\mathbb{R}^n_+$ is*

$$2^{\frac{k(k-1)}{2}} (n+1)^k = 2^{\binom{k}{2}} (n+1)^k.$$

Equations and inequations

Problem. Does there exist some $\mathbf{a} \in \overline{\mathbb{K}}^n$ satisfying $f_1(\mathbf{a}) = \cdots = f_m(\mathbf{a}) = 0$ and $g_1(\mathbf{a}) \neq 0, \ldots, g_l(\mathbf{a}) \neq 0$?

Solution. This is equivalent to the existence of $(\mathbf{a}, b) \in \overline{\mathbb{K}}^{n+1}$ satisfying

$$f_1(\mathbf{a}) = \cdots = f_m(\mathbf{a}) = g_1(\mathbf{a}) \cdots g_l(\mathbf{a}) \cdot b - 1 = 0.$$

Let z be a new variable and compute the Gröbner basis G for $\langle f_1, \ldots, f_m, g_1 \cdots g_l, z - 1 \rangle$. There exists no such solution if and only if $G = \{1\}$.

10.6 Notes and further references

Gröbner bases were invented by Bruno Buchberger during his Ph.D. research in order to solve the ideal membership problem for polynomial ideals [65]. Since then they have become a fundamental tool in computational algebra with applications in a large variety of area: statistics, combinatorics, optimization, computational biology, robotics, automated/geometric theorem proving, coding theory, to name just a few of them [66, 67]. There are quite a few excellent textbooks on Gröbner bases, each laying emphasis on different aspects of the topic depending on the intended readership; see, for example, [5, 47, 78, 207, 208].

The special structure of Gröbner bases are used as a basis for finding solutions of systems of polynomial equations. There are well-known eigenvalue methods for polynomial equations; see, e.g., [79, §2.4] and [115, 316]). The key idea for extracting solutions of zero-dimensional varieties is the fact that from the Gröbner basis G_I of an ideal I one can obtain a finite-dimensional representation of the vector space $\mathbb{K}[x_1, \ldots, x_n]/I$ and its multiplicative structure, where I is the ideal $\langle F \rangle$. In order to do this, we need to compute a basis of the vector space $\mathbb{K}[x_1, \ldots, x_n]/I$, and construct matrix representations for the multiplication operators

$$M_{x_i} : \mathbb{K}[x_1, \ldots, x_n]/I \to \mathbb{K}[x_1, \ldots, x_n]/I,$$

where $[f] \mapsto [x_i f]$ for all $[f] \in \mathbb{K}[x_1, \ldots, x_n]/I$. To do this one uses the fact that the *nonstandard monomials*, those monomials outside the ideal $\mathrm{LT}_{\succ}(I)$, can be read from the initial monomials of G_I. Recently, semidefinite programming has become quite important on the recovery of *real* solutions of equations; see, e.g., [219, 220].

In the following chapter, Chapter 11, we will consider Gröbner bases of very special ideals, namely *toric ideals*, and show how they can be employed as optimality certificates in integer programming.

10.7 Exercises

Exercise 10.7.1. Prove that there are $\binom{n+d-1}{d}$ monomials of degree d on n indeterminates.

Exercise 10.7.2. Show that the lexicographic order is a monomial order.

Exercise 10.7.3. Let $S = V(f_1, f_2, \ldots, f_k)$. Show that $\langle f_1, f_2, \ldots, f_k \rangle \subseteq I(S)$.

Exercise 10.7.4. Show that $I(F) \subseteq I(G)$ implies $V(G) \subseteq V(F)$.

Exercise 10.7.5. Let $f_1, f_2, \ldots, f_k \in \mathbb{K}[x]$ be univariate polynomials. Show that

$$\langle f_1, f_2, \ldots, f_k\rangle = \langle \gcd(f_1, f_2, \ldots, f_k)\rangle.$$

Consequently, we have for any $f \in \mathbb{K}[x]$ that $f \in \langle f_1, f_2, \ldots, f_k\rangle$ if and only if $\gcd(f_1, f_2, \ldots, f_k) \mid f$.

Exercise 10.7.6. Show that the third condition (well-ordering) for a monomial order $\succ$ can be replaced by $\mathbf{x}^{\boldsymbol{\alpha}} \succ \mathbf{x}^{\mathbf{0}}$ for all $\boldsymbol{\alpha} \in \mathbb{Z}_+^n$.

Exercise 10.7.7. Show that an order $\succ$ of monomials (equivalently of exponent vectors) is a *well-ordering* if and only if any strictly decreasing sequence of monomials terminates: $\mathbf{x}^{\alpha_1} \succ \mathbf{x}^{\alpha_2} \succ \cdots \succ \mathbf{x}^{\alpha_k} = \mathbf{x}^{\alpha_{k+1}}$.

Exercise 10.7.8. Compute the lex, graded lex, and graded reverse lex ordering for the following polynomial:

$$7xy^2z + 4x^3 + 4xyz^5 - 5y^4.$$

Exercise 10.7.9. Prove that every ascending chain of ideals $I_1 \subseteq I_2 \subseteq \cdots$ eventually stabilizes, that is, $I_n = I_{n+1} = \cdots$ holds after some index n.

Exercise 10.7.10. Prove that for every ideal $I \subseteq \mathbb{K}[x_1, \ldots, x_n]$, there exist finitely many polynomials $f_1, \ldots, f_s \in I$ such that $V(I) = V(f_1, \ldots, f_s)$.

Exercise 10.7.11. Show that the lexicographic order is a monomial order.

Exercise 10.7.12. Let $S = V(f_1, f_2, \ldots, f_k)$. Then $\langle f_1, f_2, \ldots, f_k\rangle \subseteq I(S)$.

Exercise 10.7.13. Let $f_1, f_2, \ldots, f_k \in \mathbb{K}[x]$ be univariate polynomials. Then

$$\langle f_1, f_2, \ldots, f_k\rangle = \langle \gcd(f_1, f_2, \ldots, f_k)\rangle.$$

Consequently, we have for any $f \in \mathbb{K}[x]$ that $f \in \langle f_1, f_2, \ldots, f_k\rangle$ if and only if $\gcd(f_1, f_2, \ldots, f_k) \mid f$.

Exercise 10.7.14. Compute the lex, graded lex, and graded reverse lex ordering for the following polynomial:

$$7xy^2z + 4x^3 + 4xyz^5 - 5y^4.$$

Chapter 11

Gröbner Bases in Integer Programming

We have seen the usefulness of Graver bases in integer linear programming in Chapter 3 and Chapter 4. A more refined concept is that of a *Gröbner basis*, which was introduced in the previous chapter as a key tool in computational algebraic geometry. In this chapter we come to reintroduce Gröbner bases in a framework closer to integer programming. We follow the presentations of [178, 324, 339] and most specially [155] where the fastest algorithms used today for toric Gröbner basis calculations are presented.

11.1 Toric ideals and their Gröbner bases

Polynomial ideals are important algebraic structures. We introduce a special kind of ideals for the solution of integer programming problems, so-called *toric ideals*.

Definition 11.1.1. Let $A = (\mathbf{a}_1, \ldots, \mathbf{a}_n) \subseteq \mathbb{Z}^{d\times n}$ be an integer matrix. The *toric ideal* I_A is the kernel of the polynomial map

$$\rho\colon \mathbb{K}[x_1,\ldots,x_n] \to \mathbb{K}[\mathbf{z}^{\mathbf{a}_1},\ldots,\mathbf{z}^{\mathbf{a}_n}] \subseteq \mathbb{K}[z_1^{\pm 1},\ldots,z_d^{\pm 1}]$$

induced by

$$x_i \mapsto \mathbf{z}^{\mathbf{a}_i}.$$

From $\ker(\rho) = I_A$ we obtain the quotient ring $\mathbb{K}[x_1,\ldots,x_n]/I_A$, the so-called *toric ring*.

Note that by definition of I_A, all binomials of the form $\mathbf{x}^{\mathbf{u}} - \mathbf{x}^{\mathbf{v}}$ with $\mathbf{u},\mathbf{v} \in \mathbb{Z}_+^n$ and $A\mathbf{u} = A\mathbf{v}$ belong to I_A. The following lemma shows that they do in fact generate the ideal I_A.

Lemma 11.1.2. *The toric ideal I_A is a $\mathbb{K}$-vector space spanned by the binomials*

$$\left\{\mathbf{x}^{\mathbf{u}} - \mathbf{x}^{\mathbf{v}} : A\mathbf{u} = A\mathbf{v},\ \mathbf{u},\mathbf{v} \in \mathbb{Z}_+^n\right\}$$

and therefore we have

$$I_A = \langle \mathbf{x}^{\mathbf{u}} - \mathbf{x}^{\mathbf{v}} : A\mathbf{u} = A\mathbf{v},\ \mathbf{u},\mathbf{v} \in \mathbb{Z}_+^n \rangle.$$

Proof. A binomial $\mathbf{x}^{\mathbf{u}} - \mathbf{x}^{\mathbf{v}}$ lies in I_A if and only if $A\mathbf{u} = A\mathbf{v}$. We need to show that each $f \in I_A$ is a $\mathbb{K}$-linear combination of these binomials. Fix a term order $\succ$ on $\mathbb{K}[x_1,\dots,x_n]$. Suppose there exists a polynomial $f \in I_A$ that cannot be written as a $\mathbb{K}$-linear combination of these binomials. Among all such polynomials choose one such that $\mathrm{LM}_{\succ}(f) = \mathbf{x}^{\mathbf{u}}$ is smallest with respect to $\succ$. As $f \in I_A$ and $\mathrm{LM}_{\succ}(f) = \mathbf{x}^{\mathbf{u}}$, we have

$$0 = f(\mathbf{z}^{\mathbf{a}_1},\dots,\mathbf{z}^{\mathbf{a}_n}) = \mathbf{z}^{A\mathbf{u}} + \text{other terms},$$

which implies that $\mathbf{z}^{A\mathbf{u}}$ must cancel during expansion. Thus, there exists some other monomial $\mathbf{x}^{\mathbf{v}}$ in f, clearly with $\mathbf{x}^{\mathbf{u}} \succ \mathbf{x}^{\mathbf{v}}$, such that $A\mathbf{v} = A\mathbf{u}$. Hence $f' := f - (\mathbf{x}^{\mathbf{u}} - \mathbf{x}^{\mathbf{v}})$ cannot be written as a linear combination of binomials either, because otherwise f could. As by construction $\mathrm{LM}_{\succ}(f) \succ \mathrm{LM}_{\succ}(f')$, we arrive at a contradiction for the minimality property of f.

The relation $I_A = \langle \mathbf{x}^{\mathbf{u}} - \mathbf{x}^{\mathbf{v}} : A\mathbf{u} = A\mathbf{v},\ \mathbf{u},\mathbf{v} \in \mathbb{Z}^n_+ \rangle$ is now an immediate consequence. □

Let us continue with some properties of I_A. First note that by Hilbert's basis theorem, Theorem 10.4.20, already finitely many binomials suffice to generate I_A. Note that we can restrict our attention to binomials $\mathbf{x}^{\mathbf{u}} - \mathbf{x}^{\mathbf{v}}$ with disjoint support, that is, with $\mathrm{supp}(\mathbf{u}) \cap \mathrm{supp}(\mathbf{v}) = \emptyset$. Assuming otherwise that $\gcd(\mathbf{x}^{\mathbf{u}},\mathbf{x}^{\mathbf{v}}) = \mathbf{x}^{\gamma}$, we could replace the generator $\mathbf{x}^{\mathbf{u}} - \mathbf{x}^{\mathbf{v}} \in I_A$ of I_A by $\mathbf{x}^{\mathbf{u}-\gamma} - \mathbf{x}^{\mathbf{v}-\gamma} \in I_A$ without enlarging the generated ideal. This simple observation allows us to encode such binomials $\mathbf{x}^{\mathbf{u}} - \mathbf{x}^{\mathbf{v}}$ (with disjoint support) by the vector $\mathbf{u} - \mathbf{v}$, as any vector $\mathbf{w} \in \ker(A) \cap \mathbb{Z}^n$ can be mapped *uniquely* back to a binomial with disjoint support, namely to $\mathbf{x}^{\mathbf{w}_+} - \mathbf{x}^{\mathbf{w}_-}$.

Example 11.1.3. Take $A = \begin{pmatrix} 1 & 1 & 1 & 1 \\ 0 & 1 & 2 & 3 \end{pmatrix}$. Then $\rho\colon \mathbb{K}[a,b,c,d] \to \mathbb{K}[x,y]$ and $I_A = \langle ac - b^2, bd - c^2, ad - bc, a^2d - b^3, ad^2 - c^3 \rangle$.

Interestingly, there is still no (comparably) simple way known to find such a finite generating set for a given matrix A. We will present an algorithm based on Gröbner basis techniques in Section 11.4 below.

It should be noted that Buchberger's algorithm to compute Gröbner bases (Algorithm 10.4) is *binomial friendly*. This means that polynomial division of a binomial by a set of binomials either returns 0 or a binomial and also the S-polynomial of two binomials is either 0 or a binomial. Consequently, the reduced Gröbner basis of a binomial ideal consists of binomials only. We obtain the following result:

Theorem 11.1.4. *For every term order $\succ$, there is a finite set of vectors $G_{\succ}$ in $\ker(A) \cap \mathbb{Z}^n$ such that the reduced Gröbner basis of I_A with respect to $\succ$ is $\{\mathbf{x}^{\mathbf{u}_+} - \mathbf{x}^{\mathbf{u}_-} : \mathbf{u} \in G_{\succ}\}$.*

11.2 Toric ideals and integer programming

Let us assume for the moment that we can compute a generating set for the toric ideal I_A for given matrix A and that we can therefore compute a Gröbner basis for I_A for any monomial ordering $\succ$ we are interested in. How could we use this to solve an integer linear program?

Let us consider the family of optimization problems

$$(\mathrm{IP})(\mathbf{b}) \qquad \min\left\{\mathbf{c}^\top \mathbf{x} : A\mathbf{z} = \mathbf{b},\ \mathbf{z} \in \mathbb{Z}^n_+\right\},$$

where $A \in \mathbb{Z}^{m\times n}$ is a fixed integer matrix, where $\mathbf{c} \neq \mathbf{0}$ is a fixed cost vector, and where the right-hand side vector $\mathbf{b} \in \mathbb{Z}^m$ is varying. The set of feasible solutions are the lattice points in the polyhedron

$$P_I(\mathbf{b}) := \operatorname{conv}\left\{\mathbf{z} : A\mathbf{z} = \mathbf{b},\ \mathbf{z} \in \mathbb{Z}^n_+\right\}.$$

Let us further assume that the linear program

$$(\mathrm{LP})(\mathbf{b}) \qquad \min\left\{\mathbf{c}^\top\mathbf{z} : A\mathbf{z} = \mathbf{b},\ \mathbf{z} \geq \mathbf{0}\right\}$$

is bounded for every $\mathbf{b}$. Clearly, this implies that the integer programs (IP)($\mathbf{b}$) are bounded, too. Note that, by Farkas' lemma, (LP)($\mathbf{b}$) is bounded for every $\mathbf{b}$ if and only if there exists a strictly positive vector $\mathbf{d}$ in the row-span of A. Such a vector $\mathbf{d}$ can be computed via a linear program. (How?) As $\mathbf{d}^\top\mathbf{z}$ is the same for all $\mathbf{z} \in P_I(\mathbf{b})$, the integer program (IP)($\mathbf{b}$) is equivalent to the integer program

$$\min\left\{(\mathbf{c} + \lambda\mathbf{d})^\top\mathbf{z} : A\mathbf{z} = \mathbf{b},\ \mathbf{z} \in \mathbb{Z}^n_+\right\},$$

for any $\lambda \in \mathbb{R}$. Choosing λ big enough, we can assume from here on that our cost vector $\mathbf{c}$ is nonnegative (or even positive) in every component. Moreover, for similar reasons, we may also assume that the matrix A contains only nonnegative/positive entries. If the cost vector $\mathbf{c}$ is not generic, two integer points in P_I may have the same objective function value. To avoid this, we refine the order given by the objective function to a monomial order. We choose any monomial order $\succ$ (for example, a lexicographic order) and use it as a tiebreaker in case the objective function values agree:

$$\mathbf{y} \succ_{\mathbf{c}} \mathbf{z} \Longleftrightarrow \begin{cases} \mathbf{c}^\top\mathbf{y} > \mathbf{c}^\top\mathbf{z}, & \text{or} \\ \mathbf{c}^\top\mathbf{y} = \mathbf{c}^\top\mathbf{z} & \text{and } \mathbf{y} \succ \mathbf{z}. \end{cases}$$

Note that $\succ_{\mathbf{c}}$ is a monomial order if and only if $\mathbf{c} \geq \mathbf{0}$. Thus, the following integer programs

$$(\mathrm{IP})_\succ(\mathbf{b}) \qquad \min_{\succ_{\mathbf{c}}}\left\{\mathbf{z} : A\mathbf{z} = \mathbf{b},\ \mathbf{z} \in \mathbb{Z}^n_+\right\}$$

either have no feasible solution or they have a unique optimal solution. We will consider the family of all these integer programs with fixed A and $\mathbf{c}$, but varying right-hand side vector $\mathbf{b}$. Interestingly, for this family of integer programs there exists an optimality certificate that is typically much smaller than the Graver basis of A, namely, the reduced minimal Gröbner basis of I_A with respect to $\succ_{\mathbf{c}}$. In fact, any nonreduced or nonminimal Gröbner basis of I_A with respect to $\succ_{\mathbf{c}}$ can be used.

Theorem 11.2.1. *Let $\succ$ be any fixed monomial order, $A \in \mathbb{Z}^{m\times n}$ a fixed matrix, and $\mathbf{c} \in \mathbb{R}^n_+$ a fixed cost vector. Moreover, let $G_{\succ_{\mathbf{c}}}$ be the reduced minimal Gröbner basis of I_A with respect to $\succ_{\mathbf{c}}$. Then for any right-hand side vector $\mathbf{b}$ and any nonoptimal feasible solution $\mathbf{z}_0$ there is some binomial $\mathbf{x}^{\mathbf{u}} - \mathbf{x}^{\mathbf{v}} \in G_{\succ_{\mathbf{c}}}$ such that $\mathbf{z}_0 - \mathbf{u} + \mathbf{v}$ is a better feasible solution than $\mathbf{z}_0$.*

In fact, if $\mathbf{x}^{\bar{\mathbf{z}}} :=$ remainder$(\mathbf{x}^{\mathbf{z}_0}, G_{\succ_{\mathbf{c}}})$, then $\bar{\mathbf{z}}$ is the unique optimal solution to (IP)$_\succ$($\mathbf{b}$).

Proof. Let $\mathbf{z}^*$ denote the unique optimal solution to (IP)$_\succ$($\mathbf{b}$). Then $\mathbf{z}_0 - \mathbf{z}^* \in I_A$ and $\mathrm{LM}(\mathbf{x}^{\mathbf{z}_0} - \mathbf{x}^{\mathbf{z}^*}) = \mathbf{x}^{\mathbf{z}_0}$. Thus, there exists some element $\mathbf{x}^{\mathbf{u}} - \mathbf{x}^{\mathbf{v}} \in G_{\succ_{\mathbf{c}}}$ with $\mathbf{x}^{\mathbf{u}} \,|\, \mathbf{x}^{\mathbf{z}_0}$. Let $\mathbf{x}^{\mathbf{z}_1} := \mathbf{x}^{\mathbf{z}_0} - \mathbf{x}^{\mathbf{z}_0 - \mathbf{u}}(\mathbf{x}^{\mathbf{u}} - \mathbf{x}^{\mathbf{v}}) = \mathbf{x}^{\mathbf{z}_0 - \mathbf{u} + \mathbf{v}}$. By construction, $\mathbf{z}_0 \succ \mathbf{z}_1$, $\mathbf{z}_1 \geq \mathbf{0}$, and $A\mathbf{z}_1 = A(\mathbf{z}_0 - \mathbf{u} + \mathbf{v}) = A\mathbf{z}_0 + A(-\mathbf{u} + \mathbf{v}) = \mathbf{b} + \mathbf{0} = \mathbf{b}$. This proves the first claim.

The second claim, that $\mathbf{x}^{\bar{\mathbf{z}}} :=$ remainder$(\mathbf{x}^{\mathbf{z}_0}, G_{\succ_{\mathbf{c}}})$ gives the unique optimal solution $\bar{\mathbf{z}} = \mathbf{z}^*$ to (IP)$_\succ$($\mathbf{b}$), is now immediate. □

In general, however, we do not know a generating set for the ideal I_A; so we cannot compute a Gröbner basis of it with respect to $\succ_{\mathbf{c}}$, either. Moreover, we need to find an initial feasible solution from which we start the augmentation algorithm using Gröbner basis directions.

To approach this problem, we do something similar to phase I of the simplex method in linear programming: we create a larger integer program, which has an obvious integer feasible point, and for which the toric ideal has a nice generating set to start from. Then we eliminate the extra variables.

For this assume, without loss of generality, that $\mathbf{b} \geq \mathbf{0}$ and consider the "extended integer programs"

$$(\mathrm{EIP})_{\succ}(\mathbf{b}) \qquad \min_{\succ(\mathbf{1},\mathbf{0})} \left\{ (\mathbf{s},\mathbf{z}) : \mathbf{s} + A\mathbf{z} = \mathbf{b},\ \mathbf{z} \in \mathbb{Z}_+^n,\ \mathbf{s} \in \mathbb{Z}_+^m \right\}.$$

Note that all of the programs $(\mathrm{EIP})_{\succ}(\mathbf{b})$ are feasible: they have the obvious solution $\mathbf{z} = \mathbf{0}, \mathbf{s} = \mathbf{b}$. However, an optimal solution will be of the form $\mathbf{z} = \mathbf{z}_0, \mathbf{s} = \mathbf{0}$ if the program $(\mathrm{IP})_{\succ}(\mathbf{b})$ contains some feasible solution. If $(\mathrm{IP})_{\succ}(\mathbf{b})$ is infeasible, then the extended program $(\mathrm{EIP})_{\succ}(\mathbf{b})$ has an optimal solution with $\mathbf{s} \gneq \mathbf{0}$. Putting both cases together, we see that it is sufficient to solve the extended programs.

Proposition 11.2.2 (Conti and Traverso [73]). *For $A \in \mathbb{Z}_+^{m\times n}$, the toric ideal*

$$I_{(I,A)} := \left\langle \mathbf{y}^{\mathbf{a}_1^+}\mathbf{x}^{\mathbf{a}_2^+} - \mathbf{y}^{\mathbf{a}_1^-}\mathbf{x}^{\mathbf{a}_2^-} : (I,A)\begin{pmatrix}\mathbf{a}_1\\ \mathbf{a}_2\end{pmatrix} = \mathbf{0},\ \begin{pmatrix}\mathbf{a}_1\\ \mathbf{a}_2\end{pmatrix} \in \mathbb{Z}^{m+n} \right\rangle$$

is generated by the binomials

$$\mathbf{y}^{A\mathbf{e}_j} - x_j \quad \text{for } 1 \leq j \leq n.$$

In particular, $I_A = I_{(I,A)} \cap \mathbb{K}[x_1,\ldots,x_n]$ can be computed via elimination of the indeterminates $\mathbf{y}$.

Proof. In fact, the binomials $x_i - \mathbf{y}^{A\mathbf{e}_i}$ form a Gröbner basis for $I_{(I,A)}$, for a lexicographic term order with $x_i \succ_{\mathrm{lex}} y_j$ for all pairs i and j. Thus they certainly generate the ideal. □

This completes the missing pieces. We know how to compute an initial feasible solution, we know how to get a generating set for the toric ideal I_A, from which we can compute a Gröbner basis for I_A for a term ordering $\succ_{\mathbf{c}}$, which then can be used as an optimality certificate to augment the initial feasible solution to optimality. It remains to find a generating set of the toric ideal I_A for a given matrix A.

11.3 Generating sets and toric ideals of lattice ideals

As we have mentioned before, the computation of a generating set for the toric ideal $I_A = \langle \mathbf{x}^{\mathbf{u}^+} - \mathbf{x}^{\mathbf{u}^-} : \mathbf{u} \in \ker(A) \cap \mathbb{Z}^n \rangle$ of a given integer matrix A is not trivial. In this section we show how one can transform this algebraic problem into the language of geometry. In fact, we will deal with this question for general lattice ideals

$$I(\mathcal{L}) = \langle \mathbf{x}^{\mathbf{u}^+} - \mathbf{x}^{\mathbf{u}^-} : \mathbf{u} \in \mathcal{L} \rangle$$

for any lattice $\mathcal{L} \subseteq \mathbb{Z}^n$, not just $\mathcal{L} = \ker(A) \cap \mathbb{Z}^n$. Gröbner basis computations are very sensitive to the number of indeterminates involved. In this section, we will see how this can

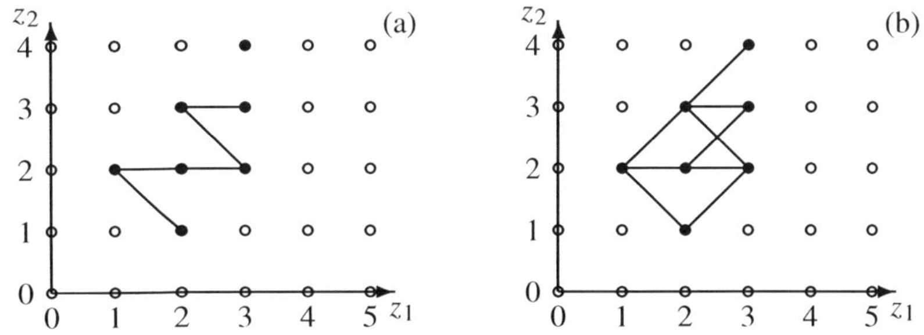

Figure 11.1. *The graphs* (a) $\mathcal{G}(\mathcal{F}_{\mathcal{L},\mathbf{b}}, M)$ *and* (b) $\mathcal{G}(\mathcal{F}_{\mathcal{L},\mathbf{b}}, M')$ *projected onto the* (z_1, z_2)*-plane.*

be avoided. We translate or (re)define some algebraic notions for lattice ideals in geometric terms.

Given a lattice $\mathcal{L} \subseteq \mathbb{Z}^n$, and a vector $\mathbf{b} \in \mathbb{Z}^n$, we define

$$\mathcal{F}_{\mathcal{L},\mathbf{b}} := \{\mathbf{z} : \mathbf{z} \equiv \mathbf{b} \ (\text{mod } \mathcal{L}),\ \mathbf{z} \in \mathbb{Z}^n_+\}.$$

For $S \subseteq \mathcal{L}$, we define $\mathcal{G}(\mathcal{F}_{\mathcal{L},\mathbf{b}}, S)$ to be the *undirected* graph with nodes $\mathcal{F}_{\mathcal{L},\mathbf{b}}$ and edges $(\mathbf{u},\mathbf{v})$ if $\mathbf{u}-\mathbf{v} \in S$ or $\mathbf{v}-\mathbf{u} \in S$ for $\mathbf{u},\mathbf{v} \in \mathcal{F}_{\mathcal{L},b}$. We will assume throughout the rest of this chapter that $\mathcal{L} \cap \mathbb{Z}^n_+ = \{\mathbf{0}\}$. Hence, $\mathcal{F}_{\mathcal{L},\mathbf{b}}$ is finite for any $\mathbf{b}$.

Definition 11.3.1. We call $S \subseteq \mathcal{L}$ a *generating set* of $\mathcal{F}_{\mathcal{L},\mathbf{b}}$ if the graph $\mathcal{G}(\mathcal{F}_{\mathcal{L},\mathbf{b}}, S)$ is connected. If S is a generating set of $\mathcal{F}_{\mathcal{L},\mathbf{b}}$ for every $\mathbf{b} \in \mathbb{Z}^n$, we call S a *generating set* or *Markov basis* of $\mathcal{L}$.

Note that connectedness of $\mathcal{G}(\mathcal{F}_{\mathcal{L},\mathbf{b}}, S)$ simply states that between each pair $\mathbf{u},\mathbf{v} \in \mathcal{F}_{\mathcal{L},b}$ there exists a path from $\mathbf{u}$ to $\mathbf{v}$ in $\mathcal{G}(\mathcal{F}_{\mathcal{L},\mathbf{b}}, S)$. Be aware of the difference between a *generating set* of a lattice and a *spanning set* of a lattice! A spanning set is any set $S \subseteq \mathcal{L}$ such that any point in $\mathcal{L}$ can be represented as a linear integer combination of the vectors in S. A generating set of $\mathcal{L}$ is a spanning set of $\mathcal{L}$, but the converse is not necessarily true.

Example 11.3.2. Let $S := \{(1,-1,-1,-3,-1,2)^\top, (1,0,2,-2,-2,1)^\top\}$, and let $\mathcal{L} \subseteq \mathbb{Z}^6$ be the lattice spanned by S. So, by definition, S is a spanning set of $\mathcal{L}$, but S is not a generating set of $\mathcal{L}$. Consider $\mathbf{b} := (2,2,4,2,4,1)^\top$, then the graph $\mathcal{G}(\mathcal{F}_{\mathcal{L},\mathbf{b}}, S)$ is not connected (see Figure 11.1a) because the point $(3,4,12,2,0,0)^\top \in \mathcal{F}_{\mathcal{L},\mathbf{b}}$ is disconnected.

Let $S' := S \cup \{(1,1,5,-1,-3,0)^\top\}$. The graph $\mathcal{G}(\mathcal{F}_{\mathcal{L},\mathbf{b}}, S')$ is now connected (see Figure 11.1b); however, S' is still not a generating set of $\mathcal{L}$ since for $\mathbf{b}' := (0,0,0,0,1,1)^\top$, we have $\mathcal{F}_{\mathcal{L},\mathbf{b}'} = \{(0,0,0,0,1,1)^\top, (0,1,3,1,0,0)^\top\}$ and the graph $\mathcal{G}(\mathcal{F}_{\mathcal{L},\mathbf{b}'}, S')$ is disconnected. The set $S'' := S' \cup \{(0,1,3,1,-1,-1)^\top\}$ is a generating set of $\mathcal{L}$.

The next lemma relates Markov bases of a lattice $\mathcal{L}$ to generating sets of the associated lattice ideal $I(\mathcal{L})$. Here, we give a proof of the crucial fact that a set $M \subseteq \mathcal{L}$ is a Markov basis of a lattice $\mathcal{L}$ if and only if $\{\mathbf{x}^{\mathbf{u}^+} - \mathbf{x}^{\mathbf{u}^-} : \mathbf{u} \in M\}$ is a generating set of the lattice ideal $I(\mathcal{L})$, that is, $I(\mathcal{L}) = I(M) := \langle \mathbf{x}^{\mathbf{u}^+} - \mathbf{x}^{\mathbf{u}^-} : \mathbf{u} \in M \rangle$ [113].

Lemma 11.3.3. *A set* $M \subseteq \mathcal{L}$ *is a Markov basis of* $\mathcal{L}$ *if and only if* $I(M) = I(\mathcal{L})$.

Proof. Assume that M is a Markov basis of $\mathcal{L}$. We show that $\mathbf{x}^{\mathbf{u}^+} - \mathbf{x}^{\mathbf{u}^-} \in I(M)$ for every $\mathbf{u} \in \mathcal{L}$ and the result follows. Let $\mathbf{u} \in \mathcal{L}$. Then, $\mathbf{u}^+, \mathbf{u}^- \in \mathcal{F}_{\mathcal{L},\mathbf{v}}$ where $\mathbf{v} = \mathbf{u}^+$. Since M is a Markov basis, there exists a path from $\mathbf{u}^+$ to $\mathbf{u}^-$ in the graph $\mathcal{G}_{\mathcal{L}}(\mathbf{v}, M)$, or more explicitly, there exists a sequence of points $(\mathbf{z}_1, \mathbf{z}_2, \ldots, \mathbf{z}_k) \subseteq \mathcal{F}_{\mathcal{L},\mathbf{v}}$ where $\mathbf{z}_1 = \mathbf{u}^+$, $\mathbf{z}_k = \mathbf{u}^-$, and $\mathbf{z}_i - \mathbf{z}_{i+1} = \delta_i \mathbf{m}_i$ for some $\mathbf{m}_i \in M$ and $\delta_i \in \{1, -1\}$ for $i = 1, \ldots, k-1$. Then, there exists $\boldsymbol{\gamma}_i \in \mathbb{Z}^n_+$ such that $\mathbf{z}_i = (\delta_i \mathbf{m}_i)^+ + \boldsymbol{\gamma}_i$, and $\mathbf{z}_{i+1} = (\delta_i \mathbf{m}_i)^- + \boldsymbol{\gamma}_i$ for every $i = 1, \ldots, k-1$. Hence, $\mathbf{x}^{\mathbf{z}_i} - \mathbf{x}^{\mathbf{z}_{i+1}} = \delta_i \mathbf{x}^{\boldsymbol{\gamma}_i}(\mathbf{x}^{\mathbf{m}_i^+} - \mathbf{x}^{\mathbf{m}_i^-})$, and therefore,

$$\mathbf{x}^{\mathbf{u}^+} - \mathbf{x}^{\mathbf{u}^-} = \sum_{i=1}^{k-1} \mathbf{x}^{\mathbf{z}_i} - \mathbf{x}^{\mathbf{z}_{i+1}} = \sum_{i=1}^{k-1} \delta_i \mathbf{x}^{\boldsymbol{\gamma}_i}(\mathbf{x}^{\mathbf{m}_i^+} - \mathbf{x}^{\mathbf{m}_i^-}) \in I(M)$$

as required.

Conversely, assume that $I(M) = I(\mathcal{L})$. Also, assume there exists $\boldsymbol{\alpha}, \boldsymbol{\beta} \in \mathcal{F}_{\mathcal{L},\mathbf{v}}$ for some $\mathbf{v} \in \mathbb{Z}^n$ such that $\boldsymbol{\alpha}$ and $\boldsymbol{\beta}$ are not connected in $\mathcal{G}_{\mathcal{L}}(\mathbf{v}, M)$. We will derive a contradiction. The binomial $\mathbf{x}^{\boldsymbol{\alpha}} - \mathbf{x}^{\boldsymbol{\beta}}$ is in $I(\mathcal{L}) = I(M)$, so we may write $\mathbf{x}^{\boldsymbol{\alpha}} - \mathbf{x}^{\boldsymbol{\beta}} = \sum_{i=1}^{d} c_i \mathbf{x}^{\boldsymbol{\gamma}_i}(\mathbf{x}^{\mathbf{m}_i^+} - \mathbf{x}^{\mathbf{m}_i^-})$, where $\mathbf{m}_i \in M$, $c_i \in \mathbb{K}$, and $\boldsymbol{\gamma}_i \in \mathbb{Z}^n_+$. Note that we allow $\mathbf{m}_i = \mathbf{m}_j$ for $i \neq j$. Now, let $I \subseteq \{1, \ldots, d\}$ be the set of $i \in \{1, \ldots, d\}$ such that the point $(\boldsymbol{\gamma}_i + \mathbf{m}_i^+)$ is in $\mathcal{F}_{\mathcal{L},\mathbf{v}}$ and $(\boldsymbol{\gamma}_i + \mathbf{m}_i^+)$ is connected to $\boldsymbol{\alpha}$ in $\mathcal{G}_{\mathcal{L}}(\mathbf{v}, M)$. Note that if $(\boldsymbol{\gamma}_i + \mathbf{m}_i^+)$ is connected to $\boldsymbol{\alpha}$ in $\mathcal{G}_{\mathcal{L}}(\mathbf{v}, M)$, then $(\boldsymbol{\gamma}_i + \mathbf{m}_i^-)$ is also connected to $\boldsymbol{\alpha}$ in $\mathcal{G}_{\mathcal{L}}(\mathbf{v}, M)$ since $(\boldsymbol{\gamma}_i + \mathbf{m}_i^+) - (\boldsymbol{\gamma}_i + \mathbf{m}_i^-) = \mathbf{m}_i \in M$. Thus, the set of monomials consisting of $\mathbf{x}^{\boldsymbol{\gamma}_i}\mathbf{x}^{\mathbf{m}_i^+}$ and $\mathbf{x}^{\boldsymbol{\gamma}_i}\mathbf{x}^{\mathbf{m}_i^-}$ for all $i \in I$, which includes $\mathbf{x}^{\boldsymbol{\alpha}}$ and not $\mathbf{x}^{\boldsymbol{\beta}}$, is disjoint from the set of monomials consisting of $\mathbf{x}^{\boldsymbol{\gamma}_i}\mathbf{x}^{\mathbf{m}_i^+}$ and $\mathbf{x}^{\boldsymbol{\gamma}_i}\mathbf{x}^{\mathbf{m}_i^-}$ for all $i \notin I$, which includes $\mathbf{x}^{\boldsymbol{\beta}}$ and not $\mathbf{x}^{\boldsymbol{\alpha}}$. Let $f(\mathbf{x}) = \sum_{i \in I} c_i \mathbf{x}^{\boldsymbol{\gamma}_i}(\mathbf{x}^{\mathbf{m}_i^+} - \mathbf{x}^{\mathbf{m}_i^-})$ and let $g(\mathbf{x}) = -\sum_{i \notin I} c_i \mathbf{x}^{\boldsymbol{\gamma}_i}(\mathbf{x}^{\mathbf{m}_i^+} - \mathbf{x}^{\mathbf{m}_i^-})$. Thus, the polynomials $f(\mathbf{x})$ and $g(\mathbf{x})$ have a disjoint set of monomials, and therefore, $f(\mathbf{x}) = \mathbf{x}^{\boldsymbol{\alpha}}$ and $g(\mathbf{x}) = \mathbf{x}^{\boldsymbol{\beta}}$ since $\mathbf{x}^{\boldsymbol{\alpha}} - \mathbf{x}^{\boldsymbol{\beta}} = f(\mathbf{x}) - g(\mathbf{x})$. However, this is impossible since $f(\mathbf{1}) = 0$ and $g(\mathbf{1}) = 0$ but $\mathbf{1}^{\boldsymbol{\alpha}} = 1$ and $\mathbf{1}^{\boldsymbol{\beta}} = 1$. □

It is this connectivity property of generating sets of lattice ideals that makes them useful in statistics. As we can reach any lattice point in $\mathcal{G}(\mathcal{F}_{\mathcal{L},\mathbf{b}}, M)$ from any other point in this graph, we may use a Markov basis of $\mathcal{L}$ to run a Monte-Carlo Markov-chain (MCMC) process to test the validity of statistical models via sampling [113]. For this reason, the notion *Markov basis* becomes more and more common to denote a generating set of a lattice ideal.

When we orient the edges of the graph $\mathcal{G}(\mathcal{F}_{\mathcal{L},\mathbf{b}}, M)$ according to a given term order $\succ$ on $\mathbb{Z}^n_+$, we can even motivate the notion of a Gröbner basis of a lattice ideal in geometric terms. The connection to integer programming will be self-evident!

For the definition of a Gröbner basis, we need a term order $\succ$ for $\mathcal{L}$. We call $\succ$ a *term order* for $\mathcal{L}$ if $\succ$ is a total well-ordering on $\mathcal{F}_{\mathcal{L},\mathbf{b}}$ for every $\mathbf{b} \in Z^n$ and $\succ$ is an additive order, that is, for $\mathbf{u}, \mathbf{v} \in \mathcal{F}_{\mathcal{L},\mathbf{b}}$ for some $\mathbf{b}$, $\mathbf{u} \succ \mathbf{v}$ implies $\mathbf{u} + \boldsymbol{\gamma} \succ \mathbf{v} + \boldsymbol{\gamma}$ for every $\boldsymbol{\gamma} \in \mathbb{Z}^n_+$.

We also need the notion of a decreasing path: a path $(\mathbf{z}^0, \ldots, \mathbf{z}^k)$ in $\mathcal{G}(\mathcal{F}_{\mathcal{L},\mathbf{b}}, G)$ is *decreasing* with respect to $\succ$ if $\mathbf{z}^i \succ \mathbf{z}^{i+1}$ for $i = 0, \ldots, k-1$. We define $\mathcal{L}_{\succ} := \{\mathbf{u} \in \mathcal{L} : \mathbf{u}^+ \succ \mathbf{u}^-\}$.

Definition 11.3.4. We call $G \subseteq \mathcal{L}_{\succ}$ a *Gröbner basis* of $\mathcal{L}$ with respect to $\succ$ if for every $\mathbf{b} \in \mathbb{Z}^n_+$ there exists a decreasing path in $\mathcal{G}(\mathcal{F}_{\mathcal{L},\mathbf{b}}, G)$ from $\mathbf{b}$ to the unique $\succ$-minimal element in $\mathcal{F}_{\mathcal{L},\mathbf{b}}$.

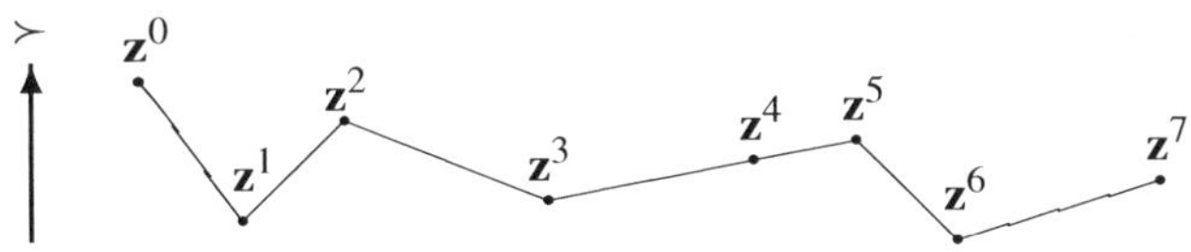

Figure 11.2. *Reduction path between* $\mathbf{z}^0$ *and* $\mathbf{z}^7$.

This definition can be rephrased to capture more of the algebraic property that the leading term ideal of $I(\mathcal{L})$ is generated by the leading terms of the Gröbner basis elements.

Lemma 11.3.5. *Let* $\succ$ *be a term order of* $\mathbb{Z}^n_+$. *A set* $G \subseteq \mathcal{L}_\succ$ *is a* $\succ$*-Gröbner basis of* $\mathcal{L}$ *if and only if, for all* $\mathbf{v} \in \mathcal{L}_\succ$, *there exists a vector* $\mathbf{u} \in G$ *such that* $\mathbf{u}^+ \leq \mathbf{v}^+$.

If $G \subseteq \mathcal{L}_\succ$ is a Gröbner basis, then G is a generating set of $\mathcal{L}$ since, given $\mathbf{u},\mathbf{v} \in \mathcal{G}(\mathcal{F}_{\mathcal{L},\mathbf{b}}, G)$ for some $\mathbf{b}$, there exists a decreasing path from $\mathbf{u}$ to the unique $\succ$-minimal element in $\mathcal{F}_{\mathcal{L},\mathbf{u}}$ and from $\mathbf{v}$ to the same element, and thus, $\mathbf{u}$ and $\mathbf{v}$ are connected in $\mathcal{G}(\mathcal{F}_{\mathcal{L},\mathbf{b}}, G)$.

Also, $G \subseteq \mathcal{L}_\succ$ is a Gröbner basis if and only if for every $\mathbf{u} \in \mathbb{Z}^n_+$, $\mathbf{u}$ is either the unique $\succ$-minimal element in $\mathcal{F}_{\mathcal{L},\mathbf{u}}$ or there exists a vector $\mathbf{v} \in G$ such that $\mathbf{u}-\mathbf{v} \in \mathcal{F}_{\mathcal{L},\mathbf{u}}$ and $\mathbf{u} \succ \mathbf{u}-\mathbf{v}$. Consequently, a Gröbner basis G is a *test set* or *optimality certificate* for the *lattice program*

$$(\mathrm{IP})^{\mathcal{L}}_{\succ}(\mathbf{v}) := \min_{\succ} \left\{ \mathbf{z} : \mathbf{z} \in \mathcal{F}_{\mathcal{L},\mathbf{v}} \right\}.$$

Any feasible integer program can be reformulated as such a lattice program. For example, the integer program $(\mathrm{IP})(\mathbf{b}) = \left\{ \mathbf{c}^\top \mathbf{z} : A\mathbf{z} = \mathbf{b}, \mathbf{z} \in \mathbb{Z}^n_+ \right\}$ is equivalent to the lattice program $(\mathrm{IP})^{\mathcal{L}}_{\succ}(\mathbf{v})$, where $\mathcal{L}$ is the lattice $\mathcal{L} = \ker(A) \cap \mathbb{Z}^n$, $\mathbf{v} \in \mathbb{Z}^n$ is any vector where $A\mathbf{v} = \mathbf{b}$, and $\succ$ is a term order $\succ_\mathbf{c}$ as defined above.

Let us rephrase the property of a Gröbner basis being an optimality certificate.

Lemma 11.3.6. *Let* $\succ$ *be a term order of* $\mathbb{Z}^n_+$. *A set* $G \subseteq \mathcal{L}_\succ$ *is a* $\succ$*-Gröbner basis of* $\mathcal{L}$ *if and only if, for all* $\mathbf{v} \in \mathcal{L}_\succ$, *there exists a vector* $\mathbf{u} \in G$ *such that* $\mathbf{u}^+ \leq \mathbf{v}^+$, *that is, we have* $\mathbf{u}^+ - \mathbf{v} \in \mathcal{F}_{\mathcal{L},\mathbf{u}^+}$ *and* $\mathbf{u}^+ \succ \mathbf{u}^+ - \mathbf{v}$.

The defining property of a Gröbner basis is very strong, so we redefine it in terms of reduction paths. A path $(\mathbf{z}^0,\ldots,\mathbf{z}^k)$ in $\mathcal{G}(\mathcal{F}_{\mathcal{L},\mathbf{b}}, G)$ is a *reduction path* with respect to $\succ$ if for no $i \in \{1,\ldots,k-1\}$, we have $\mathbf{z}^i \succ \mathbf{z}^0$ and $\mathbf{z}^i \succ \mathbf{z}^k$. For example, see Figure 11.2.

Lemma 11.3.7. *A set* $G \subseteq \mathcal{L}_\succ$ *is a Gröbner basis of* $\mathcal{L}$ *with respect to* $\succ$ *if and only if for each* $\mathbf{b} \in \mathbb{Z}^n$ *and for each pair* $\mathbf{u},\mathbf{v} \in \mathcal{F}_{\mathcal{L},\mathbf{b}}$ *there exists a reduction path in* $\mathcal{G}(\mathcal{F}_{\mathcal{L},\mathbf{b}}, G)$ *between* $\mathbf{u}$ *and* $\mathbf{v}$.

Proof. If $\mathcal{G}(\mathcal{F}_{\mathcal{L},\mathbf{b}}, G)$ contains decreasing paths from $\mathbf{u},\mathbf{v} \in \mathcal{F}_{\mathcal{L},\mathbf{b}}$ to the unique $\succ$-minimal element in $\mathcal{F}_{\mathcal{L},\mathbf{b}}$, then joining the two paths (and removing cycles if necessary) forms a reduction path between $\mathbf{u}$ and $\mathbf{v}$.

For the other direction, we assume that there is a reduction path between each pair $\mathbf{u},\mathbf{v} \in \mathcal{F}_{\mathcal{L},\mathbf{b}}$. Denote by $\mathbf{z}^*$ the unique $\succ$-minimal element in $\mathcal{F}_{\mathcal{L},\mathbf{b}}$; thus, every $\mathbf{u} \in \mathcal{F}_{\mathcal{L},\mathbf{b}}$ is connected to $\mathbf{z}^*$ by a reduction path. In particular, by the definition of a reduction path, if $\mathbf{u} \neq \mathbf{z}^*$, then the first node $\mathbf{z}^1 \neq \mathbf{u}$ in this path must satisfy $\mathbf{u} \succ \mathbf{z}^1$. Repeating this argument

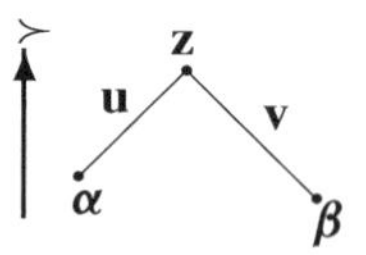

Figure 11.3. *A critical path for* $(\mathbf{u},\mathbf{v})$ *between* $\boldsymbol{\alpha}$, $\mathbf{z}$, *and* $\boldsymbol{\beta}$.

iteratively with $\mathbf{z}^1$ instead of $\mathbf{u}$, we get a decreasing path from $\mathbf{u}$ to $\mathbf{z}^*$. This follows from the fact that $\succ$ is a term order, which implies that every decreasing path must be finite. However, the only node from which the decreasing path cannot be lengthened is $\mathbf{z}^*$. □

Checking for a given $G \subseteq \mathcal{L}_\succ$ whether there exists a reduction path in $\mathcal{G}(\mathcal{F}_{\mathcal{L},\mathbf{b}},G)$ for every $\mathbf{b} \in \mathbb{Z}^n$ and for each pair $\mathbf{u},\mathbf{v} \in \mathcal{F}_{\mathcal{L},\mathbf{b}}$ involves infinitely many situations that need to be checked. In fact, far fewer checks are needed: we only need to check for a reduction path from $\boldsymbol{\alpha}$ to $\boldsymbol{\beta}$ if there exists a *critical path* from $\boldsymbol{\alpha}$ to $\boldsymbol{\beta}$. (These critical paths correspond to the construction of an S-polynomial, which are enough to be checked by Buchberger's theorem.)

Definition 11.3.8. Given $G \subseteq \mathcal{L}_\succ$ and $\mathbf{b} \in \mathbb{Z}^n$, a path $(\boldsymbol{\alpha},\mathbf{z},\boldsymbol{\beta})$ in $\mathcal{G}(\mathcal{F}_{\mathcal{L},\mathbf{b}},G)$ is a *critical path* if $\mathbf{z} \succ \boldsymbol{\alpha}$ and $\mathbf{z} \succ \boldsymbol{\beta}$.

If $(\boldsymbol{\alpha},\mathbf{z},\boldsymbol{\beta})$ is a critical path in $\mathcal{G}(\mathcal{F}_{\mathcal{L},\mathbf{b}},G)$, then $\boldsymbol{\alpha}+\mathbf{u} = \mathbf{z} = \boldsymbol{\beta}+\mathbf{v}$ for some pair $\mathbf{u},\mathbf{v} \in G$, in which case we call $(\boldsymbol{\alpha},\mathbf{z},\boldsymbol{\beta})$ a critical path for $(\mathbf{u},\mathbf{v})$ (see Figure 11.3). The point $\mathbf{z}$ plays the role of the lcm of the leading monomials of the two polynomials involved. $\boldsymbol{\alpha}$ and $\boldsymbol{\beta}$ play the role of these leading monomials. In other words, $\mathbf{x}^{\mathbf{z}} = \operatorname{lcm}(\mathbf{x}^{\boldsymbol{\alpha}},\mathbf{x}^{\boldsymbol{\beta}})$.

The following lemma will be a crucial ingredient in the correctness proofs of the algorithms presented below. It will guarantee correctness of the algorithm under consideration, since the necessary reduction paths have been constructed during the run of the algorithm. In the next lemma, we cannot assume that G is a generating set of $\mathcal{L}$, since often this is what we are trying to construct.

Lemma 11.3.9. *Let* $\boldsymbol{\alpha},\boldsymbol{\beta} \in \mathcal{F}_{\mathcal{L},\mathbf{b}}$ *for some* $\mathbf{b} \in \mathbb{Z}^n$, *and let* $G \subseteq \mathcal{L}_\succ$ *be such that there is a path between* $\boldsymbol{\alpha}$ *and* $\boldsymbol{\beta}$ *in* $\mathcal{G}(\mathcal{F}_{\mathcal{L},\mathbf{b}},G)$. *If there exists a reduction path between* $\boldsymbol{\alpha}'$ *and* $\boldsymbol{\beta}'$ *for every critical path* $(\boldsymbol{\alpha}',\mathbf{z}',\boldsymbol{\beta}')$ *in* $\mathcal{G}(\mathcal{F}_{\mathcal{L},\mathbf{b}},G)$, *then there exists a reduction path between* $\boldsymbol{\alpha}$ *and* $\boldsymbol{\beta}$ *in* $\mathcal{G}(\mathcal{F}_{\mathcal{L},\mathbf{b}},G)$.

Proof. Assume on the contrary that no such reduction path exists from $\boldsymbol{\alpha}$ to $\boldsymbol{\beta}$. Among all paths $(\boldsymbol{\alpha} = \mathbf{z}^0,\ldots,\mathbf{z}^k = \boldsymbol{\beta})$ in $\mathcal{G}(\mathcal{F}_{\mathcal{L},\mathbf{b}},G)$ choose one such that $\max_\succ\{\mathbf{z}^0,\ldots,\mathbf{z}^k\}$ is minimal. Such a minimal path exists since $\succ$ is a term order. Let j be such that $\mathbf{z}^j$ attains this maximum.

By assumption $(\mathbf{z}^0,\ldots,\mathbf{z}^k)$ is not a reduction path, and thus, $\mathbf{z}^j \succ \mathbf{z}^0$ and $\mathbf{z}^j \succ \mathbf{z}^k$, and since $\mathbf{z}^j$ is maximal, we have $\mathbf{z}^j \succ \mathbf{z}^{j-1}$ and $\mathbf{z}^j \succ \mathbf{z}^{j+1}$. Let $\mathbf{u} = \mathbf{z}^j - \mathbf{z}^{j-1}$ and $\mathbf{v} = \mathbf{z}^j - \mathbf{z}^{j+1}$. Then $(\mathbf{z}^{j-1},\mathbf{z}^j,\mathbf{z}^{j+1})$ forms a critical path. Consequently, we can replace the path $(\mathbf{z}^{j-1},\mathbf{z}^j,\mathbf{z}^{j+1})$ with the reduction path $(\mathbf{z}^{j-1} = \bar{\mathbf{z}}^0,\ldots,\bar{\mathbf{z}}^s = \mathbf{z}^{j+1})$ in the path $(\mathbf{z}^0,\ldots,\mathbf{z}^k)$ and obtain a new path between $\boldsymbol{\alpha}$ and $\boldsymbol{\beta}$ with the property that the $\succ$-maximum of the intermediate nodes is strictly less than $\mathbf{z}^j = \max_\succ\{\mathbf{z}^1,\ldots,\mathbf{z}^{k-1}\}$ (see Figure 11.4). This contradiction proves our claim. □

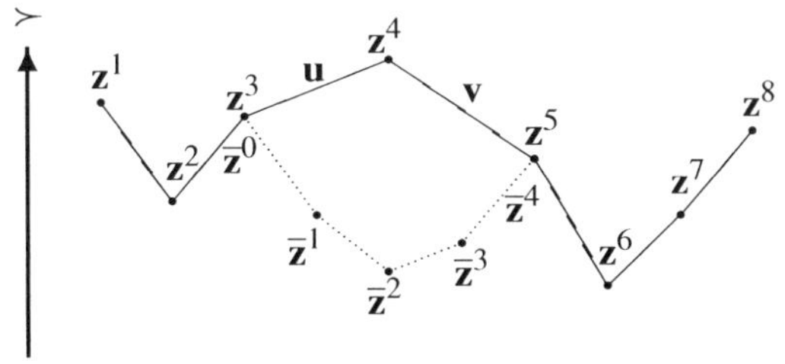

Figure 11.4. *Replacing a critical path by a reduction path.*

The following corollary is a straightforward consequence of Lemma 11.3.9, but nonetheless, it is worthwhile stating explicitly.

Corollary 11.3.10. *Let $G \subseteq \mathcal{L}_{\succ}$. If for all $\mathbf{b}' \in \mathbb{Z}^n$ and for every critical path $(\boldsymbol{\alpha}', z', \boldsymbol{\beta}')$ in $\mathcal{G}(\mathcal{F}_{\mathcal{L},\mathbf{b}'}, G)$, there exists a reduction path between $\boldsymbol{\alpha}'$ and $\boldsymbol{\beta}'$, then for all $\mathbf{b} \in \mathbb{Z}^n$ and for all $\boldsymbol{\alpha}, \boldsymbol{\beta} \in \mathcal{F}_{\mathcal{L},\mathbf{b}}$ where $\boldsymbol{\alpha}$ and $\boldsymbol{\beta}$ are connected in $\mathcal{G}(\mathcal{F}_{\mathcal{L},\mathbf{b}}, G)$, there exists a reduction path between $\boldsymbol{\alpha}$ and $\boldsymbol{\beta}$ in $\mathcal{G}(\mathcal{F}_{\mathcal{L},\mathbf{b}}, G)$.*

Combining Corollary 11.3.10 with Lemma 11.3.7, we arrive at the following result for Gröbner bases of a lattice.

Corollary 11.3.11. *A set $G \subseteq \mathcal{L}_{\succ}$ is a Gröbner basis of $\mathcal{L}$ with respect to $\succ$ if and only if G is a generating set of $\mathcal{L}$ and if for all $\mathbf{b} \in \mathbb{Z}^n$ and for every critical path $(\boldsymbol{\alpha}, z, \boldsymbol{\beta})$ in $\mathcal{G}(\mathcal{F}_{\mathcal{L},\mathbf{b}}, G)$, there exists a reduction path between $\boldsymbol{\alpha}$ and $\boldsymbol{\beta}$ in $\mathcal{G}(\mathcal{F}_{\mathcal{L},\mathbf{b}}, G)$.*

In Corollary 11.3.10 and Corollary 11.3.11, it is not necessary to check for a reduction path from $\boldsymbol{\alpha}$ to $\boldsymbol{\beta}$ for every critical path $(\boldsymbol{\alpha}, \mathbf{z}, \boldsymbol{\beta})$ in $\mathcal{G}(\mathcal{F}_{\mathcal{L},\mathbf{b}}, G)$ for all $\mathbf{b} \in \mathbb{Z}^n$. Consider the case where there exists another critical path $(\boldsymbol{\alpha}', \mathbf{z}', \boldsymbol{\beta}')$ in $\mathcal{G}(\mathcal{F}_{\mathcal{L},\mathbf{b}'}, G)$ for some $\mathbf{b}' \in \mathbb{Z}^n$ such that $(\boldsymbol{\alpha}, \mathbf{z}, \boldsymbol{\beta}) = (\boldsymbol{\alpha}' + \boldsymbol{\gamma}, \mathbf{z}' + \boldsymbol{\gamma}, \boldsymbol{\beta}' + \boldsymbol{\gamma})$ for some $\boldsymbol{\gamma} \in \mathbb{Z}^n_+ \setminus \{\mathbf{0}\}$. Then, a reduction path from $\boldsymbol{\alpha}'$ to $\boldsymbol{\beta}'$ in $\mathcal{G}(\mathcal{F}_{\mathcal{L},\mathbf{b}'}, G)$ translates by $\boldsymbol{\gamma}$ to a reduction path from $\boldsymbol{\alpha}$ to $\boldsymbol{\beta}$ in $\mathcal{G}(\mathcal{F}_{\mathcal{L},\mathbf{b}}, G)$. Thus, we only need to check for a reduction path from $\boldsymbol{\alpha}'$ to $\boldsymbol{\beta}'$.

We call a critical path $(\boldsymbol{\alpha}, \mathbf{z}, \boldsymbol{\beta})$ *minimal* if there does not exist another critical path $(\boldsymbol{\alpha}', \mathbf{z}', \boldsymbol{\beta}')$ such that $(\boldsymbol{\alpha}, \mathbf{z}, \boldsymbol{\beta}) = (\boldsymbol{\alpha}' + \boldsymbol{\gamma}, \mathbf{z}' + \boldsymbol{\gamma}, \boldsymbol{\beta}' + \boldsymbol{\gamma})$ for some $\boldsymbol{\gamma} \in \mathbb{Z}^n_+ \setminus \{\mathbf{0}\}$, or equivalently, $\min\{\alpha_i, z_i, \beta_i\} = 0$ for all $i = 1, \ldots, n$. Consequently, if there exists a reduction path between $\boldsymbol{\alpha}$ and $\boldsymbol{\beta}$ for all minimal critical paths $(\boldsymbol{\alpha}, \mathbf{z}, \boldsymbol{\beta})$, then there exists a reduction path between $\boldsymbol{\alpha}'$ and $\boldsymbol{\beta}'$ for all critical paths $(\boldsymbol{\alpha}', \mathbf{z}', \boldsymbol{\beta}')$.

Also, for each pair of vectors $\mathbf{u}, \mathbf{v} \in \mathcal{L}$, there exists a unique minimal critical path $(\boldsymbol{\alpha}^{(\mathbf{u},\mathbf{v})}, \mathbf{z}^{(\mathbf{u},\mathbf{v})}, \boldsymbol{\beta}^{(\mathbf{u},\mathbf{v})})$ determined by $\boldsymbol{\alpha}^{(\mathbf{u},\mathbf{v})} := \max\{\mathbf{u}^+, \mathbf{v}^+\}$ componentwise, $\boldsymbol{\alpha}^{(\mathbf{u},\mathbf{v})} := \mathbf{z}^{(\mathbf{u},\mathbf{v})} - \mathbf{u}$ and $\boldsymbol{\beta}^{(\mathbf{u},\mathbf{v})} := \mathbf{z}^{(\mathbf{u},\mathbf{v})} - \mathbf{v}$. So, any other critical path for $(\mathbf{u}, \mathbf{v})$ is of the form $(\boldsymbol{\alpha}^{(\mathbf{u},\mathbf{v})} + \boldsymbol{\gamma}, \mathbf{z}^{(\mathbf{u},\mathbf{v})} + \boldsymbol{\gamma}, \boldsymbol{\beta}^{(\mathbf{u},\mathbf{v})} + \boldsymbol{\gamma})$ for some $\boldsymbol{\gamma} \in \mathbb{Z}^n_+$. In algebraic terms, minimal paths correspond to the construction of S-polynomials of two binomials each of which has two terms of disjoint support.

Using minimal critical paths, we can rewrite Corollary 11.3.10 and Corollary 11.3.11, so that we only need to check for a *finite* number of reduction paths.

Lemma 11.3.12. *Let $G \subseteq \mathcal{L}_{\succ}$. If there exists a reduction path between $\boldsymbol{\alpha}^{(\mathbf{u},\mathbf{v})}$ and $\boldsymbol{\beta}^{(\mathbf{u},\mathbf{v})}$ for every pair $\mathbf{u}, \mathbf{v} \in G$, then for all $\mathbf{b} \in \mathbb{Z}^n$ and for all $\boldsymbol{\alpha}, \boldsymbol{\beta} \in \mathcal{F}_{\mathcal{L},\mathbf{b}}$ where $\boldsymbol{\alpha}$ and $\boldsymbol{\beta}$ are connected* *in* $\mathcal{G}(\mathcal{F}_{\mathcal{L},\mathbf{b}}, G)$*, there exists a reduction path between* $\boldsymbol{\alpha}$ *and* $\boldsymbol{\beta}$ *in* $\mathcal{G}(\mathcal{F}_{\mathcal{L},\mathbf{b}}, G)$*.*

Corollary 11.3.13. *A set* $G \subseteq \mathcal{L}_{\succ}$ *is a Gröbner basis of* $\mathcal{L}$ *with respect to* $\succ$ *if and only if* G *is a generating set of* $\mathcal{L}$ *and for each pair* $\mathbf{u}, \mathbf{v} \in G$*, there exists a reduction path between* $\boldsymbol{\alpha}^{(\mathbf{u},\mathbf{v})}$ *and* $\boldsymbol{\beta}^{(\mathbf{u},\mathbf{v})}$ *in* $\mathcal{G}(\mathcal{F}_{\mathcal{L},\boldsymbol{\alpha}^{(\mathbf{u},\mathbf{v})}}, G)$*.*

Note that Corollary 11.3.13 corresponds exactly to Buchberger's S-pair criterion for general polynomial ideals, Theorem 10.5.8. And as for general polynomial ideals, Corollary 11.3.13 can also be turned into an algorithm to compute a Gröbner basis of a lattice ideal $I(\mathcal{L})$ once a generating set of $I(\mathcal{L})$ is known. This is the geometric analogue to Buchberger's algorithm for the case of lattice ideals. Note, however, that it is *not* a one-to-one translation! The algorithm below deals only with vectors, which means in algebraic terms that we always divide out the gcd of the two parts of any binomial. We can do this, since the resulting binomial still lies in $I(\mathcal{L})$. Such a step is not performed by Buchberger's algorithm, as it is not valid for general polynomial ideals.

The following algorithm, Algorithm 11.1 below, called the *geometric Buchberger algorithm*, guarantees by Lemma 11.3.12 that if for a set $S \subseteq \mathcal{L}$ the points $\boldsymbol{\alpha}$ and $\boldsymbol{\beta}$ are connected in $\mathcal{G}(\mathcal{F}_{\mathcal{L},\boldsymbol{\alpha}}, S)$, then there exists a reduction path between $\boldsymbol{\alpha}$ and $\boldsymbol{\beta}$ in $\mathcal{G}(\mathcal{F}_{\mathcal{L},\boldsymbol{\alpha}}, G)$, where G denotes the set returned by the algorithm. Thus, if S is a generating set of $\mathcal{L}$, then Algorithm 11.1 returns a set G that is a Gröbner basis of $\mathcal{L}$ with respect to $\succ$ by Corollary 11.3.13.

Given a set $S \subseteq \mathcal{L}$, Algorithm 11.1 first sets $G := S$, and then directs all vectors in G according to $\succ$ such that $G \subseteq \mathcal{L}_{\succ}$. Note that at this point $\mathcal{G}(\mathcal{F}_{\mathcal{L},\mathbf{b}}, S) = \mathcal{G}(\mathcal{F}_{\mathcal{L},\mathbf{b}}, G)$ for all $\mathbf{b} \in \mathbb{Z}^n$. The algorithm then determines whether the set G satisfies Lemma 11.3.12, that is, it tries to find a reduction path from $\boldsymbol{\alpha}^{(\mathbf{u},\mathbf{v})}$ to $\boldsymbol{\beta}^{(\mathbf{u},\mathbf{v})}$ for every pair $\mathbf{u}, \mathbf{v} \in G$. If G satisfies Lemma 11.3.12, then we are done. Otherwise, no reduction path was found for some $(\mathbf{u}, \mathbf{v})$, in which case, we add a vector to G so that a reduction path exists and then again, we test whether G satisfies Lemma 11.3.12, and so on. As we "*complete* G so that it satisfies Lemma 11.3.12 in more and finally in all cases," Algorithm 11.1 is an example of a *completion procedure* [65].

To check for a reduction path, using the "maximal decreasing path" algorithm, we construct a maximal decreasing path in $\mathcal{G}(\mathcal{F}_{\mathcal{L},\boldsymbol{\alpha}^{(\mathbf{u},\mathbf{v})}}, G)$ from $\boldsymbol{\alpha}^{(\mathbf{u},\mathbf{v})}$ to some $\boldsymbol{\alpha}'$, and a maximal decreasing path in $\mathcal{G}(\mathcal{F}_{\mathcal{L},\boldsymbol{\alpha}^{(\mathbf{u},\mathbf{v})}}, G)$ from $\boldsymbol{\beta}^{(\mathbf{u},\mathbf{v})}$ to some $\boldsymbol{\beta}'$. If $\boldsymbol{\alpha}' = \boldsymbol{\alpha}'$, then we have found a reduction path from $\boldsymbol{\alpha}^{(\mathbf{u},\mathbf{v})}$ to $\boldsymbol{\beta}^{(\mathbf{u},\mathbf{v})}$. Otherwise, we add the vector $\mathbf{r}$ to G where $\mathbf{r} := \boldsymbol{\alpha}' - \boldsymbol{\beta}'$ if $\boldsymbol{\alpha}' \succ \boldsymbol{\beta}'$, and $\mathbf{r} := \boldsymbol{\beta}' - \boldsymbol{\alpha}'$ otherwise, so therefore, there is now a reduction path from $\boldsymbol{\alpha}^{(\mathbf{u},\mathbf{v})}$ to $\boldsymbol{\beta}^{(\mathbf{u},\mathbf{v})}$ in $\mathcal{G}(\mathcal{F}_{\mathcal{L},\boldsymbol{\alpha}^{(\mathbf{u},\mathbf{v})}}, G)$. Note that before we add $\mathbf{r}$ to G, since the paths from $\boldsymbol{\alpha}$ to $\boldsymbol{\alpha}'$ and from $\boldsymbol{\beta}$ to $\boldsymbol{\beta}'$ are maximal, there does not exist $\mathbf{u} \in G$ such that $\boldsymbol{\alpha}' \geq \mathbf{u}^+$ or $\boldsymbol{\beta}' \geq \mathbf{u}^+$. Therefore, there does not exist $\mathbf{u} \in G$ such that $\mathbf{r}^+ \geq \mathbf{u}^+$. This condition is needed to ensure that the completion procedure terminates. Note that in algebraic terms, the "maximal decreasing path" corresponds to the division of a monomial by a set of binomials, where *maximal* just indicates that the remainder cannot be divided any further.

ALGORITHM 11.1. Geometric Buchberger algorithm.

input a term ordering $\succ$ and a set $S \subseteq \mathcal{L}$.
output a set $G \subseteq \mathcal{L}_{\succ}$ such that if $\boldsymbol{\alpha}, \boldsymbol{\beta}$ are connected in $\mathcal{G}(\mathcal{F}_{\mathcal{L},\boldsymbol{\alpha}}, S)$, then there exists a reduction path between $\boldsymbol{\alpha}$ and $\boldsymbol{\beta}$ in $\mathcal{G}(\mathcal{F}_{\mathcal{L},\boldsymbol{\alpha}}, G)$.
$G \leftarrow \{\mathbf{u} : \mathbf{u}^+ \succ \mathbf{u}^-, \mathbf{u} \in S\} \cup \{-\mathbf{u} : \mathbf{u}^- \succ \mathbf{u}^+, \mathbf{u} \in S\}$.
$C \leftarrow \{(\mathbf{u}, \mathbf{v}) : \mathbf{u}, \mathbf{v} \in G\}$.

while $C \neq \emptyset$ **do**
  Select $(\mathbf{u},\mathbf{v}) \in C$.
  $C \leftarrow C \setminus \{(\mathbf{u},\mathbf{v})\}$.
  $\mathbf{r} \leftarrow \text{MDPA}(\boldsymbol{\alpha}^{(\mathbf{u},\mathbf{v})}, G) - \text{MDPA}(\boldsymbol{\beta}^{(\mathbf{u},\mathbf{v})}, G)$.
  if $\mathbf{r} \neq \mathbf{0}$ **then**
    if $\mathbf{r}^- \succ \mathbf{r}^+$ **then**
      $\mathbf{r} \leftarrow -\mathbf{r}$.
      $C \leftarrow C \cup \{(\mathbf{r},\mathbf{s}) : \mathbf{s} \in G\}$.
      $G \leftarrow G \cup \{\mathbf{r}\}$.
return G.

Here MDPA is the following algorithm.

ALGORITHM **11.2. Maximal decreasing path algorithm (MDPA).**

input a vector $\boldsymbol{\alpha} \in \mathbb{Z}_+^n$ and a set $G \subseteq \mathcal{L}_\succ$.
output a vector $\boldsymbol{\alpha}'$ where there is a maximal decreasing path from $\boldsymbol{\alpha}$ to $\boldsymbol{\alpha}'$ in $\mathcal{G}(\mathcal{F}_{\mathcal{L},\boldsymbol{\alpha}}, G)$.
$\boldsymbol{\alpha}' \leftarrow \boldsymbol{\alpha}$.
while there is some $\mathbf{u} \in G$ such that $\mathbf{u}^+ \leq \boldsymbol{\alpha}'$ **do**
  $\boldsymbol{\alpha}' \leftarrow \boldsymbol{\alpha}' - \mathbf{u}$.
return $\boldsymbol{\alpha}'$.

Lemma 11.3.14. *Algorithm* 11.1 *terminates and satisfies its specifications.*

Proof. Let $(\mathbf{r}^1, \mathbf{r}^2, \ldots)$ be the sequence of vectors $\mathbf{r}$ that are added to the set G during the Algorithm 11.1. Since before we add $\mathbf{r}$ to G, there does not exist $\mathbf{u} \in G$ such that $\mathbf{r}^+ \geq \mathbf{u}^+$, the sequence satisfies ${\mathbf{r}^i}^+ \not\leq {\mathbf{r}^j}^+$ whenever $i < j$. By the Gordan–Dickson lemma, such a sequence must be finite and thus, Algorithm 11.1 must terminate.

When the algorithm terminates, the set G must satisfy the property that for each $\mathbf{u},\mathbf{v} \in G$ there exists a reduction path from $\boldsymbol{\alpha}^{(\mathbf{u},\mathbf{v})}$ to $\boldsymbol{\beta}^{(\mathbf{u},\mathbf{v})}$, and therefore, by Lemma 11.3.12 there exists a reduction path between $\boldsymbol{\alpha}$ and $\boldsymbol{\beta}$ in $\mathcal{G}(\mathcal{F}_{\mathcal{L},\mathbf{b}}, G)$ for all $\boldsymbol{\alpha},\boldsymbol{\beta} \in \mathcal{F}_{\mathcal{L},\mathbf{b}}$ for all $\mathbf{b} \in \mathbb{Z}^n$ where $\boldsymbol{\alpha}$ and $\boldsymbol{\beta}$ are connected in $\mathcal{G}(\mathcal{F}_{\mathcal{L},\mathbf{b}}, G)$. Moreover, by construction, $S \subseteq G \cup -G$, and therefore, if $\boldsymbol{\alpha}$ and $\boldsymbol{\beta}$ are connected in $\mathcal{G}(\mathcal{F}_{\mathcal{L},\mathbf{b}}, S)$, then $\boldsymbol{\alpha}$ and $\boldsymbol{\beta}$ are connected in $\mathcal{G}(\mathcal{F}_{\mathcal{L},\mathbf{b}}, G)$. □

As a little *dictionary* for the translation of Buchberger's algorithm to compute Gröbner bases of general polynomial ideals to the computation of Gröbner bases of lattice ideals, we list the correspondence between (algebraic) operations on binomials and (geometric) operations of vectors below. First, let $\mathbf{u},\mathbf{v} \in \mathcal{L}$ where $\mathbf{u}^+ \succ \mathbf{u}^-$ and $\mathbf{v}^+ \succ \mathbf{v}^-$.

- The *S-polynomial* of the binomials $\mathbf{x}^{\mathbf{u}^+} - \mathbf{x}^{\mathbf{u}^-}$ and $\mathbf{x}^{\mathbf{v}^+} - \mathbf{x}^{\mathbf{v}^-}$ is the binomial

$$\frac{\mathbf{x}^{\mathbf{v}^+ \vee \mathbf{u}^+}}{\mathbf{x}^{\mathbf{v}^+}}(\mathbf{x}^{\mathbf{v}^+} - \mathbf{x}^{\mathbf{v}^-}) - \frac{\mathbf{x}^{\mathbf{v}^+ \vee \mathbf{u}^+}}{\mathbf{x}^{\mathbf{u}^+}}(\mathbf{x}^{\mathbf{u}^+} - \mathbf{x}^{\mathbf{u}^-}) = \mathbf{x}^{(\mathbf{v}^+ - \mathbf{u}^+)^+ + \mathbf{u}^-} - \mathbf{x}^{(\mathbf{v}^+ - \mathbf{u}^+)^- + \mathbf{v}^-}.$$

  Note that the least common multiple of $\mathbf{x}^{\mathbf{u}^+}$ and $\mathbf{x}^{\mathbf{v}^+}$ is simply $\mathbf{x}^{\mathbf{v}^+ \vee \mathbf{u}^+}$, where for two vectors $\mathbf{u},\mathbf{v} \in \mathbb{Z}^n$, we write $\mathbf{u} \vee \mathbf{v}$ for the componentwise maximum of $\mathbf{u}$ and $\mathbf{v}$, that is, $(\mathbf{u} \vee \mathbf{v})_i := \max\{u_i, v_i\}$. We can replace the S-polynomial with the binomial $\mathbf{x}^{(\mathbf{v}-\mathbf{u})^+} - \mathbf{x}^{(\mathbf{v}-\mathbf{u})^-}$. Thus, the *S-vector* of the vectors $\mathbf{u},\mathbf{v}$ is the vector $\mathbf{v} - \mathbf{u}$.

- If $\mathbf{x}^{\mathbf{u}^+}$ divides $\mathbf{x}^{\mathbf{v}^+}$, then $\mathbf{x}^{\mathbf{u}^+} - \mathbf{x}^{\mathbf{u}^-}$ reduces $\mathbf{x}^{\mathbf{v}^+} - \mathbf{x}^{\mathbf{v}^-}$ to the binomial $\mathbf{x}^{(\mathbf{v}^+ - \mathbf{u}^+)+\mathbf{u}^-} - \mathbf{x}^{\mathbf{v}^-}$, which we can replace with $\mathbf{x}^{(\mathbf{v}-\mathbf{u})^+} - \mathbf{x}^{(\mathbf{v}-\mathbf{u})^-}$. Thus, using just vectors, we have that if $\mathbf{u}^+ \leq \mathbf{v}^+$, then $\mathbf{u}$ reduces $\mathbf{v}$ to $\mathbf{v}-\mathbf{u}$.

- If $\mathbf{x}^{\mathbf{u}^+}$ divides $\mathbf{x}^{\mathbf{v}^-}$, then $\mathbf{x}^{\mathbf{u}^+} - \mathbf{x}^{\mathbf{u}^-}$ reduces $\mathbf{x}^{\mathbf{v}^+} - \mathbf{x}^{\mathbf{v}^-}$ to the binomial $\mathbf{x}^{\mathbf{v}^+} - \mathbf{x}^{(\mathbf{v}^- - \mathbf{u}^+)+\mathbf{u}^-}$, which we can replace with $\mathbf{x}^{(\mathbf{u}+\mathbf{v})^+} - \mathbf{x}^{(\mathbf{u}+\mathbf{v})^-}$. Thus, using just vectors, we have that if $\mathbf{u}^+ \leq \mathbf{v}^-$, then $\mathbf{u}$ reduces $\mathbf{v}$ to $\mathbf{v}+\mathbf{u}$.

11.4 Computing generating sets of lattice ideals

Let us finally turn to the problem of computing a generating set of $I(\mathcal{L})$ for given lattice $\mathcal{L} \subseteq \mathbb{Z}^n$ with $\mathcal{L} \cap \mathbb{Z}^n_+ = \{\mathbf{0}\}$. In fact, we now present the current state-of-the-art algorithm for computing a generating set of a lattice ideal. This algorithm computes a generating set of a lattice ideal via a sequence of Gröbner basis computations for a hierarchy of projections of the lattice; hence, we call this algorithm the *project-and-lift algorithm*.

Before we start, let us introduce some notation that we will need: Given a vector $\mathbf{u} \in \mathbb{Z}^n$ and a set $\sigma \subseteq \{1,\ldots,n\}$, we write $\mathbf{u}_\sigma := (u_i)_{i\in\sigma}$ and $\mathbf{u}^\sigma := (u_i)_{i\notin\sigma}$. So, $\mathbf{u}_\sigma$ is the projection of $\mathbf{u}$ onto the σ components, and $\mathbf{u}^\sigma$ is the projection of $\mathbf{u}$ onto the $\bar{\sigma} := \{1,\ldots,n\}\setminus\sigma$ components. We extend this notation to sets. Given $S \subseteq \mathbb{Z}^n$, we write $S^\sigma := \{\mathbf{u}^\sigma : \mathbf{u} \in S\}$, and $S_\sigma := \{\mathbf{u}_\sigma : \mathbf{u} \in S\}$. For the sake of brevity, we often write i meaning the set $\{i\}$ so, for example, $\mathbf{u}^i = \mathbf{u}^{\{i\}}$ and $S^i = S^{\{i\}}$.

The fundamental idea behind the project-and-lift algorithm is that we can compute a Markov basis of $I(\mathcal{L})$ given a Markov basis of $I(\mathcal{L}^i)$. Thus, analogously, given any set $\sigma \subseteq \{1,\ldots,n\}$, we can compute a Markov basis of $I(\mathcal{L}^{\sigma\setminus i})$ given a Markov basis of $I(\mathcal{L}^\sigma)$ for any $i \in \sigma$. So, starting from a given set $M \subseteq \mathcal{L}$ such that M^σ is a Markov basis of $\mathcal{L}^\sigma$ for some $\sigma \subseteq \{1,\ldots,n\}$, we compute a Markov basis of $I(\mathcal{L}^{\sigma\setminus i})$ for some $i \in \sigma$ and then set $\sigma = \sigma \setminus i$ and repeat until $\sigma = \emptyset$.

There are two cases to consider.

Case 1: We consider the situation when there exists a vector $\mathbf{a} \in \mathcal{L} \cap \mathbb{Z}^n_+$ where $a_i > 0$, that is, there exists a binomial $\mathbf{x}^\mathbf{a} - 1 \in I(\mathcal{L})$ where $a_i > 0$. Also, we require that the projection map from $\mathcal{L}$ to $\mathcal{L}^i$ is a bijection, that is, the kernel of the projection map, written $\ker^i(\mathcal{L}) := \{\mathbf{u} \in \mathcal{L} : \mathbf{u}^i = \mathbf{0}\}$, is trivial (i.e., $\ker^i(\mathcal{L}) = \{\mathbf{0}\}$).

Next, we prove the result we need to find a Markov basis of $\mathcal{L}$ given a Markov basis of $\mathcal{L}^i$ for this case.

Lemma 11.4.1. *Let $i \in \{1,\ldots,n\}$. Assume $\ker^i(\mathcal{L}) = \{\mathbf{0}\}$, and assume there exists $\mathbf{a} \in \mathcal{L} \cap \mathbb{Z}^n_+$ where $a_i > 0$. Let $M \subseteq \mathcal{L}$ such that M^i is a Markov basis of $\mathcal{L}^i$. Then, $M \cup \{\mathbf{a}\}$ is a Markov basis of $\mathcal{L}$.*

Proof. Let $\boldsymbol{\alpha}, \boldsymbol{\beta} \in \mathcal{F}_{\mathcal{L},\boldsymbol{\nu}}$ for some $\boldsymbol{\nu} \in \mathbb{Z}_n$. Since M^i is a Markov basis of $\mathcal{L}^i$, there exists a path from $\boldsymbol{\alpha}^i$ to $\boldsymbol{\beta}^i$ in $\mathcal{G}_{\mathcal{L}^i}(\boldsymbol{\nu}^i, M^i)$. Let $(\mathbf{z}_1,\mathbf{z}_2,\ldots,\mathbf{z}_k) \subseteq \mathbb{Z}^n$ such that $\boldsymbol{\alpha} = z_1$, $\boldsymbol{\beta} = z_k$, and $(\mathbf{z}^i_1,\mathbf{z}^i_2,\ldots,\mathbf{z}^i_k) \subseteq \mathcal{F}_{\mathcal{L}^i,\boldsymbol{\nu}^i}$ is a path from $\boldsymbol{\alpha}^i$ to $\boldsymbol{\beta}^i$ in $\mathcal{G}_{\mathcal{L}^i}(\boldsymbol{\nu}^i, M^i)$. Now, since $\ker^i(\mathcal{L}) = \{\mathbf{0}\}$ and $\pm(\mathbf{z}^i_j - \mathbf{z}^i_{j+1}) \in M^i$ for all $j < k$, we must have $\pm(\mathbf{z}_j - \mathbf{z}_{j+1}) \in M$ for all $j < k$. Then, there must exist a $\lambda \in \mathbb{Z}_+$ such that $\mathbf{z}_j + \lambda\mathbf{a} \geq \mathbf{0}$ for all j. Then, $(\mathbf{z}_1 + \lambda\mathbf{a}, \mathbf{z}_2 + \lambda\mathbf{a}, \ldots, \mathbf{z}_k + \lambda\mathbf{a})$ is a path from $\boldsymbol{\alpha} + \lambda\mathbf{a}$ to $\boldsymbol{\beta} + \lambda\mathbf{a}$ in $\mathcal{G}_\mathcal{L}(\boldsymbol{\nu}, M)$.

Also, $\boldsymbol{\beta}$ is connected to $\boldsymbol{\beta} + \lambda\mathbf{a}$ in $\mathcal{G}_\mathcal{L}(\boldsymbol{\nu}, M)$ by the path $(\boldsymbol{\beta}, \boldsymbol{\beta} + \mathbf{a}, \ldots, \boldsymbol{\beta} + \lambda\mathbf{a})$, and similarly, $\boldsymbol{\alpha}$ is connected to $\boldsymbol{\alpha} + \lambda\mathbf{a}$ in $\mathcal{G}_\mathcal{L}(\boldsymbol{\nu}, M)$. Thus, $\boldsymbol{\alpha}$ is connected to $\boldsymbol{\beta}$ in $\mathcal{G}_\mathcal{L}(\boldsymbol{\nu}, M)$ as required. $\square$

Lemma 11.4.1 above is equivalent to the following algebraic result. Assume that $\ker^i(\mathcal{L}) = \{\mathbf{0}\}$, and assume that there exists $\mathbf{x}^{\mathbf{a}} - 1 \in I(\mathcal{L})$ for some $\mathbf{a} \in \mathcal{L} \cap \mathbb{Z}_+^n$ where $a_i > 0$. Let $M \subseteq \mathcal{L}$ such that $I(M^i) = I(\mathcal{L}^i)$. Then, $I(M \cup \{\mathbf{a}\}) = I(\mathcal{L})$.

Now, analogously to the above, given $\sigma \subseteq \{1,\ldots,n\}$ and $i \in \sigma$, we can find a Markov basis of $\mathcal{L}^{\sigma\setminus i}$ from a Markov basis of $\mathcal{L}^\sigma$ using the following corollary of Lemma 11.4.1. Note that $\mathcal{L}^{\sigma\setminus i}$ and $\mathcal{L}^\sigma$ correspond to $\mathcal{L}$ and $\mathcal{L}^i$ respectively in Lemma 11.4.1. Also, note that $\ker^i_\sigma(\mathcal{L}) := \left\{\mathbf{u}^{\sigma\setminus i} : \mathbf{u} \in \mathcal{L}, \mathbf{u}^\sigma = \mathbf{0}\right\}$ is the kernel of the projective map from $\mathcal{L}^{\sigma\setminus i}$ to $\mathcal{L}^\sigma$.

Corollary 11.4.2. *Let $\sigma \subseteq \{1,\ldots,n\}$ and let $i \in \sigma$. Assume $\ker^i_\sigma(\mathcal{L}) = \{\mathbf{0}\}$, and assume there exists $\mathbf{a} \in \mathcal{L}$ where $\mathbf{a}_\sigma \geq \mathbf{0}$ and $a_i > 0$. Let $M \subseteq \mathcal{L}$ such that M^σ is a Markov basis of $\mathcal{L}^\sigma$. Then, $(M \cup \{\mathbf{a}\})^{\sigma\setminus i}$ is a Markov basis of $\mathcal{L}^{\sigma\setminus i}$.*

Note that using linear algebra and linear programming we can efficiently find a vector $a \in \mathcal{L} \cap \mathbb{Z}_+^n$, where $\mathbf{a}_\sigma \geq \mathbf{0}$ and $a_i > 0$, or determine that there is no such vector.

Case 2: There is no vector $\mathbf{a} \in \mathcal{L} \cap \mathbb{Z}_+^n$ where $a_i > 0$, that is, there does not exist a binomial $\mathbf{x}^{\mathbf{a}} - 1 \in I(\mathcal{L})$ where $a_i > 0$. Note that this implies that $\ker^i(\mathcal{L}) = \{\mathbf{0}\}$.

For this case, we will show that, given a set $G \subseteq \mathcal{L}$, there exists a term order $\succ$ such that G is a $\succ$-Gröbner basis of $\mathcal{L}$ if and only if G^i is a $\succ^i$-Gröbner basis of $\mathcal{L}^i$ where $\succ^i$ is the restriction of $\succ$ to the components $\{1,\ldots,n\}\setminus i$ where we treat the i-th component as if it were zero. In other words, $J(G)$ is a $\succ$-Gröbner basis of $I(\mathcal{L})$ if and only if $J(G^i)$ is a $\succ^i$-Gröbner basis of $I(\mathcal{L}^i)$.

We now define the term order that we need. It follows from Farkas' lemma in linear programming that there is no vector $\mathbf{a} \in \mathcal{L} \cap \mathbb{Z}_+^n$, where $a_i > 0$, if and only if there exists a vector $\mathbf{c} \in \mathbb{R}_+^n$ such that $c_i = 0$ and $\mathbf{c}^\top \mathbf{u} = -u_i$ for all $\mathbf{u} \in \mathcal{L}$. Now, given any term order $\succ$ on $\mathbb{Z}_+^n$, we define the term order $\succ_{\mathbf{c}}$ on $\mathbb{Z}_+^n$ as $\boldsymbol{\alpha} \succ_{\mathbf{c}} \boldsymbol{\beta}$ if $\mathbf{c}^\top\boldsymbol{\alpha} > \mathbf{c}^\top\boldsymbol{\beta}$ or $\mathbf{c}^\top\boldsymbol{\alpha} = \mathbf{c}^\top\boldsymbol{\beta}$ and $\boldsymbol{\alpha} \succ \boldsymbol{\beta}$. Note that $\succ_c$ is a term order since $\mathbf{c} \geq \mathbf{0}$. The term order $\succ_{\mathbf{c}}$ has the crucial property that $u_i \leq 0$ for every $\mathbf{u} \in \mathcal{L}_\succ$ because $\mathbf{u}^+ \succ_{\mathbf{c}} \mathbf{u}^- \Rightarrow \mathbf{c}^\top\mathbf{u}^+ \geq \mathbf{c}^\top\mathbf{u}^- \Leftrightarrow \mathbf{c}^\top\mathbf{u} \geq 0 \Leftrightarrow u_i \leq 0$. We also define the term order $\succ^i_{\mathbf{c}}$ as the term order $\succ_{\mathbf{c}}$ restricted to the components $\{1,\ldots,n\}\setminus i$ where we treat the i-th component as if it were zero. Crucially, for all $\mathbf{u} \in \mathcal{L}$, we have $\mathbf{u} \in \mathcal{L}_\succ$ if and only if $\mathbf{u}^i \in \mathcal{L}^i_{\succ^i_{\mathbf{c}}}$, that is, $\mathbf{u}^+ \succ_{\mathbf{c}} \mathbf{u}^-$ if and only if $(\mathbf{u}^i)^+ \succ^i_{\mathbf{c}} (\mathbf{u}^i)^-$. This follows since $\mathbf{c}^\top\mathbf{u} = \mathbf{c}^{i\top}\mathbf{u}^i = -u_i$.

Lemma 11.4.3. *Assume that there exists $\mathbf{c} \in \mathbb{R}_+^n$ such that $c_i = 0$ and $\mathbf{c}^\top\mathbf{u} = -u_i$ for all $\mathbf{u} \in \mathcal{L}$, and let $\succ$ be a term order on $\mathbb{Z}_+^n$. A set $G \subseteq \mathcal{L}_{\succ_{\mathbf{c}}}$ is a $\succ_{\mathbf{c}}$-Gröbner basis of $\mathcal{L}$ if and only if G^i is a $\succ^i_{\mathbf{c}}$-Gröbner basis of $\mathcal{L}^i$.*

Proof. Assume that G is a $\succ_{\mathbf{c}}$-Gröbner basis of $\mathcal{L}$. Let $\mathbf{u} \in \mathcal{L}$ such that $u^i \in \mathcal{L}^i_{\succ^i_{\mathbf{c}}}$. We must show that there exists $\mathbf{v} \in G$ such that $(\mathbf{v}^i)^+ \leq (\mathbf{u}^i)^+$ and the result follows by Lemma 11.3.5. Now, $\mathbf{u} \in \mathcal{L}_\succ$ as well, and so, since G is a $\succ_{\mathbf{c}}$-Gröbner basis of $\mathcal{L}$, there exists $\mathbf{v} \in G$ such that $\mathbf{v}^+ \leq \mathbf{u}^+$ by Lemma 11.3.5. Hence, $(\mathbf{v}^i)^+ \leq (\mathbf{u}^i)^+$ as required.

Assume that G^i is a $\succ^i_{\mathbf{c}}$-Gröbner basis of $\mathcal{L}^i$. Let $\mathbf{u} \in \mathcal{L}_{\succ_{\mathbf{c}}}$. We must show that there exists $\mathbf{v} \in G$ such that $\mathbf{v}^+ \leq \mathbf{u}^+$, and the result follows by Lemma 11.3.5. Since G^i is a $\succ^i_{\mathbf{c}}$-Gröbner basis of $\mathcal{L}^i$ and $\mathbf{u}^i \in \mathcal{L}^i_{\succ_{\mathbf{c}}}$, there exists $\mathbf{v} \in G$ such that $(\mathbf{v}^i)^+ \leq (\mathbf{u}^i)^+$. Then, since $\mathbf{v}_i^+ = 0$, we must have $\mathbf{v}^+ \leq \mathbf{u}^+$ as required. □

Lemma 11.4.3 above is equivalent to the following algebraic result. Let $\succ_{\mathbf{c}}$ be defined as above, and let $G \subseteq \mathcal{L}_{\succ_{\mathbf{c}}}$. The set $J(G)$ is a $\succ_{\mathbf{c}}$-Gröbner basis of $I(\mathcal{L})$ if and only if $J(G^i)$

is a $\succ^i_{\mathbf{c}}$-Gröbner basis of $I(\mathcal{L}^i)$.

So, given a set $M \subseteq \mathcal{L}$ such that M^i is a Markov basis of $\mathcal{L}^i$, we can compute a $G \subseteq \mathcal{L}$ such that G^i is a $\succ^i_{\mathbf{c}}$-Gröbner basis of $\mathcal{L}^i$ using the geometric Buchberger algorithm. Then, G is a $\succ_{\mathbf{c}}$-Gröbner basis of $\mathcal{L}$ and thus G is a Markov basis of $\mathcal{L}$ since Gröbner bases are Markov bases.

Now, analogously to the above, given $\sigma \subseteq \{1,\dots,n\}$ and $i \in \sigma$ where $\ker^i_\sigma(\mathcal{L}) = \{\mathbf{0}\}$, we can find a Markov basis of $\mathcal{L}^{\sigma\setminus i}$ from a Markov basis of $\mathcal{L}^\sigma$ using the following corollary of Lemma 11.4.3. First, note that, from Farkas' lemma, there is no vector $\mathbf{a} \in \mathcal{L}$ where $\mathbf{a}^\sigma \geq \mathbf{0}$ and $a_i > 0$ if and only if there exists a vector $\mathbf{c} \in \mathbb{R}^n_+$ such that $\mathbf{c}_\sigma = \mathbf{0}$ and $\mathbf{c}^\top \mathbf{u} = -u_i$ for all $\mathbf{u} \in \mathcal{L}$. Also, analogously to $\succ^i_{\mathbf{c}}$, we define $\succ^\sigma_{\mathbf{c}}$ as the term order $\succ_c$ restricted to the $\bar{\sigma}$ components and $\succ^{\sigma\setminus i}_{\mathbf{c}}$ as the term order $\succ_{\mathbf{c}}$ restricted to the $\bar{\sigma} \cup i$ components. Lastly, note that $\mathcal{L}^\sigma_{\succ^\sigma_{\mathbf{c}}} := \left\{\mathbf{u}^\sigma : \mathbf{u} \in \mathcal{L}, (\mathbf{u}^\sigma)^+ \succ^\sigma_{\mathbf{c}} (\mathbf{u}^\sigma)^-\right\}$.

Corollary 11.4.4. *Assume that there exists $\mathbf{c} \in \mathbb{R}^n_+$ such that $\mathbf{c}_\sigma = \mathbf{0}$ and $\mathbf{c}^\top \mathbf{u} = -u_i$ for all $\mathbf{u} \in \mathcal{L}$, and let $\succ$ be a term order on $\mathbb{Z}^n_+$. Let $\sigma \subseteq \{1,\dots,n\}$ and let $i \in \sigma$. Let $G \subseteq \mathcal{L}$ where $G^\sigma \subseteq \mathcal{L}^\sigma_{\succ^\sigma_{\mathbf{c}}}$. The set G is a $\succ^\sigma_{\mathbf{c}}$-Gröbner basis of $\mathcal{L}^\sigma$ if and only if $G^{\sigma\setminus i}$ is a $\succ^{\sigma\setminus i}_{\mathbf{c}}$-Gröbner basis of $\mathcal{L}^{\sigma\setminus i}$.*

So, given a set $M \subseteq \mathcal{L}$ such that M^σ is a Markov basis of $\mathcal{L}^\sigma$, we can compute a $G \subseteq \mathcal{L}$ such that G^σ is a $\succ^\sigma_{\mathbf{c}}$-Gröbner basis of $\mathcal{L}^\sigma$. Then, $G^{\sigma\setminus i}$ is a $\succ^{\sigma\setminus i}_{\mathbf{c}}$-Gröbner basis of $\mathcal{L}^{\sigma\setminus i}$ and thus $G^{\sigma\setminus i}$ is a Markov basis of $\mathcal{L}^{\sigma\setminus i}$.

We can now describe the project-and-lift algorithm. The first step is to find a set $\sigma \subseteq \{1,\dots,n\}$ such that $\ker^\sigma(\mathcal{L}) := \{\mathbf{u} \in \mathcal{L} : \mathbf{u}^\sigma = \mathbf{0}\} = \{\mathbf{0}\}$. Note that $\ker^\sigma(\mathcal{L}) = \{\mathbf{0}\}$ implies that $\ker^i_{\sigma'}(\mathcal{L}) = \{\mathbf{0}\}$ for all $\sigma' \subseteq \sigma$ and for all $i \in \sigma'$. Next, we must find a set $M \subseteq \mathcal{L}$ such that M^σ is a Markov basis of $\mathcal{L}^\sigma$. Then, we can apply the above reasoning for either case 1 or case 2 to compute a Markov basis of $\mathcal{L}^{\sigma\setminus i}$ from a Markov basis of $\mathcal{L}^\sigma$ for some $i \in \sigma$, and we do this iteratively for every $i \in \sigma$. Note that, at the beginning of each iteration of the algorithm, M^σ is a Markov basis of $\mathcal{L}^\sigma$.

ALGORITHM 11.3. Project-and-lift.

input a lattice $\mathcal{L}$.
output a Markov basis M of $\mathcal{L}$.
Find a set $\sigma \subseteq \{1,\dots,n\}$ such that $\ker^\sigma(\mathcal{L}) = \{\mathbf{0}\}$.
Find a set $M \subseteq \mathcal{L}$ such that M^σ is a Markov basis of $\mathcal{L}^\sigma$.
while $\sigma \neq \emptyset$ **do**
    Select $i \in \sigma$.
    if there exists $\mathbf{a} \in \mathcal{L}$ such that $\mathbf{a}^\sigma \geq \mathbf{0}$ and $a_i > 0$ **then**
        Set $M \leftarrow M \cup \{\mathbf{a}\}$.
    else
        Find $\mathbf{c} \in \mathbb{R}^n_+$ such that $\mathbf{c}_\sigma = \mathbf{0}$ and $\mathbf{c}^\top \mathbf{u} = -u_i$ for all $\mathbf{u} \in \mathcal{L}$.
        Using M, compute $G \subseteq \mathcal{L}$ such that G^σ is a $\succ^\sigma_{\mathbf{c}}$-Gröbner basis of $\mathcal{L}^\sigma$.
        Set $M \leftarrow G$.
    $\sigma \leftarrow \sigma \setminus i$.
return M.

We now show how to find a set $\sigma \subseteq \{1,\dots,n\}$ such that $\ker^\sigma(\mathcal{L}) = \{\mathbf{0}\}$ and how to find a set $M \subseteq \mathcal{L}$ such that M^σ is a Markov basis of $\mathcal{L}^\sigma$. Consider the special case that

$\mathcal{L} \subseteq \mathbb{Z}^n$ is an n-dimensional lattice (i.e., there are n vectors in a basis of $\mathcal{L}$). Let $B \in \mathbb{Z}^{n\times n}$ be a matrix in Hermite normal form (HNF) (see Section 2.3) that is, B is an upper triangular matrix with positive diagonal entries and nonpositive entries elsewhere such that the rows of B form a basis of $\mathcal{L}$. We can always construct such a matrix B in polynomial time from any basis of $\mathcal{L}$ using the HNF algorithm, which performs operations on the rows of B such as multiplying a row by -1 or adding some integer multiple of one row to another. The rows of the matrix B actually give a Gröbner basis of $\mathcal{L}$ with respect to the lexicographic order written $\succ_{\text{lex}}$. See Lemma 11.4.5 below. Thus, the rows of B are a Markov basis of $\mathcal{L}$.

Lemma 11.4.5. *Let $\mathcal{L} \subseteq \mathbb{Z}^n$ be an n-dimensional lattice. Let $B \in \mathbb{Z}^{n\times n}$ be a matrix in HNF such that the rows of B form a basis of $\mathcal{L}$. The rows of B give a $\succ_{\text{lex}}$-Gröbner basis of $\mathcal{L}$.*

Proof. Let $\mathbf{u} \in \mathcal{L}_{\succ_{\text{lex}}}$. Then, the first nonzero entry in $\mathbf{u}$ is positive. Assume u_i is the first nonzero entry, and let $\mathbf{b}_i$ be the i-th row of B. Since B is in HNF and B is a basis of $\mathcal{L}$, we must have that b_{ii} divides u_i. Since $b_{ij} \leq 0$ for all $j \neq i$, we must have $\mathbf{b}_i^+ \leq \mathbf{u}^+$ and, therefore, the rows of B give a $\succ_{\text{lex}}$-Gröbner basis of $\mathcal{L}$ by Lemma 11.3.5. □

Consider now the general case of a k-dimensional lattice $\mathcal{L} \subseteq \mathbb{Z}^n$. Let B be a matrix such that the rows of B form a basis of $\mathcal{L}$. Any k linearly independent columns of B then suffice to give a set $\bar{\sigma}$ such that there is a one-to-one correspondence between vectors in $\mathcal{L}^\sigma$ and vectors in $\mathcal{L}$; that is, $\ker^\sigma(\mathcal{L}) = \{\mathbf{0}\}$. Such a set σ can be found via Gaussian elimination. Now, consider the set B^σ, which is the square submatrix of B consisting of the columns indexed by $\bar{\sigma}$. We can assume that B^σ is in HNF form. Then, as discussed above, from Lemma 11.4.5, B^σ is a Markov basis of $\mathcal{L}^\sigma$ as required.

The project-and-lift algorithm does not in general compute a minimal Markov basis. See [69] for an algorithm to compute a minimal Markov basis.

The following example shows an example computation using the project-and-lift algorithm.

Example 11.4.6. Consider the lattice $\mathcal{L}$ generated by the vectors $(1,-5,-3,3)^\top$ and $(0,8,4,-5)^\top$. The set

$$M := \left\{(2,-2,-2,1)^\top,(3,1,-1,-1)^\top,(5,-1,-3,0)^\top,(1,3,1,-2)^\top\right\}$$

is a minimal Markov basis of $\mathcal{L}$. We will compute this set using the project-and-lift algorithm.

The set $M = \left\{(1,-5,-3,3)^\top,(0,8,4,-5)^\top\right\}$ is a basis of $\mathcal{L}$. Let $\sigma = \{3,4\}$. Then, $\ker \mathcal{L}\sigma = \{\mathbf{0}\}$. Note that $M^\sigma = \left\{(1,-5)^\top,(0,8)^\top\right\}$, and by Lemma 11.4.5, M^σ is a Markov basis of $\mathcal{L}^\sigma$.

Select $i := 3$. Let $\mathbf{a} = (0,8,4,-5)^\top$. Then, $\mathbf{a} \in \mathcal{L}$ and $\mathbf{a}^\sigma \geq \mathbf{0}$ and $a_i > 0$. Set $M := M \cup \{\mathbf{a}\} = \left\{(1,-5,-3,3)^\top,(0,8,4,-5)^\top\right\}$. Then, $M^{\sigma\setminus i}$ is a Markov basis of $\mathcal{L}^{\sigma\setminus i}$ by Lemma 11.4.1. Set $\sigma = \sigma \setminus i = \{4\}$.

Select $i := 4$. Let $\mathbf{c} := (\frac{1}{8},\frac{5}{8},0,0)^\top$. Then, $\mathbf{c}\cdot\mathbf{u} = -u_i$ for all $\mathbf{u} \in \mathcal{L}$ and $\mathbf{c}_\sigma = 0$. The set $\left\{(-5,1,3)^\top,(-2,2,2)^\top,(1,3,1)^\top,(3,1,-1)^\top,(8,0,-4)^\top\right\}$ is a $\succ_{\mathbf{c}}^\sigma$-Gröbner basis of $\mathcal{L}^\sigma$ where $\succ$ is a degree reverse lexicographic ordering. Set

$$M := \left\{(-5,1,3,0)^\top,(-2,2,2,-1)^\top,(1,3,1,-2)^\top,(3,1,-1,-1)^\top,(8,0,-4,-1)^\top\right\}.$$

Then, M^σ is a $\succ_{\mathbf{c}}^\sigma$-Gröbner basis of $\mathcal{L}^\sigma$. Thus, $M^{\sigma\setminus i}$ is a Markov basis of $\mathcal{L}^{\sigma\setminus i}$ by Lemma 11.4.3. Set $\sigma := \sigma \setminus i = \emptyset$.

Since $\sigma = \emptyset$, M is a Markov basis of $\mathcal{L}$, and we are done.

The set

$$M = \left\{(-5,1,3,0)^\top,(-2,2,2,-1)^\top,(1,3,1,-2)^\top,(3,1,-1,-1)^\top,(8,0,-4,-1)^\top\right\}$$

is thus a Markov basis of $\mathcal{L}$. Note that we computed one more vector than necessary: the vector $(8,0,-4,-1)^\top$ is not needed in a Markov basis of $\mathcal{L}$.

In the project-and-lift method, the order for selecting an $i \in \sigma$ is left unspecified because it is not important for the correctness of the algorithm; however, it is important in practice. As a general rule, it is better to select $i \in \sigma$ for which there exists $\mathbf{c} \in \mathbb{R}^n_+$ such that $\mathbf{c}_\sigma = \mathbf{0}$ and $\mathbf{c}^\top \mathbf{u} = -u_i$ (case 2) in preference to selecting $j \in \sigma$ for which there exists $\mathbf{a} \in \mathcal{L}$ such that $\mathbf{a}_{\bar{\sigma}} \geq \mathbf{0}$ and $a_j > 0$ (case 1).

The reason being is that case 2 is much more computationally expensive than case 1 since it involves computing a Gröbner basis of $\mathcal{L}^\sigma$ and Gröbner bases of $\mathcal{L}^\sigma$ are generally much smaller the larger σ is. Note that changing the order of selecting $i \in \sigma$ can affect whether a component $j \in \{1,\ldots,n\}$ satisfies case 1 or case 2, so it is difficult to say with certainly what is the better selection method.

11.5 Notes and further references

Several algorithms exist for computing a generating set of $I(\mathcal{L})$ given a basis of the lattice $\mathcal{L}$. The first such algorithms were given by Conti and Traverso [73] and Pottier [274] (see also [5]). Thomas [324] presented the first geometric version of Buchberger's algorithm for the toric ideal case, which was improved by Li et al. [232]. These algorithms involve introducing $n-d$ or $n-d+1$ additional indeterminates, where d is the dimension of the lattice, and then eliminating the extra indeterminates by computing a Gröbner basis for an elimination ordering. Since the performance of the Buchberger algorithm is so sensitive to the number of indeterminates, adding extra indeterminates is computationally very costly, so these algorithms are not efficient. Hoşten and Sturmfels [175] presented an algorithm that does not require introducing any additional variables. This algorithm is based upon the result that

$$I(\mathcal{L}) = (\cdots((I(S):\mathbf{x}_1^\infty):\mathbf{x}_2^\infty)\cdots):\mathbf{x}_n^\infty,$$

where S is a lattice basis of $\mathcal{L}$. Using this result, the algorithm computes a generating set of $I(\mathcal{L})$ from S via a sequence of *saturation* steps, where each individual saturation step involves a Gröbner basis computation in n indeterminates.

Bigatti, LaScala, and Robbiano [52] gave an algorithm that requires introducing one extra variable and a single Gröbner basis computation; the algorithm is based upon the related result that $I(\mathcal{L}) = I(S) : (\mathbf{x}_1 \cdot \mathbf{x}_2 \cdots \mathbf{x}_n)^\infty$.

Another way of computing a generating set of a lattice ideal is to compute a *Graver basis* of a lattice [144], which is a superset of a Markov basis of a lattice [319]. In Section 3.8, we have presented an algorithm to compute a Graver basis of a lattice. (Note that we used a project-and-lift approach there, too.) A Graver basis of a lattice is generally much larger than a minimal Markov basis of a lattice, but there are special lattices for which any Markov basis is necessarily also a Graver basis [319].

During the Buchberger algorithm, one must check whether the S-polynomial of every critical pair reduces to 0. Checking reduction to 0 is computationally expensive, so it is desirable to avoid this as much as possible. Hemmecke and Malkin [155] present three criteria that are very useful in practice for avoiding reduction to 0. The first two criteria are well-known for general Gröbner basis computations and they present their geometric variants. The third criterion is specific to saturated ideals (i.e., $I = I : \mathbf{x}^\infty$) and thus lattice ideals, and it is extremely effective in practice. The project-and-lift algorithms to compute generating sets and Gröbner bases of lattice ideals are implemented in the software package `4ti2` [1].

Besides the computational aspects, there are a variety of papers on different questions related to toric ideals, for example: truncated Gröbner bases [325], Gröbner bases for 0/1 problems [51], variation of the objective function [319], Gröbner bases in two-stage stochastic integer programming [302], or Gröbner bases for scheduling in presence of setups and correlated demands [322]. We must also remark that Blanco and Puerto have investigated applications of Gröbner techniques in the multiobjective optimization setting [54, 55]. Finally, in [156], the concept of Gröbner complexity is defined for (universal) Gröbner bases of N-fold matrices (similar to Graver complexity as in Section 4.1.1). It is shown that for any unimodular matrix A, Graver and Gröbner complexities of A and B coincide for any matrix B of suitable dimensions.

11.6 Exercises

Exercise 11.6.1. For a problem in the form

$$\max\left\{\mathbf{c}^\top\mathbf{z} : A\mathbf{z} \leq \mathbf{b},\ \mathbf{z} \geq \mathbf{0}\right\},$$

with $A \in \mathbb{Z}_+^{m\times n}$, $\mathbf{b} \in \mathbb{Z}_+^m$, and $\mathbf{c} \in \mathbb{Z}^n$, how can we compute upper bounds u_i for the variables x_i? What happens in the special case where A has a zero column?

Exercise 11.6.2. Prove that for $\mathbf{a},\mathbf{b} \in \mathcal{L}$ and $\mathbf{a} \succ \mathbf{b}$, the S-polynomial of $\mathbf{x}^{\mathbf{a}^+} - \mathbf{x}^{\mathbf{a}^-} \in I(\mathcal{L})$ and $\mathbf{x}^{\mathbf{b}^+} - \mathbf{x}^{\mathbf{b}^-} \in I(\mathcal{L})$ is

$$\mathbf{x}^{\min(\mathbf{a}^-,\mathbf{b}^-)}\left(\mathbf{x}^{(\mathbf{a}-\mathbf{b})^+} - \mathbf{z}^{(\mathbf{a}-\mathbf{b})^-}\right),$$

a monomial times the binomial corresponding to $\mathbf{a}-\mathbf{b}$.

Exercise 11.6.3. Let I be an ideal generated by the binomials corresponding to some sublattice $\mathcal{L} \subseteq \mathbb{Z}^n$:

$$I_{\mathcal{L}} := \langle \mathbf{x}^{\mathbf{a}^+} - \mathbf{x}^{\mathbf{a}^-} : \mathbf{a} \in \mathcal{L}\rangle.$$

Then a binomial $\mathbf{x}^{\mathbf{a}} - \mathbf{x}^{\mathbf{b}}$, with $\mathbf{a},\mathbf{b} \geq \mathbf{0}$, is contained in $I(\mathcal{L})$ if and only if $\mathbf{a}-\mathbf{b} \in \mathcal{L}$. (Hint: $\mathbf{x}^{\mathbf{a}}$ and $\mathbf{x}^{\mathbf{b}}$ reduce to the same standard monomial.)

Exercise 11.6.4. Let $\mathcal{L} \subseteq \mathbb{Z}^n$ be an n-dimensional integral lattice, and associate with it the lattice ideal

$$I(\mathcal{L}) := \langle \mathbf{x}^{\mathbf{a}^+} - \mathbf{x}^{\mathbf{a}^-} : \mathbf{a} \in \mathcal{L}\rangle.$$

Here we assume that the lattice $\mathcal{L}$ is generated by the columns of a nonnegative matrix $A \in \mathbb{Z}_+^{n\times n}$.

Show that if the columns $(\mathbf{a}_1,\mathbf{a}_2,\ldots,\mathbf{a}_n)$ of A generate $\mathcal{L}$, and if they are all nonnegative, then the ideal $I(\mathcal{L})$ is generated by the binomials

$$\mathbf{x}^{\mathbf{a}_i^+} - \mathbf{x}^{\mathbf{a}_i^-} = \mathbf{x}^{\mathbf{a}_i} - 1.$$

Show that this can fail if we do not assume A to be nonnegative.

Exercise 11.6.5. Let $\mathcal{L}_A \subseteq \mathbb{Z}^2$ be the two-dimensional lattice generated by the columns of $A = \left(\begin{smallmatrix} 1 & 4 \\ 4 & 3 \end{smallmatrix}\right)$. Compute all the (four) different reduced Gröbner bases for the corresponding ideal. Describe the structure of the various Gröbner basis elements. How many standard monomials are there in each case? (The standard monomials are those monomials outside the ideal $\mathrm{LT}_\succ(I)$.)

Exercise 11.6.6. (following [88].) For a given matrix A, let I_A denote the toric ideal associated with a given matrix. For a given matrix W, let $\succ_W$ be the matrix term ordering defined in Definition 10.4.3. Show that in fixed dimension one can compute the rational generating function A (Chapter 7) of the reduced minimal Gröbner basis of I_A with respect to $\succ_W$ in polynomial time.

Part V

Nullstellensatz and Positivstellensatz Relaxations

Chapter 12

The Nullstellensatz in Discrete Optimization

Hilbert's Nullstellensatz is among the classical theorems in algebraic geometry. It regards a certificate for the infeasibility of a polynomial system. Here we discuss how the Nullstellensatz can be used to certify infeasibility for combinatorial optimization problems modeled with nonlinear polynomial constraints. It turns out that from such algebraic identities either a sequence of linear algebra problems or a sequence of semidefinite programming problems can be used to solve the desired problem.

12.1 Motivation and examples

We are particularly motivated by problems of combinatorial origin. For example, in the case of graphs one can easily think about stable sets, k-colorability, and max-cut problems in terms of polynomial (nonlinear) constraints:

Theorem 12.1.1. *Let $G = (V, E)$ be a graph.*

1. *For a given positive integer k, consider the following polynomial system:*

$$x_i^2 - x_i = 0 \ \forall i \in V, \quad x_i x_j = 0 \ \forall (i, j) \in E, \ \textit{and} \ \sum_{i \in V} x_i = k.$$

   This system is feasible if and only if G has a stable set of size k.

2. *We can represent the set of cuts of G (i.e., bipartitions on V) as the 0-1 incidence vectors*

$$SG := \{\chi^F : F \subseteq E \text{ is contained in a cut of } G\} \subseteq \{0,1\}^E.$$

   Thus, the max-cut problem with nonnegative weights w_e on the edges $e \in E$ is

$$\max \left\{ \sum_{e \in E(G)} w_e x_e \ : \ x \in SG \right\}.$$

   The vectors χ^F are the solutions of the polynomial system

$$x_e^2 - x_e = 0 \, \forall \, e \in E(G) \quad \textit{and} \quad \prod_{e \in T \cap E(G)} x_e = 0 \ \forall \, T \text{ an odd cycle in } G.$$

3. *For a positive integer k, consider the following polynomial system of* $|V|+|E|$ *equations:*
$$x_i^k - 1 = 0 \ \forall i \in V \quad \text{and} \quad \sum_{s=0}^{k-1} x_i^{k-1-s} x_j^s = 0 \ \forall (i,j) \in E.$$
The graph G is k-colorable if and only if this system has a complex solution. Furthermore, when k is odd, G is k-colorable if and only if this system has a common root over $\overline{\mathbb{F}}_2$*, the algebraic closure of the finite field with two elements.*

Proof. We leave as exercises the proofs of the first two statements and concentrate on the characterization of coloring problems:

First let us argue the statement is correct over the complex numbers: If G is k-colorable, assign a k-th root of unity to each of the colors. Say the j-th color is assigned to $\beta_j = e^{2\pi j/k}$, substitute on x_l the root that corresponds to the color of the vertex labeled l. We claim that then we have a solution of the system. Clearly the equations of type $x_i^k - 1$ are all satisfied. Only the edge equations need to be verified: Take an edge ij; since we substituted roots of unity we know $x_i^k - x_j^k = 0$. We know that
$$x_i^k - x_j^k = (x_i - x_j)(x_i^{k-1} + x_i^{k-2}x_j + x_i^{k-3}x_j + \cdots + x_j^{k-1}) = 0;$$
since $x_i - x_j$ cannot be zero because the roots are different (they came from a coloring!), the other factor must then be zero.

Conversely, suppose we have a common zero for all those equations. Clearly such a solution point is made of k-th roots of unity. We claim that adjacent nodes received different roots of unity. Otherwise, say the pair ij of adjacent nodes receives the same root of unity β (think of it as a color). The equation $x_i^{k-1} + x_i^{k-2}x_j + x_i^{k-3}x_j + \cdots + x_1^{k-1} = 0$ becomes then $\beta^{k-1} + \beta^{k-1} + \cdots + \beta^{k-1} = k\beta^{k-1} = 0$, but we know that $\beta \neq 0$ a contradiction.

Now to finish the proof we need to verify that exactly the same argument we gave over the complex numbers works for $\overline{\mathbb{F}}_2$. Although $x_i^k - 1$ has only one root over $\mathbb{F}_2$ (the number 1), again it factors completely into distinct roots over $\overline{\mathbb{F}}_2$, $1, \beta_i, \ldots, \beta_{k-1}$ (this time not complex numbers). The same arguments above work, we only need to be careful in the second half of the argument since we need to have $k\beta_i^{k-1}$ be nonzero. This is fine because k is odd and by construction $\beta_i^k = 1$. □

There is a wide range of applications of polynomials to problems in combinatorics and graph theory (see, e.g., the following interesting references [11, 12, 84, 125, 128, 243, 245, 253]). The particular encodings presented in Theorem 12.1.1 have in fact interesting applications in the theory of graphs too. For example, when one considers the ideal $I(G,k)$ generated by the k-coloring polynomials one realizes from the theory of polynomial ideals (see Chapter 1 of [79]) that the dimension of the vector space $\mathbb{C}[x_1,\ldots,x_n]/I(G,k)$ equals $k!$ times the number of distinct k-colorings of the graph G. As a result we obtain that a graph G is uniquely k-colorable if and only if the dimension of the $\mathbb{C}$-vector space $\mathbb{C}[x_1,\ldots,x_n]/I(G,k)$ is $k!$ (for details see [169]). Similarly the stable set polynomials of the first statement have had several interesting applications, many introduced by Lovász [236]. There are many other combinatorial problems that can be modeled concisely by polynomial systems (see [100] and the many references therein).

It is also worth noticing that a combinatorial problem can often be modeled nonlinearly in many different ways, and in practice choosing a "good" formulation is critical for an efficient solution. We will see an example of this later on. See also the exercises for a few challenges.

Note that in the three combinatorial problems above the algebraic variety of the corresponding polynomial system has finitely many solutions (e.g., colorings, cuts, stable sets, etc.). Let $\overline{\mathbb{K}}$ be a field. For an ideal $I \subseteq \mathbb{K}[x_1,\ldots,x_n]$, when the variety $V_{\overline{\mathbb{K}}}(I)$ is finite, the ideal is called *zero dimensional* (this is the case for all of the applications considered here). When we say that a system of polynomial equations is a *combinatorial system* we mean its variety encodes a combinatorial problem (e.g., zeros represent stable sets, colorings, matchings, etc.) and it is zero dimensional.

Optimization algorithms are intimately tied to the development of infeasibility certificates. For example, in linear programming the simplex method is closely related to Farkas' lemma. Our main theme in this chapter and the next is the generalization of this important principle.

First we need the "right" *infeasibility certificates* or *theorems of alternative*, just as Farkas' lemma is a centerpiece for the development of linear programming. Two powerful infeasibility certificates for general polynomial systems will generalize the classical ones for linear optimization. The infeasibility of polynomial systems can *always* be certified by one of these particular algebraic identities (on nonlinear polynomials). In practice, to find these two infeasibility certificates, we rely either on *linear algebra* or *semidefinite programming*.

12.2 Nullstellensatz and solving combinatorial systems of equations

In this section, we revisit the problem of solving a given system of polynomial equations $f_1(\mathbf{x}) = 0, f_2(\mathbf{x}) = 0,\ldots, f_m(\mathbf{x}) = 0$ where $f_1,\ldots,f_m \in \mathbb{K}[x_1,\ldots,x_n]$, or to show that no solution $\mathbf{x}$ exists. We often abbreviate the system as $F(\mathbf{x}) = \mathbf{0}$ where $F := \{f_1,\ldots,f_m\} \subset \mathbb{K}[x_1,\ldots,x_n]$. We begin with Hilbert's Nullstellensatz. To warm up, recall Fredholm's alternative theorem from Section 1.2.

Theorem 12.2.1. *The system of linear equations $A\mathbf{x} = \mathbf{b}$ has a no solution if and only if there exists a vector $\mathbf{y}$ with the property that $\mathbf{y}^\top A = \mathbf{0}^\top$ but $\mathbf{y}^\top\mathbf{b} \neq 0$.*

The important point is that there is a stronger far-reaching generalization for *nonlinear* polynomials equations:

Theorem 12.2.2 (Hilbert's Nullstellensatz). *Consider $F = \{f_1,\ldots,f_m\} \subseteq \mathbb{K}[x_1,\ldots,x_n]$. Then the variety $\left\{\mathbf{x} \in \overline{\mathbb{K}}^n : f_1(\mathbf{x}) = 0,\ldots, f_s(\mathbf{x}) = 0\right\}$ is empty if and only if 1 belongs to the ideal $\langle F\rangle$ generated by F. Note that $1 \in \langle F\rangle$ means that there exist polynomials $\beta_1,\ldots,\beta_m$ in the ring $\mathbb{K}[x_1,\ldots,x_n]$ such that $1 = \sum_{i=1}^m \beta_i f_i$, and this polynomial identity is a certificate of infeasibility of $F(\mathbf{x}) = \mathbf{0}$.*

Note that Fredholm's theorem is simply a linear version of Hilbert's Nullstellensatz where all the polynomials are linear and the β_i's are constant. Our main goal for this chapter is relating the combinatorial optimization problems of Theorem 12.1.1 to the Hilbert Nullstellensatz. In the next chapter we will present a self-contained proof of Theorem 12.2.2.

Here we choose to focus on techniques that fit well with traditional optimization methods. The main idea is that solving a polynomial system of equations can be reduced to solving a sequence of linear algebra problems. Variants of this technique have been applied to stable sets [100, 240], vertex coloring [104, 240], satisfiability (see e.g., [71]), and cryptography (see for example [77]). We will not present it here but, in its more advanced form, our technique is strongly related to *Border bases* techniques, which can also be viewed in terms of linear algebra computations (see e.g., [191, 192, 256, 257, 316]). For the specific connection to optimization see the appendix in [106].

12.2.1 Linear algebra relaxations

There is an interesting corollary to Hilbert's Nullstellensatz: If there exist *numeric vectors* $\boldsymbol{\mu} \in \mathbb{K}^m$ such that $\sum_{i=1}^m \mu_i f_i = 1$, then already we see the polynomial system $F(\mathbf{x}) = \mathbf{0}$ must be infeasible. The crucial point here is that determining whether there exist $\boldsymbol{\mu} \in \mathbb{K}^m$ such that $\sum_{i=1}^m \mu_i f_i = 1$ is just a linear algebra problem over the field $\mathbb{K}$. There is a strong interplay among the system of nonlinear equations $F(\mathbf{x}) = \mathbf{0}$, the ideal $\langle F \rangle$, and the $\mathbb{K}$-vector space $\operatorname{span}_{\mathbb{K}}(F)$ generated by the polynomials in F (when we think of them as $\mathbb{K}$-valued vectors). Let us do an example:

Example 12.2.3. Consider the following set of polynomials:

$$F := \{f_1 := x_1^2 - 1, f_2 := 2x_1x_2 + x_3, f_3 := x_1 + x_2, f_4 := x_1 + x_3\}.$$

We can abbreviate the infeasible polynomial system of equations $f_1(\mathbf{x}) = 0, f_2(\mathbf{x}) = 0, f_3(\mathbf{x}) = 0, f_4(\mathbf{x}) = 0$ as $F(\mathbf{x}) = \mathbf{0}$. We can prove that the system $F(\mathbf{x}) = \mathbf{0}$ is infeasible if we can find $\boldsymbol{\mu} \in \mathbb{R}^4$ satisfying the following:

$$\begin{aligned} \mu_1 f_1 + \mu_2 f_2 + \mu_3 f_3 + \mu_4 f_4 &= 1, \\ \Leftrightarrow \mu_1(x_1^2 - 1) + \mu_2(2x_1x_2 + x_3) + \mu_3(x_1 + x_2) + \mu_4(x_1 + x_3) &= 1, \\ \Leftrightarrow \mu_1 x_1^2 + 2\mu_2 x_1x_2 + (\mu_2 + \mu_4)x_3 + \mu_3 x_2 + (\mu_3 + \mu_4)x_1 - \mu_1 &= 1. \end{aligned}$$

Then, equating coefficients of *monomials* on the left- and right-hand sides of the equation above gives the following *linear* system of equations (corresponding monomial in parenthesis):

$$\begin{aligned} -\mu_1 &= 1 \quad (1), & \mu_3 + \mu_4 &= 0 \quad (x_1), & \mu_3 &= 0 \quad (x_2), \\ \mu_3 + \mu_4 &= 0 \quad (x_3), & 2\mu_2 &= 0 \quad (x_1x_2), & \mu_1 &= 0 \quad (x_1^2). \end{aligned}$$

Even though $F(\mathbf{x}) = \mathbf{0}$ is infeasible, the linear system obtained above is infeasible, and so, we have not found a certificate of infeasibility of $F(\mathbf{x}) = \mathbf{0}$.

More generally, let $f_i = \sum_{\boldsymbol{\alpha} \in \mathbb{N}^n} f_{i,\boldsymbol{\alpha}} \mathbf{x}^{\boldsymbol{\alpha}}$ where only finitely many coefficients $f_{i,\boldsymbol{\alpha}}$ are nonzero $i = 1, \ldots, m$. Then, we have $\sum_{i=1}^m \mu_i f_i = 1$ if and only if $\sum_{i=1}^m \mu_i f_{i,0} = 1$ and $\sum_{i=1}^m \mu_i f_{i,\boldsymbol{\alpha}} = 0$ for all $\boldsymbol{\alpha} \in \mathbb{N}^n$ where $\boldsymbol{\alpha} \neq \mathbf{0}$.

Multiply by the μ_i's and expand by monomials on the x variables. Note that, as we saw in the example, *there is one linear equation per monomial appearing in F*. Since the polynomial is supposed to be equal to the constant polynomial 1 and because two polynomials are equal if and only if their coefficients of the same monomials are equal we obtain a system of equations with as many variables μ_i as polynomials in F, and as many equations as distinct monomials present in F.

Let us use matrix notation. Let us construct a matrix of M_F whose columns are labeled by monomials in F, in some prescribed order, and rows are labeled by the m polynomials in F. First of all, the matrix M_F has one column per monomial $\mathbf{x}^{\boldsymbol{\alpha}}$ present in F and one row per polynomial $f_i\ in F$, Thus, the entries of M_F are labeled by pairs $(f_i, \mathbf{x}^\alpha)$ where the entry labeled $(M_F)_{f_i,\mathbf{x}^\alpha}$ equals the coefficient of the monomial $\mathbf{x}^\alpha$ in the polynomial f_i. Of course, the entry can be zero.

The real variable vector $\boldsymbol{\mu}$ has one entry for each polynomial of f_i denoted $\boldsymbol{\mu}_{f_i}$. We can write a row vector $\boldsymbol{\mu} = (\mu_1, \mu_2, \ldots, \mu_m)$ and a vector $(\mathbf{0}, 1)^\top = (0, 0, \ldots, 0, 1)^\top$ with as many entries as monomials; except the last entry corresponding to the degree zero monomial, the constant term. Thus we can write our linear system as $\boldsymbol{\mu} M_F = (\mathbf{0}, 1)^\top$. Note that in the special case where $F(x) = \mathbf{0}$ is a *linear* system of equations, then Fredholm's alternative (Theorem 1.2.1) says that $F(\mathbf{x}) = \mathbf{0}$ is infeasible if and only if $\boldsymbol{\mu} M_F = (\mathbf{0}, 1)^\top$ has a solution.

Example 12.2.4. For the system F in Example 12.2.3 the matrix in question will be

$$\begin{pmatrix} -1 & 0 & 0 & 0 & 0 & 1 \\ 0 & 0 & 1 & 2 & 0 & 0 \\ 0 & 1 & 1 & 0 & 0 & 0 \\ 0 & 1 & 0 & 1 & 0 & 0 \end{pmatrix}.$$

Where the rows are labeled by the variables $\mu_1, \mu_2, \mu_3, \mu_4$ and the columns are labeled by the monomials (in ascending order from left to right) 1, x_1, x_2, x_3, x_1x_2, and x_1^2.

It is worth noticing right now that when we solve the linear system $\boldsymbol{\mu} M_F = (\mathbf{0}, 1)^\top$, we can do so over the smallest subfield of $\mathbb{K}$ containing the coefficients of the polynomials in F, which is particularly useful if such a subfield is a finite field.

Unfortunately, as the above example shows, even if the polynomial system $F(\mathbf{x}) = \mathbf{0}$ is infeasible, the linear system $\boldsymbol{\mu} M_F = (\mathbf{0}, 1)^\top$ may not be feasible. Although it did not work we can still add polynomials from the ideal $\langle F \rangle$ to the current F and try again to find a vector $\boldsymbol{\mu}$ such that $\boldsymbol{\mu} M_F = (\mathbf{0}, 1)^\top$. The message is that Hilbert's Nullstellensatz guarantees that, if $F(\mathbf{x}) = \mathbf{0}$ is infeasible, it is enough to add a finite set of polynomials from the ideal $\langle F \rangle$ to F so that eventually the *linear algebra* system $\boldsymbol{\mu} M_F = (\mathbf{0}, 1)^\top$ is feasible.

More precisely, note it is enough to add polynomials of the form $\mathbf{x}^\alpha f$ for $\mathbf{x}^\alpha$ a monomial and some polynomial $f \in F$. Why is this? If $F(x) = 0$ is infeasible, then Hilbert's Nullstellensatz says $\sum_{i=1}^m \beta_i f_i = 1$ for some $\beta_1, \ldots, \beta_m \in \mathbb{K}[x_1, \ldots, x_n]$. Let $d = \max_i\{\deg(\beta_i)\}$. Add to the set F those polynomials of the form $\mathbf{x}^\alpha f$, where $f \in F$ and $\deg(\mathbf{x}^\alpha) \leq d$. We create a new larger finite set of polynomials F' and then as a consequence a larger linear $\boldsymbol{\mu} M_{F'} = (\mathbf{0}, 1)^\top$. Note that the ideal generated by F' is the same as the ideal generated by F (e.g., set of solutions for $F'(\mathbf{x})$ is the same as the set of solutions for $F(\mathbf{x})$), but now the vector space $\mathbb{K}$-linear span of F', that is $\operatorname{span}_{\mathbb{K}}(F')$ with the monomials as basis, contains $\beta_i f_i$ for all i, and thus $1 \in \operatorname{span}_{\mathbb{K}}(F)$ or equivalently $\boldsymbol{\mu} M_{F'} = (\mathbf{0}, 1)^\top$ has a solution as a linear algebra problem.

Example 12.2.5. Consider again the polynomial system $F(\mathbf{x}) = \mathbf{0}$ from Example 12.2.3. We must add "redundant" polynomial equations to the system F (redundant as they themselves will be multiples of the f_i). In particular, we add the following polynomial equations: $x_2 f_1(\mathbf{x}) = 0$, $x_1 f_2(\mathbf{x}) = 0$, $x_1 f_3(\mathbf{x}) = 0$, and $x_1 f_4(\mathbf{x}) = 0$. Let

$$F' := \{f_1, f_2, f_3, f_4, x_2 f_1, x_1 f_2, x_1 f_3, x_1 f_4\}.$$

Note that, in particular, $\operatorname{span}_{\mathbb{K}}(F) \subset \operatorname{span}_{\mathbb{K}}(F')$.

Then, the new linear system $\boldsymbol{\mu} M_{F'} = (\mathbf{0},1)^\top$ is now as follows:

$$-\mu_1 = 1 \quad (1), \qquad \mu_3+\mu_4=0 \quad (x_1), \qquad \mu_3-\mu_5=0 \quad (x_2),$$
$$\mu_2+\mu_4=0 \quad (x_3), \qquad 2\mu_2+\mu_7=0 \quad (x_1x_2), \qquad \mu_1+\mu_7+\mu_8=0 \quad (x_1^2),$$
$$\mu_6+\mu_8=0 \quad (x_1x_3), \qquad \mu_5+2\mu_6=0 \quad (x_1^2x_2).$$

That this linear system has a solution proves that the nonlinear system $F(\mathbf{x}) = \mathbf{0}$ is infeasible. The solution is $\boldsymbol{\mu} = (-1,-\frac{2}{3},-\frac{2}{3},\frac{2}{3},-\frac{2}{3},-\frac{1}{3},\frac{4}{3},-\frac{1}{3})$, which gives the following certificate of infeasibility as given in Example 12.3.2:

$$-f_1-\frac{2}{3}f_2-\frac{2}{3}f_3+\frac{2}{3}f_4-\frac{2}{3}x_2f_1+\frac{1}{3}x_1f_2+\frac{4}{3}x_1f_3-\frac{1}{3}x_1f_4=1.$$

There is an important moral to this story: *adding new polynomial equations, multiples from the f_i, can eventually lead to a tighter linear algebra system that proves the system of nonlinear polynomials F has no solution.*

12.2.2 Nullstellensatz linear algebra algorithm (NulLA)

We now present an algorithm for determining whether a polynomial system of equations is infeasible using linear algebra relaxations. Let $F \subseteq \mathbb{K}[x_1,\ldots,x_n]$ and again let $F(\mathbf{x}) = \mathbf{0}$ be the polynomial system $f(\mathbf{x}) = 0$ for all $f \in F$. We wish to determine whether $F(\mathbf{x}) = \mathbf{0}$ has a solution over the algebraic closure of $\mathbb{K}$. Note of course that we saw another solution method in Chapters 10 and 11 on Gröbner bases.

The idea behind NulLA [100, 104] is straightforward: we check whether the linear system $\boldsymbol{\mu} M_F = (\mathbf{0},1)^\top$ is feasible (i.e., $1 \in \operatorname{span}_{\mathbb{K}}(F)$) using linear algebra over $\mathbb{K}$ and, if not, then we add polynomials from $\langle F\rangle$ to F and try again.

We add polynomials in the following systematic way: for each polynomial $f \in F$ and for each variable x_i, we add $x_i f$ to F. So, the NulLA algorithm is as follows: If $\boldsymbol{\mu} M_F = (\mathbf{0},1)^\top$ has a solution, stop; we have proved that $F(\mathbf{x}) = \mathbf{0}$ is infeasible. Otherwise for every variable x_i and every $f \in F$ add the polynomial $x_i f$ to F and repeat. Recall from above that $\boldsymbol{\mu} M_F = (\mathbf{0},1)^\top$ has a solution if and only if $1 \in \operatorname{span}_{\mathbb{K}}(F)$.

For the $\mathbb{K}$-vector subspace $V_F = \operatorname{span}_{\mathbb{K}}(F)$ of all polynomials in $\mathbb{K}[x_1,\ldots,x_n]$ generated by $\mathbb{K}$-linear combinations of the polynomials F, we define $V_F^+ := V_F + \sum_{i=1}^n x_i V_F$ where $x_i V_F := \{x_i f : f \in V_F\}$. Note that V_F^+ is also a vector subspace of $\mathbb{K}[x_1,\ldots,x_n]$ supported on particular monomials. Then, V_F^+ is precisely the linear span of V_F and $x_i V_F$ for all $i = 1,\ldots,n$. If $1 \in \operatorname{span}_{\mathbb{K}}(F)$, then $F(\mathbf{x}) = \mathbf{0}$ is infeasible and stop, otherwise replace $\operatorname{span}_{\mathbb{K}}(F)$ with $(\operatorname{span}_{\mathbb{K}}(F))^+$ and repeat. To create $(\operatorname{span}_{\mathbb{K}}(F))^+$ it suffices to add all polynomials of the form $x_i F$ to the original F. Because of the Nullstellensatz, this process must terminate if the original system F was infeasible.

If $F(\mathbf{x}) = \mathbf{0}$ is infeasible, there must exist a Nullstellensatz certificate of infeasibility $\sum_i \beta_i f_i = 1$, where $\deg(\beta_i) \leq D$, that is, the degree of the certificate polynomial is at most $\deg(F) + D$. Then we say $F(\mathbf{x}) = \mathbf{0}$ has a Nullstellensatz bound D. Although it is not stated in the Hilbert Nullstellensatz in algebraic geometry it is known that there is a universal upper bound D, which in turn bounds the number of iterations of our algorithm. It is known the bound D can be written in terms of the number of variables, number of polynomials in F, and the largest degree of a monomial present in F. We state Kollár's quantitative version of the Nullstellensatz without proof:

Lemma 12.2.6 (Kollár [199]). *Let $\mathbb{K}$ be a field. Let $f_1,\ldots,f_k$ be polynomials in $\mathbb{K}[x_1,\ldots,x_n]$ of degrees $d_1 \geq d_2 \geq \ldots d_k \geq 2$. If the polynomials have no common zero over the algebraic* *closure of* $\mathbb{K}$, *then there exist* $g_1,\dots,g_k$ *in* $\mathbb{K}[x_1,\dots,x_n]$ *such that* $g_1 f_1 + \cdots + g_k f_k = 1$ *and each* $g_i f_i$ *has degree at most* D. *Here the value of* D *is set as follows: If* $k \le n$ *we have* $D = d_1 \cdots d_k$; *if* $k > n > 1$ *we have* $D = d_1 \cdots d_{n-1} d_k$; *and if* $k > n = 1$ *we have* $D = d_1 + d_k - 1$. *This estimate for* D *is sharp for arbitrary polynomials.*

Using Kollár's important theorem (see [199]) we can state the NulLA algorithm as follows (see [100]):

ALGORITHM 12.1. NulLA algorithm.

input An initial finite set of polynomials $F \subset \mathbb{K}[x_1,\dots,x_n]$ and a bound D.
output FEASIBLE, if $F(\mathbf{x}) = \mathbf{0}$ is feasible over $\overline{\mathbb{K}}$, else NO SOLUTION.
for $k = 0,1,2,\dots,D$ **do**
  if $1 \in \text{span}_{\mathbb{K}}(F)$ **then**
    return NO SOLUTION.
  else
    replace $\text{span}_{\mathbb{K}}(F)$ by $(\text{span}_{\mathbb{K}}(F))^+$ (adding the polynomials $x_i F$ to the original F).
return FEASIBLE.

After d iterations of NulLA, the set $\text{span}_{\mathbb{K}}(F)$ contains all linear combinations of polynomials of the form $\mathbf{x}^{\boldsymbol{\alpha}} f$ where $|\boldsymbol{\alpha}| \le d$ and where f was one of the initial polynomials in F, and so, if the system is infeasible, then NulLA will find a certificate of infeasibility in at most D number of iterations. We refer to the number of iterations (the *for* loop) that NulLA takes to solve a given system of equations as the *NulLA rank* of the nonlinear system F. Note that if an infeasible system $F(x) = 0$ has a NulLA rank of r, then it has a Nullstellensatz certificate of infeasibility of degree $r + \deg(F)$.

Kollár's theorem shows that D can grow exponentially for *general* polynomial systems, but we will see now that better bounds exist for specialized systems like those arising in combinatorics. How about the combinatorial systems of equations presented in Theorem 12.1.1? Can the NulLA rank really grow exponentially? A result by D. Lazard [226] provides for ideals like those with a *linear* bound.

Lemma 12.2.7 (Lazard [226]). *Let* $f_1,\dots,f_k$ *be homogeneous polynomials of* $\mathbb{K}[x_0,\dots,x_n]$ *that generate an ideal* I, *let* d_i *be the degree of* f_i, *and assume that* $d_1 \ge d_2 \ge \cdots \ge d_k \ge 1$ *and* $k \ge n+1$. *Then the following conditions are equivalent:*

1. *The* k *projective hypersurfaces defined by* $f_1,\dots,f_k$ *have no point in common over the algebraic closure of* $\mathbb{K}$ *(in particular, they have no point in common at infinity).*

2. *The ideal* I *contains a power of the maximal ideal* $M = \langle x_0,x_1,\dots,x_n\rangle$; *namely, for some power* p, $x_i^p \in I$ *for all* x_i.

3. $M^p \subset I$ *with* $p = d_1 + d_2 + \cdots + d_{n+1} - n \le (n+1)(\max_{1\le i\le n+1}\{d_i\} - 1) + 1$.

4. *The map* $\phi\colon (\beta_1,\dots,\beta_k) \mapsto \sum \beta_i f_i$ *is surjective among all polynomials of degree* p *when, for all* i, β_i *is a homogeneous polynomial of degree* $p - d_i$.

The proof of Lemma 12.2.7 relies on advanced techniques in commutative and homological algebra, and is presented in [226, p. 169]. As a consequence of Lemma 12.2.7, when given polynomials $f_i \in \mathbb{K}[x_1,\dots,x_n]$, we can consider their homogenization $\bar{f}_i$, using an extra variable x_0 (e.g., $x^2 - x$ can be homogenized to $x^2 - xx_0$). If we are able to find a "projective" Nullstellensatz of the form

$$x_0^p = \sum \beta_i \bar{f}_i,$$

then we can substitute $x_0 = 1$ in the above equation and obtain the form of the Nullstellensatz that is more desirable for computation (e.g., $1 = \sum \beta_i' f_i$). Furthermore, the degree of β_i' is less than or equal to the degree of β_i.

We can summarize Lazard's lemma as follows (see also Brownawell [63]):

Corollary 12.2.8. *Given polynomials $f_1,\dots,f_s \in \mathbb{K}[x_1,\dots,x_n]$, where $\mathbb{K}$ is an algebraically-closed field and $d = \max\{\deg(f_i)\}$, if $f_1,\dots,f_s$ have no common zeros, and $f_1,\dots,f_s$ have no common zeros at infinity, then $1 = \sum_{i=1}^{s} \beta_i f_i$ where*

$$\deg(\beta_i) \le n(d-1).$$

Therefore, the bound on the Nullstellensatz obtained for the combinatorial ideals described in Corollary 12.2.8 is a considerable improvement on the exponential bound predicted by Kollár in Theorem 12.2.6 for general ideals. For the coloring and independent set ideals the bound is $2n$ and n, respectively.

For a given class of polynomial system of equations, it is interesting to understand the growth of the NulLA rank. For fixed NulLA rank d, it is quite interesting to characterize which systems can be solved at that rank d, because this class of systems is polynomial-time solvable.

Lemma 12.2.9. *Let $d \in \mathbb{Z}_+$ be fixed. Let $F = \{f_1, f_2,\dots, f_m\}$ be a polynomial set in $\mathbb{K}[x_1,\dots,x_n]$. Consider the finite dimensional vector space of $\mathbb{K}[x_1,\dots,x_n]$ generated by F, $\text{span}_{\mathbb{K}}(F)$; here polynomials are assumed to be encoded as vectors of coefficients indexed by all monomials of degree at most* $\deg(F)$. *Then, the first d iterations (the* for *loop) of the NulLA algorithm* (*Algorithm* 12.1) *can be computed in polynomial time in the input size of F.*

Proof. The issue here is how big is the linear system of equations that we generate at each iteration of NulLA. We will show that when d is fixed then this has size polynomial in n, m and the input coefficients. To calculate the matrix used, consider the input polynomial system $F = \{f_1,\dots,f_m\}$. We recall our assumption that, at the d-th iteration, we are working with a Nullstellensatz certificate $\sum \beta_i f_i = 1$ where every β_i, $\deg(\beta_i) < d$ for the constant d. For a given fixed positive integer d serving as a tentative degree for the Nullstellensatz certificate, the Nullstellensatz coefficients come from the solution of a system of linear equations.

We now take a closer look at the matrix equation $\boldsymbol{\mu} M_{F,d} = \mathbf{b}_{F,d}$ defining the system of linear equations. First of all, the matrix $M_{F,d}$ has one column per monomial $\mathbf{x}^\alpha$ of degree less than or equal to d on the n variables and one row per polynomial of the form $\mathbf{x}^\delta f_i$, i.e., the product of a monomial $\mathbf{x}^\delta$ of degree less than or equal to $d - \deg(f_i)$ with a polynomial $f_i \in F$. Thus, $M_{F,d} = (M_{\mathbf{x}^\delta f_i, \mathbf{x}^\alpha})$, where $M_{\mathbf{x}^\delta f_i, \mathbf{x}^\alpha}$ equals the coefficient of the monomial $\mathbf{x}^\alpha$ in the polynomial $\mathbf{x}^\delta f_i$. The real variable vector $\boldsymbol{\mu}$ has one entry for every polynomial of the form $\mathbf{x}^\delta f_i$ denoted $\boldsymbol{\mu}_{\mathbf{x}^\delta f_i}$, and the vector $\mathbf{b}_{F,d}$ has one entry for every monomial $\mathbf{x}^\alpha$ of degree less than or equal to d, where $(\mathbf{b}_{F,d})_{\mathbf{x}^\alpha} = 0$ if $\alpha \ne 0$ and $(\mathbf{b}_{F,d})_1 = 1$.

Since $\deg(\beta_i) < d$, an upper bound on the number of terms in β_i is the total number of monomials of degrees $0,\dots,d$ in n variables. In other words,

$$\#\text{ of terms in } \beta_i = \binom{n+d-1}{n-1} + \binom{n+d-2}{n-1} + \cdots + \binom{n-1}{n-1} = O(n^d).$$

Thus the matrix in question has $O(n^d)$ columns. Similarly, to calculate the number of rows we see the generators in the Nullstellensatz certificate consist of m polynomials f_i, which at the d-th iteration, each gets multiplied by a single β_i, a monomial of degree d. Since we have $O(n^d)$ such terms, there are at most $O(mn^d)$ columns on $M_{F,d}$. The matrix $M_{F,d}$ gives rise to a large linear system, but when d is constant this system is of polynomial size. The coefficients are extracted from the coefficients of polynomials in F.

Finally, from Corollary 3.2b of [296], we know that if a system of equations $A\mathbf{x} = \mathbf{b}$ has a solution, it has one of size polynomially bounded by the bit sizes of A and $\mathbf{b}$. In our case the entries of $A = M_{F,d}$ are coefficients of the terms within the f_i polynomials. The solutions of the system are of polynomially bounded bit size, since, for constant d, they are the solution to a system of $O(n^d)$ linear equations with $O(mn^d)$ variables with the same input coefficients as F. □

Note that we also see from the proof of Lemma 12.2.9 that although a linear bound is an improvement from the growth predicted by Kollár's bound, it is still far from being computationally practical. Note that then the growth of the matrices is $O(n^n)$. Fortunately we have observed in practice that the degree growth of polynomial systems for coloring problems is often *very* slow—slow enough to deal with some fairly large graphs. The interested reader can see the tables of computation in [100, 104]. Interestingly enough, we sometime can use NulLA to prove noncolorability in polynomial time. That is what we see next.

12.2.3 Detecting non-3-colorable graphs

To illustrate NulLA we consider here the problem of deciding graph 3-colorability [100, 101, 106]. We use Theorem 12.1.1 part 3 to certify graph non-3-colorability very rapidly using linear algebra over $\mathbb{F}_2$.

In the theory of discrete optimization it is important to consider the level of complexity of a problem. The class of NP-complete problems contains the problem of deciding whether a graph is 3-colorable. On the other hand, we note that the existence of the Nullstellensatz certificate for the system of equations in Theorem 12.1.1(3) with $k = 3$ is a proof that the graph is not 3-colorable. If that certificate is "short" and always easy to find then we would have an easy way to detect when the graph is not 3-colorable. This has some interesting consequences for NulLA:

Corollary 12.2.10. *If* NP $\neq$ coNP, *then there must exist an infinite family of graphs whose minimal-degree Nullstellensatz certificates of non-3-colorability grow with respect to the number of vertices and edges in the graph.*

Proof. We will prove this theorem by contradiction. Recall from complexity theory: For L an NP-Complete problem, if the complementary problem $\overline{L}$ belongs to NP, then NP = coNP. Graph 3-colorability is a well-known NP-complete problem, and non-3-colorability is its complement.

Consider a non-3-colorable graph that has been encoded as a system of polynomial equations $(x_i^3 - 1) = 0$ for $i = 1,\dots,n$, and $(x_i^2 + x_i x_j + x_j^2) = 0$ for $(i,j) \in E(G)$. Assume

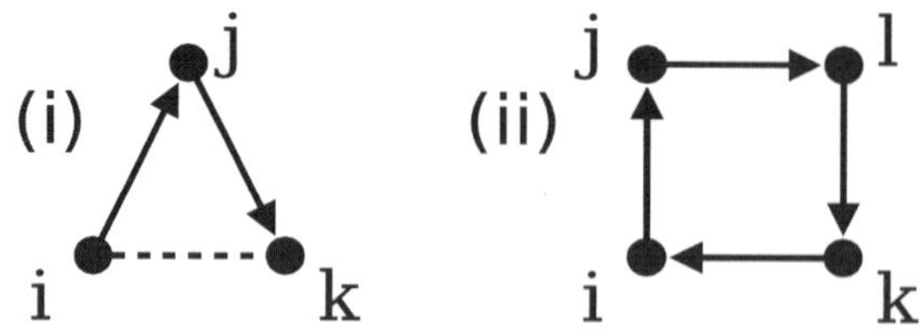

Figure 12.1. (i) *Partial* 3-*cycle,* (ii) *chordless* 4-*cycle.* [101]

that any minimal degree *not* 3-colorable Nullstellensatz certificate has $\deg(\alpha) < d$ for some constant d. We have shown in Lemma 12.2.9 that then one can compute the corresponding Nullstellensatz certificate in polynomial time. Thus 3-colorability would be in coNP, but this shows that NP = coNP, a contradiction. □

Although the NulLA rank must grow, we are interested in what happens within fixed low rank. The main theoretical result of this section is a complete combinatorial characterization of the class of non-3-colorable simple undirected graphs $G = (V,E)$ with NulLA rank one. For this we use the following system (with $\mathbb{K} = \mathbb{F}_2$) from Proposition 12.1.1:

$$J_G = \{x_i^3 + 1 = 0,\ x_i^2 + x_i x_j + x_j^2 = 0 :\ i \in V,\ \{i,j\} \in E\}. \tag{12.1}$$

Definition 12.2.11. This polynomial system has a degree one Nullstellensatz certificate of infeasibility ($D = 1$, NulLA algorithm) if and only if there exist coefficients $a_i, a_{ij}, b_{ij}, b_{ijk} \in \mathbb{F}_2$ such that

$$\sum_{i\in V}(a_i + \sum_{j\in V} a_{ij}x_j)(x_i^3+1) + \sum_{\{i,j\}\in E}\left(b_{ij} + \sum_{k\in V} b_{ijk}x_k\right)(x_i^2 + x_i x_j + x_j^2) = 1. \tag{12.2}$$

We now describe the combinatorial characterization. Let G be a graph with vertices $V = \{1,\ldots,n\}$ and edges E, and let

$$\mathrm{Arcs}(G) = \{(i,j) : i,j \in V(G), \text{ and } \{i,j\} \in E\}.$$

Our characterization involves two types of substructures on the graph G (see Figure 12.1). The first of these are *oriented partial* 3-*cycles,* which are pairs of arcs $\{(i,j),(j,k)\} \subseteq \mathrm{Arcs}(G)$, also denoted (i,j,k), in which $(k,i) \in \mathrm{Arcs}(G)$ (the vertices i,j,k induce a 3-cycle in G). The second are *oriented chordless 4-cycles,* which are sets of four arcs $\{(i,j),(j,k),(k,l),(l,i)\} \subseteq \mathrm{Arcs}(G)$, denoted (i,j,k,l), with $(i,k),(j,l) \notin \mathrm{Arcs}(G)$ (the vertices i,j,k,l induce a chordless 4-cycle).

Theorem 12.2.12. *For a given simple undirected graph $G = (V,E)$ the following two conditions are equivalent:*

1. *The polynomial system over $\mathbb{F}_2$ encoding the 3-colorability of G*

$$J_G = \{x_i^3 + 1 = 0,\ x_i^2 + x_i x_j + x_j^2 = 0 :\ i \in V,\ \{i,j\} \in E\}$$

has a degree one Nullstellensatz certificate of infeasibility.

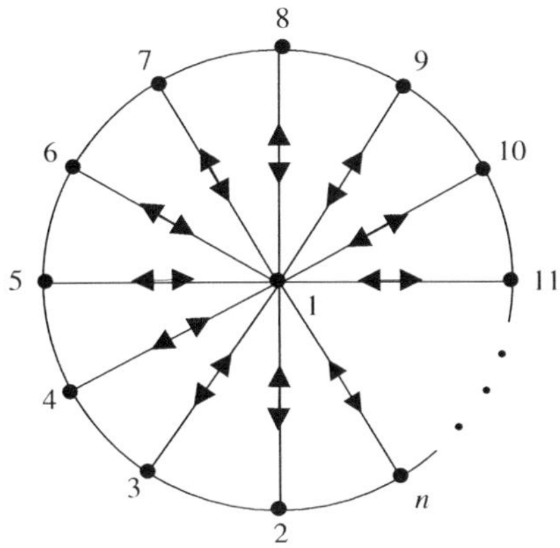

Figure 12.2. *Odd wheel.* [101]

2. *There exists a set C of oriented partial 3-cycles and oriented chordless 4-cycles from* $\mathrm{Arcs}(G)$ *such that*

   (a) $|C_{(i,j)}| + |C_{(j,i)}| \equiv 0 \pmod 2$ *for all* $\{i,j\} \in E$, *and*

   (b) $\sum_{(i,j)\in \mathrm{Arcs}(G), i<j} |C_{(i,j)}| \equiv 1 \pmod 2$,

   where $C_{(i,j)}$ *denotes the set of cycles in* C *in which the arc* $(i,j) \in \mathrm{Arcs}(G)$ *appears.*

Moreover, such graphs are non-3-colorable and can be recognized in polynomial time.

We can consider the set C in Theorem 12.2.12 as a covering of E by directed edges. From this perspective, Condition 1 in Theorem 12.2.12 means that every edge of G is covered by an even number of arcs from cycles in C. On the other hand, Condition 2 says that if $\hat{G}$ is the directed graph obtained from G by the orientation induced by the total ordering on the vertices $1 < 2 < \cdots < n$, then when summing the number of times each arc in $\hat{G}$ appears in the cycles of C, the total is odd.

Note that the 3-cycles and 4-cycles in G that correspond to the partial 3-cycles and chordless 4-cycles in C give an edge covering of a non-3-colorable subgraph of G. Also, note that if a graph G has a non-3-colorable subgraph whose polynomial encoding has a degree one infeasibility certificate, then the encoding of G will also have a degree one infeasibility certificate.

The class of graphs with encodings that have degree one infeasibility certificates includes all graphs containing odd wheels as subgraphs (e.g., a 4-clique) [240].

Corollary 12.2.13. *If a graph $G = (V, E)$ contains an odd wheel, then the encoding of 3-colorability of G from Theorem 12.2.12 has a degree one Nullstellensatz certificate of infeasibility.*

Proof. Assume G contains an odd wheel with vertices labeled as in Figure 12.2. Let

$$C := \{(i,1,i+1) : 2 \le i \le n-1\} \cup \{(n,1,2)\}.$$

Figure 12.2 illustrates the arc directions for the oriented partial 3-cycles of C. Each edge of G is covered by exactly zero or two partial 3-cycles, so C satisfies Condition 1

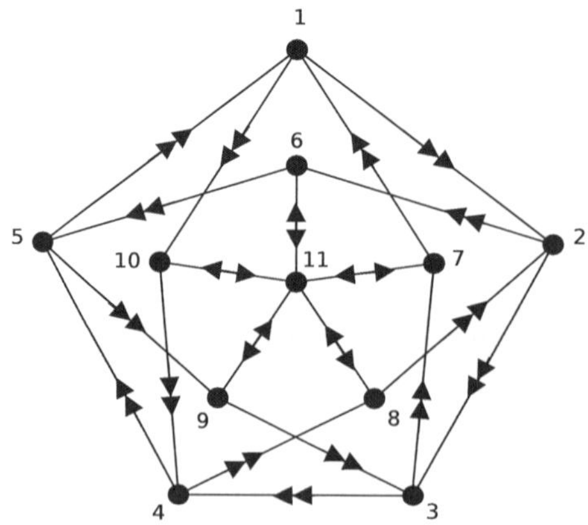

Figure 12.3. *Grötzsch graph.* [101]

of Theorem 12.2.12. Furthermore, each arc $(1,i) \in \mathrm{Arcs}(G)$ is covered exactly once by a partial 3-cycle in C, and there is an odd number of such arcs. Thus, C also satisfies Condition 2 of Theorem 12.2.12. □

A nontrivial example of a non-3-colorable graph with a degree one Nullstellensatz certificate is the Grötzsch graph.

Example 12.2.14. Consider the Grötzsch graph in Figure 12.3, which has no 3-cycles. The following set of oriented chordless 4-cycles gives a certificate of non-3-colorability by Theorem 12.2.12:

$$C := \{(1,2,3,7),(2,3,4,8),(3,4,5,9),(4,5,1,10),(1,10,11,7),\\ (2,6,11,8),(3,7,11,9),(4,8,11,10),(5,9,11,6)\}.$$

Figure 12.3 illustrates the arc directions for the 4-cycles of C. Each edge of the graph is covered by exactly two 4-cycles, so C satisfies Condition 1 of Theorem 12.2.12. Moreover, one can check that Condition 2 is also satisfied. It follows that the graph has no proper 3-coloring.

We now prove Theorem 12.2.12 using ideas from polynomial algebra. First, notice that we can simplify a degree one certificate as follows: Expanding the left-hand side of (12.2) and collecting terms, the only coefficient of $x_j x_i^3$ is a_{ij} and thus $a_{ij} = 0$ for all $i, j \in V$. Similarly, the only coefficient of $x_i x_j$ is b_{ij}, and so $b_{ij} = 0$ for all $\{i,j\} \in E$. We thus arrive at the following simplified expression:

$$\sum_{i\in V} a_i(x_i^3+1) + \sum_{\{i,j\}\in E}\left(\sum_{k\in V} b_{ijk}x_k\right)(x_i^2+x_ix_j+x_j^2) = 1. \tag{12.3}$$

Now, consider the following set F of polynomials:

$$x_i^3+1 \qquad \forall i \in V, \tag{12.4}$$

$$x_k(x_i^2+x_ix_j+x_j^2) \qquad \forall \{i,j\} \in E,\ k \in V. \tag{12.5}$$

The elements of F are those polynomials that can appear in a degree one certificate of infeasibility. Thus, there exists a degree one certificate if and only if the constant polynomial 1 is in the linear span of F; that is, $1 \in \langle F \rangle_{\mathbb{F}_2}$, where $\langle F \rangle_{\mathbb{F}_2}$ is the vector space over $\mathbb{F}_2$ generated by the polynomials in F.

We next simplify the set F. Let H be the following set of polynomials:

$$\begin{aligned}
&x_i^2x_j + x_ix_j^2 + 1 && \forall\{i,j\} \in E, && (12.6)\\
&x_ix_j^2 + x_jx_k^2 && \forall(i,j),(j,k),(k,i) \in \mathrm{Arcs}(G), && (12.7)\\
&x_ix_j^2 + x_jx_k^2 + x_kx_l^2 + x_lx_i^2 && \forall(i,j),(j,k),(k,l),(l,i) \in \mathrm{Arcs}(G), &&\\
& && (i,k),(j,l) \notin \mathrm{Arcs}(G). && (12.8)
\end{aligned}$$

If we identify the monomials $x_ix_j^2$ as the arcs (i,j), then the polynomials (12.7) correspond to oriented partial 3-cycles and the polynomials (12.8) correspond to oriented chordless 4-cycles. The following lemma says that we can use H instead of F to find a degree one certificate.

Lemma 12.2.15. *We have $1 \in \langle F \rangle_{\mathbb{F}_2}$ if and only if $1 \in \langle H \rangle_{\mathbb{F}_2}$.*

Proof. The polynomials (12.5) above can be split into two classes of equations: (i) $k = i$ or $k = j$ and (ii) $k \neq i$ and $k \neq j$. Thus, the set F consists of

$$\begin{aligned}
&x_i^3 + 1 && \forall i \in V, && (12.9)\\
&x_i(x_i^2 + x_ix_j + x_j^2) = x_i^3 + x_i^2x_j + x_ix_j^2 && \forall\{i,j\} \in E, && (12.10)\\
&x_k(x_i^2 + x_ix_j + x_j^2) = x_i^2x_k + x_ix_jx_k + x_j^2x_k && \forall\{i,j\} \in E,\ k \in V, &&\\
& && i \neq k \neq j. && (12.11)
\end{aligned}$$

Using polynomials (12.9) to eliminate the x_i^3 terms from (12.10), we arrive at the following set of polynomials, which we label F':

$$\begin{aligned}
&x_i^3 + 1 && \forall i \in V, && (12.12)\\
&x_i^2x_j + x_ix_j^2 + 1 = (x_i^3 + x_i^2x_j + x_ix_j^2) + (x_i^3 + 1) && \forall\{i,j\} \in E, && (12.13)\\
&x_i^2x_k + x_ix_jx_k + x_j^2x_k && \forall\{i,j\} \in E,\ k \in V, &&\\
& && i \neq k \neq j. && (12.14)
\end{aligned}$$

Observe that $\langle F \rangle_{\mathbb{F}_2} = \langle F' \rangle_{\mathbb{F}_2}$. We can eliminate the polynomials (12.12) as follows. For every $i \in V$, $(x_i^3 + 1)$ is the only polynomial in F' containing the monomial x_i^3 and thus the polynomial $(x_i^3 + 1)$ cannot be present in any nonzero linear combination of the polynomials in F' that equals 1. We arrive at the following smaller set of polynomials, which we label F''.

$$\begin{aligned}
&x_i^2x_j + x_ix_j^2 + 1 && \forall\{i,j\} \in E, && (12.15)\\
&x_i^2x_k + x_ix_jx_k + x_j^2x_k && \forall\{i,j\} \in E, k \in V, i \neq k \neq j. && (12.16)
\end{aligned}$$

So far, we have shown $1 \in \langle F \rangle_{\mathbb{F}_2} = \langle F' \rangle_{\mathbb{F}_2}$ if and only if $1 \in \langle F'' \rangle_{\mathbb{F}_2}$.

Next, we eliminate monomials of the form $x_ix_jx_k$. There are three cases to consider.

Case 1: $\{i,j\} \in E$ but $\{i,k\} \notin E$ and $\{j,k\} \notin E$. In this case, the monomial $x_i x_j x_k$ appears in only one polynomial, $x_k(x_i^2 + x_i x_j + x_j^2) = x_i^2 x_k + x_i x_j x_k + x_j^2 x_k$, so we can eliminate all such polynomials.

Case 2: $i,j,k \in V$, $(i,j),(j,k),(k,i) \in \text{Arcs}(G)$. Graphically, this represents a 3-cycle in the graph. In this case, the monomial $x_i x_j x_k$ appears in three polynomials:

$$x_k(x_i^2 + x_i x_j + x_j^2) = x_i^2 x_k + x_i x_j x_k + x_j^2 x_k, \tag{12.17}$$
$$x_j(x_i^2 + x_i x_k + x_k^2) = x_i^2 x_j + x_i x_j x_k + x_j x_k^2, \tag{12.18}$$
$$x_i(x_j^2 + x_j x_k + x_k^2) = x_i x_j^2 + x_i x_j x_k + x_i x_k^2. \tag{12.19}$$

Using the first polynomial, we can eliminate $x_i x_j x_k$ from the other two:

$$x_i^2 x_j + x_j x_k^2 + x_i^2 x_k + x_j^2 x_k = (x_i^2 x_j + x_i x_j x_k + x_j x_k^2) + (x_i^2 x_k + x_i x_j x_k + x_j^2 x_k),$$
$$x_i x_j^2 + x_i x_k^2 + x_i^2 x_k + x_j^2 x_k = (x_i x_j^2 + x_i x_j x_k + x_i x_k^2) + (x_i^2 x_k + x_i x_j x_k + x_j^2 x_k).$$

Now we can eliminate the polynomial (12.17). Moreover, we can use the polynomials (12.15) to rewrite the above two polynomials as follows.

$$\begin{aligned} x_k x_i^2 + x_i x_j^2 &= (x_i^2 x_j + x_j x_k^2 + x_i^2 x_k + x_j^2 x_k) + (x_j x_k^2 + x_j^2 x_k + 1) \\ &\quad + (x_i x_j^2 + x_i^2 x_j + 1), \\ x_i x_j^2 + x_j x_k^2 &= (x_i x_j^2 + x_i x_k^2 + x_i^2 x_k + x_j^2 x_k) + (x_i x_k^2 + x_i^2 x_k + 1) \\ &\quad + (x_j x_k^2 + x_j^2 x_k + 1). \end{aligned}$$

Note that both of these polynomials correspond to two of the arcs of the 3-cycle

$$(i,j),(j,k),(k,i) \in \text{Arcs}(G).$$

Case 3: $i,j,k \in V$, $(i,j),(j,k) \in \text{Arcs}(G)$ and $(k,i) \notin \text{Arcs}(G)$. We have

$$x_k(x_i^2 + x_i x_j + x_j^2) = x_i^2 x_k + x_i x_j x_k + x_j^2 x_k, \tag{12.20}$$
$$x_i(x_j^2 + x_j x_k + x_k^2) = x_i x_j^2 + x_i x_j x_k + x_i x_k^2. \tag{12.21}$$

As before we use the first polynomial to eliminate the monomial $x_i x_j x_k$ from the second:

$$\begin{aligned} &x_i x_j^2 + x_j x_k^2 + (x_i^2 x_k + x_i x_k^2 + 1) \\ &\quad = (x_i x_j^2 + x_i x_j x_k + x_i x_k^2) + (x_i^2 x_k + x_i x_j x_k + x_j^2 x_k) + (x_j x_k^2 + x_j^2 x_k + 1). \end{aligned}$$

We can now eliminate (12.20); thus, the original system has been reduced to the following one, which we label as F''':

$$x_i^2 x_j + x_i x_j^2 + 1 \qquad \forall \{i,j\} \in E, \tag{12.22}$$
$$x_i x_j^2 + x_j x_k^2 \qquad \forall (i,j),(i,k),(j,k) \in \text{Arcs}(G), \tag{12.23}$$
$$x_i x_j^2 + x_j x_k^2 + (x_i^2 x_k + x_i x_k^2 + 1) \qquad \forall (i,j),(j,k) \in \text{Arcs}(G),\ (k,i) \notin \text{Arcs}(G). \tag{12.24}$$

Note that $1 \in \langle F \rangle_{\mathbb{F}_2}$ if and only if $1 \in \langle F''' \rangle_{\mathbb{F}_2}$.

The monomials $x_i^2 x_k$ and $x_i x_k^2$ with $(k,i) \notin \mathrm{Arcs}(G)$ always appear together and only in the polynomials (12.24) in the expression $(x_i^2 x_k + x_i x_k^2 + 1)$. Thus, we can eliminate the monomials $x_i^2 x_k$ and $x_i x_k^2$ with $(k,i) \notin \mathrm{Arcs}(G)$ by choosing one of the polynomials (12.24) and using it to eliminate the expression $(x_i^2 x_k + x_i x_k^2 + 1)$ from all other polynomials in which it appears. Let $i,j,k,l \in V$ be such that $(i,j),(j,k),(k,l),(l,i) \in \mathrm{Arcs}(G)$ and $(k,i),(i,k) \notin \mathrm{Arcs}(G)$. We can then eliminate the monomials $x_i^2 x_k$ and $x_i x_k^2$ as follows:

$$\begin{aligned}x_i x_j^2 + x_j x_k^2 + x_k x_l^2 + x_l x_i^2 = {} & (x_i x_j^2 + x_j x_k^2 + x_i^2 x_k + x_i x_k^2 + 1)\\ & + (x_k x_l^2 + x_l x_i^2 + x_i^2 x_k + x_i x_k^2 + 1).\end{aligned}$$

Finally, after eliminating the polynomials (12.24), we have system H (polynomials (12.6), (12.7), and (12.8)):

$$\begin{aligned}& x_i^2 x_j + x_i x_j^2 + 1 && \forall \{i,j\} \in E,\\ & x_i x_j^2 + x_j x_k^2 && \forall (i,j),(j,k),(k,i) \in \mathrm{Arcs}(G),\\ & x_i x_j^2 + x_j x_k^2 + x_k x_l^2 + x_l x_i^2 && \forall (i,j),(j,k),(k,l),(l,i) \in \mathrm{Arcs}(G),\\ & && (i,k),(j,l) \notin \mathrm{Arcs}(G).\end{aligned}$$

The system H has the property that $1 \in \langle F''' \rangle_{\mathbb{F}_2}$ if and only if $1 \in \langle H \rangle_{\mathbb{F}_2}$, and thus, $1 \in \langle F \rangle_{\mathbb{F}_2}$ if and only if $1 \in \langle H \rangle_{\mathbb{F}_2}$ as required. □

We now establish that the sufficient condition for infeasibility $1 \in \langle H \rangle_{\mathbb{F}_2}$ is equivalent to the combinatorial parity conditions in Theorem 12.2.12.

Lemma 12.2.16. *There exists a set C of oriented partial 3-cycles and oriented chordless 4-cycles satisfying Conditions 1 and 2 of Theorem 12.2.12 if and only if $1 \in \langle H \rangle_{\mathbb{F}_2}$.*

Proof. Assume that $1 \in \langle H \rangle_{\mathbb{F}_2}$. Then there exist coefficients $c_h \in \mathbb{F}_2$ such that $\sum_{h \in H} c_h h = 1$. Let $H' := \{h \in H : c_h = 1\}$; then, $\sum_{h \in H'} h = 1$. Let C be the set of oriented partial 3-cycles (i,j,k), where $x_i x_j^2 + x_j x_k^2 \in H'$, together with the set of oriented chordless 4-cycles (i,j,l,k), where $x_i x_j^2 + x_j x_l^2 + x_l x_k^2 + x_k x_i^2 \in H'$. Now, $|C_{(i,j)}|$ is the number of polynomials in H' of the form (12.7) or (12.8) in which the monomial $x_i x_j^2$ appears, and similarly, $|C_{(j,i)}|$ is the number of polynomials in H' of the form (12.7) or (12.8) in which the monomial $x_j x_i^2$ appears. Thus, $\sum_{h \in H'} h = 1$ implies that, for every pair $x_i x_j^2$ and $x_j x_i^2$, either

1. $|C_{(i,j)}| \equiv 0 \pmod 2$, $|C_{(j,i)}| \equiv 0 \pmod 2$, and $x_i^2 x_j + x_i x_j^2 + 1 \notin H'$, or

2. $|C_{(i,j)}| \equiv 1 \pmod 2$, $|C_{(j,i)}| \equiv 1 \pmod 2$, and $x_i^2 x_j + x_i x_j^2 + 1 \in H'$.

In either case, we have $|C_{(i,j)}| + |C_{(j,i)}| \equiv 0 \pmod 2$. Moreover, since $\sum_{h \in H'} h = 1$, there must be an odd number of the polynomials of the form $x_i^2 x_j + x_i x_j^2 + 1$ in H'. That is, case 2 above occurs an odd number of times and therefore, $\sum_{(i,j) \in \mathrm{Arcs}(G), i<j} |C_{(i,j)}| \equiv 1 \pmod 2$ as required.

Conversely, assume that there exists a set C of oriented partial 3-cycles and oriented chordless 4-cycles satisfying the conditions of Theorem 12.2.12. Let H' be the set of polynomials $x_i x_j^2 + x_j x_k^2$ where $(i,j,k) \in C$ and the set of polynomials $x_i x_j^2 + x_j x_l^2 + x_l x_k^2 +$

$x_k x_i^2$ where $(i,j,l,k) \in C$ together with the set of polynomials $x_i^2 x_j + x_i x_j^2 + 1 \in H$ where $|C_{(i,j)}| \equiv 1$. Then, $|C_{(i,j)}| + |C_{(j,i)}| \equiv 0 \pmod 2$ implies that every monomial $x_i x_j^2$ appears in an even number of polynomials of H'. Moreover, since $\sum_{(i,j)\in \mathrm{Arcs}(G), i<j} |C_{(i,j)}| \equiv 1 \pmod 2$, there are an odd number of polynomials $x_i^2 x_j + x_i x_j^2 + 1$ appearing in H'. Hence, $\sum_{h\in H'} h = 1$ and $1 \in \langle H \rangle_{\mathbb{F}_2}$. □

Combining Lemmas 12.2.15 and 12.2.16, we arrive at the characterization stated in Theorem 12.2.12. That such graphs can be decided in polynomial time follows from Theorem 12.2.9.

12.2.4 Limitations of NulLA and stable sets of graphs

We have seen encouraging good behavior for NulLA for the 3-coloring problem. Unfortunately, in other combinatorial problems the NulLA rank grows rapidly in many examples (which contrasts with coloring). We show the NulLA rank growth happens in the stable set problem. This shows that the bound of Lemma 12.2.7 is tight.

Recall that a *stable set* or *independent set* in a graph $G = (V, E)$ is a subset of vertices such that no two vertices in the subset are adjacent. The maximum size $\alpha(G)$ of a stable set is called the *stability number* of G. As it is customary in optimization, we view the stable sets in terms of their *incidence vectors*. These are vectors of length $|V(G)|$, one for each stable set, filled with zeros and ones, where a one in the i-th entry says the i-th vertex is a member of associated stable set. There has been much work on understanding the description of the linear inequalities that describe the convex hull of these incidence vectors, the so-called *stable set polytope* [299], but this is difficult, perhaps even impossible, to complete (assuming P not equal NP). On the other hand, these same $0/1$ vectors can be trivially described in a small system of *quadratic* equations by Lovász, shown in Part (1) of Theorem 12.1.1.

The following theorem demonstrates *linear* growth in the minimum degree of the Nullstellensatz certificates for the nonexistence of stables sets larger than $\alpha(G)$. It also proves that the certificates are *dense* as all possible monomials of certain degree appear in the certificate. This settles an open problem in [236] where Lovász asked about finding a family of graphs with growth in the minimum degree of their Nullstellensatz certificates.

Theorem 12.2.17. *Given a graph G, a minimum-degree Nullstellensatz certificate (associated with the Lovász encoding of Theorem* 12.1.1*) for the nonexistence of a stable set of size greater than $\alpha(G)$ has degree equal to $\alpha(G)$ and contains at least one term for every stable set in G.*

The theorem establishes new lower bounds for the degree and number of terms in Nullstellensatz certificates. In earlier work, researchers in logic and complexity showed both logarithmic and linear growth in the degree of Nullstellensatz certificates over finite fields or for special instances, e.g., Nullstellensatz related to the pigeonhole principle (see [68], [180] and references therein). Our main complexity result below settles the open question posed by Lovász [236]:

Corollary 12.2.18. *There exist infinite families of graphs G_n, on n vertices, such that the degree of a minimum-degree Nullstellensatz certificate (associated with the Lovász encoding of Theorem 12.1.1) grows linearly in n and, at the same time, the number of terms in the coefficient polynomials of the Nullstellensatz certificate is exponential in n. Thus the*

NulLA rank can grow linearly and the Nullstellensatz certificate can have exponentially many terms.

Proof. We describe two infinite families explicitly. First, the disjoint union of $n/3$ triangles has exactly $4^{n/3}-1$ stable sets and the minimum degree of the Nullstellensatz certificates is $\alpha(G)=n/3$. Second, graphs with no edges have $\alpha(G)=n$, and the number of stable sets is 2^n. □

The proof of the entire theorem is not difficult but a bit long. For brevity, here we only show that, for *every* graph, there exists an explicit Nullstellensatz certificate of degree $\alpha(G)$. Proving that this is the smallest degree possible and thus that the NulLA rank grows is omitted. For more details see [100].

In what follows let S_i be the set of all stable sets of size i in the graph G. For any stable set $I \in S_i$, if I consists of the vertices $\{c_1,c_2,\dots,c_i\}$, then $x_I := x_{c_1}x_{c_2}\cdots x_{c_i}$, and we refer to the monomial x_I as a "stable set." We define $S_0 := \emptyset$, and $x_\emptyset = 1$. If we say $I\cup k \in S_{i+1}$, we explicitly mean that $I\cap k=\emptyset$, and that $x_I x_k$ is a square-free stable set monomial of degree $i+1$. If $I\cup k \notin S_{i+1}$, we explicitly mean that $I\cap k=\emptyset$ but $I\cup k$ contains at least one edge $\{k,c_j\}$. In other words, $x_I x_k$ is a square-free nonstable set monomial of degree $i+1$. In this case, let $\min_k(I)$ denote the *smallest* $c_j \in I$ such that $\{k,c_j\}\in E(G)$. Finally, let

$$P_i := \sum_{I\in S_i} x_I, \qquad \text{with} \quad P_0 := 1,$$

and

$$L_i := \frac{i\,L_{i-1}}{\alpha(G)+r-i}, \qquad \text{with} \quad L_0 := \frac{1}{\alpha(G)+r}.$$

Theorem 12.2.19. *Given a graph G, there exists a Nullstellensatz certificate of degree $\alpha(G)$ certifying the nonexistence of a stable set of size $\alpha(G)+r$ (for $r\geq 1$) such that*

$$1 = A\left(-(\alpha(G)+r)+\sum_{i=1}^{n} x_i\right)+\sum_{\{u,v\}\in E(G)} Q_{uv}x_u x_v+\sum_{k=1}^{n} Q_k(x_k^2-x_k), \tag{12.25}$$

where

$$A = -\sum_{i=0}^{\alpha(G)} L_i P_i, \qquad Q_{uv} = \sum_{i=1}^{\alpha(G)}\left(\sum_{\substack{I\in S_i:\, I\cup v\notin S_{i+1}\, and \\ \min_v(I)=u}} L_{i+1}x_{I\setminus u}\right), \quad \textit{and}$$

$$Q_k = \sum_{i=0}^{\alpha(G)}\left(\sum_{I\in S_i:\, I\cup k\in S_{i+1}} L_{i+1}x_I\right).$$

Proof. Our proof is simply the verification of equation (12.25). Let B, C, and D equal

$$1 = \underbrace{A\left(-(\alpha(G)+r)+\sum_{i=1}^{n} x_i\right)}_{B}+\underbrace{\sum_{\{u,v\}\in E(G)} Q_{uv}x_u x_v}_{C}+\underbrace{\sum_{k=1}^{n} Q_k(x_k^2-x_k)}_{D}.$$

It is easy to see that

$$-L_0P_0\big(-(\alpha(G)+r)\big) = -\frac{1}{\alpha(G)+r}(-(\alpha(G)+r)) = 1.$$

We will now show that the coefficient for every other monomial in equation (12.25) simplifies to zero. We begin by observing that every monomial in A, Q_k, or Q_{uv} is a stable set, and furthermore, that the stable set monomials in Q_k do not contain the variable x_k, and the stable set monomials in Q_{uv} contain neither x_u nor x_v. Therefore, in the expanded certificate $AB+C+D$, only three types of monomials appear: square-free stable set monomials, square-free nonstable set monomials, and stable set monomials with exactly one variable squared.

- **Square-free stable set:** Let $I = \{c_1, c_2, \ldots, c_m\}$ be any stable set of size m. The monomial x_I is created in AB in two ways: $x_{I\setminus c_k}x_{c_k}$ (formed m times, one for each c_k), or $x_I\big(-(\alpha(G)+r)\big)$. Thus, the coefficient for x_I in AB is

$$\begin{aligned}&-mL_{m-1} - L_m\big(-(\alpha(G)+r)\big)\\ &\qquad\qquad = -m\frac{L_m(\alpha(G)+r-m)}{m} + L_m(\alpha(G)+r) = mL_m.\end{aligned}$$

  The monomial x_I does not appear in C, because x_I is a stable set monomial. However, the monomial x_I is produced by $x_{I\setminus c_k}(-x_{c_k})$ in D (formed m times, one for each c_k), and the coefficient for x_I in D is $-mL_m$. Therefore, we see that

$$\underbrace{mL_m}_{\text{from } AB}\ \underbrace{-mL_m}_{\text{from } D} = 0.$$

- **Square-free nonstable set:** Let $I = \{c_1, c_2, \ldots, c_{m-1}, u\}$ be any stable set of size m, and consider the monomial x_Ix_v where $u = \min_v I$ and $\{u,v\} \in E(G)$. Now, consider all $\binom{m+1}{m}$ subsets of $\{c_1, c_2, \ldots, c_{m-1}, u, v\}$, and let M be the number of stable sets among those $\binom{m+1}{m}$ subsets. Each of those M subsets appears as a stable set monomial in A. Therefore, the monomial x_Ix_v is created M times in AB, and the coefficient for x_Ix_v in AB is $-ML_m$. The monomial x_Ix_v does not appear in D, because it is a nonstable set monomial, and it appears exactly M times in C. Therefore, the coefficient for x_Ix_v in C is ML_m, and we see that

$$\underbrace{-ML_m}_{\text{from } AB} + \underbrace{ML_m}_{\text{from } C} = 0.$$

- **Stable set with one variable squared:** Let $I = \{c_1, c_2, \ldots, c_{m-1}, k\}$ be any stable set of size m, and consider the monomial $x_{I\setminus k}x_k^2$. This monomial is created in AB by the direct product x_Ix_k, and the coefficient is $-L_m$. This monomial is not created in C, because it contains no edges, and it is created in D by $x_{I\setminus k}x_k^2$. Thus, the coefficient for x_Ix_k in D is L_m, and we see that

$$\underbrace{-L_m}_{\text{from } AB} + \underbrace{L_m}_{\text{from } D} = 0.$$

Since the constant term in $AB+C+D$ is one, and the coefficient for every other monomial is zero, equation (12.25) is a Nullstellensatz certificate of degree $\alpha(G)$. □

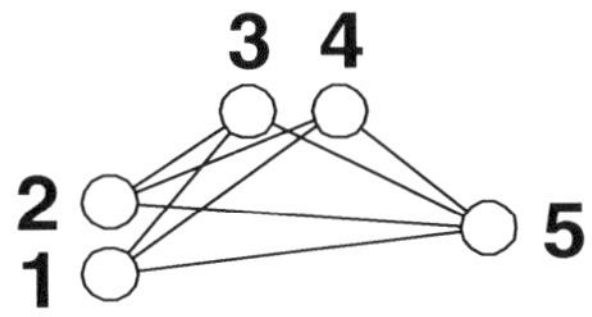

Figure 12.4. *Turán graph* $T(5,3)$.

Example 12.2.20. Here is an example of the construction and of Theorem 12.2.17. Figure 12.4 depicts the $T(5,3)$ Turán graph. It is clear that $\alpha(T(5,3)) = 2$. Therefore, we construct a certificate via Theorem 12.2.19 verifying the nonexistence of a stable set of size 3. This is in fact the *smallest* certificate.

$$
\begin{aligned}
1 = {} & \left(\frac{1}{3}x_4 + \frac{1}{3}x_2 + \frac{1}{3}\right)x_1x_3 + \left(\frac{1}{3}x_2 + \frac{1}{3}\right)x_1x_4 + \left(\frac{1}{3}x_2 + \frac{1}{3}\right)x_1x_5 + \left(\frac{1}{3}x_4 + \frac{1}{3}\right)x_2x_3 \\
& + \left(\frac{1}{3}\right)x_2x_4 + \left(\frac{1}{3}\right)x_2x_5 + \left(\frac{1}{3}x_4 + \frac{1}{3}\right)x_3x_5 + \left(\frac{1}{3}\right)x_4x_5 + \left(\frac{1}{3}x_2 + \frac{1}{6}\right)(x_1^2 - x_1) \\
& + \left(\frac{1}{3}x_1 + \frac{1}{6}\right)(x_2^2 - x_2) + \left(\frac{1}{3}x_4 + \frac{1}{6}\right)(x_3^2 - x_3) + \left(\frac{1}{3}x_3 + \frac{1}{6}\right)(x_4^2 - x_4) + \left(\frac{1}{6}\right)(x_5^2 - x_5) \\
& + \underbrace{\left(-\frac{1}{3}(x_1x_2 + x_3x_4) - \frac{1}{6}(x_1 + x_2 + x_3 + x_4 + x_5) - \frac{1}{3}\right)}_{A}(x_1 + x_2 + x_3 + x_4 + x_5 - 3).
\end{aligned}
$$

In this example, note that A, Q_i, Q_{ij} are polynomials in $\mathbb{Q}[x_1, \ldots, x_n]$, and furthermore, note that A contains one monomial for every stable set in $T(5,3)$. For example, note that the term $-\frac{1}{3}x_1x_2$ corresponds to the stable set formed by vertices 1 and 2 in $T(5,3)$. Additionally, every monomial in every coefficient is also a stable set in $T(5,3)$.

From a computational perspective, the density of the Nullstellensatz certificates represents a serious obstacle. In this case, we have demonstrated that computing Hilbert's Nullstellensatz is at least as hard as counting all possible stable sets in a graph, which is known to be #P-complete, even for graphs with low-degree vertices [117]. Furthermore, we strongly expect, based on the structure of our proof and the resulting telescopic sums, that *any* polynomial system containing 0/1 equations ($x_i^2 - x_i = 0$) will exhibit similar computational difficulties, e.g., for matching problems.

Remark 12.2.21. This also has ramifications for Gröbner bases computations, showing that any such computation can generate exponentially many intermediate monomials. On the other hand, it suggests that the natural binary encodings of stable set problems are not the most desirable for NulLA. Finally, we note that, since the degrees of the polynomials from Lovász encoding of the stable set problem are less than or equal to two, and because we have demonstrated *linear* growth in the minimum degree of the associated Nullstellensatz certificates, we have now shown that the Lazard bound on projective Nullstellensatz certificates presented in Corollary 12.2.8 is tight.

12.3 A simple proof of the Nullstellensatz

We present a simple, self-contained proof of Hilbert's Nullstellensatz. The proof we present here is based on the very nice proof by Arrondo ([15]).

Theorem 12.3.1 (Hilbert's Nullstellensatz). *Let $f_1 = 0, f_2 = 0, \dots, f_s = 0$ be a system of polynomials in $\mathbb{K}[x_1, \dots, x_m]$ with $\mathbb{K}$ a field, then the system has no root, or solution over the algebraic closure $\overline{\mathbb{K}}$, if and only if there exist polynomials $\alpha_1, \alpha_2, \dots, \alpha_s \in \mathbb{K}[x_1, \dots, x_m]$ such that $1 = \sum_{i=1}^{n} \alpha_i(x_1, \dots, x_m) f_i(x_1, \dots, x_m)$*

It is clear that the identity shows nonsolvability of the system. The difficult part of the theorem is proving that the nonexistence of solutions implies the existence of this identity between two polynomials.

Our proof proceeds by induction on the number of variables and then uses a simple version of Noether's normalization lemma and the basic theory of resultants to eliminate variables and finish the induction.

Example 12.3.2. Recall from the previous example the following set of polynomials in $\mathbb{R}[x_1, x_2, x_3]$:

$$F := \{f_1 := x_1^2 - 1, f_2 := 2x_1x_2 + x_3, f_3 := x_1 + x_2, f_4 := x_1 + x_3\}.$$

By the Nullstellensatz, the system $f_1(x) = 0, f_2(x) = 0, f_3(x) = 0, f_4(x) = 0$ is infeasible over $\mathbb{C}$ if and only if there exist polynomials $\beta_1, \beta_2, \beta_3, \beta_4 \in \mathbb{R}[x_1, x_2, x_3]$ that satisfy the polynomial identity $\beta_1 f_1 + \beta_2 f_2 + \beta_3 f_3 + \beta_4 f_4 = 1$. Here, the system is infeasible, so there exist such polynomials as follows:

$$\beta_1 = -1 - \frac{2}{3}x_2, \quad \beta_2 = -\frac{2}{3} + \frac{1}{3}x_1, \quad \beta_3 = -\frac{2}{3} + \frac{4}{3}x_1, \quad \beta_4 = \frac{2}{3} - \frac{1}{3}x_1.$$

The resulting identity provides a certificate of infeasibility of the system.

In the case that the system is infeasible a certificate of infeasibility is given by Hilbert's Nullstellensatz. It is very important to note that the coefficients of the certificate exist within the original field of coefficients $\mathbb{K}$.

12.3.1 Proof for polynomials in one variable

A constructive proof of the Hilbert Nullstellensatz in one variable uses easy properties of the greatest common divisor (remember we saw the gcd can be computed using the Euclidean algorithm in Section 10.2).

Lemma 12.3.3. *Suppose $f_1, \dots, f_s$ are univariate polynomials in $\mathbb{K}[x]$, then (when taking the* gcd *to be a monic polynomial),*

$$\gcd(f_1, f_2, \dots, f_s) = \gcd(f_1, \gcd(f_2, \dots, f_s)).$$

Proof. Note $\gcd(f_1, f_2, \dots, f_s) \mid f_i$ for $i = 1, \dots, s$. This implies $\gcd(f_1, f_2, \dots, f_s)$ divides $\gcd(f_2, \dots, f_s)$ and it also divides f_1. Therefore $\gcd(f_1, \gcd(f_2, f_3, \dots, f_s))$ is divided by $\gcd(f_1, f_2, f_3, \dots, f_s)$. Note that if $h \mid f_1$ and $h \mid \gcd(f_2, f_3, \dots, f_s)$ then $h \mid f_i$ for $i = 1 \dots s$. Thus $\gcd(f_1, \gcd(f_2, f_3, \dots, f_s))$ must divide the polynomial $\gcd(f_1, f_2, f_3, \dots, f_s)$ too. This proves they are equal. □

Remember our goal is to find solutions of the system of polynomial equations or certify that none exist. How can we determine whether a system $g_1(x) = 0, \ldots, g_r(x) = 0$ has a common root? The key point is that if a is a common root, $(x-a)$ is a common factor. Thus when $\gcd(g_1(x), \ldots, g_r(x)) \neq 1$, you have found a common solution, as any of the roots of this gcd will be a common root of the system.

Lemma 12.3.4. *If $h = \gcd(f_1, f_2, \ldots, f_s)$, there exist polynomials $a_1, a_2, \ldots, a_s$ such that $h = a_1 f_1 + \cdots + a_s f_s$.*

We already saw how to obtain such an expression for two polynomials using Lemma 12.3.3.

Proof. This is proved by induction on the number of polynomials. Say for $n = 2$, apply the extended Euclidean algorithm (see discussion before Lemma 10.2.6) to f_1, f_2. The algorithm gives an expression, $a_1 f_1 + a_2 f_2 = \gcd(f_1, f_2)$. Assume the result is true for $s-1$ or fewer polynomials. We know from Lemma 12.3.3 that $\gcd(f_1, f_2, \ldots, f_s) = \gcd(f_1, \gcd(f_2, \ldots, f_s))$. Thus by induction, $\gcd(f_2, \ldots, f_s)$ can be written as $h = \gcd(f_2, \ldots, f_s) = b_2 f_2 + \ldots + b_s f_s$, which by the case of two polynomials, gives, as desired, $\gcd(f_1, f_2, \ldots, f_s) = \gcd(f_1, h) = r_1 f_1 + sh$. This equals $r_1 f_1 + s(b_2 f_2 + \ldots + b_s f_s) = r_1 f_1 + s b_2 f_2 + \cdots + s b_s f_s$, then $r_1 = a_1, s b_2 = a_2, \ldots, s b_s = a_s$. □

The reader can easily verify the following consequence of Lemma 12.3.4.

Corollary 12.3.5. *A system of equations $f_1(x) = 0, \ldots, f_s(x) = 0$ with univariate polynomial equations has no common root if and only if one can find polynomials $a_1, a_2, \ldots, a_s$ that give the identity $1 = a_1 f_1 + \cdots + a_s f_s$.*

12.3.2 Resultants and multivariate polynomials

Now that we have the Nullstellensatz for systems of polynomials in one variable, we will use *resultants* to reduce any other system to this simpler case. The first ingredient is the resultant of two polynomials, which is a special determinant expression obtained from two polynomials in $\mathbb{K}[x]$ ($\mathbb{K}$ is any field) that decides the existence of common factors. We begin with the following property.

Lemma 12.3.6. *Let $\mathbb{K}$ be a field and let f and g be polynomials in $\mathbb{K}[x]$ such that f, g are nonzero. Let $\deg(f) = n, \deg(g) = m$, then f, g have a common factor (i.e., gcd of the polynomials is different from 1) if and only if there exist nonzero polynomials A, B in $\mathbb{K}[x]$ such that*

1. $Af + Bg = 0$,
2. $\deg(A) < m$, $\deg(B) < n$.

Proof. ($\Rightarrow$) Suppose $f = f_1 h$, $g = g_1 h$, for some common h. Let $A = g_1$ and $B = -f_1$, then $g_1 f + -f_1 g = g_1 f_1 h - g_1 f_1 h = 0$.
($\Leftarrow$) By contradiction. We assume the existence of A, B as, in the statement above, $-Af = Bg$. Suppose further that f, g have no common factor. That implies that there exist polynomials R, S in $\mathbb{K}[x]$ such that $Rf + Sg = 1$ (from the extended Euclidean algorithm). Then $B = BRf + BSg = BRf - SAf = (BR - SA)f$. This implies that $\deg(B) > n$. This contradicts the condition that $\deg(B) < n$. □

The previous lemma easily suggests how to construct the *Sylvester matrix* $\mathrm{Syl}(f,g)$ of two polynomials. This matrix will have the property that the polynomials f,g have a common root factor in $\mathbb{K}[x_1,\ldots,x_n]$ if and only if $\det(\mathrm{Syl}(f,g))=0$. Let us begin with a suggestive example:

Example 12.3.7. Suppose we wish to know whether $f=2x^3+4x-2$, $g=2x^2+3x+9$ have a common factor. From Lemma 12.3.6, they do if and only if there exist A,B with $Af+Bg=0$. In this case $A=a_1x+a_0$, $B=b_2x^2+b_1x+b_0$, with some unknown coefficients. We obtain an identity of polynomials

$$0=Af+Bg=(a_1x+a_0)(2x^3+4x-2)+(b_2x^2+b_1x+b_0)(2x^2+3x+9).$$

Recollecting terms by powers of x gives the identity

$$\begin{aligned}0=(2a_1+2b_2)x^4+(2a_0+3b_2+3b_1)x^3+(4a_1+9b_2+3b_1+2b_0)x^2\\+(-2a_1+4a_0+9b_1+3b_0)x+(-2a_0+9b_0).\end{aligned}$$

Of course, equating monomials of equal degree each of the coefficients must equal zero, thus we get a linear system of equations (in decreasing degree):

$$\begin{aligned}2a_1+2b_2&=0,\\2a_0+3b_2+2b_1&=0,\\4a_1+9b_2+3b_1+2b_0&=0,\\-2a_1+4a_0+9b_1+3b_0&=0,\\-2a_0+9b_0&=0.\end{aligned}$$

The system can be represented by a matrix $\mathrm{Syl}(f,g)$:

$$\begin{pmatrix}2&0&2&0&0\\0&2&3&2&0\\4&0&9&3&2\\-2&4&0&9&3\\0&-2&0&0&9\end{pmatrix}\begin{pmatrix}a_1\\a_0\\b_2\\b_1\\b_0\end{pmatrix}=\begin{pmatrix}0\\0\\0\\0\\0\end{pmatrix}.$$

Because $\det(\mathrm{Syl}(f,g))=1163\neq 0$, there cannot exist A,B as predicted, thus f and g have no common factor.

The example suggests that the same process that yielded the matrix in the example above can be used in general.

Definition 12.3.8. Given two nonzero univariate polynomials $f=a_nx^n+a_{n-1}x^{n-1}+\cdots+a_0$ and $g=b_mx^m+b_{m-1}x^{m-1}+b_{m-2}x^{m-2}+\cdots+b_0$ with coefficients over a field

$\mathbb{K}$, the *Sylvester matrix* $\mathrm{Syl}(f,g)$ is given by the $(m+n)\times(m+n)$ matrix

$$\begin{pmatrix} a_n & 0 & \dots & 0 & b_m & 0 & \dots & 0 \\ a_{n-1} & a_n & \ddots & 0 & b_{m-1} & b_m & \ddots & 0 \\ \vdots & a_{n-1} & \ddots & 0 & \vdots & b_{m-1} & \ddots & 0 \\ a_0 & \vdots & \ddots & a_n & b_0 & \vdots & \ddots & b_m \\ 0 & a_0 & \dots & \vdots & 0 & b_0 & \dots & \vdots \\ 0 & 0 & \dots & a_0 & 0 & 0 & \dots & b_0 \end{pmatrix}.$$

We call the determinant of the Sylvester matrix the *resultant* of f and g, denoted $\mathrm{Res}(f,g)$. Clearly $\mathrm{Res}(f,g)$ is a polynomial whose variables are the coefficients of f and g and it has integer coefficients.

Theorem 12.3.9. *The polynomials f,g have a common root factor in $\mathbb{K}[x]$ if and only if $\mathrm{Res}(f,g)=\det(\mathrm{Syl}(f,g))=0$. Moreover, there exist polynomials $\tilde{C},\tilde{D}\in\mathbb{K}[x]$, such that $\tilde{C}f+\tilde{D}g=\mathrm{Res}(f,g)$.*

Proof. First, from Lemma 12.3.6 and the rearrangement argument used in the example, one can easily see that the existence of A,B as in Lemma 12.3.6 depends on whether there is a nonzero solution to the homogeneous linear system given by $\mathrm{Syl}(f,g)$. This happens when the determinant

$$\mathrm{Res}(f,g)=\det\begin{pmatrix} a_n & 0 & \dots & 0 & b_m & 0 & \dots & 0 \\ a_{n-1} & a_n & \ddots & 0 & b_{m-1} & b_m & \ddots & 0 \\ \vdots & a_{n-1} & \ddots & 0 & \vdots & b_{m-1} & \ddots & 0 \\ a_0 & \vdots & \ddots & a_n & b_0 & \vdots & \ddots & b_m \\ 0 & a_0 & \dots & \vdots & 0 & b_0 & \dots & \vdots \\ 0 & 0 & \dots & a_0 & 0 & 0 & \dots & b_0 \end{pmatrix}$$
$$\qquad\qquad\underbrace{\hphantom{a_n\;0\;\dots\;0}}_{m}\quad\underbrace{\hphantom{b_m\;0\;\dots\;0}}_{n}$$

vanishes. To prove the second statement, when $\mathrm{Res}(f,g)=0$, take $C=0$, $D=0$; then we are done proving the second statement. When $\mathrm{Res}(f,g)\neq 0$ this implies $\det[\mathrm{Syl}(f,g)]\neq 0$. Because $\mathrm{Res}(f,g)\neq 0$ then $\gcd(f,g)=1$. Then there exist C,D such that $Cf+Dg=1$ which translates into a linear system of equations

$$\begin{pmatrix} & & \\ & \mathrm{Syl}(f,g) & \\ & & \end{pmatrix} \begin{pmatrix} c_n \\ c_{n-1} \\ \vdots \\ c_0 \\ d_m \\ d_{m-1} \\ \vdots \\ d_0 \end{pmatrix} = \begin{pmatrix} 0 \\ 0 \\ \vdots \\ \vdots \\ 0 \\ 0 \\ 0 \\ 1 \end{pmatrix}.$$

Thus, by Cramer's rule, the values of the variables c_i and d_j (which are the coefficients of the polynomials C, D) can be written in the form

$$\frac{\det(\textit{some submatrix of } \mathrm{Syl}(f,g))}{\mathrm{Res}(f,g)}.$$

Multiply the equation $Cf + Dg = 1$ by $\mathrm{Res}(f,g)$. The denominators of the coefficients disappear, hence we obtain a new polynomial identity $\tilde{C}f + \tilde{D}g = \mathrm{Res}(f,g)$, which is the desired result. □

To continue the induction we are about to use a neat additional trick of randomly rotating (or skewing) the system of equations to prepare it for the projection to one lower dimension. This is a "random projection" argument that rewrites equations in a special way easier to analyze. We need a lemma first.

Lemma 12.3.10. *Let $\mathbb{K}$ be an infinite field and $B(x_1,x_2,\dots,x_n) \in \mathbb{K}[x_1,x_2,\dots,x_n]$ be a polynomial of degree d that is not identically zero. Then there exist $z_1,z_2,\dots,z_n \in \mathbb{K}$ such that $B(z_1,z_2,\dots,z_n) \neq 0$.*

Proof. Shown by induction on n, the number of variables. If $n = 1$, we are done, because univariate polynomials have finitely many roots due to the fundamental theorem of algebra. Thus $B(x_1) \neq 0$ for infinitely many values. Suppose then the statement is true for $n-1$ variables and take $B(x_1,x_2,\dots,x_n)$ not identically zero. When expanded as powers of x_n, B looks like

$$B(x_1,\dots,x_n) = P_d(x_1,\dots,x_{n-1})x_n^d + P_{d-1}(x_1,\dots,x_{n-1})x_n^{d-1} + \cdots + P_1(x_1,\dots,x_{n-1}).$$

By hypothesis, $P_d(x_1,x_2,\dots,x_n)$ is not identically zero. Thus by induction there exist $z_1,z_2,\dots,z_{n-1} \in \mathbb{K}$ where $P_i(z_1,z_2,\dots,z_{n-1}) \neq 0$. Substitute $x_k = z_k$ for $k = 1,2,\dots,n-1$ in B to obtain $B(z_1,z_2,\dots,z_{n-1},x_n)$. This is a univariate polynomial of bounded degree, so if we compute its roots $\beta_1,\beta_2,\dots,\beta_n$, let z_n be any complex number $\neq \beta_i$. □

Lemma 12.3.11. *Let $B(x_1,x_2,\dots,x_n) \in \mathbb{K}[x_1,x_2,\dots,x_n]$ be a polynomial of degree $\leq d$, Then there exists a change of coordinates of the form*

$$x_1 = \lambda_1 y_1 + y_n, \quad x_2 = \lambda_2 y_2 + y_n, \quad \dots, \quad x_i = \lambda_i y_i + y_n,$$

$$x_n = y_n,$$

such that $B(y_1 + \lambda_1 y_n,\dots,y_{n-1} + \lambda_{n-1} y_n, y_n) = P(y_1,\dots,y_n)$ is a polynomial whose coefficient for y_n^d is nonzero.

Proof. Do the substitution with parameters $\lambda_1, \lambda_2, \ldots, \lambda_{n-1}$. We see d is the highest degree for y_n. The leading coefficient will be a polynomial C_d in $\lambda_1, \lambda_2, \ldots, \lambda_{n-1}$. From the previous lemma we can find $\lambda_1, \lambda_2, \ldots, \lambda_{n-1} \in \mathbb{K}$ that make $C_d \neq 0$. Divide by the constant $C_d(\lambda_1, \lambda_2, \ldots, \lambda_{n-1}) \neq 0$. This yields the result. □

Theorem 12.3.12 (Hilbert's Nullstellensatz, original version). *Let $f_1 = 0, f_2 = 0, \ldots, f_s = 0$ be a system of polynomial equations, $f_1, \ldots, f_s \in \mathbb{K}[x_1, \ldots, x_n]$ with $\mathbb{K}$ an algebraically closed field. Then, there is no common solution over $\mathbb{K}^n$ if and only if there exist polynomials $\alpha_1, \alpha_2, \ldots, \alpha_s \in \mathbb{K}[x_1, \ldots, x_n]$ such that*

$$\alpha_1 f_2 + \alpha_2 f_2 + \ldots + \alpha_s f_s = 1.$$

Proof. Clearly the existence of the identity $\alpha_1 f_2 + \alpha_2 f_2 + \cdots + \alpha_s f_s = 1$ is sufficient to show no solutions are possible in this case. We prove the contrapositive statement using ideals. If we denote by I the ideal generated by $f_1, \ldots, f_s$, the statement we want to prove is equivalent to the following: If I is a proper ideal of the polynomial ring ($1 \notin I$) then there is a common solution to the system. We will use induction on the number of variables to prove this. We know it is true for $n = 1$ using the properties of the gcd. Assume that the theorem holds for all proper ideals in $\mathbb{K}[x_1, \ldots, x_{n-1}]$. We will prove the statement for the proper ideal $I \subset \mathbb{K}[x_1, \ldots, x_n]$ (i.e., 1 is not in the ideal I).

From Lemma 12.3.11, we know that for *any* nonconstant polynomial $f \in I$, we can translate it in such a way that the coefficient of x_n^d is nonzero. Therefore, by scaling accordingly, we can find a $g \in I$ that is *monic* in x_n (otherwise the ideal would be in fewer variables).

Let I' be a proper ideal in $\mathbb{K}[x_1, \ldots, x_{n-1}]$ consisting of all of the polynomials in I that do not contain the variable x_n. By the induction hypothesis, there exists a point $(a_1, \ldots, a_{n-1})$ such that $f(a_1, \ldots, a_{n-1}) = 0$ for all $f \in I'$. Now we claim: The set $J = \{f(a_1, \ldots, a_n, x_n) : f \in I\}$ is a proper ideal of $\mathbb{K}[x_n]$.

If this claim is true, J is a proper ideal of $\mathbb{K}[x_n]$, and thus it is generated by a single polynomial $h(x_n)$. If h is nonconstant then, since $\mathbb{K}$ is algebraically closed, $h(x_n)$ has a root a_n. Similarly if $h(x_n)$ is the constant zero polynomial. In either case, $f(a_1, \ldots, a_{n-1}, a_n) = 0$ for all $f \in I$. Thus, for any proper ideal $I \in \mathbb{K}[x_1, \ldots, x_n]$, there exists a point $(a_1, \ldots, a_n)$ such that $f(a_1, \ldots, a_n) = 0$ for all $f \in I$. This would end the proof of Hilbert's Nullstellensatz. We just have to prove the claim.

Suppose, for the purpose of deriving a contradiction, that there exists an $f \in I$ such that $f(a_1, \ldots, a_{n-1}, x_n) = 1$. Then, we can write f as

$$f = f_0 + f_1 x_n + \ldots + f_d x_n^d,$$

where $f_i \in \mathbb{K}[x_1, \ldots, x_{n-1}]$ and

$$f_1(a_1, \ldots, a_{n-1}) = \cdots = f_d(a_1, \ldots, a_{n-1}) = 0, \quad f_0(a_1, \ldots, a_{n-1}) = 1.$$

In addition, we can similarly express the monic polynomial g as

$$g = g_0 + g_1 x_n + \cdots + g_{e-1} x_n^{e-1} + x_n^e$$

with $g_j \in \mathbb{K}[x_1, \ldots, x_{n-1}]$ for $j = 0, \ldots, (e-1)$. The resultant R of f and g with respect to the variable x_n is the polynomial in $\mathbb{K}[x_1, \ldots, x_{n-1}]$ given by the determinant of the following $(e+d) \times (e+d)$ square matrix:

$$R = \begin{vmatrix} f_0 & f_1 & \cdots & \cdots & \cdots & f_d & 0 & \cdots & \cdots & 0 \\ 0 & f_0 & f_1 & \cdots & \cdots & f_{d-1} & f_d & 0 & \cdots & 0 \\ \vdots & & \ddots & \ddots & & & \ddots & \ddots & & \vdots \\ \vdots & & & \ddots & \ddots & & & \ddots & \ddots & 0 \\ 0 & \cdots & \cdots & 0 & f_0 & f_1 & \cdots & \cdots & f_{d-1} & f_d \\ g_0 & g_1 & \cdots & \cdots & g_{e-1} & 1 & 0 & \cdots & \cdots & 0 \\ 0 & g_0 & g_1 & \cdots & \cdots & g_{e-1} & 1 & 0 & \cdots & 0 \\ \vdots & & \ddots & \ddots & \ddots & \ddots & \ddots & \ddots & & \vdots \\ \vdots & & & \ddots & \ddots & \ddots & \ddots & \ddots & \ddots & 0 \\ 0 & \cdots & \cdots & \cdots & 0 & g_0 & g_1 & \cdots & g_{e-1} & 1 \end{vmatrix}.$$

It follows from Theorem 12.3.9 that the resultant $R \in I$. Furthermore, $R \in \mathbb{K}[x_1,\dots,x_{n-1}]$, and therefore R is also in I' (since I' was defined to contain all polynomials in I that do not contain the variable x_n). Note $R(a_1,\dots,a_{n-1})$ is equivalent to evaluating all of the polynomials at $(a_1,\dots,a_{n-1})$ and then taking the determinant. This yields a lower triangular matrix, whose diagonal values are all 1. Therefore, $R(a_1,\dots,a_{n-1}) = 1$. But, according to our induction hypothesis, all polynomials in I' vanish on the point $(a_1,\dots,a_n)$. Thus, $R \notin I'$, and we have reached a contradiction. Therefore, there cannot exist a polynomial $f \in J$ that is identically 1. Therefore, $J = \{f(a_1,\dots,a_n,x_n) : f \in I\}$ is a proper ideal of $\mathbb{K}[x_n]$. □

It is easy to translate Hilbert's Nullstellensatz into the form of Theorem 12.3.1 which is more useful for computation and applications. It follows because by Theorem 12.3.12 an infeasible system must yield an identity $1 = \sum \beta_i f_i$, where the β have coefficients in the field $\overline{\mathbb{K}}$. This must mean that the linear system of equations with coefficients in $\mathbb{K}$ has a solution in $\overline{\mathbb{K}}$, so it must have a solution in $\mathbb{K}$.

To conclude, we briefly comment on a consequence of Hilbert's Nullstellensatz. Recall, for a subset of $\mathbb{K}^n$, we defined $I(S)$ as the set of all polynomials that vanish in S. Let V be a variety; it is clear that if $f^m \in I(V)$, then $f \in I(V)$. This motivates the important definition.

Definition 12.3.13. An ideal I is *radical* if $f^m \in I$ implies that $f \in I$.

Thus $I(V)$ is an example of radical ideal.

Definition 12.3.14. For an ideal I, the *radical of I* is the set

$$\sqrt{I} := \left\{ f : f^m \in I \text{ for some power } m \geq 1 \right\}.$$

It is important to note the radical gives a one-to-one correspondence between varieties and radical ideals as a consequence of the Nullstellensatz.

12.4 Notes and further references

This chapter was based on the papers [100, 101, 104] and the doctoral dissertation of Susan Margulies [240]. There are several improvements to NulLA that are based on the idea of Border bases as presented in [106, 256, 257] that we did not discuss here. Specifically, what we presented is mostly the *primal* approach were we solve a linear system to find constant multipliers $\mu \in \mathbb{K}^m$ such that $1 = \sum_{i=1}^m \mu_i f_i$ providing a certificate of (nonlinear) *infeasibility*. What we did not emphasize enough is that it is possible to do a *dual* approach too. This aims to find a vector λ with entries in $\mathbb{K}$ indexed by monomials such that $\sum_\alpha \lambda_{\mathbf{x}^\alpha} f_{i,\alpha} = 0$ for all $i = 1,\dots,m$ and $\lambda_1 = 1$ where $f_i = \sum_\alpha f_{i,\alpha}\mathbf{x}^\alpha$ for all i. The dual approach amounts to constructing linear relaxations of the set of *feasible solutions*.

12.5 Exercises

Exercise 12.5.1. Complete the proof of parts 1 and 2 of Theorem 12.1.1.

Exercise 12.5.2. Use polynomials to model the problem of deciding whether a graph contains a 3-colorable subgraph with k edges.

Exercise 12.5.3. Use polynomials to model the problem of deciding whether a graph has a Hamiltonian path.

Exercise 12.5.4. Every complex Nullstellensatz certificate for non-3-colorability of a graph has degree at least four. Moreover, in the case of a graph containing an odd wheel or a clique as a subgraph, a minimum-degree Nullstellensatz certificate for non-3-colorability has degree exactly four.

Exercise 12.5.5. Use Hilbert's Nullstellensatz to prove the following corollary: Given $h_i, g \in \mathbb{C}[x_1,\dots,x_m]$, consider the variety $V = \{\mathbf{x} : h_i(\mathbf{x}) = 0, i = 1,\dots,n\}$. Then, the polynomial $g(\mathbf{x}) = 0$ for all $\mathbf{x} \in V$ if and only if

$$g^r(x_1,\dots,x_m) = \sum_{i=1}^n \alpha_i(x_1,\dots,x_m)h_i(x_1,\dots,x_m),$$

where $\alpha_i \in \mathbb{C}[x_1,\dots,x_m]$.

Exercise 12.5.6. One can think of the Nullstellensatz as an identity between polynomials. Suppose someone gave us such an identity without further calculation. How can we verify that an identity of polynomials is really true? Suppose we have a polynomial someone else states to be constant zero. How can we check this (i.e., we wish to check $p(x_1,x_2,\dots,x_n) \equiv 0$)?

In general, checking the equation deterministically is not practical as there may be too many terms involved to verify there is always a cancellation. This exercise is about using a probabilistic approach. Using induction, prove the following theorem and its corollary.

> **Theorem 12.5.7.** *Let $\mathbb{K}$ be a field, $p(x_1,x_2,\dots,x_n)$ be a polynomial, where p is not identically zero, with n variables, largest total degree d, and let the coefficients of p be in $\mathbb{K}$. Choose $S \subseteq \mathbb{K}$ a finite subset. Then the equation $p(x_1,x_2,\dots,x_n) = 0$ has at most $d|S|^{n-1}$ solutions with $x_i \in S$.*
>
> **Corollary 12.5.8.** *Let $p(x_1,x_2,\dots,x_n)$ as in the theorem, then*
>
> $$\text{Prob}(p(x_1,x_2,\dots,x_n) = 0) \le \frac{d|S|^{n-1}}{|S|^{n-1}} = \frac{d}{|S|},$$
>
> *for $(x_1,x_2,\dots,x_n)$ a random element of S^n.*

Chapter 13

Positivity of Polynomials and Global Optimization

The Nullstellensatz deals only with the case of systems of polynomial *equations*. The inclusion of inequalities in the problem formulation poses algebraic challenges. In this case the solutions we seek use real numbers. Here we discuss other powerful identities that hold under the presence of inequalities. These are results in real algebraic geometry and we will skip their proofs (but they can be found in the nice books [58, 241]). We intend to stress ways to apply them in optimization.

Many problems in global optimization with polynomial constraints can be approached using results from real algebraic geometry. The connection between positivity of polynomials, sum of squares, and optimization can be traced back to the work of N. Shor [312]. His work went relatively unnoticed for several years, until several authors, including Lasserre, Nesterov, and Parrilo, observed, around the year 2000, that the existence of sum of squares decompositions and the search for infeasibility certificates for a semialgebraic set can be addressed via a sequence of semidefinite programs relaxations [212, 260, 270, 271]. This is a topic that today is still in high activity by researchers and we only highlight a few key ideas. This includes computational experiments; with these schemes implemented in software packages such as SOSTOOLS [278], GloptiPoly [166], and YALMIP [233].

13.1 Unconstrained optimization of polynomials and sums of squares

Let us start our discussion with the simplest special case to motivate the ideas to come.

Consider $f(x)$ to be a *univariate* polynomial. Suppose we want to minimize $f(x)$ over $x \in \mathbb{R}$. Methods from elementary calculus can tell you where to find all local minima, but to directly find a global minimum value we could rewrite the problem as a question of when polynomials are nonnegative:

$$\begin{array}{ll} \max & \gamma \\ \text{subject to} & f(x) - \gamma \geq 0 \quad \forall x \in \mathbb{R}. \end{array}$$

Indeed, minimizing $f(x)$ is the same as asking for the largest value of γ for which the polynomial $p(x) = f(x) - \gamma$ is nonnegative for all $x \in \mathbb{R}$. Now, if a polynomial is the sum of squares of other polynomials then clearly it is nonnegative. Surprisingly, it turns out that, for polynomials in one variable, being a sum of squares is the only way to be nonnegative too.

Theorem 13.1.1. *Let p be a univariate polynomial (function). Then p is nonnegative over $\mathbb{R}$ if and only if p has even degree and p is a sum of squares.*

Proof. Note that if $\deg(p)$ is odd then p cannot be nonnegative. This is clear because for large values of x, $p(x)$ is dominated by its leading term $a_n x^n$. Since n is odd, both $a_n x^n$ and $a_n(-x)^n$ cannot both be positive.

($\Leftarrow$) Suppose $p(x) = g_1^2(x) + g_2^2(x) + \cdots + g_k^2(x)$, then clearly $p(x)$ is nonnegative for all $x \in \mathbb{R}$.

($\Rightarrow$) Let $p(x) = a_n x^n + a_{n-1}x^{n-1} + \cdots + a_0$. Using the fundamental theorem of algebra, we can write

$$\begin{aligned} p(x) &= a_n x^n + a_{n-1}x^{n-1} + \cdots + a_0 \\ &= a_n \prod_j (x - r_j)^{n_j} \prod_k (x - (a_k + ib_k))^{m_k} (x - (a_k - ib_k))^{m_k} \\ &= a_n \prod_j (x - r_j)^{n_j} \prod_k \left((x - a_k)^2 + b_k^2\right)^{m_k}, \end{aligned}$$

for $r_j \in \mathbb{R}$, $a_k \pm ib_k \in \mathbb{C}$ ($a_k, b_k \in \mathbb{R}$ are the roots for $p(x)$). By nonnegativity, $n_j = 2s_j$, $a_n > 0$, we have

$$p(x) = a_n \prod_j (x - r_j)^{2s_j} \prod_k \left((x - a_k)^2 + b_k^2\right)^{m_k}.$$

This is a sum of squares by the distributive property. □

But then *how do we quickly check if p is a sum of squares?* Here is the first indication that semidefinite optimization is really tightly related to this topic. Recall that an $n \times n$ matrix Q is *positive semidefinite*, written as $Q \succeq O$, if $\mathbf{y}^\top Q \mathbf{y} \geq 0$ for all $\mathbf{y} \in \mathbb{R}^n$. Note that an application of the following theorem is to determine whether a polynomial is nonnegative by rewriting it as a sum of squares.

Theorem 13.1.2. *A univariate polynomial p with even degree $2d$ is a sum of squares if and only if there exists a symmetric positive definite matrix Q such that*

$$p = [x]_d^\top Q [x]_d, \text{ where } [x]_d = (1, x, x^2, \ldots, x^d)^\top.$$

Proof. ($\Rightarrow$) Let

$$\begin{aligned} p(x) = \sum_{k=1}^{m} q_k^2(x) &= \begin{pmatrix} q_1(x) \\ \vdots \\ q_m(x) \end{pmatrix}^\top \begin{pmatrix} q_1(x) \\ \vdots \\ q_m(x) \end{pmatrix} \\ &= \left([x]_d^\top V^\top\right)(V[x]_d) = [x]_d^\top (V^\top V)[x]_d, \end{aligned}$$

where $V[x]_d = (q_1(x), \ldots, q_m(x))^\top$, that is, the k-th row of V is made up of coefficients of the polynomial q_k. So $Q = V^\top V \succeq O$.

($\Leftarrow$) Because Q is a positive semidefinite matrix, we may use the Cholesky factorization to write Q as $Q = V^\top V$, and then reverse the process. Expanding the expression

$$p = [x]_d^\top Q[x]_d = [x]_d^\top V^\top V[x]_d$$

gives the sum of squares decomposition for p. □

Example 13.1.3. Let $p = (1+x+x^2)^2 + (3x-7)^2$. Here $d = n$, so $[x]_n = [x]_2 = (1,x,x^2)^\top$. One can write p as

$$(1+x+x^2, 3x-7)(1+x+x^2, 3x-7)^\top.$$

We then develop our matrix Q:

$$(1+x+x^2, 3x-7) = \begin{pmatrix} 1 & 1 & 1 \\ -7 & 3 & 0 \end{pmatrix} [x]_2^\top.$$

Therefore,

$$p = [x]_2 \begin{pmatrix} 1 & -7 \\ 1 & 3 \\ 1 & 0 \end{pmatrix} \begin{pmatrix} 1 & 1 & 1 \\ -7 & 3 & 0 \end{pmatrix} [x]_2^\top = [x]_2 (BB^\top)[x]_2^\top.$$

Can the same method work for polynomials in many variables? Can we apply the same procedure as in the univariate polynomial case? The multivariate polynomial $f(\mathbf{x})$, $\mathbf{x} \in \mathbb{R}^n$ of even degree $\deg(f(\mathbf{x})) = 2d$ is a nonnegative polynomial if and only if it satisfies

$$f(\mathbf{x}) \geq 0 \quad \forall \mathbf{x} \in \mathbb{R}^n. \tag{13.1}$$

Clearly, a sufficient condition in order for $f(\mathbf{x})$ to be nonnegative is

$$f(\mathbf{x}) = \sum_i q_i^2(\mathbf{x}), \tag{13.2}$$

that is, if $f(\mathbf{x})$ is reducible to a sum of squares (SOS), it is a nonnegative polynomial. Say we now wish to find a global minimum of a *multivariate* polynomial $f(\mathbf{x})$:

$$\begin{array}{rl} \min & f(\mathbf{x}) \\ \text{subject to} & \mathbf{x} \in \mathbb{R}^n. \end{array} \tag{13.3}$$

This problem is still equivalent to finding

$$\begin{array}{rl} \max & \gamma \\ \text{subject to} & f(\mathbf{x}) - \gamma \geq 0 \text{ for all } \mathbf{x} \in \mathbb{R}^n. \end{array} \tag{13.4}$$

We still have that every sums of squares polynomial can be written as a quadratic form in a set of monomials, with the corresponding matrix being positive semidefinite. The proof is identical to what we did in one variable, but the vector of monomials $\mathbf{x}$ in general depends on the degree, the number of variables, and the sparsity pattern of $p(\mathbf{x})$. If $p(\mathbf{x})$ has n variables and total degree $2d$, then $\mathbf{x}$ can always be chosen as a subset of the set of monomials of degree less than or equal to d, which has cardinality $\binom{n+d}{d}$.

Example 13.1.4. The polynomial $p(x_1,x_2)=x_1^2-x_1x_2^2+x_2^4+1$ is SOS. Among infinitely many others, $p(x_1,x_2)$ has the following decompositions:

$$\begin{aligned} p(x_1,x_2) &= \frac{3}{4}(x_1-x_2^2)^2+\frac{1}{4}(x_1+x_2^2)^2+1 \\ &= \frac{1}{9}(3-x_2^2)^2+\frac{2}{3}x_2^2+\frac{1}{288}(9x_1-16x_2^2)^2+\frac{23}{32}x_1^2. \end{aligned}$$

The polynomial $p(x_1,x_2)$ has the following representation:

$$p(x_1,x_2)=\frac{1}{6}\begin{pmatrix}1\\x_2\\x_2^2\\x_1\end{pmatrix}^{\top}\begin{pmatrix}6&0&-2&0\\0&4&0&0\\-2&0&6&-3\\0&0&-3&6\end{pmatrix}\begin{pmatrix}1\\x_2\\x_2^2\\x_1\end{pmatrix},$$

where the matrix in the expression above is positive semidefinite.

In the representation $f(\mathbf{x})=\mathbf{x}^{\top}Q\mathbf{x}$, for the right- and left-hand sides to be identical, all the coefficients of the corresponding polynomials should be equal. Since Q is simultaneously constrained by linear equations and a positive semidefiniteness condition, the problem can be easily seen to be directly equivalent to a semidefinite programming feasibility problem in the standard primal form.

Unfortunately, there is a very big obstacle to proceed as we did with a single variable: *not all nonnegative polynomials are sums of squares!* In the 1950s, Motzkin and Robinson [255, 284] exhibited polynomials that cannot be written like that, so in general, the sums of squares condition for nonnegativity is only sufficient.

Example 13.1.5. Here is Motzkin's example of a nonnegative polynomial with $n=2$ and $d=3$ that cannot be reduced to a sum of squares of polynomials:

$$f(x,y)=1+x^4y^2+x^2y^4-3x^2y^2. \tag{13.5}$$

We show that for all $(x,y)\in\mathbb{R}^2$, $f(x,y)\geq 0$ and that nevertheless $f(x,y)$ is not a sum of squares.

Proof. We know $1+x^4y^2+x^2y^4\geq 3x^2y^2$ is true, because of the inequality

$$\frac{a_1+\cdots+a_n}{n}\geq\sqrt[n]{a_1\cdot a_2\cdot\dots\cdot a_n} \tag{13.6}$$

and

$$1+x^4y^2+x^2y^4\geq 3\sqrt[3]{x^4y^2(x^2y^4)}=3x^2y^2. \tag{13.7}$$

The proof that the nonnegative polynomial $f(x,y)$ cannot be reduced to a sum of squares is by contradiction; suppose that $f=\sum s_i^2, s_i\in\mathbb{R}[x,y]$. We think of the exponent vectors of s_i as vectors in $\mathbb{Z}^2$. The *Newton polytope* of a polynomial is the convex hull of the exponent vectors of the polynomial. Clearly if we have a square of a polynomial the exponent vectors are sums of the exponent vectors of the original, thus we note that each polynomial s_i must have Newton polytope contained in the triangle $(0,0),(1,2),(2,1)$; therefore s_i is a linear combination of $1,xy^2,x^2y,xy$, but s_i^2 contains x^2y^2 and it must appear with a positive coefficient ($s_i=a+bx^2y+cxy^2+dxy$ implies that x^2y^2 has coefficient $d^2\geq 0$), but this is in contradiction with the fact that x^2y^2 appears with coefficient -3 in $f(x,y)$. □

Despite the existence of "bad" nonnegative polynomials that are not sums of squares, one can still use a different certification of nonnegativity, one that always proves when a polynomial is positive semidefinite. This is Emil Artin's solution to Hilbert's 17th problem:

Theorem 13.1.6 (Artin's theorem, 1927). *A polynomial $p(\mathbf{x}) \in \mathbb{R}[x_1,\dots,x_n]$ is positive semidefinite in $\mathbb{R}^n$ (i.e., $p(\mathbf{x}) \geq 0$ for all $\mathbf{x}$ in $\mathbb{R}^n$) if and only if p can be written as a quotient of two sums of squares of polynomials.*

In other words, p is positive semidefinite if and only if there exist sum-of-squares polynomials Q and R such that $Qp = R$. This kind of identity can be set up as a semidefinite programming problem again when we assume a bound r on the degrees of the polynomials Q, R, exactly as we did before. Exactly as we did with the Nullstellensatz in Chapter 12, we look for the existence of such an identity $Qp = R$, increasing the degrees of Q and R at each iteration. The main obstacle is that, unlike the case of the Nullstellensatz, we do not have a good general bound for the degree r that we need to use.

The proof of Artin's theorem is too technical for this book. A modern nonconstructive proof can be found, for instance, in [58]. To conclude this section we give an algorithmic proof in a special case.

Definition 13.1.7. A *positive definite* form F is a real homogeneous polynomial which is (strictly) positive in $\mathbb{R}^n \setminus \{\mathbf{0}\}$.

What we are going to see is that any polynomial f whose homogenization F is a positive definite form can be written as a quotient of two sums of squares. Note that such a decomposition is a certificate for the infeasibility of the inequality $f < 0$. To give the construction, we need a lemma. In 1928 the famous Hungarian mathematician George Pólya proved the following theorem (see [148]) which we will use later for our algorithm (and it plays a recurrent central role in effective real algebraic geometry).

Theorem 13.1.8 (Pólya's lemma). *Let $F(x_1,\dots,x_n)$ be a real homogeneous polynomial which is positive in $x_i \geq 0$, $\sum x_i > 0$. Then, for a sufficiently large integer p, the product*

$$F(x_1,\dots,x_n)\cdot(x_1+\cdots+x_n)^p$$

has all its coefficients strictly positive.

The smallest exponent p that satisfies the properties of the theorem will be called the *Pólya exponent* of F. For illustration of Pólya's lemma consider the examples in Table 13.1. Today, there are known explicit bounds on the Pólya exponent. See [277] for details. These bounds can be written in terms of the degree, the number of variables, and the size of the coefficients of F. In what follows we only use the existence of the power p.

Observe that Theorem 13.1.8 implies that F can be written in the form $F = G/H$, where G and H are homogeneous polynomials with only positive coefficients. This is a necessary and sufficient condition for F to be strictly positive in the nonnegative orthant. In a similar way, Artin's decomposition of a polynomial as a quotient of two sums of squares is necessary and sufficient to guarantee positive semidefiniteness in $\mathbb{R}^n$.

Pólya's theorem was first used by Habicht [147] to give explicit solutions to Hilbert's 17th problem in the case of positive definite homogeneous polynomials. Here, we present a different way to do this [87]. See also [283], where the author gives concrete decompositions for the same family of polynomials as a sum of even powers of linear forms in the numerator and a power of $\sum x_i^2$ in the denominator.

Table 13.1. *Some polynomials and their Pólya exponents*

Polynomial	Pólya exponent
$1000x^2 - 1999xy + 1000y^2$	3997
$50x^2 - 99xy + 50y^2$	197
$(50x^2 - 99xy + 50y^2)(x^2 + y^2)$	193
$(50x^2 - 99xy + 50y^2)(x^4 + x^2y^2 + y^4)$	187
$(x-y)^2(x+6y)^2 + y^4$	197
$5x^4 + (x-y)^2(x+6y)^2 + y^4$	44
$10x^4 + (x-y)^2(x+6y)^2 + y^4$	30
$(x-z)^2 + (y-z)^2 + (x+y)^2$	3

Here is the key idea. If we have a positive definite homogeneous polynomial F in n variables, Pólya's theorem can be applied to $F(\epsilon_1 x_1, \epsilon_2 x_2, \ldots, \epsilon_n x_n)$, where $\epsilon_i \in \{+,-\}$. In this way we have 2^n Pólya-like expressions, each of them certifying the positiveness of F inside a different orthant. We then proceed to "glue" these local certificates. The decomposition of F obtained in this way is a quotient of two sums of even powers of monomials in the "variables" $x_1, x_2, \ldots, x_n, F$.

Lemma 13.1.9. *Let $F \in \mathbb{R}[x_1, \ldots, x_n]$. Suppose that for a given variable x_i we have two identities $F \cdot A_1 = B_1$ and $F \cdot A_2 = B_2$, where A_1, B_1 are polynomials in $\mathbb{R}_+[x_i, T, F^2]$ and A_2, B_2 are polynomials in $\mathbb{R}_+[-x_i, T, F^2]$, for some arbitrary set of indeterminates T. Assume that both B_1 and B_2 have a nonzero constant term. Then we can find an expression of the form $F \cdot R = S$ where R and S are polynomials in $\mathbb{R}_+[x_i^2, T, F^2]$ and S has a nonzero constant term. Moreover* $\deg(S) \leq \deg(B_1) + \deg(B_2)$.

Proof. We can decompose $A_1 = A_{1,1} + x_i A_{1,2}$, $B_1 = B_{1,1} + x_i B_{1,2}$, $A_2 = A_{2,1} - x_i A_{2,2}$, and $B_2 = B_{2,1} - x_i B_{2,2}$ with $A_{1,1}, A_{1,2}, B_{1,1}, B_{1,2}, A_{2,1}, A_{2,2}, B_{2,1}$, and $B_{2,2} \in \mathbb{R}_+[x_i^2, T, F^2]$. Separate the two identities in the form:

$$FA_{1,1} - B_{1,1} = -x_i FA_{1,2} + x_i B_{1,2}, \quad FA_{2,1} - B_{2,1} = x_i FA_{2,2} - x_i B_{2,2}.$$

Multiplying side by side the above equations and grouping together terms with F we obtain

$$\begin{aligned} &F \cdot (A_{1,1}B_{2,1} + B_{1,1}A_{2,1} + x_i^2 A_{1,2}B_{2,2} + x_i^2 B_{1,2}A_{2,2}) \\ &\qquad = F^2(A_{1,1}A_{2,1} + x_i^2 A_{1,2}A_{2,2}) + B_{1,1}B_{2,1} + x_i^2 B_{1,2}B_{2,2}. \end{aligned}$$

By hypothesis both $B_{1,1}$ and $B_{2,1}$ have a nonzero constant term and thus $B_{1,1}B_{2,1}$ has a nonzero constant term. The constant term of $F^2 A_{1,1}A_{2,1}$ is either zero or positive, and thus the constant term of the right-hand side of the equation above is positive. From the above expression it is clear that $\deg(S) \leq \deg(B_1) + \deg(B_2)$. □

As an immediate application of the above lemma and as a preparation for the multivariate case we present a method to decompose a real univariate strictly positive polynomial F as a quotient of two sums of squares. We remark that in the univariate case, the additional condition of F being bigger than a strictly positive constant is redundant. Applying Theorem 13.1.8 to the homogenization of F and a sufficiently large exponent we have the following expression where $B_1(x)$ has only positive coefficients,

$$F(x)(x+1)^p = B_1(x).$$

With the same process applied to the polynomial $F(-x)$ we obtain

$$F(x)(1-x)^q = B_2(-x).$$

Taking $A_1 = (x+1)^p$, $A_2 = (1-x)^q$, and $T = \emptyset$ we are in the situation of Lemma 13.1.9. This will give an expression $F \cdot R = S$ with R, S polynomials in $\mathbb{R}_+[x^2, F^2]$ and thus sums of squares.

Theorem 13.1.10. *Let $F(x_1, x_2, \ldots, x_n)$ be a real strictly positive polynomial of degree d, whose homogenization is positive definite. For each $\boldsymbol{\epsilon} = (\epsilon_1, \ldots, \epsilon_n)$ in $E^n = \{+,-\}^n$, let p_ϵ be the Pólya exponent of the homogenization of F in the orthant where the sign of the i-th coordinate equals ϵ_i. Let $P = \sum_{\boldsymbol{\epsilon} \in E^n} p_{\boldsymbol{\epsilon}}$ and D be the maximum of $d+1$ and $n+1$. Then we can write:*

$$F \cdot R = S,$$

where $R, S \in \mathbb{R}_+[x_1^2, x_2^2, \ldots, x_n^2, F^2]$ and $\deg(S) \leq P + 2^n d$ *(where S is considered as a polynomial in the original variables $x_1, x_2, \ldots, x_n$ to compute* $\deg(S)$*).*

If $F \in \mathbb{R}[x_1, x_2, \ldots, x_n]$, then we can find R and S in $\mathbb{R}_+[x_1^2, x_2^2, \ldots, x_n^2, F^2]$.

Proof. Let $E^n = \{+,-\}^n$. For each $\boldsymbol{\epsilon} = (\epsilon_1, \ldots, \epsilon_n) \in E^n$ we have a Pólya expression in the corresponding orthant

$$F(x_1, x_2, \ldots, x_n) \cdot A_{\boldsymbol{\epsilon}} = B_{\boldsymbol{\epsilon}},$$

where $A_{\boldsymbol{\epsilon}}, B_{\boldsymbol{\epsilon}} \in R_+[\epsilon_1 x_1, \ldots, \epsilon_n x_n]$. Moreover, $A_{\boldsymbol{\epsilon}} = (1 + \epsilon_1 x_1 + \cdots + \epsilon_n x_n)^{p_\epsilon}$ and thus $B_{\boldsymbol{\epsilon}}$ has degree $p_{\boldsymbol{\epsilon}} + d$ and nonzero constant term. Our goal is to glue the 2^n expressions in pairs using Lemma 13.1.9. More explicitly, for each $\boldsymbol{\sigma} \in E^{n-1}$ consider the two expressions $FA_{\boldsymbol{\epsilon}} = B_{\boldsymbol{\epsilon}}$ and $FA_{\boldsymbol{\epsilon}'} = B_{\boldsymbol{\epsilon}'}$, where $\boldsymbol{\epsilon} = (\boldsymbol{\sigma}, +)$ and $\boldsymbol{\epsilon}' = (\boldsymbol{\sigma}, -)$.

We can apply Lemma 13.1.9 with $T = \{\sigma_1 x_1, \sigma_2 x_2, \ldots, \sigma_{n-1} x_{n-1}\}$. This will give $2^{(n-1)}$ expressions (one for each $\boldsymbol{\sigma} = (\sigma_1, \ldots, \sigma_{n-1})$ in E^{n-1}) where the variable x_n always appears squared. Inductively, for each $\boldsymbol{\tau} \in E^{n-2}$, we take the two expressions $FA_{\boldsymbol{\sigma}} = B_{\boldsymbol{\sigma}}$ and $FA_{\boldsymbol{\sigma}'} = B_{\boldsymbol{\sigma}'}$ with $\boldsymbol{\sigma} = (\boldsymbol{\tau}, +)$ and $\boldsymbol{\sigma}' = (\boldsymbol{\tau}, -)$ and apply Lemma 13.1.9 with $T = \{\tau_1 x_1, \tau_2 x_2, \ldots, \tau_{n-2} x_{n-2}, x_n^2\}$. This process can be continued until all the variables appear squared.

For the degrees we note that in each gluing the degrees of the expressions glued are added. The degree of the final expression will be the sum of the degrees of the 2^n equations derived from bounds to Pólya's exponent. This gives the bound $P + d2^n$. □

Example 13.1.11. Here is an example of the procedure:

$F := 134xy - 92y - 182x - 24x^3 + 412x^4 + 540x^2 + 678y^2 + 533y^4 - 200y^3 - 18x^3y + 556x^2y^2 - 344x^2y + 40xy^3 + 184xy^2 + 444.$

The reader can check that the Pólya exponent for this polynomial is two! We will apply the process described in the proof of Theorem 13.1.10. Pólya's theorem applied to F in each one of the four orthants gives four identities (we show the intermediate distributions of the terms with respect to parity of the powers of y):

(i) $F(1+x+y)^2 = F(1+2x+x^2+y^2+y(2+2x)) = 1722x^2y^2 + 1442xy^2 + 874x^3 + 904x^4 + 706x + 620x^2 + 938y^2 + 811y^4 + 444 + 548x^3y^2 + 932x^4y^2 + 930y^4x + 1169y^4x^2 + 800x^5 + 412x^6 + 533y^6 + y(474x + 460x^3 + 548x^2 + 1498xy^2 + 796 + 1064y^2 + 396x^4 + 806x^5 + 1106y^4x + 1016x^2y^2 + 1134x^3y^2 + 866y^4).$

(ii) $F(1+x-y)^2 = F(1+2x+x^2+y^2-y(2+2x)) = 2562x^2y^2+1274xy^2+874x^3+904x^4+706x+620x^2+1306y^2+1611y^4+444+1996x^3y^2+1004x^4y^2+1570y^4x+1009y^4x^2+800x^5+412x^6+533y^6-y(574x+1604x^3+884x^2+1950xy^2+980+1648y^2+1156x^4+842x^5+1026y^4x+1944x^2y^2+1090x^3y^2+1266y^4)$.

(iii) $F(1-x+y)^2 = F(1-2x+x^2+y^2+y(2-2x)) = 450x^2y^2-902xy^2-1286x^3+1000x^4-1070x+1348x^2+938y^2+811y^4+444-300x^3y^2+1004x^4y^2-402y^4x+1009y^4x^2-848x^5+412x^6+533y^6+y(-934x-324x^3+740x^2-414xy^2+796+1064y^2+564x^4-842x^5+866y^4-1026y^4x+120x^2y^2-1090x^3y^2)$.

(iv) $F(1-x-y) = 716x^2y^2-628xy^2-564x^3+436x^4-626x+722x^2+770y^2+733y^4+444-538x^3y^2-573y^4x-412x^5-y(-408x-350x^3+1018x^2-56xy^2+536+878y^2+394x^4+533y^4+596x^2y^2)$.

Applying Lemma 13.1.9, with y as the distinguished variable to the pairs (i)–(ii) and (iii)–(iv) and grouping terms as in Lemma 13.1.9 we get (notice the expressions are presented now arranged by parity of powers of the variable x):

(i)–(ii) $F(7508x^2y^6+18160x^2y^2+6544x^4+4952x^2+6684y^2+10090y^4+20348x^4y^2+26700y^4x^2+5832x^6+7752y^6+8562x^4y^4+6056x^6y^2+824x^8+1066y^8+888+x(14264y^2+5640x^2+3188+22568x^2y^2+22380y^4+6964x^4+14416x^4y^2+19768y^4x^2+13160y^6+3248x^6)) = 20679124x^2y^6+6723652x^2y^2+2421240x^4+1048996x^2+1776416y^2+4654924y^4+12531112x^4y^2+16893804y^4x^2+3380292x^6+6653476y^6+24185356x^4y^4+12976984x^6y^2+3666432x^8y^4+5547524x^6y^6+1476284x^{10}y^2+4580433x^4y^8+2295630y^{10}x^2+169744x^{12}+284089y^{12}+6F^2x^2y^2+F^2+6F^2x^2+6F^2y^2+F^2x^4+F^2y^4+17274928x^6y^4+20296116x^4y^6+7688272x^8y^2+13108438x^2y^8+2726496x^8+5276765y^8+1384896x^{10}+2387282y^{10}+197136+x(3711592y^2+1651552x^2+626928+10260368x^2y^2+10310324y^4+3070608x^4+14459768x^4y^2+22021832y^4x^2+14337360y^6+3621212y^{10}+9347744x^6y^4+12882352x^4y^6+3862096x^8y^2+9701716x^2y^8+659200x^{10}+12F^2y^2+2166576x^8+4F^2+4F^2x^2+22429140x^4y^4+22897032x^2y^6+10878896x^6y^2+10718648y^8+3153936x^6)$.

(iii)–(iv) $F(8408x^2y^2+4572x^4+4836x^2+4020y^2+5134y^4+5584x^4y^2+5430y^4x^2+2520x^6+3198y^6+888-x(7456y^2+5268x^2+3028+6972x^2y^2+6162y^4+3696x^4+3584x^4y^2+4402y^4x^2+3198y^6+824x^6) =$
$5298888x^2y^6+5057552x^2y^2+3019356x^4+1588900x^2+1185008y^2+2676988y^4+6822948x^4y^2+6963772y^4x^2+3189648x^6+3371312y^6+6633028x^4y^4+5042212x^6y^2+F^2+3F^2x^2+3F^2y^2+3085788x^6y^4+3212416x^4y^6+1829476x^8y^2+1989123x^2y^8+1741568x^8+2332333y^8+529008x^{10}+852267y^{10}+197136-x(2915800y^2+2437788x^2+753024+6141928x^2y^2+4529144y^4+3340724x^4+6443844x^4y^2+7068028y^4x^2+4175308y^6+852267y^{10}+2123228x^6y^4+2840400x^4y^6+967052x^8y^2+2057377x^2y^8+169744x^{10}+3F^2y^2+1014096x^8+3F^2+F^2x^2+5639176x^4y^4+4866908x^2y^6+3468572x^6y^2+2264079y^8+2550240x^6)$.

Finally, applying again Lemma 13.1.9 with x as the distinguished variable, we get the following expression from which F is decomposed as a quotient of two sums of squares.

$F(716381628672x^2y^6+80945077792x^2y^2+1309788013600x^{10}y^4 40286070144x^4+8570997312x^2+4739888256y^2+24573722112y^4+318423130400x^4y^2+317633301040y^4x^2+99294039872x^6+68743283680y^6+1060688074256x^4y^4+672316077216x^6y^2+143469890432x^{14}y^2+1982702974736x^8y^4+2814822717360x^6y^6+768832371904x^{10}y^2+2303389497104x^4y^8+990053073384y^{10}x^2+125406307008x^{12}+116435744860y^{12}+1863486418736x^6y^4+1983138377456x^4y^6+886558335136x^8y^2+1032502294416x^2y^8+505612381472x^{12}y^4+157752051648x^8+122024198048y^8+169540742272x^{10}+144755788904y^{10}+40199723460x^2y^{16}+2507919676288x^6y^8+1687333237608x^4y^{10}+2293828274592x^8y^6+621063650084x^2y^{12}+1282298534376x^6y^{10}+1423499576064x^8y^8+379456896112x^8y^{10}+292967217500x^6y^{12}+338955778976x^{10}y^8+145245087972x^4y^{14}+1051819997200x^{10}y^6+235854251632x^2y^{14}+208393538448x^{12}y^6+60652193280x^{14}+735231433668x^4y^{12}+1817033244y^{18}+20512201152x^{16}y^2+60546876076y^{14}+17483777024x^{16}+18186081524y^{16}+1958166784x^{18}+350113536+(39456x^2+109168x^4+60284y^4+128196x^2y^8+89728x^8y^2+76044y^6+227652x^4y^6+395200x^2y^6+48644y^8+11536x^{10}+6396y^{10}+208940x^6y^4+551252x^4y^4+342344x^6y^2+451140y^4x^2+430776x^4y^2+238456x^2y^2+1776+72800x^8+18696y^2+$

$122640\,x^6)F^2 + 85194907968\,x^{14}y^4 + 430307879168\,x^{12}y^2) =$

$(153448078017504\,x^2y^6 + 11011029982720\,x^2y^2 + 851952864568912\,x^{10}y^4 + 5511262925968\,x^4 +$
$992116096128\,x^2 + 583803281664\,y^2 + 3550450975360\,y^4 + 53477285280880\,x^4y^2 +$
$53285071815840\,y^4x^2 + 16742340875856\,x^6 + 12247813101568\,y^6 + 228076657290496\,x^4y^4 +$
$144275999563072\,x^6y^2 + 177419113096224\,x^{14}y^2 + 818449513424880\,x^8y^4 +$
$1160297022426176\,x^6y^6 + 316404880090816\,x^{10}y^2 + 943080312658456\,x^4y^8 +$
$400790099977488\,y^{10}x^2 + 51946321935360\,x^{12} + 51812546094456\,y^{12} + 537983815619840\,x^6y^4 +$
$571695232269920\,x^4y^6 + 255064333134736\,x^8y^2 + 295172928849576\,x^2y^8 + 613138233076304\,x^{12}y^4 +$
$33884952842288\,x^8 + 27834477697840\,y^8 + 48891319043696\,x^{10} + 44539310679888\,y^{10} +$
$132667868133349\,x^2y^{16} + 1609709073374408\,x^6y^8 + 1074117047816472\,x^4y^{10} +$
$1484929624125296\,x^8y^6 + 391920773628424\,x^2y^{12} + 1494743643985512\,x^6y^{10} +$
$1679981404535176\,x^8y^8 + 1210562461740080\,x^8y^{10} + 922584991406952\,x^6y^{12} +$
$1101811041447776\,x^{10}y^8 + 463071536349080\,x^4y^{14} + 1257591653271104\,x^{10}y^6 +$
$274626206521816\,x^2y^{14} + 697775379415696\,x^{12}y^6 + 40598074878080\,x^{14} + 854507981216656\,x^4y^{12} +$
$11022901919929\,y^{18} + 76241258273152\,x^{16}y^2 + 44002353922224\,y^{14} + 22913533222784\,x^{16} +$
$26786486470753\,y^{16} + 8809248338688\,x^{18} + 296435588380336\,x^{14}y^4 + 40113876079218\,y^{18}x^2 +$
$5607817144761\,y^{20}x^2 + 25101305044221\,x^4y^{18} + 20172563339904\,x^{18}y^2 + 2384556922240\,x^{20}y^2 +$
$63315154652095\,x^6y^{16} + 128071716208936\,x^{10}y^{12} + 11885395021040\,x^{18}y^4 +$
$106193345637816\,x^8y^{14} + 75244284164400\,x^{14}y^8 + 113979811644760\,x^{12}y^{10} +$
$36042080253584\,x^{16}y^6 + 156928941649179\,y^{16}x^4 + 87953191492976\,x^{16}y^4 + 351913378400864\,x^6y^{14} +$
$234657785839200\,x^{14}y^6 + 525673478728032\,x^8y^{12} + 427778646436656\,x^{12}y^8 +$
$557728278696952\,x^{10}y^{10} + 38862602496 + 281083319515232\,x^{12}y^2 + 2697191817931\,y^{20} +$
$242119679763\,y^{22} + 2064497141504\,x^{20} + 201691178752\,x^{22} + (3\,y^6 + 45\,x^4y^2 + 19\,y^4 + 21\,x^2 + 9\,y^2 +$
$1 + 90\,x^2y^2 + 57\,y^4x^2 + 35\,x^4 + 7\,x^6)F^4 + (1011238152\,x^2y^6 + 214583120\,x^2y^2 + 213172104\,x^{10}y^4 +$
$1043554136\,x^4 + 27650008\,x^2 + 14240800\,y^2 + 60356936\,y^4 + 628009464\,x^4y^2 + 633987416\,y^4x^2 +$
$191801384\,x^6 + 125837552\,y^6 + 1474515760\,x^4y^4 + 929987432\,x^6y^2 + 932892160\,x^8y^4 +$
$1345813968\,x^6y^6 + 355155544\,x^{10}y^2 + 1110516186\,x^4y^8 + 486373732\,y^{10}x^2 + 54326496\,x^{12} +$
$45155610\,y^{12} + 1635596488\,x^6y^4 + 1755400200\,x^4y^6 + 771165424\,x^8y^2 + 932373358\,x^2y^8 +$
$214189576\,x^8 + 155761526\,y^8 + 142279872\,x^{10} + 112787062\,y^{10} + 436152642\,x^6y^8 + 292702206\,x^4y^{10} +$
$386752696\,x^8y^6 + 107817366\,x^2y^{12} + 7129248\,x^{14} + 5113602\,y^{14} + 1182816 + 65068392\,x^{12}y^2)F^2).$

13.2 Positivstellensätze for semidefinite programming relaxations

In Section 12.2.2 we generated a sequence of linear algebra problems to decide the solvability of a system of polynomial equations. Here we explain the variation of the same principle that deals with systems of inequalities and equations. We call the solution set of a finite system of polynomial equations and inequalities, $f_1, \dots, f_s, g_1, \dots, g_k \in \mathbb{K}[x_1, \dots, x_n]$, a *basic semialgebraic set*. Note that convex polyhedra correspond to the particular case where all the constraint polynomials have degree one. The basic question is, when is a given semialgebraic set empty?

Does there exist $\mathbf{x} \in \mathbb{R}^n$ such that

$$f_1(\mathbf{x}) = 0, \ \dots, \ f_s(\mathbf{x}) = 0, \ g_1(\mathbf{x}) \geq 0, \ \dots, \ g_k(\mathbf{x}) \geq 0?$$

Now we show a way to set up this semialgebraic feasibility problem as a sequence of semidefinite programs terminating with a feasible solution (most of this material is inspired by [222, 271]).

As inspiration recall the classical Farkas' lemma at the foundation of linear programming (see Section 1.2).

Theorem 13.2.1 (Farkas' lemma). *Let $A \in \mathbb{R}^{m\times n}$, $\mathbf{b} \in \mathbb{R}^m$, $C \in \mathbb{R}^{k\times n}$, and $\mathbf{d} \in \mathbb{R}^k$. The set of solutions*

$$\{\mathbf{x} : \mathbf{x} \in \mathbb{R}^n \text{ such that } A\mathbf{x} + \mathbf{b} = \mathbf{0}, C\mathbf{x} + \mathbf{d} \geq \mathbf{0}\}$$

is empty if and only if there exist $\boldsymbol{\lambda} \in \mathbb{R}^m_+$ *and* $\boldsymbol{\mu} \in \mathbb{R}^k$ *such that* $\boldsymbol{\mu}^\top A + \boldsymbol{\lambda}^\top C = \mathbf{0}^\top$ *and* $\boldsymbol{\mu}^\top \mathbf{b} + \boldsymbol{\lambda}^\top \mathbf{d} = -1$.

Just as Fredholm's theorem is the primitive version of the Nullstellensatz for linear *equations*, Farkas' lemma is the precursor on *linear inequalities* for a more general theorem that extends the Nullstellensatz for systems of nonlinear polynomial inequalities. Although not widely known in optimization, it turns out this generalization gives certificates of infeasibility of *nonlinear systems of polynomial equations and inequalities* over the reals. The following result appears in this form in [58] and is due to Stengle [315].

Given a set of polynomial inequalities $G = \{g_1(\mathbf{x}) \geq 0, \ldots, g_s(\mathbf{x}) \geq 0\}$ in $\mathbb{R}[x_1, \ldots, x_n]$, we define the *semialgebraic set*

$$K(G) = \left\{ \mathbf{x} \in \mathbb{R}^n : g_i(\mathbf{x}) \geq 0 \;\forall\, i = 1, \ldots, s \right\}.$$

This is the set of points that are nonnegative for all the inequalities in G. Following page 86 in Section 4.2 of [58], we define the *cone* or *preorder* generated by G (see [241]) as

$$\operatorname{cone}(G) := \left\{ \sum_{\alpha \in \{0,1\}^n} s_\alpha \mathbf{g}^\alpha : s_\alpha \in \mathbb{K}[x_1, \ldots, x_n] \text{ is SOS} \right\},$$

where $\mathbf{g}^\alpha := \prod_{i=1}^m g_i^{\alpha_i}$ and a polynomial $s(\mathbf{x}) \in \mathbb{K}[x_1, \ldots, x_n]$ will be called SOS if it can be written as a *sum of squares* of other polynomials, that is, $s(\mathbf{x}) = \sum_i q_i^2(\mathbf{x})$ for some $q_i(\mathbf{x}) \in \mathbb{K}[x_1, \ldots, x_n]$. If $s(\mathbf{x})$ is SOS, then clearly $s(\mathbf{x}) \geq 0$ for all $\mathbf{x} \in \mathbb{R}^n$. The sum in the definition of cone(G) is finite, with a total of 2^m terms, corresponding to the subsets of $\{g_1, \ldots, g_m\}$ including the empty set. The following result appears in this form in [58] and is due to Stengle [315].

Theorem 13.2.2 (Positivstellensatz and its relatives). *Let* $K(G)$ *be the semialgebraic set given by G and its preorder* cone(G). *We have the following equivalent results.*

1. $K(G) = \emptyset$ *if and only if* $-1 \in$ cone(G).
2. $f(\mathbf{x}) = 0$ *for all* $\mathbf{x} \in K(G)$ *if and only if there exists* $m \in \mathbb{Z}_{\geq 0}$ *such that* $-f^{2m} \in$ cone(G).
3. $f(\mathbf{x}) \geq 0$ *for all* $\mathbf{x} \in K(G)$ *if and only if there exists* $m \in \mathbb{Z}_{\geq 0}$ *and* $p, q \in$ cone(G) *such that* $pf = f^{2m} + q$.
4. $f(\mathbf{x}) > 0$ *for all* $\mathbf{x} \in K(G)$ *if and only if there exists* $p, q \in$ cone(G) *such that* $pf = 1 + q$.

Example 13.2.3. Consider the polynomial system $\{f = 0,\, g \geq 0\}$, where

$$f := x_2 + x_1^2 + 2 = 0, \qquad g := x_1 - x_2^2 + 3 \geq 0.$$

By the Positivstellensatz, there are no solutions $(x_1, x_2) \in \mathbb{R}^2$ if and only if there exist polynomials $\beta, s_1, s_2 \in \mathbb{R}[x_1, x_2]$ that satisfy

$$\beta \cdot f + s_1 + s_2 \cdot g = -1 \quad \text{where } s_1 \text{ and } s_2 \text{ are SOS.}$$

Here, the system is infeasible, so there exist such polynomials as follows:

$$s_1 = \tfrac{1}{3} + 2\left(x_2 + \tfrac{3}{2}\right)^2 + 6\left(x_1 - \tfrac{1}{6}\right)^2, \; s_2 = 2, \text{ and } \beta = -6.$$

The resulting identity provides a certificate of infeasibility of the system.

Now we describe an algorithm (originally presented in [270, 271]) and illustrate it with an example, on how we can use SDPs to decide the feasibility of a system of polynomial inequalities. Exactly as we did for the Nullstellensatz case, we can look for the existence of a Positivstellensatz certificate of bounded degree D. Once we assume that the degree D is fixed we can obtain a reformulation as a semidefinite programming problem. We formalize this description in the following algorithm:

ALGORITHM 13.1. The Positivstellensatz method.

input A polynomial system $\{f_i(\mathbf{x}) = 0, g_i(\mathbf{x}) \geq 0\}$ and a Positivstellensatz bound D.
output FEASIBLE, if $\{f_i(\mathbf{x}) = 0, g_i(\mathbf{x}) \geq 0\}$ is feasible over $\mathbb{R}$, else INFEASIBLE.
for $d = 0, 1, 2, \ldots, D$ **do**
  if there exist $\beta_i, s_\alpha \in \mathbb{K}[x_1, \ldots, x_n]$ such that

$$-1 = \sum_i \beta_i f_i + \sum_{\alpha \in \{0,1\}^n} s_\alpha \mathbf{g}^\alpha,$$

  with s_α SOS, $\deg(\beta_i f_i) \leq d$, $\deg(s_\alpha \mathbf{g}^\alpha) \leq d$ **then**
    return INFEASIBLE.
return FEASIBLE.

Notice that the membership test in the main loop of the algorithm is, by the results described at the beginning of this section, equivalent to a finite-sized semidefinite program. Similarly to the Nullstellensatz case, the number of iterations (i.e., the degree of the certificates) serves as a quantitative measure of the hardness in proving infeasibility of the system. Unlike the case of the Nullstellensatz the degree bounds known for the Positivstellensatz certificates are not as sharp and well-understood (see [234]).

Example 13.2.4. Consider the polynomial system $\{f = 0, g \geq 0\}$ from Example 13.2.3, where $f := x_2 + x_1^2 + 2 = 0$ and $g := x_1 - x_2^2 + 3 \geq 0$. At the d-th iteration of Algorithm 13.1 applied to the polynomial problem $\{f = 0, g \geq 0\}$, one asks whether there exist polynomials $\beta, s_1, s_2 \in \mathbb{K}[x_1, \ldots, x_n]$ such that $\beta f + s_1 + s_2 \cdot g = -1$ where s_1, s_2 are SOS and $\deg(s_1), \deg(s_2 \cdot g), \deg(\beta \cdot f) \leq d$. For each fixed positive integer d this can be tested by a (possibly large) semidefinite program.

Solving this for $d = 2$, we have $\deg(s_1) \leq 2$, $\deg(s_2) = 0$ and $\deg(\beta) = 0$, so s_2 and β are constants and

$$\begin{aligned} s_1 = \mathbf{z}^\top Q \mathbf{z} &= \begin{pmatrix} 1 \\ x_1 \\ x_2 \end{pmatrix}^\top \begin{pmatrix} Q_{11} & Q_{12} & Q_{13} \\ Q_{12} & Q_{22} & Q_{23} \\ Q_{13} & Q_{23} & Q_{33} \end{pmatrix} \begin{pmatrix} 1 \\ x_1 \\ x_2 \end{pmatrix} \\ &= Q_{11} + 2Q_{12}x_1 + 2Q_{13}x_2 + Q_{22}x_1^2 + 2Q_{23}x_1x_2 + Q_{33}x_2^2, \end{aligned}$$

where $\mathbf{z} = (1, x_1, x_2)^\top$ and $Q \in \mathbb{R}^{3\times 3}$ is a symmetric positive semidefinite matrix. Thus, the certificate for $D = 2$ is $\beta f + \mathbf{z}^\top Q \mathbf{z} + s_2 \cdot g = -1$ where $Q \succeq O$ and $s_2 \geq 0$. If we expand the left-hand side and equate coefficients on both sides of the equation, we arrive at the following SDP:

$$\begin{aligned} 2\beta + Q_{11} + 3s_2 = -1 \quad (1), &\qquad 2Q_{12} + s_2 = 0 \quad (x_1), \\ \beta + 2Q_{13} = 0 \quad (x_2), &\qquad \beta + Q_{22} = 0 \quad (x_1^2), \\ 2Q_{23} = 0 \quad (x_1x_2), &\qquad Q_{33} - s_2 = 0 \quad (x_2^2), \end{aligned}$$

where $Q \succeq O$ and $s_2 \geq 0$. This SDP has a solution as follows:

$$Q = \begin{pmatrix} 5 & -1 & 3 \\ -1 & 6 & 0 \\ 3 & 0 & 2 \end{pmatrix}, \; s_2 = 2, \text{ and } \beta = -6.$$

The resulting identity, which is the same as the one given in Example 13.2.3, proves the inconsistency of the system.

How about the problem of optimizing a polynomial over a semialgebraic set? There is an well-known straightforward way to optimize if we know how to solve the feasibility problem: By considering the emptiness of the sublevel sets of the objective function one can reduce optimization problems to the above feasibility problems. But there is another alternative.

We now wish to solve:

$$\begin{array}{lll} \text{maximize} & f(\mathbf{x}) & \\ \text{subject to} & g_1(\mathbf{x}) \geq 0, & h_1(\mathbf{x}) = 0, \\ & \vdots & \vdots \\ & g_m(\mathbf{x}) \geq 0, & h_k(\mathbf{x}) = 0, \end{array} \tag{13.8}$$

where functions $f(\cdot)$, $g_j(\cdot)$ for $j \in \{1,\ldots,m\}$, $h_i(\cdot)$ for $i \in \{1,\ldots,k\}$ are polynomials with coefficients in $\mathbb{R}$.

In many applications one wishes to obtain an optimal value of a polynomial within a basic closed semialgebraic set. Lasserre [212] also had the idea to use theorems in real algebraic geometry that help reduce the problem to an SDP computation. These are specializations of the Positivstellensatz for compact semialgebraic sets where the representation is somewhat simplified. The key theorems of Schmüdgen and Putinar (with further improvements by others) give the existence of simpler certificates that can be obtained from semidefinite programming.

We talked about the preorder generated by a set of polynomial inequalities $G = \{g_1(\mathbf{x}) \geq 0,\ g_2(\mathbf{x}) \geq 0,\ \ldots,\ g_k(\mathbf{x}) \geq 0\}$. To state the new theorems let us define the *quadratic module* of a semialgebraic set $M(G)$ which is a simpler subset of the preorder: This is the set of all linear combinations (with SOS coefficients)

$$\sigma_0 + \sum_{i=1}^{k} \sigma_i g_i(\mathbf{x}),$$

where σ_i is the sum of squares of polynomials. The reader can check $M_G \subset \text{cone}(G)$ (although sometimes the inclusion turns into equality). Again the quadratic module $M(G)$ is a convex set within the space of polynomials and this happens even when we restrict the degrees of the coefficients.

First of all we have Schmüdgen's theorem [295] (see [305] for an algorithmic version, again tied to Pólya's theorem).

Theorem 13.2.5 (Schmüdgen's theorem). *Let* $G = \{g_1(x) \geq 0,\ g_2(x) \geq 0,\ \ldots,\ g_k(x) \geq 0\}$ *be a set of polynomial inequalities. Suppose that the semialgebraic set* $K(G) = \{x \in \mathbb{R}^n : g_i(x) \geq 0 \text{ for all } i\}$ *is compact. Given any* $f(x) \in \mathbb{R}[x_1,\ldots,x_n]$ *such that* $f(\bar{x}) > 0$ *for all* $\bar{x} \in K(G)$*, then* f *belongs to the preorder* cone(G).

This theorem really requires that f is strictly positive over $K(G)$ (see exercises). One can always wonder, *how many summands are really needed in such a representation?* M. Putinar [282], with improvements by T. Jacobi and A. Prestel [182], achieve an improvement of Schmüdgen's result saying that only m terms suffice (see [262] for a more quantitative algorithmic version).

Theorem 13.2.6 (Putinar's theorem). *Consider again $G, K(G), f$ as in Theorem 13.2.5. Suppose that in addition the quadratic module $M(G)$ contains the polynomial $N - \sum_{i=1}^{n} x_i^2$ for some positive integer N. Then, the polynomial f belongs to the quadratic module $M(G)$.*

Just as we did in the case of minimizing polynomials over $\mathbb{R}^n$, Lasserre [213] realized that one can use these two theorems to find the minimum of a strictly positive polynomial f over a semialgebraic set $K(G)$. For this, as before we want to find the largest number λ such that $f - \lambda \geq 0$ over $K(G)$. Earlier, in Section 13.1, we looked for the maximum λ such that $f - \lambda$ was a sum of squares. Here we aim to find the smallest λ that makes sure f belongs to either the preorder cone(G) or to the quadratic module $M(G)$. As before we can implement the problem of finding λ by restricting ourselves to elements of $M(G)$ or cone(G) where the summands have degree at most m.

13.3 Approximating the integer hull of combinatorial problems

A central theme of discrete optimization has been the understanding of the facets that describe the integer hull of a combinatorial problem. The traditional modeling of combinatorial optimization problems often uses 0/1 incidence vectors. The set S of solutions of a combinatorial problem (e.g., the stable sets, traveling salesman tours) is often computed through the (implicit) convex hull of such lattice points. Just as in the stable set and max-cut examples in Proposition 12.1.1, the incidence vectors can be seen as the *real* solutions to a system of polynomial equations: $f_1(\mathbf{x}) = f_2(\mathbf{x}) = \cdots = f_m(\mathbf{x}) = 0$, where $f_1, \ldots, f_m \in \mathbb{R}[x_1, \ldots, x_n]$.

As we will see now, one can recover quite a bit of information about the integer hull of combinatorial problems from a sequence of combinatorially controlled SDPs. This kind of approach was pioneered in the lift-and-project method of Balas, Ceria, and Cornuéjols [19], the matrix-cut method of Lovász and Schrijver [237], and the linearization technique of Sherali–Adams [310]. See also [223] and references therein for a very extensive survey. In this last section we point to more recent developments by Gouveia et al. [141, 142] who considered a sequence of semidefinite relaxations of the convex hull of real solutions of a polynomial system encoding a combinatorial problem. They called these approximations *theta bodies* because, for stable sets of graphs, the first theta body in this hierarchy is exactly Lovász's theta body of a graph [236].

Let us start with an historically important example that gives the name to the method: Given an undirected finite graph $G = (V, E)$, consider the set S_G of characteristic vectors of stable sets of G. The convex hull of S_G, denoted by STAB(G), is the *stable set polytope*. As we already mentioned, the vanishing ideal of S_G is given by $I_G := \langle x_i^2 - x_i : i \in V,\ x_i x_j : \{i, j\} \in E \rangle$ which is a real radical zero-dimensional ideal in $\mathbb{K}[x_1, \ldots, x_n]$. In [236], Lovász introduced a semidefinite relaxation, TH(G), of the polytope STAB(G), called the *theta body* of G. There are multiple descriptions of TH(G), but the one in [237, Lemma 2.17], for instance, shows that TH(G) can be defined completely in terms of the polynomial system

I_G. It is easy to show that $\mathrm{STAB}(G) \subseteq \mathrm{TH}(G)$, and remarkably, we have that $\mathrm{STAB}(G) = \mathrm{TH}(G)$ if and only if the graph is *perfect*. We will now explain how the case of stable sets can be generalized to construct theta bodies for many other combinatorial problems.

We will construct an approximation of the convex hull of a finite set of points S, denoted conv(S), by a sequence of convex bodies recovered from "degree truncations" of the defining polynomial systems. In what follows, I will be a radical polynomial ideal. A polynomial f is *nonnegative modulo* I, written as $f \geq 0 \bmod I$, if $f(\mathbf{s}) \geq 0$ for all $\mathbf{s} \in V_{\mathbb{R}}(I)$. More strongly, the polynomial f is a *sum of squares* (SOS) *modulo* I if there exists $h_j \in \mathbb{K}[x_1,\ldots,x_n]$ such that $f \equiv \sum_{j=1}^{t} h_j^2 \pmod{I}$ for some t or, equivalently, $f - \sum_{j=1}^{t} h_j^2 \in I$. If, in addition, each h_j has degree at most k, then we say that f is k*-SOS* modulo I. The ideal I is k*-SOS* if *every* polynomial that is nonnegative modulo I is k-SOS modulo I. If every polynomial of degree at most d that is nonnegative modulo I is k-SOS modulo I, we say that I is (d,k)*-SOS*.

Note that conv($V_{\mathbb{R}}(I)$), the convex hull of $V_{\mathbb{R}}(I)$, is described by the linear polynomials f such that $f \geq 0$ modulo I. A certificate for the nonnegativity of f modulo I is the existence of an SOS-polynomial $\sum_{j=1}^{t} h_j^2$ that is congruent to f modulo I. One can now investigate the convex hull of S through the hierarchy of nested closed convex sets defined by the semidefinite programming relaxations of the set of $(1,k)$-SOS polynomials.

Definition 13.3.1. Let $I \subseteq \mathbb{K}[x_1,\ldots,x_n]$ be an ideal, and let k be a positive integer. Let $\Sigma_k \subset \mathbb{K}[x_1,\ldots,x_n]$ be the set of all polynomials that are k-SOS modulo I.

1. The k-th *theta body* of I is

$$\mathrm{TH}_k(I) := \{\mathbf{x} \in \mathbb{R}^n : f(\mathbf{x}) \geq 0 \text{ for every linear } f \in \Sigma_k\}.$$

2. The ideal I is TH_k*-exact* if the k-th theta body $\mathrm{TH}_k(I)$ coincides with the closure of conv($V_{\mathbb{R}}(I)$).

3. The *theta-rank* of I is the smallest k such that $\mathrm{TH}_k(I)$ coincides with the closure of conv($V_{\mathbb{R}}(I)$).

Example 13.3.2. Consider the ideal $I = \langle x^2y - 1\rangle \subset \mathbb{R}[x,y]$. Then we have $\mathrm{conv}(V_{\mathbb{R}}(I)) = \{(p_1,p_2) \in \mathbb{R}^2 : p_2 > 0\}$, and any linear polynomial that is nonnegative over $V_{\mathbb{R}}(I)$ is of the form $\alpha + \beta y$, where $\alpha, \beta \geq 0$. Since $\alpha y + \beta \equiv (\sqrt{\alpha}xy)^2 + (\sqrt{\beta})^2 \bmod I$, I is $(1,2)$-SOS and TH_2-exact.

Example 13.3.3. For the case of the stable sets of a graph G, one can see that

$$\mathrm{TH}_1(I_G) = \left\{ \mathbf{y} \in \mathbb{R}^n : \begin{array}{l} \exists M \succeq O, M \in \mathbb{R}^{(n+1)\times(n+1)} \text{ such that} \\ M_{00} = 1, \\ M_{0i} = M_{i0} = M_{ii} = y_i \ \forall\, i \in V, \\ M_{ij} = 0 \ \forall\, \{i,j\} \in E. \end{array} \right\}$$

It is known that $\mathrm{TH}_1(I_G)$ is precisely *Lovász's theta body of* G. The ideal I_G is TH_1-exact precisely when the graph G is perfect.

By definition, $\mathrm{TH}_1(I) \supseteq \mathrm{TH}_2(I) \supseteq \cdots \supseteq \mathrm{conv}(V_{\mathbb{R}}(I))$. As seen in Example 13.3.2, conv($V_{\mathbb{R}}(I)$) may not always be closed and so the theta-body sequence of I can converge, if at all, only to the closure of conv($V_{\mathbb{R}}(I)$). The good news for combinatorial optimization are that there is plenty of good behavior for problems arising with a finite set of possible solutions.

Example 13.3.4. For the max-cut problem we saw earlier, the defining vanishing ideal is $I(SG) = \langle x_e^2 - x_e$ for all $e \in E$, $\prod x_e$ for all $e \in T$ an odd cycle in $G\rangle$. In this case one can prove that the ideal $I(SG)$ is TH_1-exact if and only if G is a bipartite graph. In general the theta rank of $I(SG)$ is bounded above by the size of the max-cut in G. There is no constant k such that $\mathrm{TH}_k(I(SG)) = \mathrm{conv}(SG)$, for all graphs G. Other formulations of max-cut are studied in [142].

Recall that when $S \subset \mathbb{R}^n$ is a finite set, its vanishing ideal $I(S)$ is zero dimensional and real radical (see [241, Section 12.5] for a definition of the real radical). In what follows, we say that a finite set $S \subset \mathbb{R}^n$ is *exact* if its vanishing ideal $I(S) \subseteq \mathbb{R}[x_1, x_2, \ldots, x_n]$ is TH_1-exact.

Theorem 13.3.5 (see [141]). *For a finite set $S \subset \mathbb{R}^n$, the following are equivalent:*

1. *S is exact.*
2. *There is a finite linear inequality description of* $\mathrm{conv}(S)$ *in which for every inequality $g(\mathbf{x}) \geq 0$, g is 1-SOS modulo $I(S)$.*
3. *There is a finite linear inequality description of* $\mathrm{conv}(S)$ *such that for every inequality $g(\mathbf{x}) \geq 0$, every point in S lies either on the hyperplane $g(\mathbf{x}) = 0$ or on a unique parallel translate of it.*
4. *The polytope* $\mathrm{conv}(S)$ *is affinely equivalent to a compressed lattice polytope (every reverse lexicographic triangulation of the polytope is unimodular with respect to the defining lattice).*

Example 13.3.6. The vertices of the following 0/1-polytopes in $\mathbb{R}^n$ are exact for every n: (1) hypercubes, (2) (regular) cross polytopes, (3) hypersimplices (includes simplices), (4) joins of 2-level polytopes, and (5) stable set polytopes of perfect graphs on n vertices.

More strongly, one can say the following.

Proposition 13.3.7. *Suppose $S \subseteq \mathbb{R}^n$ is a finite point set such that for each facet F of* $\mathrm{conv}(S)$ *there is a hyperplane H_F such that $H_F \cap \mathrm{conv}(S) = F$ and S is contained in at most $t+1$ parallel translates of H_F. Then $I(S)$ is TH_t-exact.*

In [141] the authors show that theta bodies can be computed explicitly as projections to the feasible set of a semidefinite program. These SDPs are constructed using the *combinatorial moment matrices* introduced by [222].

13.4 Notes and further references

This chapter was based on the presentation given in [106], which is also deeply inspired in the paper [271]. See also [212, 260, 270] for the origins of the theory. This is a topic that today is still in high activity by researchers and we only highlighted a few ideas. For a more detailed exposition of this topic we recommend [218] and the wonderful forthcoming book (in this same series) devoted solely to this theme [57].

It is worth mentioning that there are variations of the method that can not only determine whether a semialgebraic set is empty but also find explicit solutions. See, e.g.,

[165, 220]. Similarly we saw in Section 13.1 that things do not always go so well with the SDP programming approximation when the domain is not compact. Today there are several alternative techniques to deal with these kinds of problems (see, e.g., [109]).

We described the search for Positivstellensatz infeasibility certificates formulated as a semidefinite programming problem. There is an alternative interpretation, obtained by dualizing the corresponding semidefinite programs. This is the exact analogue of the construction presented in [106], and is closely related to the approach via truncated moment sequences developed by Lasserre [212, 218].

13.5 Exercises

Exercise 13.5.1. Use the techniques outlined to compute or estimate the global minimum of bivariate polynomials such as $x^4+x^2+y^6-3x^2y^2$.

Exercise 13.5.2. Let A be a real, symmetric $n \times n$ matrix. Prove that A is PSD if and only if all the coefficients of its characteristic polynomial

$$p(\lambda) = \det(\lambda I - A) = \lambda^n + a_{n-1}\lambda^{n-1} + \cdots + a_1\lambda + a_0 \tag{13.9}$$

alternate in sign ($a_i(-1)^{n-i} > 0$).

Exercise 13.5.3. Prove that a quotient of two sums of squares of polynomials can always be rewritten as sum of squares of rational functions.

Exercise 13.5.4. The *Newton polytope* of a polynomial q is the convex hull of the vectors of exponents (seen as lattice points) in the monomials of q. Give a proof of the following clever result of Reznick: If $p(\mathbf{x}) = \sum q_i^2(\mathbf{x})$, then Newton($q_i$) is contained in $\frac{1}{2}$Newton($p(\mathbf{x})$) (a useful fact to write sums of squares, when possible, because it reduces the size of the PSD matrix Q).

Exercise 13.5.5. Consider the inequality $g = (1-x^2)^3 \geq 0$. The reader can verify that its semialgebraic set is $K(g) = [-1,1]$. Show that while $(1-x) > 0$ on $K(g)$, it is impossible that $1-x \in \text{cone}(g)$. Which shows the Schmüdgen's result does require strict positivity of polynomials.

Chapter 14

Epilogue

This book was just an invitation to a trend of new algebraic ideas for optimization problems. Sadly, we cannot discuss in depth some other important work, but we wish to at least point the reader to other exciting parts of this universe. We briefly mention many results but provide no details.

14.1 Algebraic and topological ideas in linear optimization

One can also find useful algebraic and geometric insights in the theory of linear optimization. Consider the usual primal and dual linear programming problems

$$\max\left\{\mathbf{c}^\top\mathbf{x} : A\mathbf{x} = \mathbf{b},\ \mathbf{x} \geq \mathbf{0}\right\}, \tag{14.1}$$

$$\min\left\{\mathbf{b}^\top\mathbf{y} : A^\top\mathbf{y} - \mathbf{s} = \mathbf{c},\ \mathbf{s} \geq \mathbf{0}\right\}, \tag{14.2}$$

where A is an $m \times n$ matrix. The primal-dual interior point methods are among the most computationally successful algorithms for linear optimization (see [285, 327]). The interior point methods follow the *central path*. The primal-dual central path is actually a portion of an algebraic curve given by the following system of equations (this observation was first made by Bayer and Lagarias [42, 43]).

$$\left\{(\mathbf{x},\mathbf{y},\mathbf{y}) : A\mathbf{x} = \mathbf{b},\ A^\top\mathbf{y} - \mathbf{s} = \mathbf{c},\ x_i s_i = \lambda \, \forall i = 1, 2, \ldots, n\right\}. \tag{14.3}$$

The system of linear and quadratic equations has several properties: For all $\lambda > 0$, the system of polynomial equations has a unique real solution $(\mathbf{x}^*(\lambda), \mathbf{y}^*(\lambda), \mathbf{s}^*(\lambda))$ with the properties $\mathbf{x}^*(\lambda) > \mathbf{0}$ and $\mathbf{s}^*(\lambda) > \mathbf{0}$. The point $\mathbf{x}^*(\lambda)$ is the optimal solution of the *logarithmic barrier function* for (14.1), which is defined as

$$f_\lambda(\mathbf{x}) := \mathbf{c}^\top\mathbf{x} + \lambda \sum_{i=1}^{n} \log x_i.$$

Any limit point $(\mathbf{x}^*(0), \mathbf{y}^*(0), \mathbf{s}^*(0))$ of these solutions for $\lambda \to 0$ is the unique solution of the complementary slackness constraints and thus yields an optimum point.

In optimization the central path is only followed *approximately* by interior point methods with some kind of Newton steps. Similarly, the central path is typically considered going from the *analytic center* within the polyhedral single cell, with $s_i \geq 0$, to the optimal solution of the linear programs. But the usual central path is just a part of an explicit real algebraic curve that extends beyond a single feasible region (given by sign constraints on variables). Instead of studying the problem with only constraints $s_i \geq 0$, one can ask for all feasible programs arising from any set of sign conditions $s_i \epsilon_i 0$, $\epsilon_i \in \{\leq, \geq\}$, $1 \leq i \leq m$. There are at most $\binom{m-1}{n}$ such feasible sign vectors. The input data $(A, \mathbf{b})$ is said to be in *general position* if this number is attained. Then the central curve passes through all the vertices of a hyperplane arrangement.

The traditional primal and dual central parts are portions of the projections of the above quadratic and linear equations to one single set of variables and one can ask what are the new defining equations. In [107] the authors computed the degree, arithmetic genus, and defining prime ideal of the central curve and their primal and dual projections. These invariants are expressed in terms of the matroid of the input matrix A. This is another evidence of the strong relation between linear programming and matroids (see, e.g., [18]).

In practical computations, interior point methods follow a piecewise-linear approximation to the central path. One way to estimate the number of Newton steps needed to reach the optimal solution is to bound the *total curvature* of the exact central path. The intuition is that curves with small curvature are easier to approximate with fewer line segments. This idea has been investigated by various authors (see, e.g., [254, 313, 329, 337]), and has yielded interesting results. For example, Vavasis and Ye [329] found that the central path contains no more than n^2 turning points. This finding led to an interior-point algorithm whose running time depends only on the constraint matrix A. Thus, in a way, curvature can be regarded as the continuous analogue of the role played by the diameter in the simplex method.

In [107, 108] the authors obtained bounds for the total curvature in terms of the degree of the Gauss maps of the (algebraic) central curve. From this one can get a formula for the average total curvature over the different polyhedral cells. For practical applications the more relevant quantity is not the average total curvature but rather the curvature in a specific feasible region. This has been investigated by Deza, Terlaky, and Zinchenko in a series of papers [110–112]. They conjectured that the curvature of a polytope, defined as the largest possible total curvature of the associated central path with respect to the various cost vectors, is no more than $2\pi m$, where m is the number of facets of the polytope. They name their conjecture the *continuous Hirsch conjecture* to suggest the similarity with the well-known problem for the simplex method (more on this below). Although the average value for the curvature for bounded cells is known to be linear, we do not have a polynomial bound for the total curvature in a single cell.

The classic simplex method of Dantzig and its variations are one of the most common algorithms for solving linear programs. It can be viewed as a family of combinatorial local search algorithms on the graph of a convex polyhedron. More precisely, the search is done over a finite graph, the one-skeleton of the polyhedron or *graph of the polyhedron*, which is composed of the zero- and one-dimensional faces of the feasible region (called *vertices* and *edges*). The search moves from a vertex of the one-skeleton to a better neighboring one according to some *pivot rule* which selects an improving neighbor. Geometrically the simplex method traces a path on the graph of the polytope. Despite great effort of analysis, it remains open whether there is always a polynomial bound on the shortest path between two vertices in the skeleton. The *diameter* of the graph of a polytope is the length of the longest shortest path among all possible pairs of vertices. Geometric and topological tools are useful on bounding the diameter.

It has been pointed out repeatedly that the proofs of known upper bounds use only very limited properties of polytopes and often hold for (topological) simplicial complexes. For example, Klee and Kleinschmidt [197] showed that some bounds for fixed number of variables hold for the ridge graphs of all pure simplicial complexes and more general objects. The *combinatorial-topological* approach to the diameter problem has a long history (see, e.g., [239]); Adler, Dantzig, and Murty [6, 7], Kalai [184], Provan and Billera [280], and others abstracted the notion of the graph of a polytope. This has become an important direction of research (see [17, 85, 130, 132, 133, 246, 251] and the many references therein). The objects one wishes to abstract are graphs of simple polyhedra. A d-dimensional polyhedron P is *simple* if each of its vertices is contained in exactly d of the n facets of P. It is well-known that it is enough to consider simple polyhedra, since the largest diameter of d-polyhedra with n facets must be achieved by a simple d-polyhedra with n facets.

A key idea in a recent exciting paper of Eisenbrand et al. [124] is working with new topological abstractions they called *base abstraction and connected layer families*. Eisenbrand et al. were able to prove that their abstraction is a reasonable generalization because it satisfies some of the known upper bounds on the diameter of polytopes. Larman proved in [211] that for a d-dimensional polytope P with n facets the diameter is no more than $2^{d-3}n$ (shortly after, this bound was improved by Barnette [26]). This bound shows that in fixed dimension the diameter must be linear in the number of facets. The best general bound of $O(n^{1+\log d})$ was obtained by Kalai and Kleitman [185]. The authors of [124] proved that the Larman bound and the Kalai-Kleitman bounds hold again for their abstractions. But a great novelty in the Eisenbrand et al.'s approach is that there are abstraction graphs with diameter greater than $\Omega(n^2/\log n)$, for a certain constant c (although they are far from being polyhedra).

Of course, when talking about upper bounds we must mention that W. Hirsch conjectured in 1957 that the diameter of the graph of a polyhedron defined by n inequalities in d dimensions is at most $n-d$. Finally, in 2010 the *Hirsch conjecture* was disproved by a clever, explicit geometric-topological construction due to Francisco Santos. He proved the following theorem (see all details in the paper [288]):

Theorem 14.1.1. *There is a* 43*-dimensional polytope with* 86 *facets and of diameter at least* 44. *There is an infinite family of non-Hirsch polytopes with n facets and diameter* $\sim (1+\epsilon)n$ *for some* $\epsilon > 0$. *This holds true even in fixed dimension.*

This still leaves open the possibility that the diameter of convex polyhedra is bounded by a polynomial on the number of facets and the dimension.

14.2 Other generating functions techniques in integer programming

There is another style of generating function to capture all the information about a family of integer linear programs, which is "dual" to the Barvinok framework presented in Part III. We can construct a different kind of generating function, the *partition generating function* defined by a given $m \times n$ integral constraint matrix A and integral vectors $\mathbf{c}, \mathbf{b}$ (as we did for Gröbner bases, $\mathbf{b}$ can be thought of as a varying parameter). The following simple lemma illustrates the connection to generating functions related to *vector partition functions* (see

[62, 81, 318, 333] and the references therein).

Lemma 14.2.1. *Let $\mathbf{a}_i$ denote the columns of a $d \times n$ integral matrix A.*

(a) *The case of 0/1 integer programs: In the monomial expansion of the polynomial*

$$\prod_{j=1}^{n}(1+\mathbf{z}^{\mathbf{a}_j}t^{c_j}) = \sum_{\textit{feasible } \mathbf{b}} \left(\sum_{\alpha \in P \cap \mathbb{Z}^n} t^{\mathbf{c}^\top \alpha} \right) \mathbf{z}^{\mathbf{b}}$$

there is a monomial of the form $t^\beta \mathbf{z}^\mathbf{b}$ if and only if there is a 0/1 vertex $\mathbf{v}$ in the polyhedron $P = \{\mathbf{x} : A\mathbf{x} = \mathbf{b},\ \mathbf{0} \le \mathbf{x} \le \mathbf{1}\}$ that has cost value $\mathbf{c}^\top \mathbf{v} = \beta$.

(b) *The case of general integer programs: In the multivariate Laurent series expansion defined by the rational function*

$$\frac{1}{\prod_{j=1}^{n}(1-\mathbf{z}^{\mathbf{a}_j}t^{c_j})} = \sum t^\beta \mathbf{z}^\mathbf{b}$$

there is a monomial of the form $t^\beta \mathbf{z}^\mathbf{b}$ if and only if $P = \{\mathbf{x} : A\mathbf{x} = \mathbf{b},\ \mathbf{x} \ge \mathbf{0}\}$ has a lattice point of cost value β.

Example 14.2.2. Consider the one-parameter family of knapsack problems:

$$\max\{x_1 + 2x_2 + x_3 : 3x_1 + 5x_2 + 17x_3 = b, x_1, x_2, x_3 \in \{0,1\}\}.$$

The family is fully encoded, for *any* right-hand side b, inside

$$\left(1+z^3 t\right)\left(1+z^5 t^2\right)\left(1+z^{17} t\right).$$

This is a short representation of any of the optimal values. For example, in the expanded monomial form

$$1 + z^{17}t + z^5t^2 + z^{22}t^3 + z^3t + z^{20}t^2 + z^8t^3 + z^{25}t^4$$

we see that for $b = 22$ there is a feasible solution with objective value 3.

Suppose now that we can choose $x_1, x_2, x_3 \in \mathbb{Z}_+$ and consider the expression

$$\frac{1}{(1-z^3t)(1-z^5t^2)(1-z^{17}t)}.$$

Its multivariate Taylor series expansion is $1 + z^3t + z^5t^2 + z^6t^2 + z^8t^3 + z^9t^3 + z^{10}t^4 + z^{11}t^4 + z^{12}t^4 + z^{13}t^5 + z^{14}t^5 + (t^6+t^5)z^{15} + z^{16}t^6 + (t^6+t)z^{17} + (t^7+t^6)z^{18} + z^{19}t^7 + (t^8+t^7+t^2)z^{20} + (t^8+t^7)z^{21} + (t^8+t^3)z^{22} + (t^9+t^8+t^3)z^{23} + (t^9+t^8)z^{24} + (t^{10}+t^9+t^4)z^{25} + (t^{10}+t^9+t^4)z^{26} + (t^{10}+t^9+t^5)z^{27} + (t^{11}+t^{10}+t^5)z^{28} + \cdots$.

In this example, observe that we have $t^{10}z^{28}$ because we have for $b = 28$ just one knapsack solution $x_1 = 6,\ x_2 = 2,\ x_3 = 0$ of objective function value 10. Note also that if we set $t = 1$, we count the lattice points instead of recording the objective function values of points: $1 + z^3 + z^5 + z^6 + z^8 + z^9 + z^{10} + z^{11} + z^{12} + z^{13} + z^{14} + 2z^{15} + z^{16} + 2z^{17} + 2z^{18} + z^{19} + 3z^{20} + 2z^{21} + 2z^{22} + 3z^{23} + 2z^{24} + 3z^{25} + 3z^{26} + 3z^{27} + 3z^{28} + \cdots$.

Thus we can summarize what we have done by the following observation: Let $\phi_A(\mathbf{b})$ be the coefficient of $\mathbf{z}^{\mathbf{b}}$ of the function

$$f(\mathbf{z}) = \frac{1}{(1-\mathbf{z}^{\mathbf{a}_1})\cdots(1-\mathbf{z}^{\mathbf{a}_n})}$$

expanded as a power series centered at $\mathbf{z}=\mathbf{0}$ or, similarly, for the binary case, the coefficient in the expansion of the polynomial

$$\prod_{j=1}^{n}(1+\mathbf{z}^{\mathbf{a}_j}t^{c_j}) = \sum_{\text{feasible } \mathbf{b}} \left(\sum_{\alpha\in P\cap\mathbb{Z}^n} t^{\mathbf{c}^\top\alpha}\right)\mathbf{z}^{\mathbf{b}}.$$

Finding the optimal value of an integer program or finding the number of lattice points inside a polyhedron, in the binary or general integer versions, is the same as computing the coefficient of the monomial $\mathbf{z}^{\mathbf{b}}$. Several interesting ways have been used for calculating such values. First, one can study these generating functions as complex meromorphic functions with poles over a hyperplane arrangement and try to recover the values of residues. This is shown in the work by Vergne and her collaborators [23, 321, 333]. An alternative method to compute the coefficients of power series expansions relates to multivariate complex integration (see work by Beck [44], Lasserre and Zeron [221], Pemantle and Wilson [273]). It follows from the study of complex residues and careful application of Cauchy integral formula that the coefficient $\phi_A(\mathbf{b})$ of $\mathbf{z}^{\mathbf{b}}$ is equal to:

$$\phi_A(\mathbf{b}) = \frac{1}{(2\pi i)^m}\int_{|z_1|=\epsilon_1}\cdots\int_{|z_m|=\epsilon_m}\frac{z_1^{-b_1-1}\cdots z_m^{-b_m-1}}{(1-\mathbf{z}^{\mathbf{a}_1})\cdots(1-\mathbf{z}^{\mathbf{a}_n})}\,d\mathbf{z}.$$

Here $0<\epsilon_1,\ldots,\epsilon_m<1$ are different numbers such that we can expand all the terms $\frac{1}{1-\mathbf{z}^{M_k}}$ into power series about 0. This method has yet to be applied in optimization; so far it has only been applied for combinatorial problems.

Lasserre [216] also looked at the existence of nonnegative integral solutions for systems of the form $A\mathbf{x}=\mathbf{b}$ using the same rational generating function setting and manages to reduce the feasibility computation to solving a (large) linear program which arises by comparing the coefficients of a certain polynomial identity coming from the rational function. See the interesting monograph by Lasserre [217] for more on this topic.

Calculations with series expansions can be made very fast. One technique is the well-known *fast Fourier transform* technique. Yu. Nesterov [261] showed that this technique, applied to the partition generating function of the n-item knapsack problem, gives a pseudopolynomial complexity that improves upon standard dynamic programming by a factor of n.

14.3 Variations on Gomory's group relaxations

A "classical" algebraic method for solving integer linear optimization problems is Gomory's *group relaxation*, introduced by Gomory in [139, 140]. We briefly introduce it and then discuss some generalizations and variations of an algebraic and geometric nature.

Given a general integer program of the form

$$\min\left\{\mathbf{c}^\top\mathbf{x} : A\mathbf{x}=\mathbf{b},\ \mathbf{x}\geq\mathbf{0},\ \mathbf{x}\in\mathbb{Z}^n\right\}, \tag{14.4}$$

its group relaxation is obtained by dropping nonnegativity restrictions on all the basic variables in the optimal solution of its linear relaxation (see [296, Section 24.2].

Gomory's group relaxation of (14.4) is the problem

$$\min\left\{\tilde{\mathbf{c}}^\top \mathbf{x}_N : A_B^{-1} A_N \mathbf{x}_N \equiv A_B^{-1}\mathbf{b} \ (\mathrm{mod}\ \mathbf{1}),\ \mathbf{x}_N \geq \mathbf{0},\ \mathbf{x}_N \in \mathbb{Z}^{n-m}\right\}, \tag{14.5}$$

where B and N denote the index sets for the basic and nonbasic columns of A corresponding to the optimal solution of the usual linear relaxation of (14.4). The vector $\mathbf{x}_N$ denotes the nonbasic variables and the cost vector $\tilde{\mathbf{c}} = \mathbf{c}_N - \mathbf{c}_B A_B^{-1} A_N$, where $\mathbf{c} = (\mathbf{c}_B, \mathbf{c}_N)$ is partitioned according to B and N. The notation $A_B^{-1} A_N \mathbf{x}_N \equiv A_B^{-1}\mathbf{b}$ (mod $\mathbf{1}$) indicates that $A_B^{-1} A_N \mathbf{x}_N - A_B^{-1}\mathbf{b}$ is a vector of integers. Problem (14.5) is called a *group relaxation* of (14.4) since it can be written in the canonical form

$$\min\left\{\tilde{\mathbf{c}}^\top \mathbf{x}_N : \sum_{j\in N} g_j x_j \equiv g_0 \ (\mathrm{mod}\ G),\ \mathbf{x}_N \geq \mathbf{0},\ \mathbf{x}_N \in \mathbb{Z}^{n-m}\right\}, \tag{14.6}$$

where G is a finite Abelian group and $g_j \in G$. Problem (14.6) can be viewed as a shortest path problem in a graph on $|G|$ nodes, which immediately furnishes pseudopolynomial algorithms for solving it. Once the optimal solution $\mathbf{x}_N^*$ of (14.5) is found, it can be uniquely *lifted* to a vector $\mathbf{x}^* = (\mathbf{x}_B^*, \mathbf{x}_N^*) \in \mathbb{Z}^n$ such that $A\mathbf{x}^* = \mathbf{b}$. If $\mathbf{x}_B^* \geq \mathbf{0}$ then $\mathbf{x}^*$ is the optimal solution of (14.4). Otherwise, $\mathbf{c}^\top \mathbf{x}^*$ is a lower bound for the optimal value of (14.4).

Note that here and below we use the typical notation that separates the indices of a vector $\mathbf{x}$ into those of a set S, namely, $\mathbf{x}_S$, and those of its complement $\mathbf{x}_{\bar{S}}$; with similar separation for matrices and their respective multiplications.

Wolsey introduced in [335] the notion of *extended* group relaxations of (14.4). Here one allows an index set B of columns that is only linearly independent, but not necessarily a basis. Thus it is allowed to drop the nonnegativity restrictions on fewer basic variables $\mathbf{x}_B$. If (14.5) does not solve (14.4), then one could resort to other extended relaxations to solve the problem. At least one of these extended group relaxations (in the worst case (14.4) itself) is guaranteed to solve the integer program (14.4).

There are several results on group relaxations that can be obtained using *Gröbner bases* of *toric ideals* presented in Chapter 11. One considers the entire family of integer programs of the form (14.4) as the right-hand side vector $\mathbf{b}$ varies. Surprisingly, the regular triangulations we introduced in Section 6.3 play a role in the algebraic study of integer programming. For this let $\Delta_\mathbf{c}$ be the regular triangulation of the column vectors of the matrix A with respect to the height vector $\mathbf{c}$. We now define a group relaxation of (14.4) with respect to each face B of $\Delta_\mathbf{c}$. Here B is again an index set of columns of A, linearly independent, but not necessarily a basis. Let N denote the complement of B. Then the group relaxation of the integer program (14.4) with respect to the face B of $\Delta_\mathbf{c}$ is the optimization problem:

$$\min\left\{\tilde{\mathbf{c}}_N^\top \mathbf{x}_N : A_B \mathbf{x}_B + A_N \mathbf{x}_N = \mathbf{b},\ \mathbf{x}_N \geq \mathbf{0},\ (\mathbf{x}_B, \mathbf{x}_N) \in \mathbb{Z}^n\right\}. \tag{14.7}$$

One can prove that the optimal solution $\mathbf{x}_N^*$ of (14.7) can be uniquely lifted to the solution $(\mathbf{x}_B^*, \mathbf{x}_N^*)$ of $A\mathbf{x} = \mathbf{b}$. The formulation of (14.7) shows that $\mathbf{x}_B^*$ is an integer vector. The group relaxation (14.7) solves the original integer program (14.4) if and only if $\mathbf{x}_B^*$ is also non-negative.

In this way we obtain relaxations indexed by the faces of a regular triangulation of the cone generated by A. Thus the simplicial complex given by triangulation $\Delta_\mathbf{c}$ encodes all the optimal bases of the linear programs arising from the coefficient matrix A and cost vector $\mathbf{c}$ (see [317]). The key point is that these "triangulation" group relaxations are precisely all the bounded feasible group relaxations of all programs in the family. One can read off a lot of information from the faces of the regular triangulation. Results include Wolsey's extended relaxations, a characterization of unimodular problems, and an extension of the notion of *total dual integrality*. See [173], [171], [172], [317, Sections 8 and 12.D], [339], [267] and the references therein.

Finally, the connection of integrality conditions and commutative algebra, mostly about the combinatorics of set packing and set covering problems discussed in [75], has also been explored in [116, 135, 136] and other references by the same authors.

The geometry of numbers makes a clean appearance in a different variation of Gomory's group relaxation. There is a fairly recent variation of the original ideas of Gomory proposed by Andersen, Louveaux, Weismantel, and Wolsey [13]. They introduced a model that relaxes Gomory's group relaxation even further, by dropping the integrality constraint for all the nonbasic variables:

$$\min\left\{\tilde{\mathbf{c}}^\top \mathbf{x}_N : A_B^{-1}A_N\mathbf{x}_N \equiv A_B^{-1}\mathbf{b} \pmod{\mathbf{1}},\ \mathbf{x}_N \geq \mathbf{0},\ \mathbf{x}_N \in \mathbb{R}^{n-m}\right\}. \tag{14.8}$$

The model has received significant attention in recent years for developing the theory behind cutting planes derived from *multiple* rows of the optimal simplex tableaux. Valid inequalities for the convex hull of the feasible region of (14.8) can be obtained using *maximal lattice-free convex sets*, i.e., convex sets containing no integer point in their interior that are maximal with respect to set inclusion. It is known (see, e.g., [40]) that the maximal lattice-free convex sets are polyhedra whose recession cones are not full dimensional. We refer to the survey [72] and the literature cited within.

If we fix the facet normals of a family of polyhedra, we can also determine the maximal lattice-free polyhedra within that family. This gives a powerful theory of arithmetic–topological nature, which is explored in [25, 291–293] and from a more algebraic point of view in [272].

14.4 Connections to matrix analysis and representation theory

An interesting approach to combinatorial optimization problems uses the theory of representations of the symmetric group. This work was started by Barvinok and Vershik in the papers [37, 38], then later continued and expanded in [27, 28, 36, 41, 264].

Let S_n denote the symmetric group of order n and let $f: S_n \to \mathbb{R}$. The theory developed concerns those combinatorial optimization problems which can be formulated in one of the following ways:

1. Find the optimum value $\max_{\pi\in S_n} f(\pi)$.
2. Find $\max_{\pi\in S_n - A} f(\pi)$ for some specified $A \subseteq S_n$
3. Decide whether, for a given $c \in \mathbb{R}$, there exists $\pi \in S_n$ satisfying $f(\pi) = c$.
4. Find the value of the sum $\sum_{\pi : f(\pi)=c} \mu(\pi)$ where $\mu : S_n \to \mathbb{R}$ is a given function.

These include well-known combinatorial optimization problems, such as the assignment problem, the quadratic assignment problem, the traveling salesman problem, and others.

For solving the above problems the key focus is the evaluation of the sums $\sum_{\pi\in S_n} f^m(\pi)$, $m \in \mathbb{Z}_+$, and the exponential sums (generating functions) $S_\mu(f;t) = \sum_{\pi\in S_n} \exp\{tf(\pi)\}\mu(\pi)$, $t \in \mathbb{R}$. The methods are based on the theory of representations of finite symmetric groups which is used for the fast evaluation of $S_\mu(f;t)$. Another key component is the analysis of the convex hull of the orbit of a point relative to the action of a permutation group. With some exceptions the convex hull of an orbit of a general point is a complicated polyhedron with many facets. Examples of orbit polytopes are the famous Birkhoff polytope (related to the assignment problem) or the permutahedron [338]. Many problems of combinatorial optimization may be posed as linear programming problems on such polytopes in the case of the symmetric group or smaller permutation groups. The papers [41, 264] investigate the structure of such polytopes for general finite groups.

What kind of result can one obtain? The results obtained include approximation algorithms for the problem (1); description of some subsets $A \subseteq S_n$ for which a polynomial algorithm for (2) exists; pseudopolynomial algorithms for (3). Take, for example, [36] where the authors investigated the quadratic assignment problem in the following form: Given an array of n^4 parameters c_{kl}^{ij}, $i,j,k,l = 1,\ldots,n$, one seeks an n-permutation σ that maximizes the function

$$f(\sigma) = \sum_{i,j=1}^{n} c_{\sigma(i)\sigma(j)}^{ij}.$$

Denote by $\overline{f}$ the average objective function value (taken over all n-permutations σ) and consider the normalized objective function

$$f_0(\sigma) := f(\sigma) - \overline{f}.$$

Letting τ be an optimal permutation, Barvinok and Stephen were interested in the quality of the value $f_0(\sigma)$ of a random permutation σ relative to $f_0(\tau)$. They can prove, for example, that for every $\alpha > 1$, there exists some $\delta = \delta(\alpha) > 0$ such that for all sufficiently large n the inequality

$$f_0(\sigma) \geq \frac{\alpha}{n^2} f_0(\tau)$$

holds with probability at least δn^{-2}.

To conclude, let us recall that algebraic properties of matrices and their decompositions have been used for some time to give speed ups in various algorithms regarding (realizable) matroids and matchings [235], but they continue to be explored in innovative ways (see, e.g., [31, 150]).

We hope that the reader is convinced of the interest of the algebraic and geometric techniques we outlined in this book. There is so much ahead still to be done, and we hope you are enticed to join the party!

Bibliography

[1] 4ti2 team. 4ti2—A software package for algebraic, geometric and combinatorial problems on linear spaces. Available at `http://www.4ti2.de`, 2012.

[2] K. Aardal. Lattice reformulation of integer programming problems. In *Integer points in polyhedra—geometry, number theory, representation theory, algebra, optimization, statistics*, volume 452 of *Contemporary Mathematics*, pages 1–13. American Mathematical Society, Providence, RI, 2008. ISBN 978-0-8218-4173-0 doi: 10.1090/conm/452/08768. URL `http://dx.doi.org/10.1090/conm/452/08768`.

[3] K. Aardal and A. K. Lenstra. Hard equality constrained integer knapsacks. *Math. Oper. Res.*, 29(3):724–738, 2004. ISSN 0364-765X. doi: 10.1287/moor.1040.0099. URL `http://dx.doi.org/10.1287/moor.1040.0099`.

[4] K. Aardal and L. A. Wolsey. Lattice based extended formulations for integer linear equality systems. *Math. Program. Ser. A*, 121:337–352, 2010. ISSN 0025-5610. doi: 10.1007/s10107-008-0236-7. URL `http://dx.doi.org/10.1007/s10107-008-0236-7`.

[5] W. W. Adams and P. Loustaunau. *An Introduction to Gröbner Bases*, volume 3 of *Graduate Studies in Mathematics*. American Mathematical Society, Providence, RI, 1994. ISBN 0-8218-3804-0.

[6] I. Adler and G. B. Dantzig. Maximum diameter of abstract polytopes. *Math. Program. Stud.*, 1:20–40, 1974. ISSN 0303-3929. Pivoting and extensions.

[7] I. Adler, G. B. Dantzig, and K. Murty. Existence of A-avoiding paths in abstract polytopes. *Math. Program. Stud.*, 1:41–42, 1974. ISSN 0303-3929. Pivoting and extensions.

[8] R. K. Ahuja, T. L. Magnanti, and J. B. Orlin. *Network Flows: Theory, Algorithms, and Applications*. Prentice-Hall, Inc., Englewood Cliffs, New Jersey, 1993.

[9] M. Ajtai. Worst-case complexity, average-case complexity and lattice problems. In *Proceedings of the International Congress of Mathematicians, Vol.* III *(Berlin, 1998), Doc. Math.* Extra Vol. III, pages 421–428 (electronic), 1998.

[10] M. Ajtai, R. Kumar, and D. Sivakumar. A sieve algorithm for the shortest lattice vector problem. In *STOC* '01: *Proceedings of the Thirty-third Annual ACM Symposium on Theory of Computing*, pages 601–610, ACM, New York, 2001. ISBN 1-58113-349-9. doi: 10.1145/380752.380857.

[11] N. Alon. Combinatorial Nullstellensatz. *Combin., Probab., and Comput.*, 8:7–29, 1999.

[12] N. Alon and M. Tarsi. Colorings and orientations of graphs. *Combinatorica*, 12: 125–134, 1992.

[13] K. Andersen, Q. Louveaux, R. Weismantel, and L. Wolsey. Inequalities from two rows of a simplex tableau. In M. Fischetti and D. Williamson, editors, *Proceedings of 12th Integer Programming and Combinatorial Optimization (IPCO 2007)*, Lecture Notes in Computer Science, pages 1–15, Springer, Berlin/Heidelberg, 2007.

[14] D. L. Applegate, R. E. Bixby, V. Chvátal, and W. J. Cook. *The Traveling Salesman Problem: A Computational Study*, Princeton Series in Applied Mathematics. Princeton University Press, Princeton, NJ, 2006. ISBN 13: 978-0-691-12993-8; ISBN 10: 0-691-12993-2.

[15] E. Arrondo. Another elementary proof of the Nullstellensatz. *Amer. Math. Monthly*, 113(2):169–170, 2006.

[16] M. Aschenbrenner and R. Hemmecke. Finiteness theorems in stochastic integer programming. *Found. Comput. Math.*, 7:183–227, 2007.

[17] D. Avis and S. Moriyama. On combinatorial properties of linear program digraphs. In *Polyhedral Computation*, volume 48 of *CRM Proceedings and Lecture Notes*, pages 1–13. Amer. Math. Soc., Providence, RI, 2009. ISBN 0-8218-47633-7.

[18] A. Bachem and W. Kern. *Linear Programming Duality: An Introduction to Oriented Matroids*. Universitext. Springer–Verlag, Berlin, 1992. ISBN 3-540-55417-3.

[19] E. Balas, S. Ceria, and G. Cornuéjols. A lift-and-project cutting plane algorithm for mixed 0-1 programs. *Math. Program. Ser. A*, 58(3):295–324, 1993. ISSN 0025-5610.

[20] V. Baldoni, N. Berline, M. Köppe, and M. Vergne. Intermediate sums on polyhedra: Computation and real Ehrhart theory. *Mathematika*, FirstView Article pp. 1–22. Published online: Jan 2012. doi: 10.1112/S0025579312000101.

[21] V. Baldoni, N. Berline, J. A. De Loera, M. Köppe, and M. Vergne. How to integrate a polynomial over a simplex. *Math. Comp.*, 80(273):297–325, 2011. doi: 10.1090/S0025-5718-2010-02378-6.

[22] V. Baldoni, N. Berline, J. A. De Loera, M. Köppe, and M. Vergne. Computation of the highest coefficients of weighted Ehrhart quasi-polynomials of rational polyhedra. *Found. Comput. Math.*, 2011. doi: 10.1007/s10208-011-9106-4. Published online 12 November 2011.

[23] W. Baldoni-Silva, J. A. De Loera, and M. Vergne. Counting integer flows in networks. *Found. Comput. Math.*, 4(3):277–314, 2004. ISSN 1615-3375. doi: 10.1007/s10208-003-0088-8. URL `http://dx.doi.org/10.1007/s10208-003-0088-8`.

[24] W. Banaszczyk, A. E. Litvak, A. Pajor, and S. J. Szarek. The flatness theorem for nonsymmetric convex bodies via the local theory of Banach spaces. *Math. Oper. Res.*, 24(3):728–750, 1999. ISSN 0364-765X. doi: 10.1287/moor24.3.728. URL `http://dx.doi.org/10.1287/moor.24.3.728`.

[25] I. Bárány, R. Howe, and H. E. Scarf. The complex of maximal lattice free simplices. In G. Rinaldi and L. A. Wolsey, editors, *Proceedings of the 3rd Conference on Integer Programming and Combinatorial Optimization*, pages 1–10, CORE, Louvain-la-Neuve, Belgium 1993.

[26] D. Barnette. An upper bound for the diameter of a polytope. *Discrete Math.*, 10: 9–13, 1974.

[27] A. I. Barvinok. The method of Newton sums in problems of combinatorial optimization. *Diskret. Mat.*, 2(1):3–15, 1990. ISSN 0234-0860. doi: 10.1515/dma.1991.1.4.349. URL `http://dx.doi.org/10.1515/dma.1991.1.4.349`.

[28] A. I. Barvinok. Combinatorial complexity of orbits in representations of the symmetric group. In *Representation theory and dynamical systems*, volume 9 of *Adv. Soviet Math.*, pages 161–182. AMS, Providence, RI, 1992.

[29] A. I. Barvinok. Exponential integrals and sums over convex polyhedra (in Russian). *Funktsional. Anal. i Prilozhen.*, 26(2):64–66, 1992. Translated in Funct. Anal. Appl. 26 (1992), no. 2, pp. 127–129.

[30] A. I. Barvinok. Polynomial time algorithm for counting integral points in polyhedra when the dimension is fixed. *Math. Oper. Res.*, 19:769–779, 1994.

[31] A. I. Barvinok. New algorithms for linear k-matroid intersection and matroid k-parity problems. *Math. Program. Ser. A*, 69:449–470, 1995. ISSN 0025-5610. doi: 10.1007/BF01585571. URL `http://dx.doi.org/10.1007/BF01585571`.

[32] A. I. Barvinok. *A Course in Convexity*, volume 54 of *Graduate Studies in Mathematics*. American Mathematical Society, Providence, RI, 2002. ISBN 0-8218-2968-8.

[33] A. I. Barvinok. Computing the Ehrhart quasi-polynomial of a rational simplex. *Math. Comp.*, 75(255):1449–1466 (electronic), 2006.

[34] A. I. Barvinok. *Integer Points in Polyhedra*. Zurich Lectures in Advanced Mathematics. European Mathematical Society (EMS), Zürich, Switzerland, 2008. ISBN 978-3-03719-052-4.

[35] A. I. Barvinok and J. E. Pommersheim. An algorithmic theory of lattice points in polyhedra. In L. J. Billera, A. Björner, C. Greene, R. E. Simion, and R. P. Stanley, editors, *New Perspectives in Algebraic Combinatorics*, volume 38 of *Math. Sci. Res. Inst. Publ.*, pages 91–147. Cambridge Univ. Press, Cambridge, 1999.

[36] A. I. Barvinok and T. Stephen. The distribution of values in the quadratic assignment problem. *Math. Oper. Res.*, 28(1):64–91, 2003. ISSN 0364-765X. doi: 10.1287/moor.28.1.64.14262. URL `http://dx.doi.org/10.1287/moor.28.1.64.14262`.

[37] A. I. Barvinok and A. M. Vershik. Convex hulls of orbits of representations of finite groups, and combinatorial optimization. *Funktsional. Anal. i Prilozhen.*, 22(3):66–67, 1988. ISSN 0374-1990. doi: 10.1007/BF01077628. URL `http://dx.doi.org/10.1007/BF01077628`.

[38] A. I. Barvinok and A. M. Vershik. Methods of representations theory in combinatorial optimization problems. *Izv. Akad. Nauk SSSR Tekhn. Kibernet.*, 6:64–71, 205, 1988 (in Russian). ISSN 0002-3388. English translation in *Soviet J. Comput. Systems Sci.* 27:1–7, 1989.

[39] A. I. Barvinok and K. Woods. Short rational generating functions for lattice point problems. *J. Amer. Math. Soc.*, 16(4):957–979, 2003.

[40] A. Basu, G. Conforti, G. Cornuéjols, and G. Zambelli. Maximal lattice-free convex sets in linear subspaces. *Math. Oper. Res.*, 35:704–720, 2010.

[41] B. Baumeister, C. Haase, B. Nill, and A. Paffenholz. On permutation polytopes. *Adv. in Math.* , 222:431–452, 2009.

[42] D. A. Bayer and J. C. Lagarias. The nonlinear geometry of linear programming. I. Affine and projective scaling trajectories. *Trans. Amer. Math. Soc.*, 314(2):499–526, 1989. ISSN 0002-9947. doi: 10.2307/2001396. URL `http://dx.doi.org/10.2307/2001396`.

[43] D. A. Bayer and J. C. Lagarias. The nonlinear geometry of linear programming. II. Legendre transform coordinates and central trajectories. *Trans. Amer. Math. Soc.*, 314(2):527–581, 1989. ISSN 0002-9947. doi: 10.2307/2001397. URL `http://dx.doi.org/10.2307/2001397`.

[44] M. Beck. Counting lattice points by means of the residue theorem. *Ramanujan J.*, 4 (3):299–310, 2000.

[45] M. Beck and F. Sottile. Irrational proofs for three theorems of Stanley. *European J. Combin.*, 28(1):403–409, 2007.

[46] M. Beck, C. Haase, and F. Sottile. Formulas of Brion, Lawrence, and Varchenko on rational generating functions for cones, Math. Intell., 31(1): 9–17 2009. ISSN 0343-6993. URL `http://dx.doi.org/10.1007/s00283-008-9013-y`.

[47] T. Becker, V. Weispfenning, and H. Kredel. *Gröbner Bases: A Computational Approach to Commutative Algebra*. Springer, New York, 1993.

[48] M. Bellare and P. Rogaway. The complexity of approximating a nonlinear program. In Pardalos [269, pp. 16–24].

[49] Y. Berstein and S. Onn. The Graver complexity of integer programming. *Ann. Combin.*, 13:289–296, 2009.

[50] D. Bertsimas and R. Weismantel. *Optimization over Integers*. Dynamic Ideas, Belmont, MA, 2005.

[51] D. Bertsimas, G. Perakis, and S. Tayur. A new algebraic geometry algorithm for integer programming. *Management Sci.*, 46:999–1008, 2000.

[52] A. M. Bigatti, R. LaScala, and L. Robbiano. Computing toric ideals. *J. Symbol. Comput.*, 27:351–365, 1999.

[53] J. R. Birge and F. Louveaux. *Introduction to Stochastic Programming*. Springer, New York, 1997.

[54] V. Blanco and J. Puerto. Partial Gröbner bases for multiobjective integer linear optimization. *SIAM J. Discrete Math.*, 23(2):571–595, 2009. ISSN 0895-4801. doi: 10.1137/070698051. URL http://dx.doi.org/10.1137/070698051.

[55] V. Blanco and J. Puerto. Some algebraic methods for solving multiobjective polynomial integer programs. *J. Symbol. Comput.*, 46(5):511–533, 2011. ISSN 0747-7171. doi: 10.1016/j.jsc.2010.10.003. URL http://dx.doi.org/10.1016/j.jsc.2010.10.003.

[56] V. Blanco and J. Puerto. A new complexity result on multiobjective linear integer programming using short rational generating functions. *Optimiz. Lett.*, 6:537–543, 2012. ISSN 1862-4472. doi: 10.1007/s11590-011-0279-1. URL http://dx.doi.org/10.1007/s11590-011-0279-1.

[57] G. Blekherman, P. Parrilo, and R. Thomas, editors. *Semidefinite Optimization and Convex Algebraic Geometry*. Volume 13 of *SIAM–MOS Series on Optimization*. SIAM, Philadelphia, PA, 2012.

[58] J. Bochnak, M. Coste, and M.-F. Roy. *Géométrie algébrique réelle*, volume 12 of *Ergebnisse der Mathematik und ihrer Grenzgebiete (3) [Results in Mathematics and Related Areas (3)]*. Springer-Verlag, New York, 1987. ISBN 3-540-16951-2.

[59] A. Bockmayr and F. Eisenbrand. Cutting planes and the elementary closure in fixed dimension. *Math. Oper. Res.*, 26(2):304–312, 2001. ISSN 0364-765X. doi: 10.1287/moor.26.2.304.10555. URL http://dx.doi.org/10.1287/moor.26.2.304.10555.

[60] C. J. Brianchon. Théorème nouveau sur les polyèdres. *J. École Polytech.*, 15:317–319, 1837.

[61] M. Brion. Points entiers dans les polyédres convexes. *Ann. Sci. École Norm. Sup.*, 21(4):653–663, 1988.

[62] M. Brion and M. Vergne. Residue formulae, vector partition functions and lattice points in rational polytopes. *J. Amer. Math. Soc.*, 10:797–833, 1997.

[63] W. Brownawell. Bounds for the degrees in the Nullstellensatz. *Ann. Math.*, 126(3):577–591, 1987.

[64] W. Bruns, B. Ichim, and C. Söger. Normaliz. Algorithms for Rational Cones and Affine Monoids. http://www.math.uos.de/normaliz.

[65] B. Buchberger. Gröbner bases: An algorithmic method in polynomial ideal theory. In N. K. Bose, editor, *Multidimensional Systems Theory*, pages 184–232. D. Reidel, Hingham, MA, 1985.

[66] B. Buchberger and M. Kauers. Groebner Bases. *Scholarpedia*, 5(10):7763, 2010. URL http://www.scholarpedia.org/article/Groebner_basis.

[67] B. Buchberger and F. Winkler, editors. *Gröbner bases and applications—Proceedings of* 33 *Years of Gröbner Bases*. London Mathematical Society Lecture Note Series 251. Cambridge University Press, Cambridge, UK, 1998.

[68] S. Buss and T. Pitassi. Good degree bounds on Nullstellensatz refutations of the induction principle. In *IEEE Conference on Computational Complexity*, pages 233–242, IEEE, Piscataway, NJ, 1996.

[69] M. Caboara, M. Kreuzer, and L. Robbiano. Efficiently computing minimal sets of critical pairs. *J. Symbol. Comput.*, 38:1169–1190, 2004.

[70] J. W. S. Cassels. *An Introduction to the Geometry of Numbers*. Classics in Mathematics. Springer-Verlag, Berlin, 1997. ISBN 3-540-61788-4. Corrected reprint of the 1971 edition.

[71] M. Clegg, J. Edmonds, and R. Impagliazzo. Using the Gröbner basis algorithm to find proofs of unsatisfiability. In *STOC '96: Proceedings of the Twenty-Eighth Annual ACM Symposium on Theory of Computing*, pages 174–183, ACM, New York, 1996. ISBN 0-89791-785-5. doi: 10.1145/237814.237860. URL `http://doi.acm.org/10.1145/237814.237860`.

[72] M. Conforti, G. Cornuéjols, and G. Zambelli. Corner polyhedra and intersection cuts. *Surveys Oper. Res. Management Sci.*, 16:105–120, 2011.

[73] P. Conti and C. Traverso. Buchberger algorithm and integer programming. In *Applied Algebra, Algebraic Algorithms and Error-Correcting Codes (New Orleans, LA, 1991)*, volume 539 of *Lecture Notes in Comput. Sci.*, pages 130–139. Springer, Berlin, 1991.

[74] W. Cook, M. Hartmann, R. Kannan, and C. McDiarmid. On integer points in polyhedra. *Combinatorica*, 12(1):27–37, 1992. ISSN 0209-9683. doi: 10.1007/BF01191202. URL `http://dx.doi.org/10.1007/BF01191202`.

[75] G. Cornuéjols. *Combinatorial Optimization: Packing and Covering*, volume 74 of *CBMS-NSF Regional Conference Series in Applied Mathematics*. SIAM, Philadelphia, PA, 2001. ISBN 0-89871-481-8. doi: 10.1137/1.9780898717105. URL `http://dx.doi.org/10.1137/1.9780898717105`.

[76] G. Cornuéjols, R. Urbaniak, R. Weismantel, and L. A. Wolsey. Decomposition of integer programs and of generating sets. In R. Burkard and G. J. Woeginger, editors, *Proceedings of the 5th European Symposium on Algorithms*, pages 92–103, Springer-Verlag, London. 1997. ISBN 3-540-63397-9.

[77] N. Courtois, A. Klimov, J. Patarin, and A. Shamir. Efficient algorithms for solving overdefined systems of multivariate polynomial equations. In *Advances in Cryptology—EUROCRYPT 2000 (Bruges)*, volume 1807 of *Lecture Notes in Computer Science*, pages 392–407, Springer, Berlin, 2000.

[78] D. A. Cox, J. B. Little, and D. O'Shea. *Ideals, Varieties, and Algorithms: An Introduction to Computational Algebraic Geometry and Commutative Algebra*. Springer-Verlag, New York, 1992.

[79] D. A. Cox, J. Little, and D. O'Shea. *Using Algebraic Geometry*, volume 185 of *Graduate Texts in Mathematics*. Second Edition. Springer, New York, 2005. ISBN 0-387-20706-6.

[80] D. Dadush, C. Peikert, and S. Vempala. Enumerative lattice algorithms in any norm via M-ellipsoid coverings. In *FOCS* 2011*, IEEE 52nd Annual Symposium on Foundations of Computer Science*, pages 580–589, Oct. 2011. IEEE, Piscataway, NJ, 2011. doi: 10.1109/FOCS.2011.31.

[81] C. De Concini and C. Procesi. *Topics in Hyperplane Arrangements, Polytopes and Box-Splines*. Universitext. Springer, New York, 2011. ISBN 978-0-387-78962-0.

[82] E. de Klerk. The complexity of optimizing over a simplex, hypercube or sphere: a short survey. *Central European J. Oper. Res.*, 16(2):111–125, Jun 2008. doi: 10.1007/s10100-007-0052-9. URL `http://dx.doi.org/10.1007/s10100-007-0052-9`.

[83] E. de Klerk, M. Laurent, and P. A. Parrilo. A PTAS for the minimization of polynomials of fixed degree over the simplex. *Theoret. Comput. Sci.*, 361:210–225, 2006.

[84] J. A. De Loera. Gröbner bases and graph colorings. *Beiträge zur Algebra und Geometrie*, 36(1):89–96, 1995.

[85] J. A. De Loera and S. Klee. Transportation problems and simplicial polytopes that are not weakly vertex-decomposable. *Math. Oper. Res.*, 2012. (to appear).

[86] J. A. De Loera and S. Onn. All linear and integer programs are slim 3-way transportation programs. *SIAM J. Optim.*, 17(3):806–821, 2006.

[87] J. A. De Loera and F. Santos. An effective version of Pólya's theorem on positive definite forms. *J. Pure Appl. Algebra*, 108(3):231–240, 1996. ISSN 0022-4049. doi: 10.1016/0022-4049(95)00042-9. URL `http://dx.doi.org/10.1016/0022-4049(95)00042-9`.

[88] J. A. De Loera, D. C. Haws, R. Hemmecke, P. Huggins, B. Sturmfels, and R. Yoshida. Short rational functions for toric algebra and applications. *J. Symbol. Comput.*, 38(2):959–973, 2004.

[89] J. A. De Loera, D. C. Haws, R. Hemmecke, P. Huggins, and R. Yoshida. Three kinds of integer programming algorithms based on Barvinok's rational functions. In *Integer Programming and Combinatorial Optimization, 10th IPCO Proceedings*, volume 3064 of *Lecture Notes in Computer Science*, pages 244–255. Springer-Verlag, Berlin, 2004. ISBN 3-540-22113-1.

[90] J. A. De Loera, R. Hemmecke, J. Tauzer, and R. Yoshida. Effective lattice point counting in rational convex polytopes. *J. Symbol. Comput.*, 38(4):1273–1302, 2004.

[91] J. A. De Loera, D. C. Haws, R. Hemmecke, P. Huggins, J. Tauzer, and R. Yoshida. LattE, version 1.2. Available from URL `http://www.math.ucdavis.edu/~latte/`, 2005.

[92] J. A. De Loera, D. C. Haws, R. Hemmecke, P. Huggins, and R. Yoshida. A computational study of integer programming algorithms based on Barvinok's rational functions. *Discrete Optim.*, 2:135–144, 2005.

[93] J. A. De Loera, R. Hemmecke, M. Köppe, and R. Weismantel. Integer polynomial optimization in fixed dimension. *Math. Oper. Res.*, 31(1):147–153, 2006.

[94] J. A. De Loera, R. Hemmecke, M. Köppe, and R. Weismantel. FPTAS for mixed-integer polynomial optimization with a fixed number of variables. In *Proceedings of the 17th Annual ACM-SIAM Symposium on Discrete Algorithms, Miami, FL, January 22–24, 2006*, pages 743–748, ACM, New York, SIAM, Philadelphia, 2006. ISBN 0-898-71605-5.

[95] J. A. De Loera, R. Hemmecke, M. Köppe, and R. Weismantel. FPTAS for optimizing polynomials over the mixed-integer points of polytopes in fixed dimension. *Math. Program. Ser. A*, 118:273–290, 2008. doi: 10.1007/s10107-007-0175-8.

[96] J. A. De Loera, R. Hemmecke, S. Onn, and R. Weismantel. *N*-fold integer programming. *Discrete Optim.*, 5(2):231–241, 2008. ISSN 1572-5286. doi: 10.1016/j.disopt.2006.06.006. URL `http://www.sciencedirect.com/science/article/pii/S1572528607000230`. In Memory of George B. Dantzig.

[97] J. A. De Loera, D. C. Haws, and M. Köppe. Ehrhart polynomials of matroid polytopes and polymatroids. *Discrete Comput. Geom.*, 42(4):670–702, 2009. doi: 10.1007/s00454-008-9080-z.

[98] J. A. De Loera, R. Hemmecke, and M. Köppe. Pareto optima of multicriteria integer linear programs. *INFORMS J. Comput.*, 21(1):39–48, Winter 2009. doi: 10.1287/ijoc.1080.0277.

[99] J. A. De Loera, R. Hemmecke, S. Onn, U. G. Rothblum, and R. Weismantel. Convex integer maximization via Graver bases. *J. Pure Appl. Algebra*, 213:1569–1577, 2009.

[100] J. A. De Loera, J. Lee, S. Margulies, and S. Onn. Expressing combinatorial problems by systems of polynomial equations and Hilbert's Nullstellensatz. *Combin. Probab. Comput.*, 18(4):551–582, 2009. ISSN 0963-5483. doi: 10.1017/S0963548309009894. URL `http://dx.doi.org/10.1017/S0963548309009894`.

[101] J. A. De Loera, C. J. Hillar, P. N. Malkin, and M. Omar. Recognizing graph theoretic properties with polynomial ideals. *Electron. J. Combin.*, 17(1):Research Paper 114, 26, 2010. ISSN 1077-8926. URL `http://www.combinatorics.org/Volume_17/Abstracts/v17i1r114.html`.

[102] J. A. De Loera, J. Rambau, and F. Santos. *Triangulations: Structures for Algorithms and Applications*, volume 25 of *Algorithms and Computation in Mathematics*, 1st edition. Springer, Berlin, 2010. ISSN 1431-1550.

[103] J. A. De Loera, B. Dutra, M. Köppe, S. Moreinis, G. Pinto, and J. Wu. A users guide for latte integrale v1.5. Available from URL `http://www.math.ucdavis.edu/~latte/`, 2011.

[104] J. A. De Loera, J. Lee, P. N. Malkin, and S. Margulies. Computing infeasibility certificates for combinatorial problems through Hilbert's Nullstellensatz. *J. Symbol. Comput.*, 46(11):1260–1283, 2011. ISSN 0747-7171. doi: 10.1016/j.jsc.2011.08.007. URL `http://dx.doi.org/10.1016/j.jsc.2011.08.007`.

[105] J. A. De Loera, B. Dutra, M. Köppe, S. Moreinis, G. Pinto, and J. Wu. Software for exact integration of polynomials over polyhedra. eprint arXiv:1108.0117 [math.MG], 2012. *Comput. Geom. Theory Appl.*, to appear.

[106] J. A. De Loera, P. N. Malkin, and P. Parrilo. Computation with polynomial equations and inequalities arising in combinatorial optimization. In J. Lee and S. Leyffer, editors, *Nonlinear Mixed Integer Optimization*, volume 154 of *IMA Volumes in Mathematics and its Applications*, pages 447–482. Springer, New York, 2012. ISBN 1-461-41926-6.

[107] J. A. De Loera, B. Sturmfels, and C. Vinzant. The central curve in linear programming. *Found. Comput. Math.*, 12:509–540, 2012.

[108] J.-P. Dedieu, G. Malajovich, and M. Shub. On the curvature of the central path of linear programming theory. *Found. Comput. Math.*, 5(2):145–171, 2005. ISSN 1615-3375. doi: 10.1007/s10208-003-0116-8. URL `http://dx.doi.org/10.1007/s10208-003-0116-8`.

[109] J. Demmel, J. Nie, and V. Powers. Representations of positive polynomials on noncompact semialgebraic sets via KKT ideals. *J. Pure Appl. Algebra*, 209(1):189–200, 2007. ISSN 0022-4049. doi: 10.1016/j.jpaa.2006.05.028. URL `http://dx.doi.org/10.1016/j.jpaa.2006.05.028`.

[110] A. Deza, T. Terlaky, and Y. Zinchenko. Polytopes and arrangements: diameter and curvature. *Oper. Res. Lett.*, 36(2):215–222, 2008. ISSN 0167-6377. doi: 10.1016/j.orl.2007.06.007. URL `http://dx.doi.org/10.1016/j.orl.2007.06.007`.

[111] A. Deza, T. Terlaky, and Y. Zinchenko. A continuous d-step conjecture for polytopes. *Discrete Comput. Geom.*, 41(2):318–327, 2009. ISSN 0179-5376. doi: 10.1007/s00454-008-9096-4. URL `http://dx.doi.org/10.1007/s00454-008-9096-4`.

[112] A. Deza, T. Terlaky, and Y. Zinchenko. Central path curvature and iteration-complexity for redundant Klee-Minty cubes. In *Advances in Applied Mathematics and Global Optimization*, volume 17 of *Advances in Mechanics and Mathematics*, pages 223–256. Springer, New York, 2009. ISBN 0-387-75714-7.

[113] P. Diaconis and B. Sturmfels. Algebraic algorithms for sampling from conditional distributions. *Ann. Statist.*, 26:363–3097, 1998.

[114] P. Diaconis, R. L. Graham, and B. Sturmfels. Primitive partition identities. In D. Miklós, V. T. Sós, and D. Szőnyi, editors, *Combinatorics, Paul Erdős is Eighty*, volume 2 of *Bolyai Society Mathematical Studies*, pages 173–192. Budapest János Bolyai Mathematical Society, 1996. ISBN 9-638-02273-6.

[115] A. Dickenstein and I. Emiris, editors. *Solving Polynomial Equations: Foundations, Algorithms, and Applications*, volume 14 of *Algorithms and Computation in Mathematics*. Springer-Verlag, Heidelberg, 2005.

[116] L. A. Dupont and R. H. Villarreal. Algebraic and combinatorial properties of ideals and algebras of uniform clutters of TDI systems. *J. Comb. Optim.*, 21(3):269–292, 2011. ISSN 1382-6905. doi: 10.1007/s10878-009-9244-7. URL `http://dx.doi.org/10.1007/s10878-009-9244-7`.

[117] M. Dyer and C. Greenhill. On Markov chains for independent sets. *J. Algorithms*, 35:17–49, 2000.

[118] M. Dyer and R. Kannan. On Barvinok's algorithm for counting lattice points in fixed dimension. *Math. Oper. Res.*, 22:545–549, 1997.

[119] M. Ehrgott and X. Gandibleux. A survey and annotated bibliography of multiobjective combinatorial optimization. *OR Spektrum*, 22:425–460, 2000.

[120] F. Eisenbrand. On the membership problem for the elementary closure of a polyhedron. *Combinatorica*, 19(2):297–300, 1999. ISSN 0209-9683. doi: 10.1007/s004930050057. URL `http://dx.doi.org/10.1007/s004930050057`.

[121] F. Eisenbrand. Fast integer programming in fixed dimension. In *Algorithms—ESA 2003*, volume 2832 of *Lecture Notes in Comput. Sci.*, pages 196–207. Springer, Berlin, 2003. doi: 10.1007/978-3-540-39658-1_20. URL `http://dx.doi.org/10.1007/978-3-540-39658-1_20`.

[122] F. Eisenbrand. Integer programming and algorithmic geometry of numbers. In M. Jünger, T. Liebling, D. Naddef, W. Pulleyblank, G. Reinelt, G. Rinaldi, and L. Wolsey, editors, 50 *Years of Integer Programming* 1958–2008, pages 505–560. Springer-Verlag, 2010. ISBN 978-3-540-68274-5.

[123] F. Eisenbrand and G. Shmonin. Parametric integer programming in fixed dimension. *Math. Oper. Res.*, 33(4):839–850, 2008. doi: 10.1287/moor.1080.0320. URL `http://mor.journal.informs.org/content/33/4/839.abstract`.

[124] F. Eisenbrand, N. Hähnle, A. Razborov, and T. Rothvoß. Diameter of polyhedra: limits of abstraction. *Math. Oper. Res.*, 35(4):786–794, 2010.

[125] S. Eliahou. An algebraic criterion for a graph to be four-colourable. In *International Seminar on Algebra and Its Applications (Spanish) (México City,* 1991*)*, volume 6 of *Aportaciones Matemáticas. Notas de Investigación*, pages 3–27. Sociedad. Matematica. Mexicana, México City, 1992. ISBN 9-083-62796-X.

[126] V. A. Emelichev and V. A. Perepelitsa. On the cardinality of the set of alternatives in discrete many-criterion problems. *Discrete Math. Appl.*, 2:461–471, 1992.

[127] J. Figueira, S. Greco, and M. Ehrgott, editors. *Multiple Criteria Decision Analysis. State of the Art Surveys.* Springer, New York, 2005. ISBN 0-387-23067-X.

[128] K. Fischer. Symmetric polynomials and Hall's theorem. *Disc. Math.*, 69(3):225–234, 1988. ISSN 0012-365X.

[129] A. Frank and É. Tardos. An application of simultaneous Diophantine approximation in combinatorial optimization. *Combinatorica*, 7(1):49–65, 1987. ISSN 0209-9683. doi: 10.1007/BF02579200. URL `http://dx.doi.org/10.1007/BF02579200`.

[130] K. Fukuda, S. Moriyama, and Y. Okamoto. The Holt-Klee condition for oriented matroids. *European J. Combin.*, 30(8):1854–1867, 2009. ISSN 0195-6698. doi: 10.1016/j.ejc.2008.12.012. URL `http://dx.doi.org/10.1016/j.ejc.2008.12.012`.

[131] M. R. Garey and D. S. Johnson. *Computers and Intractibility: A Guide to the Theory of NP-completeness.* W. H. Freeman and Co., San Francisco, 1979. ISBN 0-716-71044-7.

[132] B. Gärtner. The random-facet simplex algorithm on combinatorial cubes. *Random Struct. Algor.*, 20(3):353–381, 2002. ISSN 1042-9832. doi: 10.1002/rsa.10034. URL http://dx.doi.org/10.1002/rsa.10034. Probabilistic methods in combinatorial optimization.

[133] B. Gärtner, J. Matoušek, L. Rüst, and P. Škovroň. Violator spaces: structure and algorithms. *Disc. Appl. Math.*, 156(11):2124–2141, 2008. ISSN 0166-218X. doi: 10.1016/j.dam.2007.08.048. URL http://dx.doi.org/10.1016/j.dam.2007.08.048.

[134] J. Gathen and J. Gerhard. *Modern Computer Algebra* 2nd edition. Cambridge University Press, Cambridge, 1999. ISBN 0-521-64176-4.

[135] I. Gitler and R. H. Villarreal. *Graphs, Rings and Polyhedra*, volume 35 of *Aportaciones Mat. Textos*. Sociedad Matemáticas Mexicana, Mexico, 2011. ISBN 978-607-02-1630-5.

[136] I. Gitler, E. Reyes, and R. H. Villarreal. Blowup algebras of square-free monomial ideals and some links to combinatorial optimization problems. *Rocky Mountain J. Math.*, 39(1):71–102, 2009. ISSN 0035-7596. doi: 10.1216/RMJ-2009-39-1-71. URL http://dx.doi.org/10.1216/RMJ-2009-39-1-71.

[137] R. Gollmer, U. Gotzes, and R. Schultz. A note on second-order stochastic dominance constraints induced by mixed-integer linear recourse. *Math. Program.*, 126:179–190, 2011.

[138] R. E. Gomory. An algorithm for integer solutions to linear programs. In *Recent Advances in Mathematical Programming*, pages 269–302. McGraw-Hill, New York, 1963.

[139] R. E. Gomory. On the relation between integer and non-integer solutions to linear programs. *Proc. Nat. Acad. Sci.*, 53:260–265, 1965.

[140] R. E. Gomory. Some polyhedra related to combinatorial problems. *Linear Algebra Appl.*, 2:451–458, 1969.

[141] J. Gouveia, P. A. Parrilo, and R. R. Thomas. Theta bodies for polynomial ideals. *SIAM J. Optim.*, 20(4):2097–2118, 2010. ISSN 1052-6234. doi: 10.1137/090746525. URL http://dx.doi.org/10.1137/090746525.

[142] J. Gouveia, M. Laurent, P. Parrilo, and R. Thomas. A new semidefinite programming hierarchy for cycles in binary matroids and cuts in graphs. *Math. Program.*, 133:203–225, 2012. ISSN 0025-5610. doi: 10.1007/s10107-010-0425-z. URL http://dx.doi.org/10.1007/s10107-010-0425-z.

[143] J. P. Gram. Om rumvinklerne i et polyeder. *Tidsskrift Math. (Copenhagen)*, 3(4):161–163, 1874.

[144] J. E. Graver. On the foundations of linear and integer linear programming I. *Math. Program.*, 8:207–226, 1975.

[145] M. Grötschel, L. Lovász, and A. Schrijver. *Geometric Algorithms and Combinatorial Optimization*, volume 2 of *Algorithms and Combinatorics*. Springer, Berlin, 1988.

[146] P. M. Gruber and C. G. Lekkerkerker. *Geometry of numbers*, 2nd edition, volume 37 of *North-Holland Mathematical Library*. North-Holland Publishing Co., Amsterdam, 1987. ISBN 0-444-70152-4.

[147] W. Habicht. Über die Zerlegung strikte definiter Formen in Quadrate. *Comment. Math. Helv.*, 12:317–322, 1940. ISSN 0010-2571.

[148] G. H. Hardy, J. E. Littlewood, and G. Pólya. *Inequalities*. Cambridge Mathematical Library. Cambridge University Press, Cambridge, 1988. ISBN 0-521-35880-9. Reprint of the 1952 edition.

[149] M. E. Hartmann. *Cutting Planes and the Complexity of the Integer Hull*. Ph.D. thesis, Cornell University, Department of Operations Research and Industrial Engineering, Ithaca, NY, 1989.

[150] N. J. A. Harvey. Algebraic algorithms for matching and matroid problems. *SIAM J. Comput.*, 39(2):679–702, 2009. ISSN 0097-5397. doi: 10.1137/070684008. URL http://dx.doi.org/10.1137/070684008.

[151] A. S. Hayes and D. C. Larman. The vertices of the knapsack polytope. *Discrete Appl. Math.*, 6:135–138, 1983.

[152] S. Heinz. Complexity of integer quasiconvex polynomial optimization. *J. Complex.*, 21(4):543–556, 2005. ISSN 0885-064X. doi: http://dx.doi.org/10.1016/j.jco.2005.04.004.

[153] R. Hemmecke. On the computation of Hilbert bases of cones. In *Mathematical Software, ICMS* 2002, pp. 307–317. World Scientific, Hackensack, NJ, 2002. ISBN 9-812-38048-5.

[154] R. Hemmecke. Test sets for integer programs with $\mathbb{Z}$-convex objective. eprint arXiv:math/0309154, 2003.

[155] R. Hemmecke and P. N. Malkin. Computing generating sets of lattice ideals and Markov bases of lattices. *J. Symbol. Comput.*, 44:1463–1476, 2009.

[156] R. Hemmecke and K. A. Nairn. On the Gröbner complexity of matrices. *J. Pure Appl. Algebra*, 213:1558–1563, 2009.

[157] R. Hemmecke and R. Schultz. Decomposition of test sets in stochastic integer programming. *Math. Program. Ser. B*, 94(2–3):323–341, 2003.

[158] R. Hemmecke and R. Weismantel. Representation of sets of lattice points. *SIAM J. Optim.*, 18:133–137, 2007.

[159] R. Hemmecke, M. Köppe, and R. Weismantel. A polynomial-time algorithm for optimizing over N-fold 4-block decomposable integer programs. In F. Eisenbrand and F. B. Shepherd, editors, *Integer Programming and Combinatorial Optimization*, volume 6080 of *Lecture Notes in Computer Science*, pages 219–229. Springer, Berlin, Heidelberg, 2010. doi: 10.1007/978-3-642-13036-6_17. ISBN 3-642-13036-4.

[160] R. Hemmecke, S. Onn, and R. Weismantel. A polynomial oracle-time algorithm for convex integer minimization. *Math. Program. Ser. A*, 126:97–117, 2011. ISSN 0025-5610. doi: 10.1007/s10107-009-0276-7. URL http://dx.doi.org/10.1007/s10107-009-0276-7.

[161] R. Hemmecke, M. Köppe, and R. Weismantel. Graver basis and proximity techniques for block-structured separable convex integer minimization problems. eprint arXiv:1207.1149v1, 2012.

[162] M. Henk and R. Weismantel. Test sets of the knapsack problem and simultaneous Diophantine approximation. In R. Burkard and G. J. Woeginger, editors, *Proceedings of the 5th European Symposium on Algorithms*, pages 92–103, Springer, Berlin, 1997. ISBN 3-540-63397-9.

[163] M. Henk, M. Köppe, and R. Weismantel. Integral decomposition of polyhedra and some applications in mixed integer programming. *Math. Program. Ser. B*, 94(2–3): 193–206, 2003. doi: 10.1007/s10107-002-0315-0.

[164] P. Henrici. *Applied and Computational Complex Analysis. Vol. 1. Power Series—Integration—Conformal Mapping—Location of Zeros*, Wiley Classics Library. John Wiley & Sons, New York, 1988. ISBN 0-471-60841-6. Reprint of the 1974 original. A Wiley-Interscience Publication.

[165] D. Henrion and J. Lasserre. Detecting global optimality and extracting solutions in GloptiPoly. In *Positive Polynomials in Control*, volume 312 of *Lecture Notes in Control and Inform. Sci.*, pages 293–310. Springer, Berlin, 2005.

[166] D. Henrion and J.-B. Lasserre. GloptiPoly: global optimization over polynomials with MATLAB and SeDuMi. *ACM Trans. Math. Software*, 29(2):165–194, 2003.

[167] J. L. Higle and S. Sen. *Stochastic Decomposition: A Statistical Method for Large Scale Stochastic Linear Programming*. Kluwer, Dordrecht, 1996.

[168] R. Hildebrand and M. Köppe. A new Lenstra-type algorithm for quasiconvex polynomial integer minimization with complexity $2^{O(n \log n)}$. eprint arXiv:1006.4661 [math.OC], 2012.

[169] C. Hillar and T. Windfeldt. An algebraic characterization of uniquely vertex colorable graphs. *J. Combin. Theory Ser. B*, 98:400–414, 2008.

[170] D. S. Hochbaum and J. G. Shanthikumar. Convex separable optimization is not much harder than linear optimization. *J. ACM*, 37:843–862, October 1990. ISSN 0004-5411. doi: http://doi.acm.org/10.1145/96559.96597. URL `http://doi.acm.org/10.1145/96559.96597`.

[171] S. Hoşten and R. R. Thomas. Standard pairs and group relaxations in integer programming. *J. Pure Appl. Algebra*, 139:133–157, 1999.

[172] S. Hoşten and R. R. Thomas. The associated primes of initial ideals of lattice ideals. *Math. Res. Lett.*, 6:83–97, 1999.

[173] S. Hoşten and R. R. Thomas. Gomory integer programs. *Math. Program. Ser. B*, 96: 271–292, 2003.

[174] R. A. Horn and C. R. Johnson. *Matrix Analysis*. Cambridge University Press, Cambridge, UK, 1990. ISBN 0-521-38632-2. Corrected reprint of the 1985 original.

[175] S. Hoşten and B. Sturmfels. An implementation of Gröbner bases for integer programming. In E. Balas and J. Clausen, editors, *Proceedings of the 4th Conference on Integer Programming and Combinatorial Optimization*, pages 267–276, Springer, New York, 1995. ISBN 3-540-59408-6.

[176] S. Hoşten and B. Sturmfels. Computing the integer programming gap. *Combinatorica*, 27(3):367–382, 2007. ISSN 0209-9683. doi: 10.1007/s00493-007-2057-3. URL `http://dx.doi.org/10.1007/s00493-007-2057-3`.

[177] S. Hoşten and S. Sullivant. Finiteness theorems for Markov bases of hierarchical models. *J. Combin. Theory Ser. A*, 114:311–321, 2007.

[178] S. Hoşten and R. Thomas. Gröbner bases and integer programming. In *Gröbner bases and applications (Linz, 1998)*, volume 251 of *London Mathematical Society Lecture Note Series*, pages 144–158. Cambridge University Press, Cambridge, 1998. ISBN 0-521-63298-6.

[179] M. N. Huxley. Integer points in plane regions and exponential sums. In *Number Theory*, Trends in Mathematics, pages 157–166. Birkhäuser, Basel, 2000. ISBN 3-764-36259-6.

[180] P. Impagliazzo, P. Pudlák, and J. Sgall. Lower bounds for the polynomial calculus and the Gröbner basis algorithm. *Computat. Complex.*, 8:127–144, 1999.

[181] H. Isermann. Proper efficiency and the linear vector maximum problem. *Oper. Res.*, 22:189–191, 1974.

[182] T. Jacobi and A. Prestel. Distinguished representations of strictly positive polynomials. *J. Reine Angew. Math.*, 532:223–235, 2001. doi: 10.1515/crll.2001.023. URL `http://dx.doi.org/10.1515/crll.2001.023`.

[183] D. S. Johnson, M. Yannakakis, and Ch. H. Papadimitriou. On generating all maximal independent sets. *Inform. Proc. Lett.*, 27:119–123, 1988.

[184] G. Kalai. Upper bounds for the diameter and height of graphs of convex polyhedra. *Discrete Comput. Geom.*, 8(4):363–372, 1992. ISSN 0179-5376. doi: 10.1007/BF02293053. URL `http://dx.doi.org/10.1007/BF02293053`.

[185] G. Kalai and D. J. Kleitman. A quasi-polynomial bound for the diameter of graphs of polyhedra. *Bull. Amer. Math. Soc. (N.S.)*, 26(2):315–316, 1992. ISSN 0273-0979. doi: 10.1090/S0273-0979-1992-00285-9. URL `http://dx.doi.org/10.1090/S0273-0979-1992-00285-9`.

[186] P. Kall and S. W. Wallace. *Stochastic Programming*. Wiley, Chichester, 1994. ISBN 0-471-95108-0.

[187] R. Kannan. Improved algorithms for integer programming and related lattice problems. In *STOC '83: Proceedings of the Fifteenth Annual ACM Symposium on Theory of Computing*, pages 193–206, ACM, New York, 1983. ISBN 0-89791-099-0. doi: 10.1145/800061.808749.

[188] R. Kannan. Minkowski's convex body theorem and integer programming. *Math. Oper. Res.*, 12(3):415–440, 1987. ISSN 0364-765X. doi: 10.1287/moor.12.3.415.

[189] R. Kannan. Lattice translates of a polytope and the Frobenius problem. *Combinatorica*, 12(2):161–177, 1992.

[190] R. Kannan and A. Bachem. Polynomial algorithms for computing the Smith and Hermite normal forms of an integer matrix. *SIAM J. Comput.*, 8(4):499–507, November 1979.

[191] A. Kehrein and M. Kreuzer. Characterizations of border bases. *J. Pure Appl. Algebra*, 196(2–3):251–270, 2005. ISSN 0022-4049.

[192] A. Kehrein, M. Kreuzer, and L. Robbiano. An algebraist's view on border bases. In A. Dickenstein and I. Emiris, editors, *Solving Polynomial Equations: Foundations, Algorithms, and Applications*, volume 14 of *Algorithms and Computation in Mathematics*, chapter 4, pages 160–202. Springer-Verlag, Heidelberg, 2005. ISBN 3-540-24326-7.

[193] L. G. Khachiyan. A polynomial algorithm in linear programming. *Dokl. Akad. Nauk SSSR*, 244(5):1093–1096, 1979. ISSN 0002-3264.

[194] L. G. Khachiyan. Convexity and complexity in polynomial programming. In Z. Ciesielski and C. Olech, editors, *Proceedings of the International Congress of Mathematicians, August* 16–24, 1983, *Warszawa*, pages 1569–1577, North-Holland, New York, 1984. ISBN 0-444-86659-0.

[195] A. Khinchin. A quantitative formulation of Kronecker's theory of approximation (in Russian). *Izv. Akad. Nauk SSR Ser. Mat.*, 12:113–122, 1948.

[196] A. G. Khovanskiĭ. *Fewnomials*, volume 88 of *Translations of Mathematical Monographs*. American Mathematical Society, Providence, RI, 1991. ISBN 0-8218-4547-0. Translated from the Russian by Smilka Zdravkovska.

[197] V. Klee and P. Kleinschmidt. The d-step conjecture and its relatives. *Math. Oper. Res.*, 12(4):718–755, 1987. ISSN 0364-765X. doi: 10.1287/moor.12.4.718. URL `http://dx.doi.org/10.1287/moor.12.4.718`.

[198] M. Kochol. Constructive approximation of a ball by polytopes. *Math. Slovaca*, 44 (1):99–105, 1994. URL `http://dml.cz/dmlcz/136602`.

[199] J. Kollár. Sharp effective Nullstellensatz. *J. Amer. Math. Soc.*, 1(4):963–975, 1988.

[200] M. Köppe. A primal Barvinok algorithm based on irrational decompositions. *SIAM J. Discrete Math.*, 21(1):220–236, 2007. doi: 10.1137/060664768.

[201] M. Köppe. LattE macchiato, version 1.2-mk-0.9.3, an improved version of De Loera et al.'s LattE program for counting integer points in polyhedra with variants of Barvinok's algorithm. URL `http://www.math.ucdavis.edu/~mkoeppe/latte/`, 2008.

[202] M. Köppe. On the complexity of nonlinear mixed-integer optimization. In *Mixed-Integer Nonlinear Optimization*, volume 154 of *IMA Volumes in Mathematics and its Applications*, pages 533–557. Springer, New York, 2011. ISBN 978-1-4614-1926-6.

[203] M. Köppe and S. Verdoolaege. Computing parametric rational generating functions with a primal Barvinok algorithm. *Electronic J. Combin.*, 15:1–19, 2008. Article #R16.

[204] M. Köppe, S. Verdoolaege, and K. M. Woods. An implementation of the Barvinok–Woods integer projection algorithm. In Information Theory and Statistical Learning (ITSL 2008), Las Vegas, CSREA Press, 2008. ISBN 1-601-32079-5.

[205] M. Köppe, C. T. Ryan, and M. Queyranne. Rational generating functions and integer programming games. *Oper. Res.*, 59(6):1445–1460, November/December 2011. doi: 10.1287/opre.1110.0964. URL `http://or.journal.informs.org/content/59/6/1445.abstract`.

[206] B. Korte and J. Vygen. *Combinatorial Optimization, Theory and Algorithms*, 5th edition, volume 21 of *Algorithms and Combinatorics*. Springer, Heidelberg, 2012. ISBN 978-3-642-24487-2. doi: 10.1007/978-3-642-24488-9. URL `http://dx.doi.org/10.1007/978-3-642-24488-9`.

[207] M. Kreuzer and L. Robbiano. *Computational Commutative Algebra 1*. Springer, Berlin/Heidelberg, 2000. ISBN 3-540-67733-X.

[208] M. Kreuzer and L. Robbiano. *Computational Commutative Algebra 2*. Springer, Berlin/Heidelberg, 2005. ISBN 3-540-28296-3.

[209] B. Krishnamoorthy and G. Pataki. Column basis reduction and decomposable knapsack problems. *Discrete Optim.*, 6(3):242–270, 2009. ISSN 1572-5286. doi: 10.1016/j.disopt.2009.01.003. URL `http://dx.doi.org/10.1016/j.disopt.2009.01.003`.

[210] T. Kudo and A. Takemura. A lower bound for the Graver complexity of the incidence matrix of a complete bipartite graph. eprint arXiv:1102.4674v1, 2011.

[211] D. G. Larman. Paths of polytopes. *Proc. London Math. Soc.*, 20(3):161–178, 1970. ISSN 0024-6115.

[212] J. B. Lasserre. Global optimization with polynomials and the problem of moments. *SIAM J. Optim.*, 11(3):796–817 (electronic), 2001. ISSN 1052-6234.

[213] J. B. Lasserre. Semidefinite programming vs. LP relaxations for polynomial programming. *Math. Oper. Res.*, 27(2):347–360, 2002. ISSN 0364-765X. doi: 10.1287/moor.27.2.347.322. URL `http://dx.doi.org/10.1287/moor.27.2.347.322`.

[214] J. B. Lasserre. Integer programming, Barvinok's counting algorithm and Gomory relaxations. *Oper. Res. Letters*, 32:133–137, 2003.

[215] J. B. Lasserre. Generating functions and duality for integer programs. *Discrete Optim.*, 1:167–187, 2004.

[216] J. B. Lasserre. A discrete Farkas lemma. *Disc. Optim.*, 1(1):67–75, 2004. ISSN 1572-5286. doi: 10.1016/j.disopt.2004.04.002. URL `http://dx.doi.org/10.1016/j.disopt.2004.04.002`.

[217] J. B. Lasserre. *Linear and Integer Programming vs. Linear Integration and Counting. A Duality Viewpoint*. Springer Series in Operations Research and Financial Engineering. Springer, New York, 2009. ISBN 978-0-387-09413-7. doi: 10.1007/978-0-387-09414-4. URL `http://dx.doi.org/10.1007/978-0-387-09414-4`.

[218] J. B. Lasserre. *Moments, Positive Polynomials and their Applications*, volume 1 of *Imperial College Press Optimization Series*. Imperial College Press, London, 2010. ISBN 978-1-84816-445-1; 1-84816-445-9.

[219] J. B. Lasserre, M. Laurent, and P. Rostalski. Semidefinite characterization and computation of zero-dimensional real radical ideals. *Found. Comput. Math.*, 8(5):607–647, 2008. ISSN 1615-3375.

[220] J. B. Lasserre, M. Laurent, and P. Rostalski. A unified approach to computing real and complex zeros of zero-dimensional ideals. In M. Putinar and S. Sullivant, editors, *Emerging Applications of Algebraic Geometry*, volume 149 of *IMA Volumes in Mathematics and its Applications*, pages 125–155. Springer, New York, 2009. ISBN 0-387-09685-X.

[221] J. B. Lasserre and E. Zeron. Counting integral points in a convex rational polytope. *Math. Oper. Res.*, 28:853–870, 2003.

[222] M. Laurent. Semidefinite representations for finite varieties. *Math. Program. Ser. A*, 109:1–26, 2007. ISSN 0025-5610.

[223] M. Laurent. Sums of squares, moment matrices and optimization over polynomials. In *Emerging Applications of Algebraic Geometry*, volume 149 of *IMA Volumes in Mathematics and its Applications*, pages 157–270. Springer, New York, 2009. ISBN 0-387-09685-X.

[224] J. Lawrence. Polytope volume computation. *Math. Comp.*, 57(195):259–271, 1991.

[225] J. Lawrence. Rational-function-valued valuations on polyhedra. In *Discrete and Computational Geometry (New Brunswick, NJ, 1989/1990)*, volume 6 of *DIMACS Series in Discrete Mathematics and Theoretical Computer Science*, pages 199–208. American Mathematical Society, Providence, RI, 1991. ISBN 0-821-86595-1.

[226] D. Lazard. Algèbre linéaire sur $\mathbb{K}[x_1, \ldots, x_n]$ et élimination. *Bulletin SMF*, 105: 165–190, 1977.

[227] J. Lee, S. Onn, and R. Weismantel. On test sets for nonlinear integer maximization. *Oper. Res. Lett.*, 36:439–443, 2008.

[228] J. Lee, S. Onn, and R. Weismantel. Nonlinear optimization over a weighted independence system. In A. Goldberg and Y. Zhou, editors, *Algorithmic Aspects in Information and Management*, volume 5564 of *Lecture Notes in Computer Science*, pages 251–264. Springer-Verlag, Berlin, 2009. doi: 10.1007/978-3-642-02158-9_22.

[229] J. Lee, S. Onn, L. Romanchuk, and R. Weismantel. The quadratic Graver cone, quadratic integer minimization, and extensions. *Math. Program Ser. B*, published online 13 October 2012. doi: 10.1007/s10107-012-0605-0.

[230] A. K. Lenstra, H. W. Lenstra, Jr., and L. Lovász. Factoring polynomials with rational coefficients. *Math. Ann.*, 261(4):515–534, 1982. ISSN 0025-5831. doi: 10.1007/BF01457454. URL `http://dx.doi.org/10.1007/BF01457454`.

[231] H. W. Lenstra, Jr. Integer programming with a fixed number of variables. *Math. Oper. Res.*, 8:538–548, 1983.

[232] Q. Li, Y. Guo, T. Ida, and J. Darlington. The minimised geometric Buchberger algorithm: an optimal algebraic algorithm for integer programming. In *Proceedings of the 1997 International Symposium on Symbolic and Algebraic Computation*, pages 331–338. ACM Press, New York, 1997. ISBN 0-897-91875-4.

[233] J. Löfberg. YALMIP: A toolbox for modeling and optimization in MATLAB, In *IEEE International Symposium on Computer-Aided Control Systems Design*, Taipei, Taiwan, 2004, page 284. ISBN 0-780-38635-3. URL `http://control.ee.ethz.ch/~joloef/yalmip.php`.

[234] H. Lombardi. Une borne sur les degrés pour le théorème des zéros réel effectif. In *Real Algebraic Geometry (Rennes, 1991)*, volume 1524 of *Lecture Notes in Math.*, pages 323–345. Springer, Berlin, 1992. ISBN 3-540-55992-2. doi: 10.1007/BFb0084631. URL `http://dx.doi.org/10.1007/BFb0084631`.

[235] L. Lovász. On determinants, matchings, and random algorithms. In *Fundamentals of computation theory: Proceedings of the Conference on Algebraic, Arithmetic, and Categorical Methods in Computation Theory, Berlin/Wendisch-Rietz, 1979*, volume 2 of *Mathematical Research*, pages 565–574. Akademie-Verlag, Berlin, 1979.

[236] L. Lovász. Stable sets and polynomials. *Discrete Math.*, 124(1–3):137–153, 1994. ISSN 0012-365X.

[237] L. Lovász and A. Schrijver. Cones of matrices and set-functions and 0-1 optimization. *SIAM J. Optim.*, 1(2):166–190, 1991. ISSN 1052-6234. doi: 10.1137/0801013. URL `http://dx.doi.org/10.1137/0801013`.

[238] D. Maclagan. Antichains of monomial ideals are finite. *Proc. Amer. Math. Soc.*, 129 (6):1609–1615, 2001.

[239] P. Mani and D. W. Walkup. A 3-sphere counterexample to the W_v-path conjecture. *Math. Oper. Res.*, 5(4):595–598, 1980. ISSN 0364-765X. doi: 10.1287/moor.5.4.595. URL `http://dx.doi.org/10.1287/moor.5.4.595`.

[240] S. Margulies. *Computer Algebra, Combinatorics, and Complexity: Hilbert's Nullstellensatz and NP-Complete Problems*. Ph.D. thesis, UC Davis, 2008.

[241] M. Marshall. *Positive polynomials and sums of squares*, volume 146 of *Mathematical Surveys and Monographs*. American Mathematical Society, Providence, RI, 2008. ISBN 13: 978-0-8218-4402-1; ISBN 10: 0-8218-4402-4.

[242] Yu. V. Matiyasevich. Enumerable sets are Diophantine. *Dokl. Akad. Nauk SSSR*, 191:279–282, 1970. (Russian); English translation in Soviet Math. Dokl., 11: 354–357, 1970.

[243] Yu. V. Matiyasevich. A criteria for colorability of vertices stated in terms of edge orientations. *Discrete Anal.*, 26:65–71, 1974.

[244] Yu. V. Matiyasevich. *Hilbert's tenth problem*. The MIT Press, Cambridge, MA, USA, 1993. ISBN 0-262-13295-8.

[245] Yu. V. Matiyasevich. Some algebraic methods for calculation of the number of colorings of a graph. *Zapiski Nauchnykh Seminarov POMI*, 293:193–205, 2001.

[246] J. Matoušek and T. Szabó. RANDOM EDGE can be exponential on abstract cubes. *Adv. Math.*, 204(1):262–277, 2006. ISSN 0001-8708. doi: 10.1016/j.aim.2005.05.021. URL http://dx.doi.org/10.1016/j.aim.2005.05.021.

[247] D. Micciancio. The shortest vector in a lattice is hard to approximate to within some constant. *SIAM J. Comput.*, 30(6):2008–2035, 2001. doi: 10.1137/S0097539700373039.

[248] D. Micciancio and S. Goldwasser. *Complexity of Lattice Problems. A Cryptographic Perspective.* The Kluwer International Series in Engineering and Computer Science, volume 671. Kluwer Academic Publishers, Boston, MA, 2002. ISBN 0-7923-7688-9. doi: 10.1007/978-1-4615-0897-7. URL http://dx.doi.org/10.1007/978-1-4615-0897-7.

[249] D. Micciancio and P. Voulgaris. Faster exponential time algorithms for the shortest vector problem. In *Proceedings of the* 21*st ACM-SIAM Symposium on Discrete Algorithms (SODA)*, SIAM, Philadelphia. ACM, New York, 2010. ISBN 0-898-71698-5.

[250] D. Micciancio and P. Voulgaris. A deterministic single exponential time algorithm for most lattice problems based on Voronoi cell computations. In *Proceedings of the* 42*nd ACM Symposium on Theory of Computing STOC* 2010, pages 351–358, 2010. ISBN 978-1-4503-0050-6. doi: 10.1145/1806689.1806739. URL http://doi.acm.org/10.1145/1806689.1806739.

[251] J. Mihalisin and V. Klee. Convex and linear orientations of polytopal graphs. *Disc. Comput. Geom.*, 24(2-3):421–435, 2000. ISSN 0179-5376. doi: 10.1007/s004540010046. URL http://dx.doi.org/10.1007/s004540010046.

[252] P. Mirchandani and H. Soroush. The stochastic multicommodity flow problem. *Networks*, 20:121–155, 1990.

[253] M. Mnuk. Representing graph properties by polynomial ideals. In *Computer Algebra in Scientific Computing (ASC* 2001*)*, Proceedings of the Fourth International Workshop on Computer Algebra in Scientific Computing, Konstanz, Springer-Verlag, Berlin, pages 431–444, 2001.

[254] R. D. C. Monteiro and T. Tsuchiya. A strong bound on the integral of the central path curvature and its relationship with the iteration-complexity of primal-dual path-following LP algorithms. *Math. Program. Ser. A*, 115(1):105–149, 2008. ISSN 0025-5610. doi: 10.1007/s10107-007-0141-5. URL http://dx.doi.org/10.1007/s10107-007-0141-5.

[255] T. S. Motzkin. The arithmetic-geometric inequality. In *Inequalities (Proc. Sympos. Wright-Patterson Air Force Base, Ohio, 1965)*, pages 205–224. Academic Press, New York, 1967.

[256] B. Mourrain. A new criterion for normal form algorithms. In 13*th International Symposium AAECC*-13, volume 1719 of *Lecture Notes in Comput. Sci.*, pages 430–443. Springer, New York, 1999. ISBN 3-540-66723-7.

[257] B. Mourrain and P. Trébuchet. Stable normal forms for polynomial system solving. *Theoret. Comput. Sci.*, 409(2):229–240, 2008. ISSN 0304-3975.

[258] K. Murota, H. Saito, and R. Weismantel. Optimality criterion for a class of nonlinear integer programs. *Oper. Res. Lett.*, 32:468–472, 2004.

[259] G. L. Nemhauser and L. A. Wolsey. *Integer and Combinatorial Optimization*. Wiley, Chichester, 1988. ISBN 0-471-82819-X.

[260] Yu. Nesterov. Squared functional systems and optimization problems. In J. B. G. Frenk et al., editors, *High Performance Optimization*, pages 405–440. Kluwer Academic, Boston, 2000. ISBN 0-792-36013-3.

[261] Yu. Nesterov. Fast Fourier Transform and its applications to integer knapsack problems. CORE Discussion Paper No. 2004/64, Catholic University of Louvain (UCL), Center for Operations Research and Econometrics (CORE), September 2004.

[262] J. Nie and M. Schweighofer. On the complexity of Putinar's Positivstellensatz. *J. Complex.*, 23(1):135–150, 2007. ISSN 0885-064X. doi: 10.1016/j.jco.2006.07.002. URL `http://dx.doi.org/10.1016/j.jco.2006.07.002`.

[263] C. D. Olds, A. Lax, and G. P. Davidoff. *The Geometry of Numbers*, volume 41 of *Anneli Lax New Mathematical Library*. Mathematical Association of America, Washington, DC, 2000. ISBN 0-88385-643-3. Appendix I by Peter D. Lax.

[264] S. Onn. Geometry, complexity and combinatorics of permutation polytopes, *J. Combin. Theory Ser. A*, 64:31–49, 1993.

[265] S. Onn. *Nonlinear Discrete Optimization: An Algorithmic Theory*. Zurich Lectures in Advanced Mathematics. European Mathematical Society, Zürich, Switzerland, 2010. ISBN 3-037-19093-0.

[266] S. Onn and U. G. Rothblum. Convex combinatorial optimization. *Discrete Comput. Geom.*, 32:549–566, 2004.

[267] E. O'Shea and A. Sebö. Characterizations of total dual integrality. In M. Fischetti and D. Williamson, editors, *Proceedings of the 12th Integer Programming and Combinatorial Optimization Conference (IPCO 2007)*, Lecture Notes in Computer Science, volume 4513, pages 1–15, Springer, Berlin/Heidelberg, 2007. ISBN 3-540-72792-2.

[268] I. Pak. On sampling integer points in polyhedra. In S. Smale, F. Cucker, and J. M. Rojas, editors, *Foundations of Computational Mathematics: Proceedings of the Smalefest 2000*, page 319–324. World Scientific, River Edge, NJ, 2002. ISBN 981-02-4845-8.

[269] P. M. Pardalos, editor. *Complexity in Numerical Optimization*. World Scientific, River Edge, NJ, 1993. ISBN 9-810-21415-4.

[270] P. A. Parrilo. *Structured Semidefinite Programs and Semialgebraic Geometry Methods in Robustness and Optimization*. Ph.D. thesis, California Institute of Technology, May 2000.

[271] P. A. Parrilo. Semidefinite programming relaxations for semialgebraic problems. *Math. Program. Ser. B*, 96:293–320, 2003. ISSN 0025-5610.

[272] I. Peeva and B. Sturmfels. Generic lattice ideals. *J. Amer. Math. Soc*, 11:363–373, 1998.

[273] R. Pemantle and M. Wilson. Asymptotics of multivariate sequences, Part I: smooth points of the singular variety. *J. Combin. Theory Ser. A*, 97:129–161, 2001.

[274] L. Pottier. Minimal solutions of linear Diophantine systems: Bounds and algorithms. In R. V. Book, editor, *Rewriting Techniques and Applications, Lecture Notes in Computer Science 488*, pages 162–173. Springer, Berlin, 1991. ISBN 3-540-53904-2.

[275] L. Pottier. The Euclidean algorithm in dimension *n*. In *Proceedings of the 1996 International Symposium on Symbolic and Algebraic Computation*, ISSAC '96, pages 40–42, ACM, New York, 1996. ISBN 0-89791-796-0. doi: 10.1145/236869.236894. URL `http://doi.acm.org/10.1145/236869.236894`.

[276] W. Powell and H. Topaloglu. Dynamic-programming approximations for stochastic time-staged integer multicommodity-flow problems. *INFORMS J. Comput.*, 18:31–42, 2006.

[277] V. Powers and B. Reznick. A new bound for Pólya's theorem with applications to polynomials positive on polyhedra. *J. Pure Appl. Algebra*, 164(1-2):221–229, 2001. ISSN 0022-4049. doi: 10.1016/S0022-4049(00)00155-9. URL `http://dx.doi.org/10.1016/S0022-4049(00)00155-9`. Effective methods in algebraic geometry (Bath, 2000).

[278] S. Prajna, A. Papachristodoulou, S. Seiler, and P. Parrilo. *SOSTOOLS: Sum of squares optimization toolbox for MATLAB*, 2004. URL `http://www.cds.caltech.edu/sostools`

[279] A. Prékopa. *Stochastic Programming*. Kluwer, Dordrecht, 1995. ISBN 0-792-33482-5.

[280] S. Provan and L. Billera. Decompositions of simplicial complexes related to diameters of convex polyhedra. *Math. Oper. Res.*, 5(4):576–594, 1980. ISSN 0364-765X. doi: 10.1287/moor.5.4.576. URL `http://dx.doi.org/10.1287/moor.5.4.576`.

[281] A. V. Pukhlikov and A. G. Khovanskii. A Riemann–Roch theorem for integrals and sums of quasi-polynomials over virtual polytopes. *St. Petersburg Math. J.*, 4(4): 789–812, 1993.

[282] M. Putinar. Positive polynomials on compact semi-algebraic sets. *Indiana Univ. Math. J.*, 42(3):969–984, 1993. doi: 10.1512/iumj.1993.42.42045. URL `http://dx.doi.org/10.1512/iumj.1993.42.42045`.

[283] B. Reznick. Uniform denominators in Hilbert's seventeenth problem. *Math. Z.*, 220 (1):75–97, 1995. ISSN 0025-5874. doi: 10.1007/BF02572604. URL `http://dx.doi.org/10.1007/BF02572604`.

[284] R. M. Robinson. Some definite polynomials which are not sums of squares of real polynomials. In *Selected questions of algebra and logic (collection dedicated to the memory of A. I. Mal'cev) (Russian)*, pages 264–282. Izdat. "Nauka" Sibirsk. Otdel., Novosibirsk, 1973.

[285] C. Roos, T. Terlaky, and J.-P. Vial. *Interior Point Methods for Linear Optimization*. Springer, New York, 2006. Revised edition of *Theory and Algorithms for Linear Optimization*, Wiley, Chichester, 1997. ISBN 978-0387-26378-6; 0-387-26378-0.

[286] A. Ruszczyński. Some advances in decomposition methods for stochastic linear programming. *Ann. Oper. Res.*, 85:153–172, 1999.

[287] C. Ryan, A. X. Jiang, and K. Leyton-Brown. Symmetric games with piecewise linear utilities. In *Proceedings of the Behavioral and Quantitative Game Theory: Conference on future directions*, BQGT '10, article 41, ACM, New York, 2010. ISBN 978-1-60558-919-0. doi: 10.1145/1807406.1807447. URL `http://doi.acm.org/10.1145/1807406.1807447`.

[288] F. Santos. A counterexample to the Hirsch conjecture. Ann. of Math., 176(1), 383–412 (2012).

[289] F. Santos and B. Sturmfels. Higher Lawrence configurations. *J. Combin. Theory Ser. A*, 10:151–164, 2003.

[290] Y. Sawaragi, H. Nakayama, and T. Tanino, editors. *Theory of Multiobjective Optimization*. Academic Press, Orlando, 1985. ISBN 0-126-20370-9.

[291] H. E. Scarf. Production sets with indivisibilities, Part I: Generalities. *Econometrica*, 49:1–32, 1981.

[292] H. E. Scarf. Production sets with indivisibilities, Part II: The case of two activities. *Econometrica*, 49(2):395–423, 1981.

[293] H. E. Scarf. Neighborhood systems for production sets with indivisibilities. *Econometrica*, 54:507–532, 1986.

[294] H. E. Scarf. Test sets for integer programs. *Math. Program. Ser. A*, 79:355–368, 1997.

[295] K. Schmüdgen. The K-moment problem for compact semi-algebraic sets. *Math. Ann.*, 289(2):203–206, 1991.

[296] A. Schrijver. *Theory of Linear and Integer Programming*. John Wiley, New York, 1986. ISBN 0-471-90854-1.

[297] A. Schrijver. *Combinatorial Optimization: Polyhedra and Efficiency*. Springer-Verlag, Berlin, 2003.

[298] A. Schrijver. *Combinatorial Optimization. Polyhedra and Efficiency. Vol. A*, volume 24 of *Algorithms and Combinatorics*. Springer-Verlag, Berlin, 2003. ISBN 3-540-44389-4. Paths, flows, matchings, Chapters 1–38.

[299] A. Schrijver. *Combinatorial Optimization. Polyhedra and Efficiency. Vol. B*, volume 24 of *Algorithms and Combinatorics*. Springer-Verlag, Berlin, 2003. ISBN 3-540-44389-4. Matroids, trees, stable sets, Chapters 39–69.

[300] A. Schrijver. *Combinatorial Optimization. Polyhedra and Efficiency. Vol. C*, volume 24 of *Algorithms and Combinatorics*. Springer-Verlag, Berlin, 2003. ISBN 3-540-44389-4. Disjoint paths, hypergraphs, Chapters 70–83.

[301] R. Schultz. On structure and stability in stochastic programs with random technology matrix and complete integer recourse. *Math. Program.*, 70:73–89, 1995.

[302] R. Schultz, L. Stougie, and M. H. van der Vlerk. Solving stochastic programs with integer recourse by enumeration: A framework using Gröbner basis reductions. *Math. Program. Ser. A*, 83:229–252, 1998.

[303] A. S. Schulz and R. Weismantel. An oracle–polynomial time augmentation algorithm for integer programming. In *Proceedings of the 10th Annual ACM–SIAM Symposium on Discrete Algorithms*, pages 967–968, SIAM, Philadelphia, ACM, New York, 1999. ISBN 0-898-71434-6.

[304] A. S. Schulz, R. Weismantel, and G. M. Ziegler. 0/1 integer programming: optimization and augmentation are equivalent. In P. Spirakis, editor, *Algorithms – ESA* 95, pages 473–483. Volume 979 of Lecture Notes in Computer Science, Springer, Berlin, 1995. ISBN 3-540-60313-1.

[305] M. Schweighofer. An algorithmic approach to Schmüdgen's Positivstellensatz. *J. Pure Appl. Algebra*, 166(3):307–319, 2002. ISSN 0022-4049. doi: 10.1016/S0022-4049(01)00041-X. URL http://dx.doi.org/10.1016/S0022-4049(01)00041-X.

[306] A. Sebö. Hilbert bases, Carathéodory's theorem and combinatorial optimization. In R. Kannan and W. R. Pulleyblank, editors, *Proceedings of the 1st Conference on Integer Programming and Combinatorial Optimization*, pages 431–455, University of Waterloo Press, Waterloo, Ontario. 1990. ISBN 0-888-98099-X.

[307] I. V. Sergienko and V. A. Perepelitsa. Finding the set of alternatives in discrete multi-criterion problems. *Cybernetics*, 3:673–683, 1991.

[308] P. D. Seymour. Decomposition of regular matroids. *J. Combin. Theory Ser. B*, 28: 305–359, 1980.

[309] G. C. Shephard. An elementary proof of Gram's theorem for convex polytopes. *Canad. J. Math.*, 19:1214–1217, 1967.

[310] H. D. Sherali and W. P. Adams. A hierarchy of relaxations between the continuous and convex hull representations for zero-one programming problems. *SIAM J. Discrete Math.*, 3(3):411–430, 1990. ISSN 0895-4801.

[311] V. N. Shevchenko. On the number of extreme points in linear programming. *Kibernetika*, 2:133–134, 1981. In Russian.

[312] N. Shor. Class of global minimum bounds of polynomial functions. *Cybernetics*, 23 (6):731–734, 1987.

[313] G. Sonnevend, J. Stoer, and G. Zhao. On the complexity of following the central path of linear programs by linear extrapolation. II. *Math. Program. Ser. B*, 52(3):527–553 (1992), 1991. ISSN 0025-5610. doi: 10.1007/BF01582904. URL http://dx.doi.org/10.1007/BF01582904.

[314] R. P. Stanley. *Enumerative Combinatorics*, volume I. Cambridge University Press, Cambridge, UK, 1997. ISBN 0-521-66351-2.

[315] G. Stengle. A Nullstellensatz and a Positivstellensatz in semialgebraic geometry. *Math. Ann.*, 207:87–97, 1974. ISSN 0025-5831.

[316] H. J. Stetter. *Numerical Polynomial Algebra*. SIAM, Philadelphia, PA, 2004. ISBN 0-89871-557-1.

[317] B. Sturmfels. *Gröbner Bases and Convex Polytopes*. AMS, Providence, RI, 1996. ISBN 0-821-80487-1.

[318] B. Sturmfels. On vector partition functions. *J. Combin. Theory Ser. A*, 72(2):302–309, 1995. ISSN 0097-3165. doi: 10.1016/0097-3165(95)90067-5. URL `http://dx.doi.org/10.1016/0097-3165(95)90067-5`.

[319] B. Sturmfels and R. R. Thomas. Variation of cost functions in integer programming. *Math. Program.*, 77:357–388, 1997.

[320] B. Sturmfels, R. Weismantel, and G. M. Ziegler. Gröbner bases of lattices, corner polyhedra and integer programming. *Contributions to Algebra and Geometry*, 36: 281–298, 1995.

[321] A. Szenes and M. Vergne. Residue formulae for vector partitions and Euler-MacLaurin sums. *Adv. Appl. Math.*, 30(1-2):295–342, 2003. ISSN 0196-8858. doi: 10.1016/S0196-8858(02)00538-9. URL `http://dx.doi.org/10.1016/S0196-8858(02)00538-9`.

[322] S. Tayur, R. R. Thomas, and N. R. Natraj. An algebraic geometry algorithm for scheduling in presence of setups and correlated demands. *Math. Program.*, 69:369–401, 1995.

[323] R. R. Thomas. *Gröbner basis methods for integer programming*. Ph.D. thesis, Cornell University, Ithaca, NY, 1994.

[324] R. R. Thomas. A geometric Buchberger algorithm for integer programming. *Math. Oper. Res.*, 20:864–884, 1995.

[325] R. R. Thomas and R. Weismantel. Truncated Gröbner bases for integer programming. *Applic. Algebra Eng., Commun. Comput.*, 8:241–257, 1997.

[326] R. Urbaniak, R. Weismantel, and G. M. Ziegler. A variant of the Buchberger algorithm for integer programming. *SIAM J. Discrete Math.*, 10:96–108, 1997.

[327] R. J. Vanderbei. *Linear Programming: Foundations and Extensions*. 3rd edition. International Series in Operations Research & Management Science, 114. Springer, New York, 2008. ISBN 978-0-387-74387-5. doi: 10.1007/978-0-387-74388-2. URL `http://dx.doi.org/10.1007/978-0-387-74388-2`.

[328] S. A. Vavasis. Polynomial time weak approximation algorithms for quadratic programming. In Pardalos [269, pp. 490–497].

[329] S. A. Vavasis and Y. Ye. A primal-dual interior point method whose running time depends only on the constraint matrix. *Math. Program. Ser A*, 74(1):79–120, 1996. ISSN 0025-5610. doi: 10.1016/0025-5610(95)00043-7. URL `http://dx.doi.org/10.1016/0025-5610(95)00043-7`.

[330] S. Verdoolaege. `barvinok`. URL `http://freshmeat.net/projects/barvinok/`, 2007.

[331] S. Verdoolaege and K. M. Woods. Counting with rational generating functions. *J. Symbol. Comput.*, 43(2):75–91, 2008. ISSN 0747-7171. doi: 10.1016/j.jsc.2007.07.007. URL `http://dx.doi.org/10.1016/j.jsc.2007.07.007`.

[332] S. Verdoolaege, R. Seghir, K. Beyls, V. Loechner, and M. Bruynooghe. Counting integer points in parametric polytopes using Barvinok's rational functions. *Algorithmica*, 48(1):37–66, 2007. doi: 10.1007/s00453-006-1231-0.

[333] M. Vergne. Residue formulae for Verlinde sums, and for number of integral points in convex rational polytopes. In *European Women in Mathematics (Malta, 2001)*, pages 225–285. World Scientific Publishing, River Edge, NJ, 2003. ISBN 9-812-38190-2.

[334] R. Weismantel. Test sets of integer programs. *Math. Methods Oper. Res.*, 47:1–37, 1998.

[335] L. Wolsey. Extensions of the group theoretic approach in integer programming. *Management Sci.*, 18:74–83, 1971.

[336] K. Woods. *Rational Generating Functions and Lattice Point Sets*. Ph.D. thesis, University of Michigan, 2004.

[337] G. Zhao and J. Stoer. Estimating the complexity of a class of path-following methods for solving linear programs by curvature integrals. *Appl. Math. Optim.*, 27(1):85–103, 1993. ISSN 0095-4616. doi: 10.1007/BF01182599. URL `http://dx.doi.org/10.1007/BF01182599`.

[338] G. M. Ziegler. *Lectures on Polytopes*. Number 152 in Graduate Texts in Mathematics. Springer, New York, 1995. ISBN 0-387-94329-3.

[339] G. M. Ziegler. Gröbner bases and integer programming. In *Some Tapas of Computer Algebra*, volume 4 of *Algorithms and Computation in Mathematics*, pages 168–183. Springer, Berlin, 1999. ISBN 3-540-63480-0.

Index

0/1 linear integer minimization, 100
3-colorability, 245
`4ti2`, 80, 83

absolute convergence, 107, 130
affine combination, 3
affine hull, 18
affine hyperplane, 4
algebraic closure, 193, 238, 242, 256
algebraic variety, 239
algorithm
    ϵ-approximation, 157
    approximation, 157
    Barvinok's, 129, 134, 146
    Buchberger's, 210, 218
    Euclidean, 256
    geometric Buchberger, 226
    Gram–Schmidt orthogonalization, 50
    Graver proximity, 100
    incremental polynomial time, 141
    Lenstra's, 53, 79, 144
    LLL, 50
    Nullstellensatz linear algebra, 242
    polynomial total time, 141
    polynomial-delay, 141
    polynomial-space polynomial-delay, 141
    Pottier's, 71
    regular triangulation, 121
    weak approximation, 158
all-primal Barvinok decomposition, 153
analytic center, 282
analytic function, 137
appoximation scheme
    weak, 158
approximate continuous convex optimization, 93
approximate continuous convex optimization oracle, 101
approximate shortest vector, 133
approximation algorithm, 157
    weak, 158
approximation scheme, 157
arithmetic mean, 163
Artin's theorem, 269
ascending chain, 215
assignment problem, 287
augmentation algorithms, 93
augmentation step, 92
    Graver-best, 66, 97, 99
augmentation vector, 63, 96
auxiliary objective function, 95

`barvinok`, 135
Barvinok's algorithm, 129, 134, 146
    for integer linear optimization, 142
    homogenized, 153
Barvinok's signed decomposition, 132
    all-primal, 153
    dual, 152
    primal, 152
Barvinok's theorem, 134, 135
basic feasible solution, 19
basic semialgebraic set, 187, 273
basic solution, 19, 151
basic variables, 19
basis, 19, 168
    biorthonormal, 133
    Gröbner, 208, 217, 286
    integral, 176
    Markov, 221, 228
    of a lattice, 34
Bernoulli–l'Hôpital rule, 106
binary encoding scheme, 21
binary search, 143
binomial ideal, 218
binomials, 217
biorthonormal basis, 133
Birkhoff polytope, 288
bisection algorithm, 163
bit length, 21
bit-scaling technique, 66, 100

Blichfeldt's theorem, 41
block-structured integer programs, 77
Boolean operations, 150, 164, 184
Boolean operations lemma, 148
bounded set, 16
branch and bound algorithm, 176
branching on hyperplanes, 54
brick, 81, 90, 92
Brion's theorem, 108, 118, 120, 124
Buchberger's algorithm, 210, 218
  geometric version, 226
Buchberger's S-pair criterion, 210
Buchberger's theorem, 210
building block, 81

Carathéodory's theorem, 5, 97, 168
Cauchy integral formula, 285
central curve, 282
central path, 281
certificate of infeasibility, 239
chamber, 151
characteristic polynomial, 280
Cholesky factor, 23
circuit, 97
colorable, 238
combinatorial moment matrix, 279
combinatorial parity conditions, 251
combinatorial system of equations, 239
commutative algebra, 92
comparison oracle, 66, 85, 99
comparison point, 181
complementary slackness, 21, 281
completion algorithm, 92, 226
complex analysis, 106, 137
complex integration, 285
complex number, 238
complex plane, 106
compressed lattice polytope, 279
computational complexity, 21, 58
cone, 5, 274
  finitely generated, 5, 12, 13
  index descent, 133
  pointed, 15
  pointed rational, 130
  polyhedral, 5
  recession, 16, 17
  simplicial, 120, 130
  tangent, 118
  unimodular, 112
conformal (orthant-compatible), 97
conic combination, 3
continuous function, 198
continuous Hirsch conjecture, 282
continuous measure, 89
continuous relaxation, 93, 97
convex combination, 3
convex hull, 3
convex position, 120
convex set, 3
convex subdivision, 122
counting, 105
counting oracle, 143
critical path, 224
cut, 237

De Morgan formula, 118
decision making under uncertainty, 89
decomposition
  primal Barvinok, 152
  signed, 113
decomposition algorithm, 89
decomposition of polyhedra, 117
decomposition tree, 134
decreasing path, 222
degree, 193, 194
Delaunay triangulations, 122
derivative, 198
Descartes' rule of signs, 199, 213
determinant of a lattice, 40
diameter of a polytope, 282
Dickson's lemma, 207
dictionary, 227
dictionary order, 203
differential operator, 109, 159
digging algorithm, 145, 146
dimension, 18
directed augmentation, 66, 96
directed graph, 87
directed path, 87
discrete measure, 89
discretization, 165
division algorithm, 194
domain of convergence, 106, 115
dual basis, 133
dual linear programming, 19
duality
  strong, 20
  weak, 20
duality trick, 152

dynamic programming problem, 86, 99, 285

ϵ-approximation algorithm, 157
elementary square matrices, 6
elimination ideal, 212
elimination of variables, 212
elimination theorem, 212
ellipsoid, 22
ellipsoid method, 22, 25
Euclidean algorithm, 256
evaluation oracle, 93
evaluation problem, 160
exact set, 279
explicit enumeration, 147
exponent vector, 203, 268
exponential integrals, 115, 176
exponential sum, 114, 287
extended Euclidean algorithm, 196, 257
extended group relaxation, 286
extended integer program, 220
extreme point, 18

face, 18
factoring, 176
Farkas' lemma, 9, 21, 219, 229, 239, 273
fast Fourier transform, 285
feasibility oracle, 143
finite field, 238
finitely generated cone, 5, 12, 13
finitely generated ideal, 208
fixed dimension, 134
flatness constant, 151
flatness theorem, 54, 151
formal Laurent series, 129
Fourier–Motzkin elimination, 6, 7
FPTAS (fully polynomial-time approximation scheme), 158
Fredholm's alternative theorem, 5, 239, 274
fully polynomial-time approximation scheme (FPTAS), 158, 182
 for a maximization problem, 158
 for maximizing nonnegative polynomials, 159
 for a minimization problem, 182, 187
function
 analytic, 137
 continuous, 198
 rational, 105, 106
 separable convex, 63, 79, 85, 98
 separable convex p-piecewise affine-linear, 87
fundamental parallelepiped, 39, 109, 130
fundamental theorem of algebra, 169

gaps, 149, 152
Gaussian elimination, 6, 201, 203
gcd (greatest common divisor), 194, 195, 203
general position, 282
generating function, 105, 108, 129, 287
 Boolean operation, 147
 evaluation, 135
 integer projection, 149
 intermediate, 176
 mixed-integer, 176
 output-sensitive enumeration, 141
 positively weighted, 136, 147, 149, 165
 projection theorem, 149
 rational, 132
 specialization, 135
 substitution, 135
generating set, 221
generic, 122, 219
geometric Buchberger algorithm, 226
geometric improvement, 67
geometric series, 105, 110, 130
geometry of numbers, 29
global criterion, 179, 181, 185
global mixed-integer polynomial optimization, 157
global nonconvex polynomial optimization, 176
Gomory's group relaxation, 154
Gomory–Chvátal closure, 56
Gordan–Dickson lemma, 43, 44, 90, 207, 227
Gröbner basis, 208, 217, 286
 of a lattice ideal, 222
 reduced, 212, 218
 reduced minimal, 219
Gröbner complexity, 233
Grötzsch graph, 248
graded lexicographic order, 204
graded reverse lexicographic order, 204
Gram–Brianchon theorem, 118
Gram–Schmidt orthogonalization algorithm, 50

graph, 221, 237, 286
  of a polyhedron, 282
Graver-based dynamic programming, 86, 96, 100
Graver basis, 63, 64, 219, 232
  for stochastic IPs (integer programs), 91
  length bound, 94
Graver-best augmentation step, 66, 97, 99
Graver complexity, 80, 82, 102
Graver proximity algorithm, 100
Graver test set, 63
greatest common divisor (gcd), 194, 195, 203, 256
  of univariate polynomials, 195
grid approximation, 165
grid problem, 165, 171
group relaxation, 285

H-representation, 10
Hadamard product, 147, 184
half-open decomposition, 124, 152, 153
half-open polyhedron, 124
half-open simplicial cone, 130
half-open triangulation, 127
heights, 121
Hermite normal form (HNF), 35, 95, 231
Hilbert basis, 45, 47, 83, 154
  element enumeration, 149
Hilbert's 10th problem, 58
Hilbert's 17th problem, 269
Hilbert's basis theorem, 208, 218
Hilbert's Nullstellensatz, 213, 237, 239, 256
Hirsch conjecture, 283
HNF (Hermite normal form), 35, 95, 231
Hochbaum–Shanthikumar's proximity, 97
Hölder's inequality, 159
holes, 180
homogeneous polynomial, 243
homogenization, 14, 130, 269
homogenized Barvinok algorithm, 153
hyperplane, 4
hyperplane arrangement, 282

ideal, 183, 201
  finitely generated, 208
  generated by a set, 202
  monomial, 206
  radical, 262, 278
  toric, 217, 286
  vanishing, 202
ideal intersection, 213
ideal membership, 202, 212
ideal point, 181
ILF (integer linear feasibility problem), 54
ILP (integer linear optimization problem), 53
inapproximability, 58
inclusion-exclusion principle, 112, 117, 124
incomputable, 58
indecomposable lattice point, 47
independent set, 252
index of a cone, 111, 130
index of a sublattice, 40
indicator function, 41, 117, 124
infeasibility certificate, 239
infinite geometric series, 106
initial feasible solution, 220
integer hull, 55, 151
integer linear feasibility problem (ILF), 54
integer linear optimization problem (ILP), 53
integer linear program in fixed dimension, 79
integer programming game, 190
integer projection, 136
integral, 176
integral basis, 44
integral generating set, 44
integral polyhedron, 56
intermediate generating function, 176
intersection lemma, 147, 148, 186
irrational decomposition, 152
iterated bisection, 143

Kannan's partitioning theorem, 151
Khinchin's flatness theorem, 54, 151
knapsack, 285
knapsack problem, 284
k-SOS ideal, 278
k-th theta body, 278

ℓ_1-norm, 83, 93, 96, 99, 101
Laplace transform, 115
large scale problem, 89

largest term, 204
LattE integrale, 135
lattice, 34
  sublattice, 40
lattice basis
  LLL-reduced, 133
lattice-point-free convex bodies, 151
lattice program, 223
lattice width, 54
lattice width direction, 54
Laurent expansion, 106
Laurent polynomial, 130
Laurent series, 106, 284
Lawrence–Khovanskii–Pukhlikov theorem, 131
layers, 81
leading coefficient, 204
leading monomial, 204, 207
leading term, 194, 204
Lenstra's algorithm, 53, 79, 144
lexicographic order, 141, 181, 203
LF (linear feasibility problem), 22
lift-and-project method, 277
lifted configuration, 121
limit point, 166
lineality space, 15, 17
linear feasibility problem (LF), 22
linear optimization problem (LP), 21
linearization technique, 277
Lipschitz constant, 171
LLL (Lenstra–Lenstra–Lovász) algorithm, 50
LLL-reduced lattice basis, 133
logarithmic barrier function, 281
Lovász encoding, 252
lower envelope, 121
LP (linear optimization problem), 21

M-ellipsoid coverings, 50
Maclagan's theorem, 91
Markov basis, 221, 228
matrix term order, 204
matrix-cut method, 277
matroid, 282
max-cut problem, 237
maximal decreasing path algorithm, 226
maximal lattice-free convex set, 287
meromorphic function, 115, 130
minimal critical path, 225
Minkowski sum, 12, 14, 47
Minkowski's first theorem, 31, 42
Minkowski–Hlawka theorem, 43
mixed-integer generating function, 176
mixed-integer summation technique, 176
modulo nonpointed polyhedra, 120
moment, 89
moment curve, 139
monic polynomial, 261
monomial, 105
monomial ideal, 206
monomial map, 136
monomial order, 203
monomial substitution, 144, 150, 184
Monte-Carlo Markov-chain, 222
Moore–Bellman–Ford's algorithm, 88
multicommodity network flow problem, 77
multicriterion integer linear programming problem, 179, 181
multiepigraph, 183
multiexponent notation, 129
multigraph, 184
multiplicity, 198
multistage stochastic integer linear programs, 101
multivariate division algorithm, 205

N-fold 4-block decomposable, 77
N-fold 4-block decomposable integer program, 93
N-fold 4-block decomposable matrix, 77
N-fold integer program, 77, 96, 99
N-fold matrix, 79
Nash equilibrium, 190
Newton polytope, 268, 280
Noether's normalization lemma, 256
nonbasic variable, 19
nonnegative modulo an ideal, 278
nonstandard monomial, 214
normal form
  Smith, 38, 133
normal form algorithm, 71
NP-complete, 245
NP-hard, 58
NulLA (Nullstellensatz linear algebra algorithm), 242
NulLA rank, 243, 253
Nullstellensatz, 213, 237, 239, 256
Nullstellensatz linear algebra algorithm (NulLA), 242

objective function, 108
    auxiliary, 95
odd cycle, 237
odd wheel, 247
one-to-one projection, 152, 184
optimality certificate, 63, 65, 96, 223
optimization oracle, 93
oracle, 141, 148
    comparison, 66, 85, 99
    counting, 143
    feasibility, 143
    optimization, 93
    separation, 22
orbit polytope, 288
oriented chordless 4-cycle, 246
oriented partial 3-cycle, 246
outcome vector, 179
output-polynomial time, 75
output-sensitive complexity analysis, 141
overcounting, 112

Pólya exponent, 269
Pólya's lemma, 269
Pareto optimum, 179, 180, 184
Pareto strategy, 179, 180, 184
partition generating function, 283
permutahedron, 288
phase I, 68, 96, 100, 220
Pick's theorem, 34
piecewise affine linear, 86, 96
pivot, 201
pivot rule, 282
point configuration, 120
pointed cone, 132
pointed polyhedron, 17, 130
pointed rational cone, 130
pointed rational polyhedron, 130
polar, 10
    of a cone, 11, 130
    of a set, 10
pole, 106, 135
polygon, 31
    simple, 32
polyhedral cone, 5
polyhedral norm, 181, 185
polyhedron, 4
    pointed, 17, 130
    pointed rational, 130
    simple, 120, 283
polynomial
    monic, 261
polynomial map, 217
polynomial system, 237
polynomial-space polynomial-delay enumeration algorithm, 148, 182, 185
polynomial-space polynomial-delay prescribed-order enumeration algorithm, 150, 181
polynomial-time approximation scheme (PTAS), 157
polytopal subdivision, 121
polytope, 4
positive definite form, 269
positive semidefinite (PSD) matrix, 26
positive semidefinite (PSD) polynomial, 269
positively weighted generating function, 136, 147, 149, 165
Positivstellensatz, 274
Pottier's algorithm, 71
preorder, 274
primal Barvinok decomposition, 152
primal-dual interior point algorithm, 281
probability measure, 89
project-and-lift, 73, 230
project-and-lift algorithm, 228
projection, 164, 180, 185
projection theorem, 182, 184
proximity, 93, 97
proximity-scaling technique, 93, 97
PSD (positive semidefinite) matrix, 26
pseudonorm, 187
pseudopolynomial, 97
PTAS (polynomial-time approximation scheme), 157
Putinar's theorem, 277

quadratic assignment problem, 287
quadratic module, 276
quotient ring, 217
quotient rule, 160

radical ideal, 262, 278
Radon's lemma, 5
range of the objective function, 173
rational function, 105, 106
rational generating function, 132
real roots, 198
recession cone, 16, 17

reduced Gröbner basis, 212, 218
reduced lattice basis, 51
reduced minimal Gröbner basis, 219
reduction path, 223
regular subdivision, 122
regular triangulation, 286
regular triangulation algorithm, 121
remainder of polynomial division, 204, 209
removable singularity, 135
representation, 287
representation theorem for cones, 15
residue, 285
residue technique, 160
residue techniques, 162
resolution of polyhedra, 15
resultant, 256, 257, 259
reverse lexicographic triangulation, 279
Rolle's theorem, 198
root, 194
    multiplicity of, 197
root of unity, 238
row reduction, 201

S-polynomial, 209, 218, 224, 227
S-vector, 227
sample, 154, 177
saturated ideal, 233
saturation, 92, 232
scenario, 89
Schmüdgen's theorem, 276
SDP (semidefinite program), 27
selecting a Pareto optimum, 185
semialgebraic set, 274
semidefinite optimization problem (SDP), 27, 239
semidefinite programming, 27, 239, 275
semigroup, 110
separable convex function, 63, 79, 85, 93, 98
separation oracle, 22
separation problem, 26
series expansion, 145
set covering, 287
set packing, 287
shelling, 124
shortest path, 88
shortest path problem, 286
shortest vector, 49
    approximate, 133
shortest vector problem (SVP), 49
sign variation, 199
sign-compatible, 44
signed decomposition, 113
simple polygon, 32
simple polyhedron, 120, 283
simplex, 120
simplicial complex, 286
simplicial cone, 120, 130
singularity, 106
    removable, 135
slack variable, 17
small-gaps theorem, 151
Smith normal form, 38, 133
SOS (sum of squares), 274
    modulo an ideal, 278
spanning set, 221
specialization, 184
square-free, 254
stability number, 252
stable set, 237, 252
stable set polytope, 252, 277
stochastic integer multicommodity flow problem, 78
stochastic integer program with second-order dominance relations, 78
strong duality, 20
Sturm sequence, 200
subdeterminant, 55, 97
subdivision
    convex, 122
    regular, 122
sublattice, 40
substitution, 136
sum of squares (SOS), 274
    modulo an ideal, 278
summation formula, 105
summation method for optimizing polynomials, 159, 161
sums of squares (SOS), 265
superadditivity, 98
support, 97
supported Pareto outcome, 180
supporting hyperplane, 18
Sylvester matrix, 258
symbolic differentiation, 109
symmetric groups, 288
system of linear equations, 5
system of polynomial equations, 193

TH_k-exact, 278
tangent cone, 118
Taylor expansion, 137
term, 194
term order, 203
  graded lexicographic, 204
  graded reverse lexicographic, 204
  lexicographic, 141, 181, 203
  matrix, 204
test set, 63, 223
theorem
  Artin's, 269
  fundamental
    of algebra, 169
  Gram–Brianchon, 118
  Hilbert's basis, 208, 218
  Minkowski's first, 31, 42
  Minkowski–Hlawka, 43
  Schmüdgen's, 276
  small-gaps, 151
  Weyl–Minkowski, 4, 9, 26
theorem of the alternative, 239
theta body, 277
  k-th, 278
theta-rank, 278
three-way transportation problems, 79
tiling, 109
Todd polynomial, 137
toric ideal, 217, 286
toric ring, 217
total curvature, 282
total degree, 204
total dual integrality, 287
total order, 203
totally unimodular, 57
totally unimodular matrix, 77, 79
transportation matrix, 83, 101
transportation polytope, 79
transportation problem, 79
traveling salesman problem, 287
triangulation, 111, 121, 130
  of a cone, 120
  of a polytope, 120
  of a vector configuration, 123
  regular, 286
  reverse lexicographic, 279
truncated Gröbner basis, 233
truncated Taylor series, 139
Turán graph, 255
two-stage stochastic integer optimization problem, 77
two-stage stochastic integer programming, 88
type, 81, 87

uniform convergence, 107, 130
unimodular, 150, 279, 287
unimodular cone, 112, 133
unimodular matrix, 34, 36, 152
universality theorem, 79

V-representation, 10
valid inequality, 18
valuation, 131
vanishing ideal, 202
variety, 201, 239, 262
  zero-dimensional, 214, 239
vector configuration, 123
vector partition function, 283
vertex, 18
volume, 23
von Neumann, 20
Voronoi cell, 50

weak approximation algorithm, 158
weak approximation scheme, 158
weak composition, 95
weak duality, 20
well-ordering, 203
Weyl–Minkowski theorem, 4, 9, 26
width, 54

zero divisor, 131
zero-dimensional variety, 214, 239